Organocatalytic Enantioselective Conjugate Addition Reactions
A Powerful Tool for the Stereocontrolled Synthesis of Complex Molecules

RSC Catalysis Series

Series Editor:
Professor James J Spivey, *Louisiana State University, Baton Rouge, USA*

Advisory Board:
Krijn P de Jong, *University of Utrecht, The Netherlands*, James A Dumesic, *University of Wisconsin-Madison, USA*, Chris Hardacre, *Queen's University Belfast, Northern Ireland*, Enrique Iglesia, *University of California at Berkeley, USA*, Zinfer Ismagilov, *Boreskov Institute of Catalysis, Novosibirsk, Russia*, Johannes Lercher, *TU München, Germany*, Umit Ozkan, *Ohio State University, USA*, Chunshan Song, *Penn State University, USA*

Titles in the Series:
1: Carbons and Carbon Supported Catalysts in Hydroprocessing
2: Chiral Sulfur Ligands: Asymmetric Catalysis
3: Recent Developments in Asymmetric Organocatalysis
4: Catalysis in the Refining of Fischer-Tropsch Syncrude
5: Organocatalytic Enantioselective Conjugate Addition Reactions: A Powerful Tool for the Stereocontrolled Synthesis of Complex Molecules

How to obtain future titles on publication:
A standing order plan is available for this series. A standing order will bring delivery of each new volume immediately on publication.

For further information please contact:
Book Sales Department, Royal Society of Chemistry, Thomas Graham House, Science Park, Milton Road, Cambridge, CB4 0WF, UK
Telephone: +44 (0)1223 420066, Fax: +44 (0)1223 420247, Email: books@rsc.org
Visit our website at http://www.rsc.org/Shop/Books/

Organocatalytic Enantioselective Conjugate Addition Reactions

A Powerful Tool for the Stereocontrolled Synthesis of Complex Molecules

Edited by

Jose L. Vicario, Dolores Badía, Luisa Carrillo and Efraim Reyes
Department of Organic Chemistry II, University of the Basque Country, Bilbao, Spain

RSC Publishing

RSC Catalysis Series No. 5

ISBN: 978-1-84973-024-2
ISSN: 1757-6725

A catalogue record for this book is available from the British Library

© J. L. Vicario, D. Badía, L. Carrillo and E. Reyes 2010

Published by The Royal Society of Chemistry,
Thomas Graham House, Science Park, Milton Road,
Cambridge CB4 0WF, UK

Registered Charity Number 207890

For further information see our website at www.rsc.org

Preface

Asymmetric organocatalysis has experienced an impressive rebirth in the last few years, with a plethora of new methodologies developed for carrying out enantioselective transformations which up to that moment were only available under transition-metal catalysis. Given the operational and economical advantages associated with this methodology, together with the fact that organocatalysts are environmentally friendly, robust and very often commercially available reagents, many research groups worldwide have engaged in active research in this field. In this context, the conjugate addition reaction has attracted a particular interest by the synthetic organic chemists, which has led to an extraordinarily high number of reports dealing with enantioselective versions of this transformation carried out using small organic molecules as catalysts. Organocatalysts operating by very different mechanistic profiles have demonstrated their performance when applied to this reaction obtaining in many cases outstanding levels of chemical efficiency and stereoselectivity. The conjugate addition reaction is also a particularly interesting reaction because of its wide synthetic versatility due to the broad spectrum of donors and acceptors that may be employed on one hand and also because of its potential to be applied in tandem, domino or cascade reactions, resulting in an extremely powerful approach for the preparation of molecules of high complexity in a single step starting from readily available starting materials.

This book intends to cover a very hot topic in a rather comprehensive way, including the most recent research made in this field until 2009, presenting the advances in this field organized according to the mechanistic pathway involved in the activation of the reagents participating in the reaction by the organic catalyst. We hope that this book will provide a good state-of-the-art view to all organic chemists working in this field and also to all who wish to start projects in this area.

RSC Catalysis Series No. 5
Organocatalytic Enantioselective Conjugate Addition Reactions: A Powerful Tool for the Stereocontrolled Synthesis of Complex Molecules
By Jose L. Vicario, Dolores Badía, Luisa Carrillo and Efraim Reyes
© J. L. Vicario, D. Badía, L. Carrillo and E. Reyes 2010
Published by the Royal Society of Chemistry, www.rsc.org

Contents

RSC Catalysis Series No. 5
Organocatalytic Enantioselective Conjugate Addition Reactions: A Powerful Tool for the
Stereocontrolled Synthesis of Complex Molecules
By Jose L. Vicario, Dolores Badıa, Luisa Carrillo and Efraim Reyes
© J. L. Vicario, D. Badıa, L. Carrillo and E. Reyes 2010
Published by the Royal Society of Chemistry, www.rsc.org

CHAPTER 1
Introduction

1.1 Organocatalysis: an Emerging Field

Chemistry is the science of matter and the changes it undergoes during chemical reactions. Behind this definition one can identify the ability of chemists to change the matter at will, in order to obtain new materials or compounds with novel or improved properties. As humans evolved toward modern societies, more and more complex chemical entities were required to satisfy their needs toward higher standards of life; chemists needed to develop new methods and reactions that allowed the preparation of the target compounds in a more efficient way. These two concepts, "efficiency" and "complexity", entail the major principles behind the activity of chemists dedicated to organic synthesis: a complex molecule, with many different types of functionalities located in different positions and with a certain three-dimensional orientation of all its atoms, has to be built up starting from simple and readily available starting materials by carrying out a set of transformations in the most efficient (economic) way. This means that reactions and synthetic methodologies have to be available for the organic chemist to carry out these transformations in the simplest way, with the highest possible yield and selectivity and with the minimum generation of waste. In this context, asymmetric synthesis represents the highest level with regard to selectivity control in a chemical reaction. If a chiral compound has to be prepared as a single enantiomer, the chemist has to control the exact trajectories in which the reagents approach to each other, which means that an exquisite control of all the events taking place in the reaction vessel has to be achieved.

The agrochemical and pharmaceutical industries are fields in which chirality and stereochemical control are of special relevance. Drug chirality is now a major theme in the design, discovery, development, launching and marketing of

RSC Catalysis Series No. 5
Organocatalytic Enantioselective Conjugate Addition Reactions: A Powerful Tool for the Stereocontrolled Synthesis of Complex Molecules
By Jose L. Vicario, Dolores Badía, Luisa Carrillo and Efraim Reyes
© J. L. Vicario, D. Badía, L. Carrillo and E. Reyes 2010
Published by the Royal Society of Chemistry, www.rsc.org

new drugs and, therefore, stereochemistry is an essential dimension to be taken into account. The thalidomide case in the 1960s is a paradigmatic example of this behavior. This drug was prescribed in Europe in a racemic form to pregnant women to alleviate sickness but, while one of the enantiomers had sedative and antiemetic activities, the opposite enantiomer had teratogenic effects. This tragedy led to a new awareness of the importance of stereoselective pharmacodynamics and pharmacokinetics, enabling the differentiation of the relative contributions of enantiomers to overall drug action. When one enantiomer is responsible for the activity of interest, its paired enantiomer could be inactive, be an antagonist of the active enantiomer or have a separate activity that could be desirable or undesirable. Considering these possibilities, there appear to be major advantages in using enantiomerically pure drugs, such as a reduction of the total administered dose, enhanced therapeutic window and a more precise estimation of dose–response relationships. These factors have led to an increasing preference for single enantiomers in both industry and regulatory authorities. Regulatory control of chiral drugs began in the United States with the publication in 1992 of formal guidelines on the development of chiral drugs and was followed in the European Union in 1994 by a document entitled *Investigation of Chiral Active Substances.* Applicants must recognize the occurrence of chirality in new drugs, attempt to separate the stereoisomers, assess the contribution of the various stereoisomers to the activity of interest and make a rational selection of the stereoisomeric form that is proposed for marketing.

Among the strategies available in the synthetic organic chemist's toolbox for controlling the stereochemical outcome of a given reaction, catalysis has become the option of choice in the last 20 years. The true advantages displayed by catalytic enantioselective methods compared to the chiral auxiliary methodology justify this situation: catalytic methods constitute a more direct and atom-economic approach than the use of chiral auxiliaries because the requirement of stoichiometric quantities of the chirality source and the need for additional synthetic steps for the attachment/removal of the auxiliary are avoided. As a consequence of this, a huge number of different research groups worldwide have engaged in the development of enantioselective versions of almost all the organic reactions known up to date. In this context, the catalytic methodologies typically employed for the enantioselective preparation of chiral compounds have been considered for many years to lie in two main categories, namely metal catalysis and enzymatic methods. In fact, metal-catalyzed reactions have prevailed for many years above the enzymatic methods, reaching an exceptional level of sophistication. Proof of the paramount importance of transition-metal-mediated enantioselective transformations is the 2001 Nobel Prize in chemistry, which was awarded to William R. Knowles, Ryoji Noyori and K. Barry Sharpless for the development of important metal-catalyzed enantioselective reductions and oxidations. On the other hand, enzymatic methodologies, although a well-known method even used by living cells to produce their own metabolites, normally chiral compounds in enantiomerically pure form, have been considered for many years as a more limited strategy, due to the extremely high specificity shown by enzymes toward the substrate structure. This situation has been overcome in recent years with the advances in biotechnology, which allow the preparation of modified enzymes from genetically modified organisms.

However, in the last decade a new branch of knowledge has opened up in the interface between these two methodologies. The fact that small chiral organic molecules that do not contain metal atoms in the structure (organocatalysts) can catalyze certain chemical transformations in a stereoselective way has experienced an impressive rebirth in the chemical community, which has led to the coining of a new term, "asymmetric organocatalysis", to define this new methodological alternative for carrying out a reaction in a catalytic and enantioselective way. Organocatalysts are (purely) organic molecules, composed of (exclusively) hydrogen, carbon, silicon, nitrogen, phosphorous, oxygen, sulfur, selenium, tellurium and halides and arise as the "opposite" to metal complexes, in which the metal is the main thing responsible for the catalytic activity in the reaction. However, there are some organocatalysts which do incorporate some metal elements in their structure (for example, several chiral DMAP derivatives containing a ferrocene unit) but in which the metal atom does not participate in the catalytic cycle, remaining as a simple architectural element of the catalyst.

Asymmetric organocatalysis has become a very rapidly growing field of research, as a result of both the novelty of the concept and the high efficiencies and selectivities attained by many organocatalytic transformations.[1,2] One of the main advantages of this methodology is the fact that organocatalysts and the intermediate species participating in the catalytic cycle are usually inert toward water or oxygen, and therefore reactions do not require special care with regard to the use of an inert atmosphere or dry solvents. This turns into a high operational simplicity when running the reaction, compared with the typical specific equipment and expertise required when carrying out most of the transition metal-catalyzed reactions. In addition, the fact that the presence of hazardous metals is precluded in the reaction scheme makes this methodology even more interesting from the environmental point of view and this is of special relevance in the production of chemicals for human consumption which do not tolerate the presence of metal contaminants even at the trace level. All these practical and economical advantages, together with the fact that organocatalysts are very often commercially available reagents, have contributed significantly to the rapid growth of the field, leading many research groups to engage in the development of novel organocatalytic procedures for performing transformations which were typically run using transition-metal catalysis. This has made organocatalysis a rapidly changing field, which has experienced an impressive growth especially in the last few years, as can be seen from the evolution of the publications in this area as shown in Figure 1.1, with several very important research groups involved. As a consequence of this, organocatalysis has also become a very competitive field, which in some cases has led to multiple publications appearing almost simultaneously and covering similar reactions or reporting similar

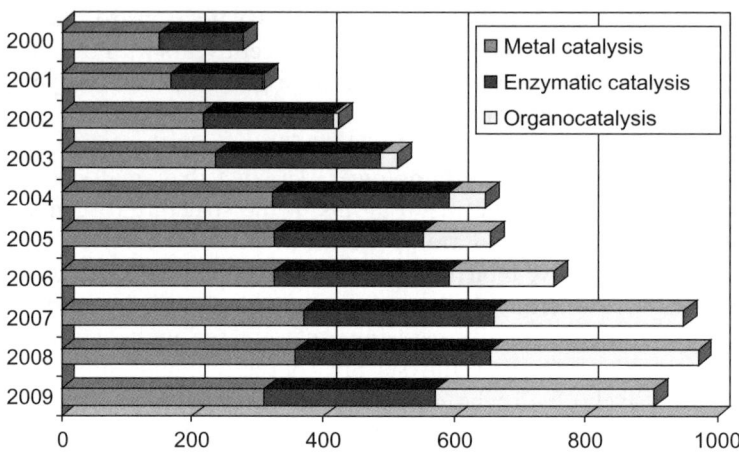

Figure 1.1 Evolution of publications in the field of asymmetric catalysis (2000–2009).[4]

results. This competitiveness and the rush for publication has also in some cases resulted in a rather superficial treatment of the experimental results obtained and a lack of a deep research into the consequences of these results and the knowledge which could be acquired thereof. In fact, many organocatalytic reactions like enamine and iminium catalysis or the reactions *via* H-bonding interactions or with *N*-heterocyclic carbenes proceed *via* mechanisms which are closely related to those operating in many enzymatic processes, although this connection has very often been underestimated during the research process.[3] A global vision on the field, and possible new advances and future directions, could have been complicated by such a fragmented way of reporting the huge amount of information gathered in many different laboratories worldwide.

However, despite the impressive advance gained in this field in the last years, many issues still remain unsolved. Some reactions have remained elusive to organocatalysis and, therefore, transition-metal catalysis stands as the only available synthetic tool. Another issue which has to be underlined is the fact that organocatalytic reactions typically require very high catalyst loadings (5–20 mol%), which at the moment is very far from the extremely high efficiencies achieved in this context by transition-metal catalytic methodologies and makes the application of the methodology to large-scale synthesis much more difficult. In fact, the application of organocatalytic asymmetric methodologies to industrial production is still to be developed. Finally, the application of organocatalytic reactions in total syntheses of complex molecules also remains rather unexplored,[5] although there is a significant tendency observed in the last years which indicates that chemists working in total synthesis are starting to seriously consider organocatalytic methodologies as a potent alternative in their chemical reactions toolkit.

1.2 Historical Evolution of Asymmetric Organocatalysis

The possibility that small organic molecules could catalyze organic transformations in a stereoselective way can even be considered as a key element in the origins of life, as it is widely accepted that the source of homochirality in biological systems should be attributed to the presence of enantiomerically enriched α-amino acids in meteorites,[6] and especially once it was demonstrated that these α-amino acids were able to catalyze aldol reactions in a stereoselective way under prebiotic conditions conducing to sugar-type products.[7]

However, the first attempts in the field were carried out when trying to understand the mechanism of several enzymatic transformations, the small chiral organic molecule being intended to operate as a model mimicking the enzyme behavior. For example, in 1908, Georg Bredig, who was indeed interested in the origin of the enzyme activity, found that the thermal decarboxylation of optically enriched camphorcarboxylic acid in the presence of (+)- or (−)-*limonene* proceeded with an enhancement of optical purity of the final product[8] and, as an extension of this work, he subsequently studied the same decarboxylation reaction in the presence of natural alkaloids like nicotine or quinidine.[9] It was also G. Bredig who reported the first enantioselective C–C bond-forming reaction under metal-free conditions, when he demonstrated that the addition of HCN to benzaldehyde in the presence of quinine or quinidine also proceeded with some degree of enantioselection (Scheme 1.1).[10] Although the observed ee was too low for preparative purposes, the proof of principle was clearly established for future advances in the field.

In the same line with these pioneering reports, the German chemist Wolfgang Langebeck made several decisive contributions to the field, in particular, coining for the first time the term "organic catalyst" to define those reactions promoted exclusively by organic compounds.[11] In fact, the origins of the research by Langebeck were strongly multidisciplinary; he dedicated most of his efforts to the identification and explanation of enzymatic processes by using simple amino acids or small peptides in order to mimic the behavior of natural enzymes.[12] In 1949, Langebeck published his book "*Organic Catalysts and Their Relations with Enzymes*",[11] in which, for example, mechanistic

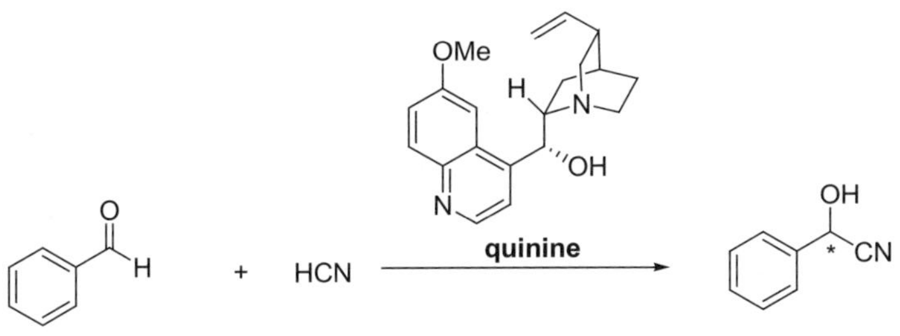

Scheme 1.1   Enantioselective addition of HCN to benzaldehyde reported by Bredig.

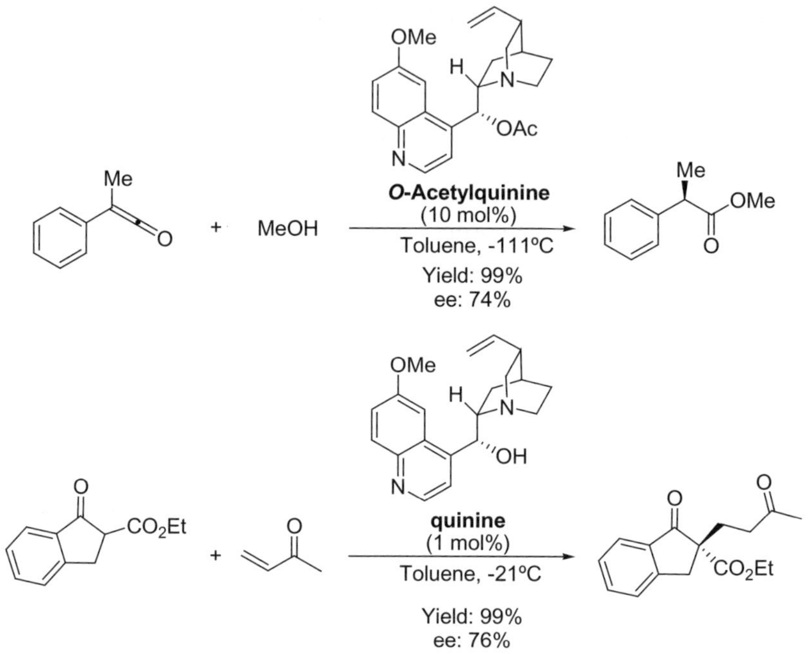

Scheme 1.2 The enantioselective addition of MeOH to methyl phenyl ketene reported by Pracejus and the quinine-catalyzed enantioselective Michael reaction reported by Wynberg.

implications related to the way organic catalysts and enzymes participate in promoting a reaction were discussed.

Probably the first organocatalytic asymmetric reaction furnishing high levels of enantioselection was the addition of methanol to methyl phenyl ketene in the presence of *O*-acetylquinine reported by Pracejus in 1960 (Scheme 1.2).[13] This reaction also formed the basis for the extension of this chemistry to the use of cinchona alkaloid-based catalysts in other reactions, by considering their implication in the reaction as a chiral Lewis base and it is in this context in which the first report on organocatalytic conjugate additions can be found in the literature. Initially, Bergstrom and Langström reported the Michael addition of β-keto esters to acrolein using 2-hydroxymethylquinuclidine as catalyst and, although the enantioselectivity was not determined, the authors reported that the final products had optical activity.[14] Subsequently, Wynberg and co-workers carried out extensive research in the use of cinchona alkaloids as promoters for conjugate additions (see the first example reported by the group in Scheme 1.2)[15] and additionally observing that natural cinchona alkaloids were superior catalysts to those in which the C-9 OH group had been modified, which led to the proposal that the cinchona alkaloids played the role of bifunctional catalysts in which the quinuclidine tertiary amine and the free

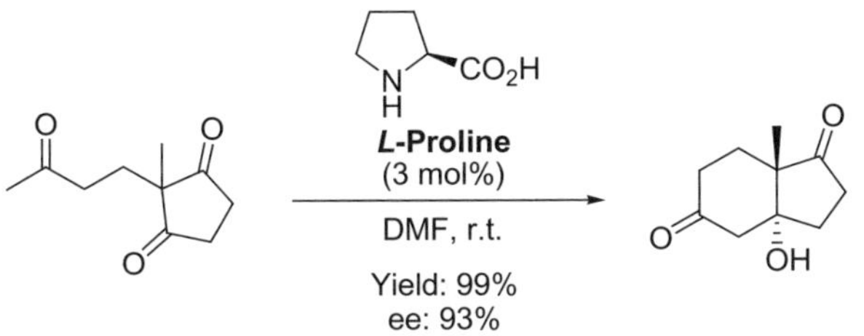

Scheme 1.3   The Hajos–Parrish–Eder–Sauer–Wiechert reaction.

9-OH groups were participating in the activation of the reagents and facilitating their orientation for achieving high stereocontrol. These works also indicated that H-bonding interactions could be a good element for the activation of molecules without the need for metal-centered Lewis acids.

One of the milestones in the development of organocatalysis is the intra-molecular aldol reaction catalyzed by proline developed independently by two industrial research groups at Hoffmann-La Roche[16] and Schering[17] (Scheme 1.3). This reaction, also known as the Hajos–Parrish–Eder–Sauer–Wiechert reaction, was reported in 1971 and is based on the foundations of stoichio-metric enamine chemistry by Stork and the mechanistic conclusions driven by Langebeck himself on some enzymatic reactions, and outlines for the first time the reversible formation of a nucleophilic enamine as the key intermediate participating in the catalytic cycle.

Some other very important events in the historic development of asymmetric organocatalysis appeared between 1980 and the late 1990s, such as the devel-opment of the enantioselective alkylation of enolates using cinchona-alkaloid-based quaternary ammonium salts under phase-transfer conditions[18] or the use of chiral Brønsted acids by Inoue[19] or Jacobsen[20] for the asymmetric hydro-cyanation of aldehydes and imines respectively. These initial reports acted as the launching point for a very rich chemistry that was extensively developed in the following years, such as the enantioselective catalysis by H-bonding acti-vation or the asymmetric phase-transfer catalysis. The same would apply to the development of enantioselective versions of the Morita–Baylis–Hillman reac-tion,[21] to the use of polyamino acids for the epoxidation of enones, also known as the Julià epoxidation[22] or to the chemistry by Denmark in the phosphor-amide-catalyzed aldol reaction.[23]

It was in 2000 that Barbas and List reported their well-known proline-cat-alyzed enantioselective intermolecular aldol reaction (Scheme 1.4),[24] as the culmination of a research which started in the 1990s with the use of aldolase antibodies as catalysts for the aldol reaction.[25] Trying to provide a mechanistic rationale for understanding these reactions, and with the evidence of enamine intermediates participating in the reaction in hand, they developed the proline-catalyzed intermolecular aldol reaction in an attempt to mimic the enzyme's behavior. Another important landmark in this context was the introduction of the iminium catalysis concept by MacMillan, related to the enantioselective

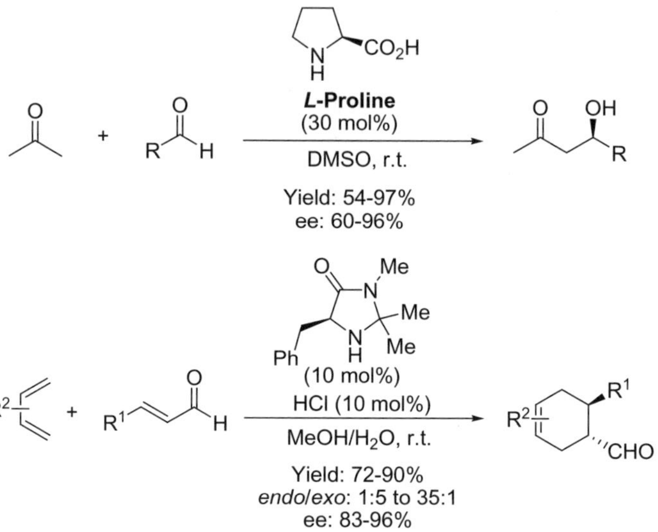

Scheme 1.4 Landmarks in the development of aminocatalysis by Barbas and List and by MacMillan, respectively.

amine-catalyzed Diels–Alder reaction, which was also reported in 2000.[26] It was MacMillan himself who, in this Diels–Alder paper, coined the term "organocatalysis" as a substitute of the usually employed "metal-free catalysis". These two seminal papers settled the start point for the development of aminocatalysis as it is known nowadays.

Finally, other important contributions that should be cited as very relevant with regard to the evolution of the organocatalysis field are those related to the use of stable *N*-heterocyclic carbenes as catalysts. The basis of this chemistry also finds its roots at the very early beginnings of organic synthesis, in particular in the well-known cyanide-catalyzed benzoin condensation by Liebig and Whöler,[27] and

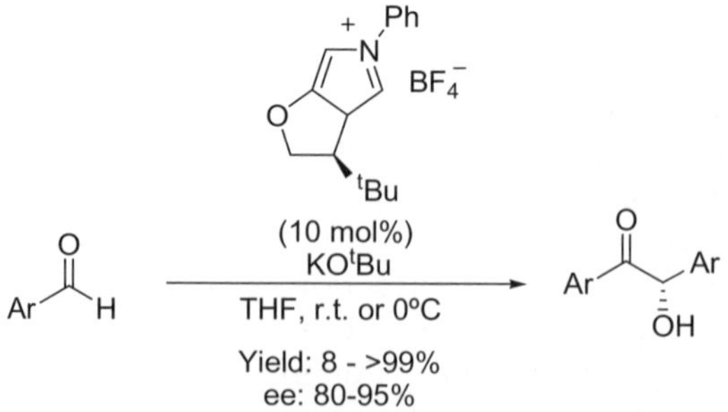

Scheme 1.5 The enantioselective *N*-heterocyclic carbene catalyzed benzoin condensation reported by Enders.

the related thiazolium salt-catalyzed benzoin condensation by Ugai.[28] This is another example of chemists imitating nature in the sense that nucleophilic acylation reactions like the benzoin condensation are biochemical processes catalyzed by transketolase enzymes in the presence of thiamine as a coenzyme, which involve the participation of *N*-heterocyclic carbene intermediates. In spite of these very early precedents the first organocatalytic enantioselective reaction promoted by chiral carbenes furnishing high levels of enantioselection had to wait until the twenty-first century, with the first example of a highly efficient benzoin condensation reported by Enders in 2002 (Scheme 1.5).[29]

1.3 Types of Organocatalysts

Organocatalysts are usually classified into different groups according to the mechanistic pathway involved in the corresponding catalytic cycle and the type of interaction/chemical reaction involved in the activation of the reagents. In this sense, organocatalysts are typically classified into two generic groups, according to the interaction between the substrate and the catalyst, namely "covalent catalysis" and "non-covalent catalysis". This classification was already outlined by Langebeck in his book "*Organic Catalysts and Their Relations with Enzymes*" published in 1949, but it is still applicable to the modern organocatalytic reactions developed very recently. Some examples are given in Figure 1.2.

Catalysts activating the substrate of a given reaction by forming a covalent bond are among the most widely used and studied types of organocatalysts to date. This method of activation implies that reversible chemical reactions have to be available for the attaching and detaching of the catalyst to the substrate/final product in order to allow activation and catalyst turnover. The strong nature of the substrate-to-catalyst interaction allows an effective and well-defined influence of the catalyst in the stereochemical outcome of the reaction, although it might make the turnover difficult, which usually results in the need for rather high catalyst loadings in order to achieve good conversion and also to rather long reaction times in many cases. Chiral amines (aminocatalysts) belong to this group of organocatalysts, participating in many types of reactions by activating the substrates by the reversible formation of azomethine compound (an enamine,[30] iminium ion[31] or iminium-radical cation, also known as SOMO catalysis[32]). Another important class of these catalysts are *N*-heterocyclic carbenes,[33] which participate by the formation of the Breslow intermediate, several tertiary amines or analogues, which participate as Lewis bases in reactions *via* the corresponding ylides[34] or acylammonium salts[35] and trialkylphosphines and trialkylamines, with the same behavior in the Morita–Baylis–Hillman reaction.[36] The oxidation reactions *via* chiral oxiranes also belong to this type of organocatalysis.[37]

With regard to catalysts interacting with the reagents participating in a given reaction by weaker interactions, the most important and relevant activation mechanism is that involving the formation of catalyst-substrate complexes by

ORGANOCATALYSIS

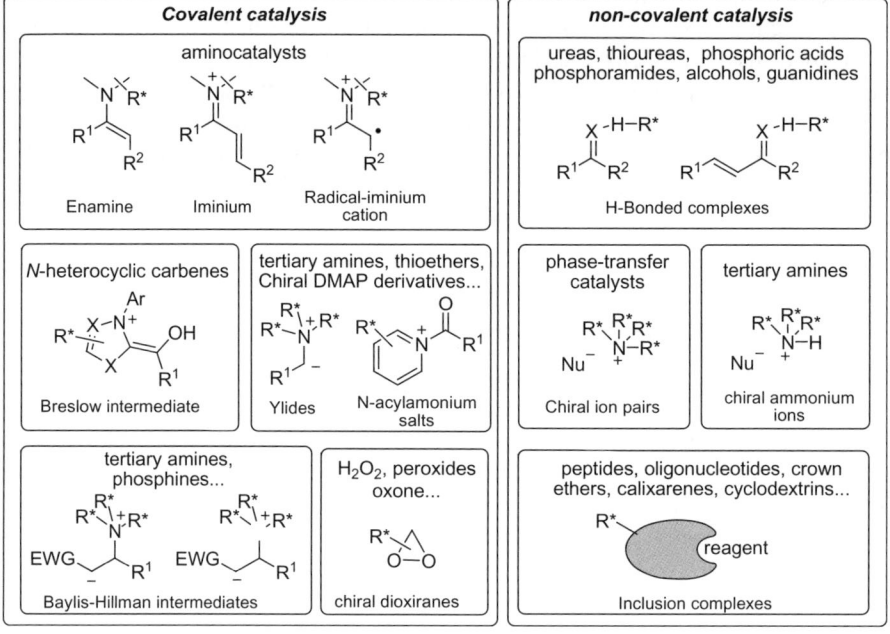

Figure 1.2   Classification of organocatalysts into "covalent" and "non-covalent" catalysis.

the formation of hydrogen bonds.[38] Other important organocatalytic methodologies are also included in this group such as those involving the formation of chiral ion pairs in phase-transfer catalysis[39] and the use of chiral tertiary amines as bases in the activation of a nucleophile by deprotonation, forming a chiral ammonium salt.[40] Also in this context, the use of peptides,[41] chiral crown ethers or other molecules of small molecular weight participating in the activation of the substrate by the formation of supramolecular host-guest complexes as catalysts should be included in this group of organocatalysts.

There is also an alternative classification of the different organocatalytic methodologies outlined by List,[42] in which organocatalysts are characterized according to their acid/base reactivity (Figure 1.3). In this sense, organocatalysts are classified as Lewis bases, which participate in the reaction by attaching to the substrate *via* nucleophilic addition or substitution, Brønsted bases, which activate one of the reagents by exerting their deprotonation, Brønsted acids, which are involved in the activation of the substrate by protonation/H-bonding interactions, and Lewis bases, which also initiate the catalytic cycle by interacting with the nucleophile through Lewis acid/base interactions. According to this classification, aminocatalysts, *N*-heterocyclic carbenes and those amines, thioethers, phosphines and related compounds participating in the formation of ylides or in Morita–Baylis–Hillman type reactions should be considered as

ORGANOCATALYSIS

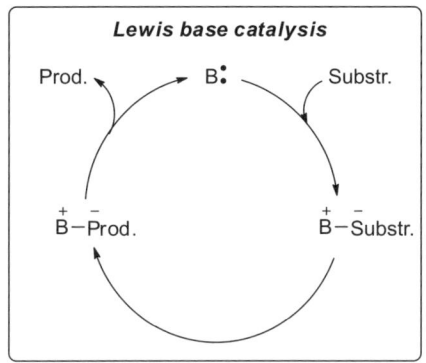

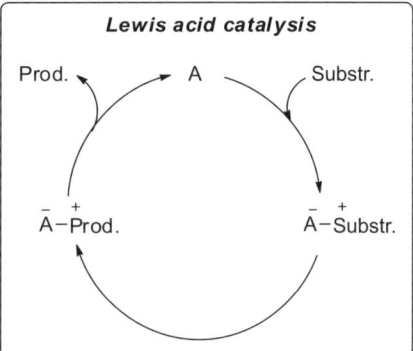

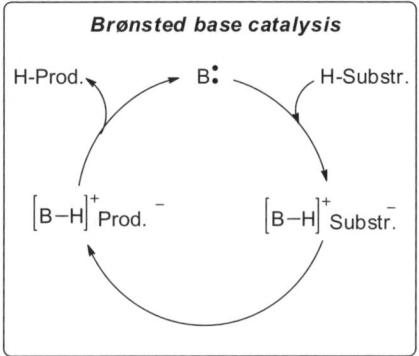

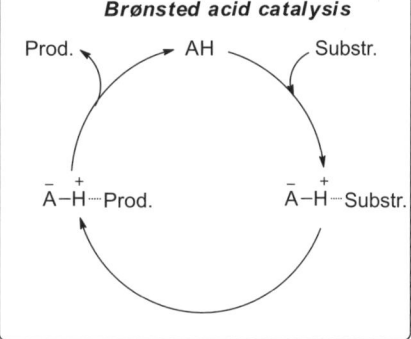

Figure 1.3   Classification of organocatalysts according to their acid/base reactivity.

Lewis bases and those participating by H-bonding interactions would belong to the Brønsted acids group. On the other hand, chiral amines or guanidines exerting their influence by activating a pro-nucleophile by deprotonation would be classified as Brønsted bases, and phase-transfer catalysis, the chiral dioxiranes involved in the oxidation of olefins, chiral *N*-oxides[43] and Denmark's phosphoramides[44] among many others would be included in the family of Lewis acid catalysis. The place that those catalysts forming inclusion complexes should occupy would vary depending on the interactions dominating the formation of these supramolecular assemblies.

1.4   The Conjugate Addition Reaction: a Very Convenient Platform for Asymmetric Organocatalysis

The conjugate addition reaction is regarded as one of the most powerful methods for the preparation of complex molecules.[45] In this transformation, a

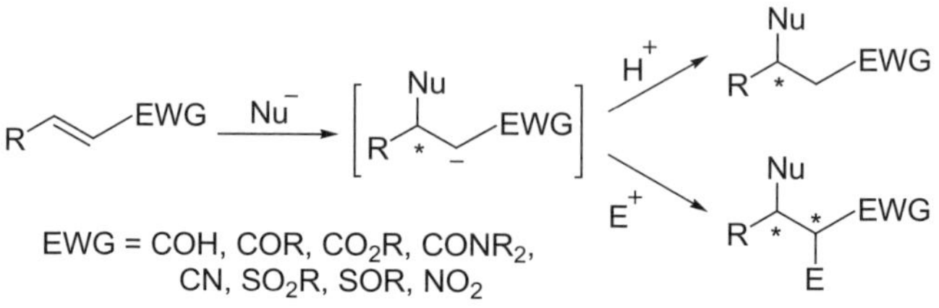

EWG = COH, COR, CO$_2$R, CONR$_2$,
CN, SO$_2$R, SOR, NO$_2$

Scheme 1.6   General scheme for the conjugate addition reaction.

nucleophile (also referred to as a donor) adds to the β-carbon of an electron-deficient olefin (usually known as an acceptor) giving a stabilized carbanion intermediate, which, after protonation or subsequent treatment with another electrophile, furnishes the final addition product (Scheme 1.6). The synthetic versatility of this reaction relies mainly upon the broad spectrum of donors and acceptors that may be employed, as has been demonstrated by the huge number of examples in which it has been applied as a key strategic transformation in total synthesis. The nucleophiles can be either carbon- or heteroatom-centered and the acceptors are usually α,β-unsaturated carbonyl compounds (aldehydes, ketones, esters, amides, *etc.*), although other activating groups like nitro, sulfonate, sulfoxide, phosphate or phosphonate have also been successfully employed.

In the last few years, many different organocatalytic versions of this important transformation have been reported, using a wide variety of different nucleophiles and conjugate acceptors. The range of different electron-withdrawing groups amenable to use for the activation of the olefin toward conjugate addition mean that many different activation mechanisms can apply to catalyze this reaction using organic molecules. For example, α,β-unsaturated aldehydes can be activated under iminium catalysis, by nitroalkenes, by H-bonding, *etc.* On the other hand, the rich profile of available Michael donors also opens up an unlimited field for the development of new applications, also allowing the use of different organocatalysts operating by their own mechanistic pathway (aldehydes can be activated *via* enamine catalysis, simple esters by phase-transfer catalysis, *etc.*). Moreover, the potential of the conjugate addition reaction to be applied in tandem, domino or cascade reactions has also been considered by some authors, leading to extremely effective novel synthetic methodologies for the preparation of molecules of high complexity in a single step.[46] Organocatalytic conjugate reactions[47] and cascade processes have been reviewed quite recently.[48]

1.5   On the Organization of This Book

The aim of this book is to provide a picture on the "state of the art" around the methodologies reported until 2009 for carrying out enantioselective conjugate additions using organocatalysis as the methodological approach to achieve stereocontrol. The book has been organized into different chapters according to the classification of organocatalysts with regard to the mechanism operating in the activation of the reagents during the conjugate addition process. For this reason, enamine activation, iminium activation, H-bonding-mediated reactions, phase-transfer catalysis and other miscellaneous organocatalytic reactions have been covered in their corresponding chapters from Chapter 2 to Chapter 6. It should be mentioned that several transformations like the Morita–Baylis–Hillman or related reactions have not been considered in this book. The reason for this is that, even though these processes indeed involve a conjugate addition process in their mechanism, these reactions do not end up in a final product in which formally a nucleophile has been incorporated at the β-position of the electron-withdrawing group which activates the olefin.

It also has to be pointed out that, in many cases, bifunctional catalysts have been employed, which operate by two or more mechanistic profiles in a synergistic form. In these cases, we have selected the way of activation which, by our understanding, was crucial for catalytic activity and we have accordingly classified the methodology in the corresponding chapter. Other authors' points of view may vary from ours, leading to a different classification of a given catalyst/reaction. Finally, we have dedicated a final chapter to cascade reactions initiated by organocatalytic conjugate additions, independently from the activation mode of the catalyst employed, although inside the chapter all these methodologies have also been organized according to mechanistic considerations.

As was pointed out earlier, a huge number of contributions from many different laboratories worldwide can be found in the literature with many examples of very useful and effective protocols for carrying out organocatalytic enantioselective conjugate addition reactions. We have made a great effort in compiling most of them into this book but it might be that a particular relevant report has been involuntarily omitted. The authors would like to apologize in advance if this has been the case.

References and Notes

1. For three reference works on asymmetric organocatalysis see (a) P. I. Dalko, *Enantioselective Organocatalysis*, Wiley-VCH, Weinheim, 2007; (b) A. Berkessel and H. Gröger, *Asymmetric Organocatalysis*, VCH, Weinheim, Germany, 2004; (c) M. T. Reetz, B. List, S. Jaroch and H. Weinmann, *Organocatalysis*, Ernst Schering Foundation Symposium Proceedings, Springer-Verlag, Berlin, 2007.
2. For some general reviews on asymmetric organocatalysis see (a) S. Bertelsen and K. A. Jørgensen, *Chem. Soc. Rev.*, 2009, **38**, 2178; (b) C. Palomo, M. Oiarbide and R. López, *Chem. Soc. Rev.*, 2009, **38**, 632; (c) D. W. C. MacMillan, *Nature*, 2008, **455**, 304; (d) P. Melchiorre, M. Marigo, A. Carlone and G. Bartoli, *Angew. Chem. Int. Ed.*, 2008, **47**, 6138; (e) A. Dondoni

and A. Massi, *Angew. Chem. Int. Ed.*, 2008, **47**, 4638; (f) Special issue on organocatalysis, *Chem. Rev.*, 2007, **107**(12); (g) Special issue on organocatalysis, *Acc. Chem. Res.*, 2004, **37**(8); (h) H. Pellissier, *Tetrahedron*, 2007, **63**, 9267; (i) B. List and J.-W. Yang, *Science*, 2006, **313**, 1584; (j) M. J. Gaunt, C. C. C. Johansson, A. McNally and N. C. Vo, *Drug Discovery Today*, 2007, **2**, 8; (k) P. I. Dalko and L. Moisan, *Angew. Chem. Int. Ed.*, 2004, **43**, 5138; (l) B. List, *Chem. Commun.*, 2006, 819; (m) P. I. Dalko and L. Moisan, *Angew. Chem. Int. Ed.*, 2001, **40**, 3726.

3. For an authoritative essay see (a) C. F. Barbas III, *Angew. Chem. Int. Ed.*, 2008, **47**, 42 For a perspective see; (b) D. Enders and A. A. Narine, *J. Org. Chem.*, 2008, **73**, 7857.

4. Source: Scifinder Scholar search under terms "asymmetric catalysis" and next "metal catalysis", "enzyme catalysis" and "organocatalysis" or papers reported in English.

5. For a review see R. M. de Figueiredo and M. Christmann, *Eur. J. Org. Chem.*, 2007, 2575.

6. (a) S. Pizzarello and A. L. Weber, *Science*, 2004, **303**, 1151; (b) J. R. Crowin and S. Pizzarello, *Science*, 1997, **275**, 951.

7. See for example (a) M. Klussmann, A. J. R. White, A. Armstrong and D. G. Blackmond, *Angew. Chem. Int. Ed.*, 2006, **45**, 7985; (b) M. Klussmann, H. Iwamura, S. Matthew, D. Wells, U. Pandya, A. Armstrong and D. G. Blackmond, *Nature*, 2006, **441**, 621 and references therein. See also; (c) J. Casas, M. Engquist, I. Ibrahem, B. Kaynak and A. Córdova, *Angew. Chem. Int. Ed.*, 2005, **44**, 1343; (d) L. E. Orgel, *Science*, 2000, **290**, 1306.

8. G. Bredig and R. W. Balcom, *Ber. Deusch. Chem. Ger.*, 1908, **41**, 740.

9. G. Bredig and K. Fajans, *Ber. Deusch. Chem. Ger.*, 1908, **41**, 752.

10. G. Bredig and P. S. Fiske, *Biochem. Z.*, 1913, **46**, 7.

11. W. Langebeck, *Die Organische Katalysatoren und ihre Beziehungen zu den Fermenten*, Springer-Verlag, Berlin, 1949.

12. (a) W. Langenbeck, *Angew. Chem.*, 1928, **41**, 740; (b) W. Langenbeck, *Angew. Chem.*, 1932, **45**, 97.

13. H. Pracejus, *Justus Liebig Ann. Chem.*, 1960, **634**, 9.

14. B. Langström and G. Bergson, *Acta Chem. Scand.*, 1973, **27**, 3118.

15. (a) H. Wynberg and R. Helder, *Tetrahedron Lett.*, 1975, **16**, 4057; (b) K. Hermann and H. Wynberg, *J. Org. Chem.*, 1979, **44**, 2238; (c) R. Helder, R. Arends, W. Bolt, H. Hiemstra and H. Wynberg, *Tetrahedron Lett.*, 1977, **18**, 2181; (d) H. Hiemstra and H. Wynberg, *J. Am. Chem. Soc.*, 1981, **103**, 417. For a review see; (e) H. Wynberg, *Top. Stereochem.*, 1986, **16**, 87.

16. (a) Z. G. Hajos and D. R. Parrish, German Patent DE 2102623, 1971; (b) Z. G. Hajos and D. R. Parrish, *J. Org. Chem.*, 1974, **39**, 1615.

17. (a) U. Eder, G. Sauer and R. Wiechert, *German Patent* DE 2014757, 1971; (b) U. Eder, G. Sauer and R. Wiechert, *Angew. Chem. Int. Ed. Engl.*, 1971, **10**, 496.

18. (a) U.-H. Dolling, P. Davis and E. J. J. Grabowski, *J. Am. Chem. Soc.*, 1984, **106**, 446; (b) R. S. E. Conn, A. V. Lovell, S. Karady and L. M. Weinstock, *J. Org. Chem.*, 1986, **51**, 4710.

19. J. Oku and S. Inoue, *J. Chem. Soc. Chem. Commun.*, 1981, 229.

20. (a) M. S. Sigman and E. N. Jacobsen, *J. Am. Chem. Soc.*, 1998, **120**, 4901 See also; (b) P. Vachal and E. N. Jacobsen, *J. Am. Chem. Soc.*, 2002, **124**, 10012.

21. (a) K. Morita, Z. Suzuki and H. Hirose, *Bull. Chem. Soc. Jpn.*, 1968, **41**, 2815; (b) A. B. Baylis and A. D. Hillman, *German Patent* DE 2155113, 1974. *Chem. Abstr.*, 1972, 77, 34174q.

22. S. Julià, J. Masana and J. C. Vega, *Angew. Chem. Int. Ed.*, 1980, **19**, 929.

23. S. E. Denmark, S. B. D. Winter, X. Su and K.-T. Wong, *J. Am. Chem. Soc.*, 1996, **118**, 7404.

24. B. List, R. A. Lerner and C. F. Barbas III, *J. Am. Chem. Soc.*, 2000, **122**, 2395.

25. For an account see C. F. Barbas III, *Angew. Chem. Int. Ed.*, 2008, **47**, 42.

26. K. A. Ahrendt, C. J. Borths and D. W. C. MacMillan, *J. Am. Chem. Soc.*, 2000, **122**, 4243.

27. F. Wöhler and J. Liebig, *Ann. Pharm.*, 1832, **3**, 249.

28. T. Ugai, S. Tanaka and S. Dokawa, *J. Pharm. Soc. Jpn.*, 1943, **63**, 296.

29. D. Enders and U. Kallfass, *Angew. Chem. Int. Ed.*, 2002, **41**, 1743.

30. Reviews: (a) S. Mukherjee, J. W. Yang, S. Hoffmann and B. List, *Chem. Rev.*, 2007, **107**, 5471; (b) S. Sulzer-Mossé and A. Alexakis, *Chem. Commun.*, 2007, 3123; (c) T. Kano and K. Maruoka, *Chem. Commun.*, 2008, 5465.

31. Reviews: (a) A. Erkkilä, I. Majander and P. M. Pihko, *Chem. Rev.*, 2007, **107**, 5416; (b) G. Lelais and D. W. C. MacMillan, *Aldrichimica Acta*, 2006, **39**, 79.

32. Highlights: (a) S. Bertelsen, M. Nielsen and K. A. Jørgensen, *Angew. Chem. Int. Ed.*, 2007, **46**, 7356; (b) P. Renaud and P. Leong, *Science*, 2008, **322**, 55; (c) P. Melchiorre, *Angew. Chem. Int. Ed.*, 2009, **48**, 1360.

33. Reviews: (a) D. Enders, O. Niemeier and A. Henseler, *Chem. Rev.*, 2007, **107**, 5606; (b) N. Marion, S. Diez-Gonzalez and S. P. Nolan, *Angew. Chem. Int. Ed.*, 2007, **46**, 2988.

34. Reviews: (a) M. J. Gaunt and C. C. C. Johanson, *Chem. Rev.*, 2007, **107**, 5596; (b) E. M. McGarrigle, E. L. Myers, O. Illa, M. A. Shaw, S. L. Riches and V. K. Aggarwal, *Chem. Rev.*, 2007, **107**, 5841; (c) S. France, D. J. Guerin, S. J. Miller and T. Lectka, *Chem. Rev.*, 2003, **103**, 2985.

35. Review: R. P. Wurz, Chem. Rev., 2007, 107, 5570.

36. Reviews: (a) V. Declerck, J. Martinez and F. Lamaty, *Chem. Rev.*, 2009, **109**, 1; (b) P. R. Krishna, R. Sachwani and P. S. Reddy, *Synlett*, 2008, 2897; (c) D. Basavaiah, K. V. Rao and R. J. Reddy, *Chem. Soc. Rev.*, 2007, **36**, 1581; (d) Y.-L. Shi and M. Shi, *Eur. J. Org. Chem.*, 2007, 2905; (e) G. Masson, C. Housseman and J. Zhu, *Angew. Chem. Int. Ed.*, 2007, **46**, 4614.

37. Reviews: (a) M. Frohn and Y. Shi, *Synthesis*, 2000, 1979; (b) S. E. Denmark and Z. Wu, *Synlett*, 1999, **847**.
38. Reviews: (a) X. H. Yu and W. Wang, *Chem. Asian J.*, 2008, **3**, 516; (b) S. J. Connon, *Chem. Commun.*, 2008, 2499; (c) H. Miyabe and Y. Takemoto, *Bull. Chem. Soc. Jpn.*, 2008, **81**, 785; (d) A. G. Doyle and E. N. Jacobsen, *Chem. Rev.*, 2007, **107**, 5173; (e) M. S. Taylor and E. N. Jacobsen, *Angew. Chem. Int. Ed.*, 2006, **45**, 1520; (f) S. J. Connon, *Chem. Eur. J.*, 2006, **12**, 5418; (g) P. R. Schreiner, *Chem. Soc. Rev.*, 2003, **32**, 289.
39. Reviews: (a) S. Jew and H. Park, *Chem. Commun.*, 2009, 7090; (b) T. Hashimoto and K. Maruoka, *Chem. Rev.*, 2007, **107**, 5656; (c) T. Ooi and K. Maruoka, *Aldrichimica Acta*, 2007, **40**, 77; (d) T. Ooi and K. Maruoka, *Angew. Chem. Int. Ed.*, 2007, **46**, 4222; (e) M. J. O'Donnell, *Acc. Chem. Res.*, 2004, **37**, 506; (f) B. Lygo and B. Andrews, *Acc. Chem. Res.*, 2004, **37**, 518; (g) K. Maruoka and T. Ooi, *Chem. Rev.*, 2003, **103**, 3013; (h) D. Martyres, *Synlett*, 1999, 1508; (i) A. Nelson, *Angew. Chem. Int. Ed.*, 1999, **38**, 1583.
40. (a) C. Palomo, M. Oiarbide and R. Lopez, *Chem. Soc. Rev.*, 2009, **38**, 632; (b) J. Shen and C.-H. Tan, *Org. Biomol. Chem.*, 2008, **6**, 3229.
41. Reviews: (a) E. A. Colby, S. M. Davie, Y. X. Mennen and S. J. Miller, *Chem. Rev.*, 2007, **107**, 5759; (b) E. R. Jarvo and S. J. Miller, *Tetrahedron*, 2002, **58**, 2481.
42. J. Seayad and B. List, *Org. Biomol. Chem.*, 2005, **3**, 719.
43. (a) A. Malkov and P. Kocovsky, *Eur. J. Org. Chem.*, 2007, 29; (b) J. Chen and N. Takenaka, *Chem. Eur. J.*, 2009, **15**, 7268.
44. Review: S. E. Denmark, Chimia, 2008, 62, 37.
45. (a) P. Perlmutter, *Conjugate Addition Reactions in Organic Synthesis*, Pergamon Press, Oxford, 1992. For general reviews on asymmetric conjugate additions see; (b) J. Christoffers, *Chem. Eur. J.*, 2003, **9**, 4862; (c) T. Hayashi and K. Yamasaki, *Chem. Rev.*, 2003, **103**, 2829; (d) N. Krause and A. Hoffmann-Röder, *Synthesis*, 2001, 171; (e) M. P. Sibi and S. Manyem, *Tetrahedron*, 2000, **56**, 8033; (f) J. Leonard, E. Díez-Barra and S. Merino, *Eur. J. Org. Chem.*, 1998, 2051; (g) N. Krause, *Angew. Chem. Int. Ed.*, 1998, **37**, 283; (h) B. E. Rossiter and N. M. Swingle, *Chem. Rev.*, 1992, **92**, 771; (i) J. d'Angelo, D. Desmaele, F. Dumas and A. Guingant, *Tetrahedron: Asymmetry*, 1992, **3**, 459.
46. For some selected reviews see (a) K. C. Nicolaou and J. S. Chen, *Chem. Soc. Rev.*, 2009, **38**, 2993; (b) A. Padwa, *Chem. Soc. Rev.*, 2009, **38**, 3072; (c) J. Poulin, C. M. Grisé-Bard and L. Barriault, *Chem. Soc. Rev.*, 2009, **38**, 3092; (d) S. Perreault and T. Rovis, *Chem. Soc. Rev.*, 2009, **38**, 3149; (e) N. Ismabery and R. Lavila, *Chem. Eur. J.*, 2008, **14**, 8444; (f) C. J. Chapman and C. G. Frost, *Synthesis*, 2007, 1; (g) H. Miyabe and Y. Takemoto, *Chem. Eur. J.*, 2007, **13**, 7280; (h) N. T. Patil and Y. Yamamoto, *Synlett*, 2007, 994; (i) J. Zhu and H. Bienayme ed., *Multicomponent Reactions*, Wiley-VCH, Weinheim, 2005; (j) D. Tejedor and F. Garcia-Tellado, *Chem. Soc. Rev.*, 2007, **36**, 484; (k) L. F. Tietze, G. Brasche and K. Gerike, *Domino Reactions in Organic Chemistry*,

Wiley-VCH, Weinheim, 2006; (l) K. C. Nicolaou, D. J. Edmonds and P. G. Bulger, *Angew. Chem. Int. Ed.*, 2006, **45**, 7134; (m) A. Doemling, *Chem. Rev.*, 2006, **106**, 17; (n) H.-C. Guo and J.-A. Ma, *Angew. Chem. Int. Ed.*, 2006, **45**, 354; (o) D. J. Ramon and M. Yus, *Angew. Chem. Int. Ed.*, 2005, **44**, 1602.

47. (a) J. L. Vicario, D. Badia and L. Carrillo, *Synthesis*, 2007, 2065; (b) D. Almaçi, D. A. Alonso and C. Nájera, *Tetrahedron: Asymmetry*, 2007, **18**, 299; (c) S. B. Tsogoeva, *Eur. J. Org. Chem.*, 2007, 1701.

48. (a) X. Yu and W. Wang, *Org. Biomol. Chem.*, 2008, **6**, 2037; (b) D. Enders, C. Grondal and M. R. M. Huettl, *Angew. Chem. Int. Ed.*, 2007, **46**, 1570; (c) G. Guillena, D. J. Ramon and M. Yus, *Tetrahedron: Asymmetry*, 2007, **18**, 693.

CHAPTER 2

Enantioselective Conjugate Addition Reactions via Enamine Activation

2.1 Introduction

Primary or secondary amines can be employed as catalysts in Michael-type reactions because of their ability to activate enolizable aldehydes or ketones toward their participation as nucleophiles in conjugate addition reactions *via* reversible formation of an enamine intermediate species (Scheme 2.1).[1] After the conjugate addition step takes place, an iminium ion is generated, which, upon hydrolysis, releases the product and releases the amine catalyst for its participation on a subsequent catalytic cycle. Two important issues are behind this breakthrough concept: first of all, the known reversibility of the reaction taking place in the formation of enamine/iminium ions allows the turnover of the amine reagent, allowing the use of this compound in substoichiometric amounts. Secondly, the condensation of the amine with the starting carbonyl compound results in the initial formation of an iminium ion with an increased α-C–H acidity because of the lower energy of its LUMO frontier orbital. This induces a fast deprotonation leading to the formation of an enamine intermediate in which a significant raising of the energy of its HOMO frontier orbital results in the formation of an active nucleophile suitable to react with other electrophiles like, for example, with electron-poor alkenes as in the case of the Michael reaction.

The foundations of this concept (enamine activation) lie in the fundamental studies by Stork and Robinson covering the stoichiometric use of enamine nucleophiles for the formation of C–C bonds. The Hajos–Parrish–Eder–Sauer–Wiechert reaction reported in 1971 (Scheme 2.2),[2] which consisted of a

RSC Catalysis Series No. 5
Organocatalytic Enantioselective Conjugate Addition Reactions: A Powerful Tool for the Stereocontrolled Synthesis of Complex Molecules
By Jose L. Vicario, Dolores Badía, Luisa Carrillo and Efraim Reyes
© J. L. Vicario, D. Badía, L. Carrillo and E. Reyes 2010
Published by the Royal Society of Chemistry, www.rsc.org

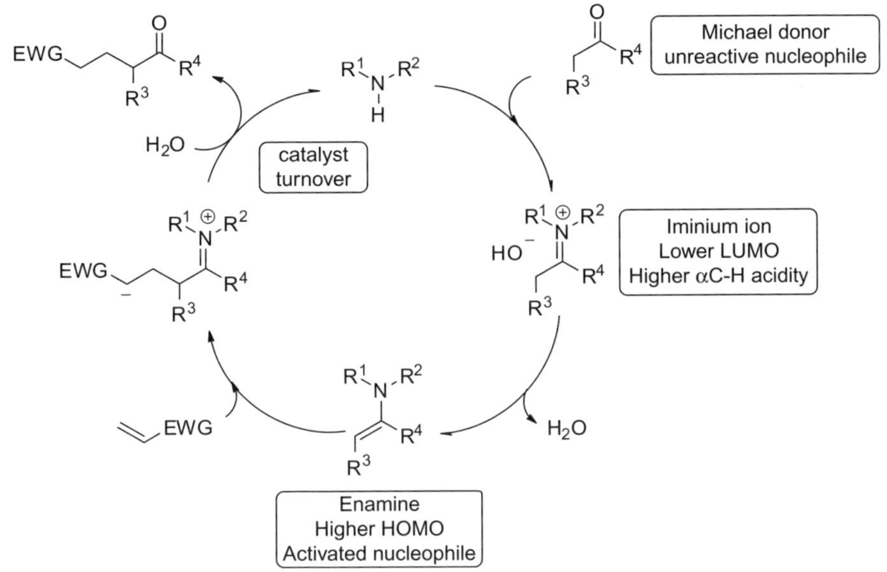

Scheme 2.1 Catalytic cycle for the Michael reaction proceeding *via* enamine activation.

proline-catalyzed enantioselective intramolecular aldol reaction used for the preparation of a valuable chiral intermediate employed in the synthesis of progesterone derivatives, was the first example in which this kind of chemistry crossed the barrier and entered into the field of asymmetric catalysis. However, this chemistry remained completely unattended for many years until 2000, when Barbas and List reported their well-known proline-catalyzed enantioselective intermolecular aldol reaction (Scheme 2.2).[3] In fact, this was the culmination of a research which started in the 1990s with the use of aldolase antibodies as catalysts for the aldol reaction, continued with the efforts directed to provide a mechanistic rationale for understanding these reactions and, with the evidence of enamine intermediates participating in the reaction in hand, ended up with the development of the proline-catalyzed intermolecular aldol reaction in an attempt to mimic the enzyme's behavior.[4]

Once these pioneering proline-catalyzed reactions were understood from the mechanistic point of view, the enamine activation concept was extended to a wide variety of reactions, especially those involving the use of different electrophiles able to engage in reactions with the intermediate enamine reagent formed. This statement particularly applies to the Michael reaction and, in the following years, a plethora of new reactions and methods were developed for carrying out the amine-catalyzed conjugate addition reaction of aldehydes or ketones to highly electrophilic α,β-unsaturated carbonyl compounds or related species like nitroalkenes or vinyl sulfones among others.[5] Obviously special emphasis was placed on carrying out the corresponding stereocontrolled

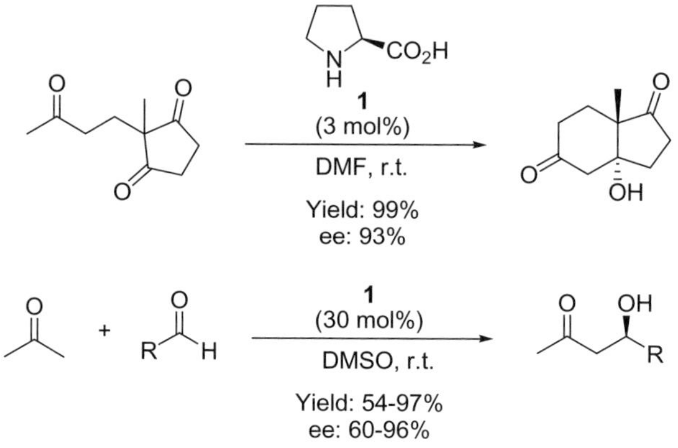

Scheme 2.2 First approaches toward the enamine activation concept.

versions of this important reaction. In addition, the initial exceptional results provided by proline were also very soon extended to other chiral amine catalysts like, for example, α,α-diarylprolinol silyl ethers,[6] which nowadays can be considered as one of the "privileged" classes of organocatalysts because of the exceptional performance furnished in many reactions, including conjugate additions.

The most remarkable feature of this strategy is that previous activation of the nucleophile (*e.g.* formation of enolate or silyl enol ether species) is avoided because it is the catalyst itself that activates the Michael donor reagent. This results in a higher atom-efficiency of the process and also allows carrying out the reaction under neutral and smooth conditions. An additional outstanding advantage of the enamine activation strategy applied to the Michael reaction is the capacity to employ unmodified aldehydes as donor reagents without the need for prior derivatization to the corresponding enolate or silyl enol ether, avoiding the extremely problematic formation of these activated reactive derivatives. On the other hand, it has to be mentioned that this methodology is limited to the use of Michael donors capable of forming enamines (aldehydes or ketones) and the lower reactivity of such nucleophiles very often limits the application of this reaction to the use of rather activated Michael acceptors.

2.1.1 Factors Influencing the Stereocontrol in Michael Additions Proceeding *via* Enamine Activation

The mechanism proposed for the secondary amine-catalyzed conjugate addition to electron-poor olefins has been outlined in Scheme 2.1. As can be seen in this catalytic cycle, all the steps involving enamine formation and hydrolysis are in dynamic equilibrium, therefore the irreversible C–C bond-formation step being the relevant process with regard to stereochemical control. As a consequence of this, it is postulated that it should be at this point in which the catalyst must exert its stereochemical influence by controlling the approach of the electrophile by one of the two diastereotopic faces of the enamine intermediate. This stereodiscrimination can be carried out by steric bias or, alternatively, by introducing a convenient stereodirecting element (a hydrogen bond donor or a positively or negatively charged substituent), which interacts with the electrophile and therefore directs its attack from a preferential position (Scheme 2.3). On the other hand, related to the simple diastereoselection, the reaction is typically proposed to occur *via* a synclinal acyclic transition state as initially proposed by Seebach and Golinski.[7] This synclinal approach is proposed to be stabilized by the electrostatic interactions which should arise between the partially positive nitrogen of the enamine and the negatively charged oxygen atoms at the electron-withdrawing group which activates the Michael acceptor that would develop as the reaction proceeds forward.

In this context, controlling the geometry of the formed enamine intermediate shows up as the first very important issue that has to be controlled if a highly stereoselective reaction is desired. Condensation of a secondary amine with an aldehyde or a ketone can lead to the formation of two diastereoisomers, *Z* or *E*, in varying proportions depending on the nature of the carbonyl reagent and the

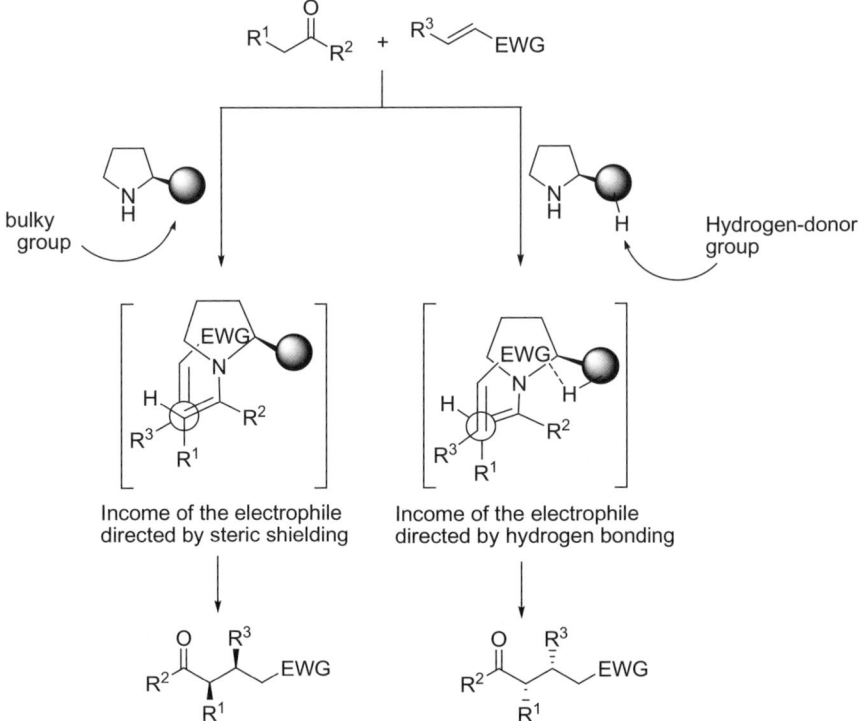

Scheme 2.3 Proposed mechanistic models for the Michael reaction under enamine activation.

secondary amine. These two diastereoisomers can interconvert through the formation of an intermediate iminium species (Scheme 2.4). However, due to the steric constraints imposed by the amine catalyst, which is usually a pyrrolidine-based heterocycle, the preferential formation of the less sterically encumbered *E* diastereoisomer is normally guaranteed. Moreover, the possibility that the enamine exists as a mixture of rotamers is also relevant to the stereochemical outcome of the reaction because the formation of a mixture of conformers of equal reactivity toward the Michael addition would eventually lead to a different shielding of the two faces of the nucleophile and therefore to the formation of mixtures of stereoisomers. Alternatively, when stereodirecting elements such as hydrogen donors are placed at the catalyst structure, the possible preferential formation of a more sterically congested conformation should also be considered if this results in a more reactive conformer because of the interaction with the stereodirecting element.

In general, better results are obtained when aldehydes are employed as Michael donors with respect to the use of ketones. Aldehydes condense much faster with the secondary amine catalyst to furnish the corresponding enamines and, in addition, the geometry and conformation control is much easier because, in this case, the preferential formation of the *s*-trans conformer is to be expected because of the large difference in size between the substituents of the enamine moiety ($R^1 = H$ in Scheme 2.4). On the other hand, the lower catalytic activity demonstrated by secondary amines when employing ketones as Michael donors is mainly derived from the inherent difficulties associated to the formation of such sterically congested enamine intermediates, which adds to the additional issue of geometry control of the enamine at the conjugate

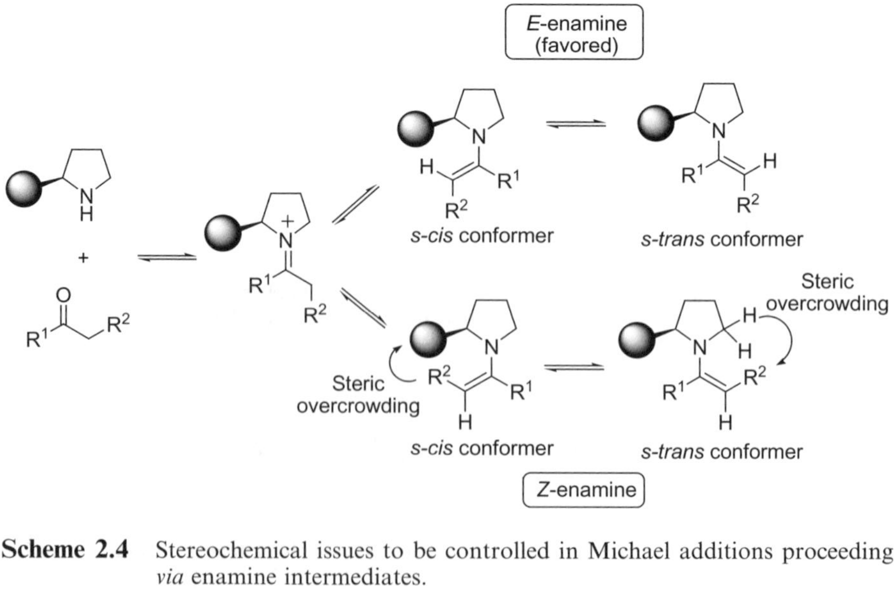

Scheme 2.4 Stereochemical issues to be controlled in Michael additions proceeding *via* enamine intermediates.

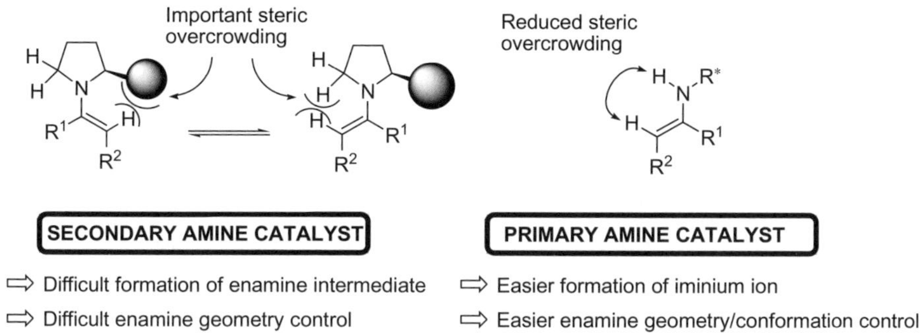

Figure 2.1 Primary *versus* secondary amine catalysts in the enamine activation of ketones.

addition step and the regioselectivity during its formation. This is not a problem when symmetrically substituted ketones (*e.g.* cyclohexanone) are employed because of the chemical equivalence of the two possible regioisomers. On the contrary, the use of unsymmetrically substituted ketones as Michael donors in which substituents of different size and nature are placed at the R^1 position makes this issue of enormous relevance in the design of a useful chiral amine catalyst. A solution to this problem has been provided more recently with the discovery that primary amines can operate as highly active and efficient aminocatalysts in several Michael transformations proceeding *via* enamine activation.[8] The condensation reaction between a ketone and a primary amine is sterically more feasible, delivering the corresponding imine, which will exist in equilibrium with the corresponding enamine. In addition, the geometry control on this intermediate is also easier because of the higher differences in steric bulk between the coplanar substituents across the conjugate N–C=C moiety (Figure 2.1).

2.2 Nitroalkenes as Michael Acceptors

As has already been mentioned, the low reactivity of enamine nucleophiles needs a highly electrophilic Michael acceptor for the reaction to proceed with good conversions in an acceptable time. In this context, the Michael reaction of aldehydes and ketones with nitroalkenes can be regarded as one of the most studied transformations in which the enamine activation concept has been applied. This reaction furnishes highly functionalized adducts with remarkable potential in organic synthesis, due to the synthetic versatility of the nitro group and the presence of the carbonyl moiety from the donor reagent.

2.2.1 Ketones as Michael Donors

The first attempts in the field were carried out in the context of the proline-catalyzed Michael reaction of ketones with nitrostyrene, which was initially

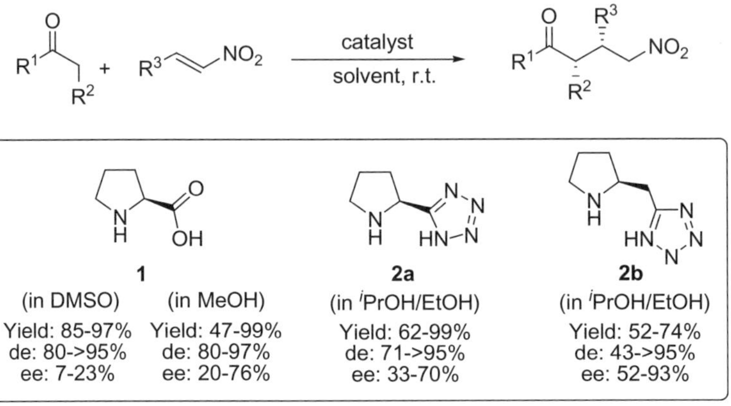

Scheme 2.5   Enantioselective Michael reactions of ketones with nitroalkenes catalyzed by proline and proline-tetrazole analogues.

reported by List, and was followed by improved versions of the initial reaction design by Enders (Scheme 2.5).[9] In all these cases, the obtained yields and diastereoselectivities were remarkably high, although moderate enantioselectivities were obtained. Soon afterwards, Ley developed the pyrrolidine-tetrazole catalyst **2a**, in which the carboxylate group had been replaced by the tetrazole moiety, which is considered an isostere of carboxylic acid due to their similar pK_a, and which was thought to have improved solubility in the reaction solvent, which was thought to be the main problem of the reaction.[10] In a subsequent report, it was found that the **2b** homologue was significantly more efficient in the same transformation, providing enantioselectivities higher than 90% ee for the reaction between six-membered cyclic ketones and different nitrostyrenes.[11] However, the reaction using acyclic ketones proceeded in all cases with rather low degrees of stereoselection.

Two different models were proposed by Ley for the **2b**-catalyzed reaction which should also be of application for the cases of proline **1** and proline-tetrazole catalysts **2a**, both of them in good agreement with the observed absolute configuration of the final Michael adducts (Figure 2.2).[11] One proposal involved the possibility of the tetrazole moiety acting as a bulky substituent which directed the income of the electrophile by the less hindered face of the enamine intermediate in the most stable pseudo-*trans* conformation. Alternatively, the formation of a hydrogen-bonded transition state was also proposed, in this case with the participation of the pseudo-*cis* enamine conformer. This second pathway was afterwards estimated to be the energetically most favored one by DFT calculations.[12]

This catalyst design in which the stereochemical outcome of the reaction is controlled by hydrogen bonding interactions with an acidic site at the catalyst has been applied to other differently modified chiral secondary amines, generally consisting of a 2-substituted pyrrolidine motif. Some representatives are

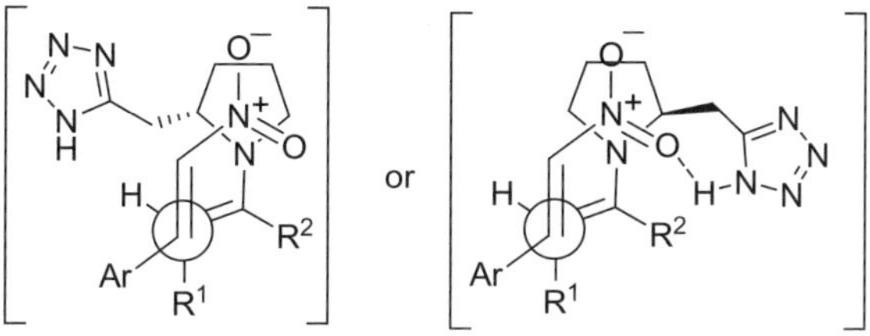

Figure 2.2 Proposed models for the **2b**-catalyzed Michael reaction between ketones and nitrostyrenes.

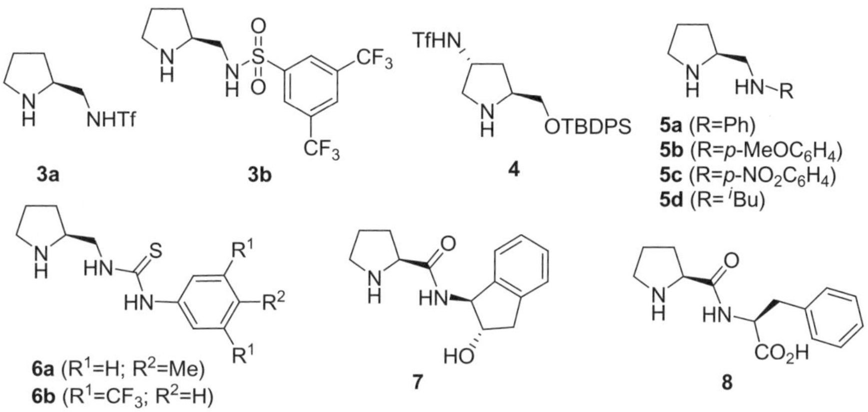

Figure 2.3 Some catalysts containing stereodirecting elements able to interact *via* hydrogen bonding with the nitroalkane electrophile.

shown in Figure 2.3, which include pyrrolidine sulfonamides **3a**,[13] **3b**[14] and **4**,[15] or pyrrolidine-aniline derivatives **5a–c**.[16] All these catalysts have demonstrated an outstanding performance with cyclohexanone or other related six-membered cyclic ketones as Michael donors but rather poor results were obtained with other cyclic ketones such as cyclopentanone or cycloheptanone or when acyclic ketones were employed as nucleophiles. Alternatively, the introduction of two potential hydrogen donors in the structure as a strategy to achieve a higher degree of stereocontrol in the reaction with the formation of a more rigid transition state has been explored by several authors,[17] as is the case of pyrrolidine-thioureas **6a**[17a] and **6b**,[17b] in which the known ability of the nitro group to engage selectively in a double hydrogen bond with the thiourea functionality guarantees a strong interaction between the intermediate enamine nucleophile and the nitroalkane acceptor, therefore favoring the formation of a very conformationally rigid transition state. In a similar context, Nájera has applied this concept for the development of aminoindole-containing derivative **7**, which was identified as the most effective catalyst from a wide array of

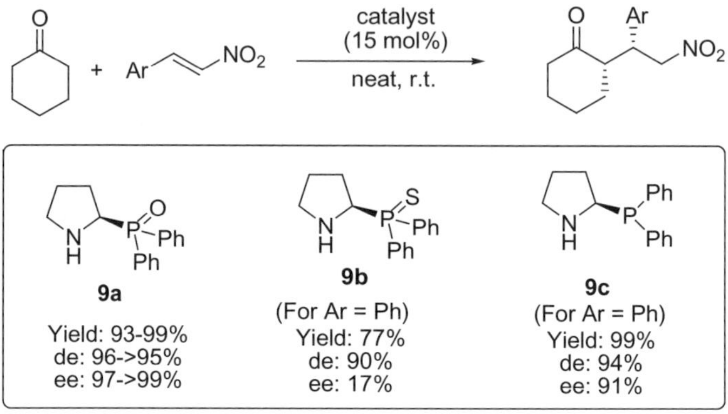

Scheme 2.6 Enantioselective Michael reactions of ketones with nitroalkenes catalyzed by **9a–c**.

different aminoalcohol-derived prolinamides tested in the Michael reaction of 3-pentanone with nitrostyrenes.[18] Moreover, in this case the interaction between the nucleophile and the electrophile by the formation of a hydrogen-bonded network was also supported by computational studies. The same concept lies behind the use of peptidyl-type catalyst **8** developed by Tsogoeva.[19]

An alternative reaction design has also been recently applied to this reaction by using a newly designed catalyst **9a** incorporating a diphenylphosphine oxide moiety in which the very polar P=O bond assists the attack of the electrophile by engaging in hydrogen bonding with the nitro moiety *via* an intercalating water molecule (Scheme 2.6).[20] Previous DFT calculations supported the formation of such a hydrogen-bonded transition state incorporating bridging water molecules between the enamine nucleophile and the Michael acceptor, which was afterwards confirmed in the reaction between cyclohexanone with different nitrostyrenes. Additional proof supporting the initial hypothesis was provided with the use of other related catalysts **9b** and **9c**, one containing a less polar P=S bond, and another incorporating a diphenylphosphine moiety (without the P=O polar bond), both of them furnishing lower diastereo- and enantioselectivities for the same model reaction, which was especially remarkable in the case of the P=S-containing catalyst **9b**.

An alternative catalyst design consists of the introduction of bulky groups at the pyrrolidine ring, which would exert their stereochemical influence *via* steric shielding of one of the diastereotopic faces of the enamine intermediate. In this context, a wide variety of different proline derivatives have been employed in this transformation, including prolinol silyl ether **10a**,[21a] homoprolinol silyl ether **11**,[21b] prolinal dithioacetal **12a**[22] and modified pyrrolidine **13** (Figure 2.4).[23] All these catalysts have been fundamentally tested in the reaction of cyclohexanone with nitrostyrenes, providing satisfactory results, although the use of other ketones as Michael donors usually led to poorer results. On the other hand, a series of different 1,2-diamines derived from proline-like **14a–c**,[24]

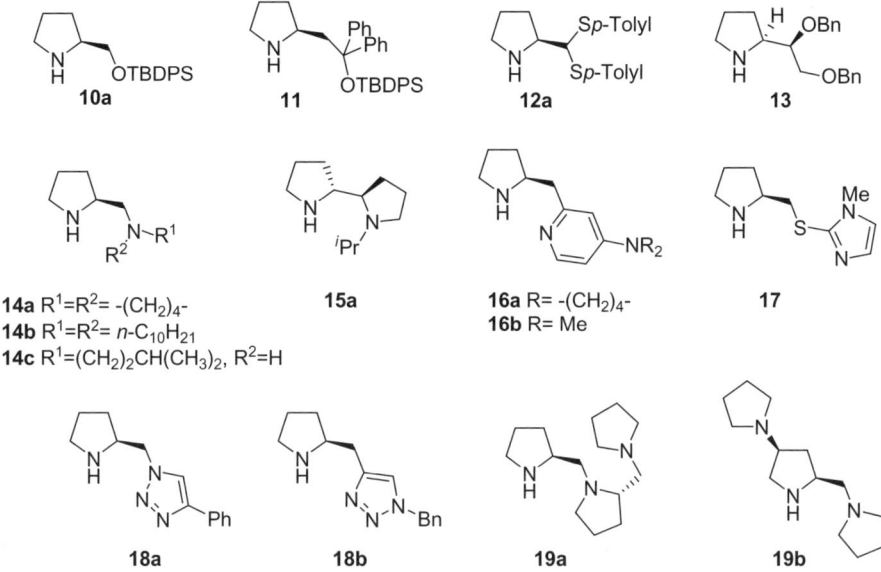

Figure 2.4 Other secondary amine catalysts employed in the enantioselective Michael addition of ketones to nitroalkenes.

and 2-(*N-iso*-propylpyrrolidin-2-yl)pyrrolidine (iPBP) **15a**,[25] pyrrolidine-pyridine derivatives **16a** and **16b**,[26] imidazolyl- and triazolyl-containing pyrrolidines **17**[27] and **18a–b**,[28] and triamines **19a** and **19b**[29] have also been successfully employed in the same context, trying to reach to more rigid reaction intermediates by the formation of additional interactions between the lateral amino moiety and the electrophile. As in the previous examples, the application of these newly designed catalysts was especially focused on the use of cyclohexanone or other cyclic six-membered ring ketones as Michael donors.

However, the activation mechanism operating when these diamine and triamine catalysts were employed deserves special attention. For these compounds, the incorporation of a Brønsted acid co-catalyst in the reaction scheme led to a significant improvement in both the yield and the enantioselectivity. A possible explanation for this might be the ability of the Brønsted acid to accelerate the reaction in which the enamine intermediate is being formed, therefore increasing the concentration of the nucleophile in the reaction medium. This would imply that the amine-containing lateral substituents of the pyrroline ring in the catalyst would exert their stereochemical influence by simple steric shielding as in the previous example shown (Figure 2.5). However, an alternative pathway can also be proposed in which the possible formation of a protonated enamine intermediate could take place, therefore directing the income of the nitroalkane electrophile by hydrogen bonding. Both possibilities account for the absolute configuration observed in the major enantiomer obtained.

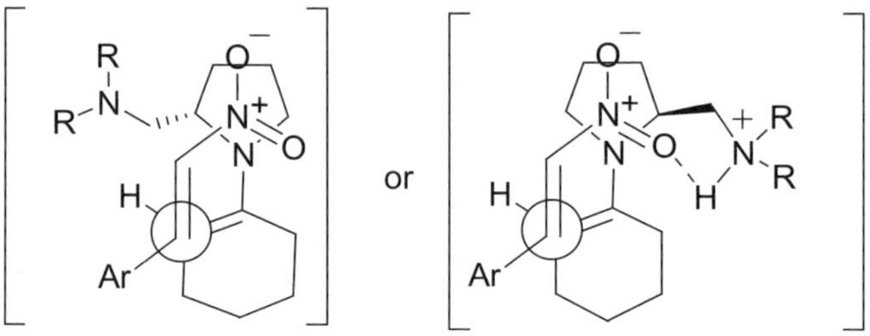

Figure 2.5 Proposed models for Michael reaction between cyclohexanone and nitrostyrenes catalyzed by pyrrolidine-diamines **14a–c** in the presence of a Brønsted acid co-catalyst.

As can be seen in almost all the examples presented up to this moment, while the reaction using cyclohexanone usually proceeds in a satisfactory way, changing to acyclic ketones as Michael donors usually leads to a dramatic drop in stereoselectivity, which is interpreted in terms of a poor geometry control exerted by the catalyst in the formation of the enamine intermediate. A singular situation appears when α-heterosubstituted ketones are employed, because, in this case, the presence of the α-heteroatom can facilitate the regio- and diastereocontrolled formation of the enamine *via* secondary interactions with the catalyst. Alexakis has studied in detail the use of its iPBP catalyst **15a** in the Michael reaction of α-hydroxy- and α-aminoketones to nitroolefins explaining the high stereocontrol achieved, by assuming the preferential formation of a Z-enamine assisted by intramolecular hydrogen bonding with the amine-containing side chain of the catalyst (Scheme 2.7).[30] This change in enamine geometry would also account for the inversion of the simple diastereoselection from *syn* to *anti* in these particular cases, which is also supported by the fact that the reaction using methoxyacetone, which is unable to interact with the catalyst *via* hydrogen bonding, furnished the corresponding *syn* major diastereoisomer.

Nevertheless, as was pointed out before, a straightforward solution to the rather limited substrate scope of the reaction with regard to the ketone reagent and also a good way to overcome the lack of reactivity of ketones toward enamine activation has been the use of primary amines as organocatalysts. In fact, literature examples indicate that primary amines are much more active catalysts for the Michael addition of both cyclic and acyclic ketones to nitroalkenes compared to the same reaction using a secondary amine catalyst like most of the proline-based derivatives already presented before.

In this context, a wide variety of different chiral primary amines has been developed and tested in this reaction with different results (Scheme 2.8). The first example of a primary amine-catalyzed Michael reaction of ketones with nitroalkenes was reported by Córdova,[31a] who found that alanine-derived amide **20a** was an excellent catalyst for this transformation. However, in this report, the reaction scope was also mainly focused on the use of cyclohexanone, with a single example of an acyclic enone (2-butanone) providing the Michael

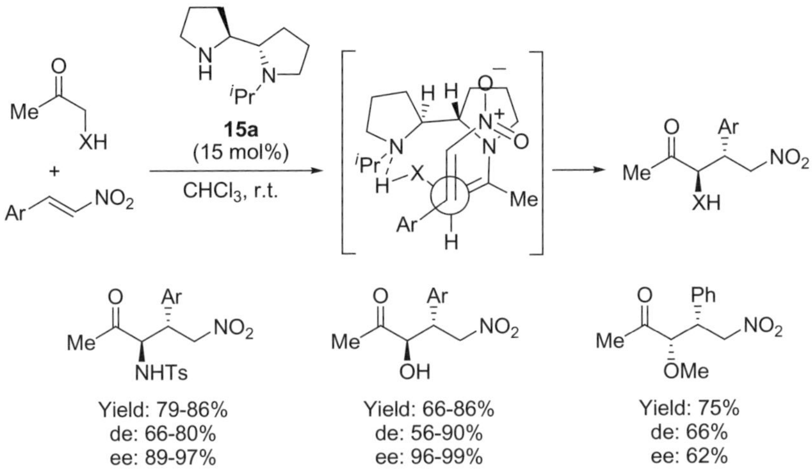

Scheme 2.7 Enantioselective Michael reactions of α-hydroxy- and α-aminoketones with nitroalkenes catalyzed by **15a**.

adduct in moderate yield and stereoselectivity. The optimized conditions found for the reaction required the inclusion of water and a Brønsted acid co-catalyst for the reaction to proceed satisfactorily, which was interpreted by the authors by assuming that interconversion of the imine/enamine/iminium intermediates present in the catalytic cycle was facilitated by these additives. Interestingly, when designing the catalyst, it was found that the presence of a primary amide moiety was a key structural feature for achieving high stereoselectivity, which pointed toward the existence of a well-organized transition state in which hydrogen-bonding interactions between the NH-moiety and the nitro group would play an important role. Following this line, the same authors developed dipeptides **20b** and **20c** as more efficient catalysts,[31b] in which multiple hydrogen bonding in the transition state could be possible by the additional incorporation of a carboxylic acid moiety. This concept was also the inspiration for the development of phenylglycine-based diamide **21**[32] and bispidine-derived catalyst **22a**,[33] the latter showing to be extremely competitive in the Michael reaction of acyclic ketones such as acetone or 2-propanone with a wide variety of nitroalkenes. In this later case, the addition of a Brønsted acid co-catalyst such as 3,3′,5,5′-tetrabromobiphenol (TBBP) was found to be important for achieving good results. Molecular modeling studies also supported the inter-action between the bispidine moiety and the nitroalkane *via* hydrogen bonding in the transition state.

A particularly synergistic situation is found when chiral primary amines containing a thiourea moiety in their structure are used as catalysts. In this case, after the enamine formation step takes place, the thiourea moiety finds a direct way for interacting with the nitro group due to the effective formation of two hydrogen bonds between the NH groups and the nitro moiety, leading to a

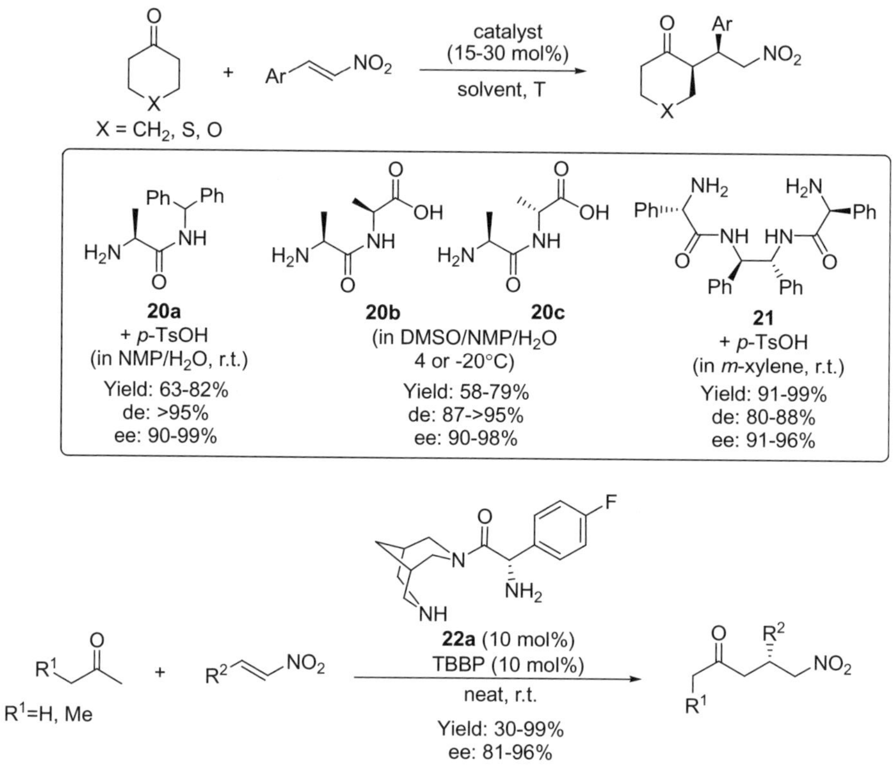

Scheme 2.8   Enantioselective Michael reactions of ketones with nitrostyrenes catalyzed by several primary amines.

well-ordered transition state. Some catalysts have been developed in this context but, perhaps, Jacobsen thiourea **23a** has shown up as the most efficient one in terms of stereoselectivity and substrate scope, showing its excellent performance for a wide variety of differently substituted ketones and also behaving well when alkyl-substituted nitroalkenes were employed as electrophiles, which are a kind of Michael acceptor and which usually provide rather poor results (Scheme 2.9).[34] Other related catalysts developed in this reaction also include the thioureas **24** and **25a** developed by Tsogoeva,[35] aminosugar containing derivative **25b**,[36] aspartate-*tert*-butyl ester based thiourea **26**[37] and dihydroabietic amine-based thiourea **25c**.[38] It has to be pointed out that a common feature shared by all these catalyst systems is their remarkably improved performances with respect to other proline-derived catalysts, especially in the case in which acyclic ketones are employed as Michael donors.

Other catalysts such as those depicted in Scheme 2.10 have also been developed following the same concept. In particular, primary amine-sulfonamides **27a** and **27b**, in which the sulfonamide N–H group plays the role of the hydrogen-donor group, have been successfully applied to the reaction of

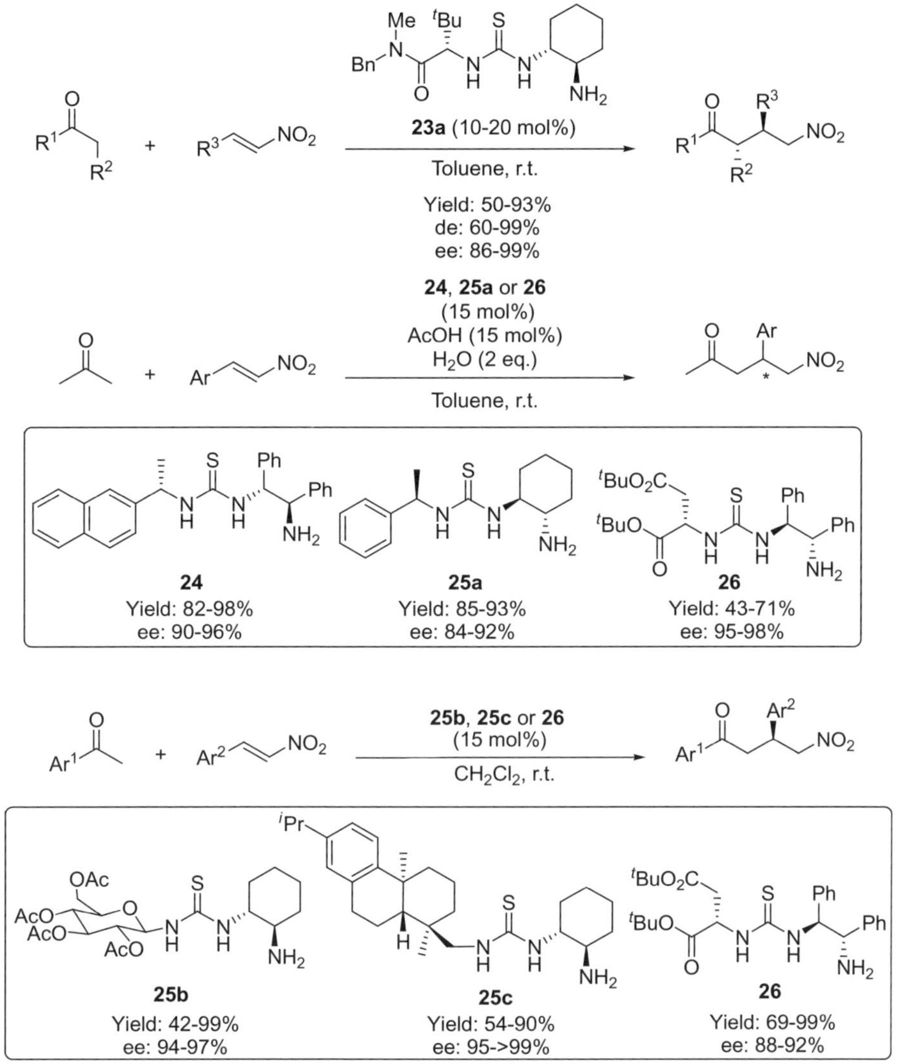

Scheme 2.9 Enantioselective Michael reactions of ketones with nitrostyrenes catalyzed by primary amine-thiourea catalysts.

acetone and aryl methyl ketones with nitrostyrenes.[39] Alternatively, 9-epi-amino cinchona alkaloid derivative **28a** has also been employed in the presence of a Brønsted acid co-catalyst in the reaction of cyclic and acyclic ketones with nitrostyrene providing good results.[40] In this case, the combination of the catalyst with the Brønsted acid leads to the formation of an intermediate enamine in which the tertiary amine moiety present at the cinchonine skeleton becomes selectively protonated leading to the corresponding ammonium salt

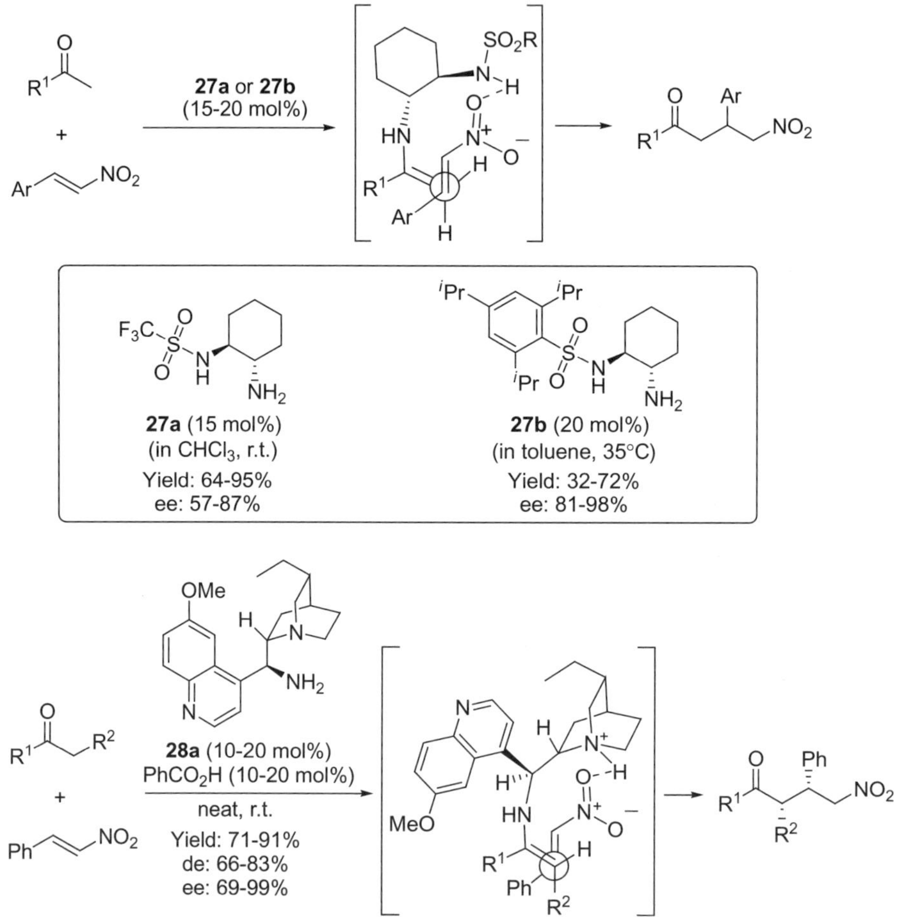

Scheme 2.10 Enantioselective Michael addition of ketones to nitroalkenes catalyzed by primary amine-sulfonamide catalysts **27a–b** and by 9-epi amino cinchona alkaloid **28a**.

and therefore playing the role of the N–H acidic site in terms of interaction with the nitro group *via* hydrogen bonding.

In all the examples presented up to this point, it has become clear that very subtle changes to the catalyst structure very often lead to significant differences in its performance in terms of yields and stereoselectivities in the same reaction, which means that a library of structurally diverse catalysts has to be available for optimization of the reaction conditions, and this very often leads to their preparation *via* multistep synthesis. A possible alternative to this situation might be the modulation of the catalyst by incorporation of achiral additives that self-assemble with the original pre-catalyst leading to a supramolecular architecture with improved performance. Working under this hypothesis, it has

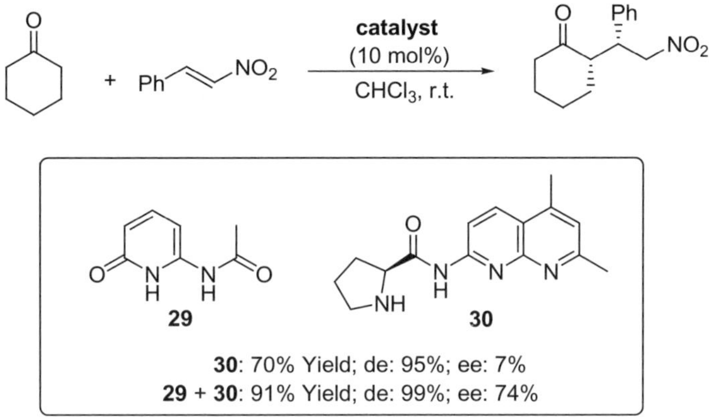

Scheme 2.11   Effect of the incorporation of achiral additive **29** in the **30**-catalyzed enantioselective Michael addition of cyclohexanone to nitrostyrene.

been shown that the incorporation of a pyridinone additive **29** to the Michael reaction between cyclohexanone and nitrostyrene mediated by prolinamide **30** resulted in a dramatic improvement of the enantioselectivity, thus transforming an unselective catalyst into a more effective one (Scheme 2.11).[41] The interaction between the host pre-catalyst molecule and the additive by hydrogen-bonding was proven by ^{1}H-NMR titration methods, confirming the original hypothesis. Another report has also shown that a chiral primary amine together with a thiourea incorporated as an additive can be jointly used as catalysts in the Michael reaction of ketones with nitroalkenes, also in this case demonstrating that self-assembly between the two catalysts must occur in order to obtain the excellent results observed, which was also supported by several experiments which ruled out the possibility of a simple synergistic effect.[42]

2.2.2   Aldehydes as Michael Donors

The enantioselective Michael addition of naked aldehyde donors to nitroalkenes has grown up very rapidly after the mechanistic interpretation of the enamine activation concept was established and therefore it has become a very powerful transformation illustrated by several important applications in the synthesis of valuable chiral compounds. The high ability of the aldehyde reagents to condense with the amine catalyst to deliver the required enamine intermediate relatively easily, together with the high electrophilicity of the nitroalkane counterpart, make this combination of reagents the ideal platform for the development of an enantioselective conjugate addition process. In fact, the higher reactivity of the aldehydes with respect to the corresponding ketone donor counterparts in the condensation reaction with the catalyst leads to a faster and much more effective reaction. In addition, the big difference in steric

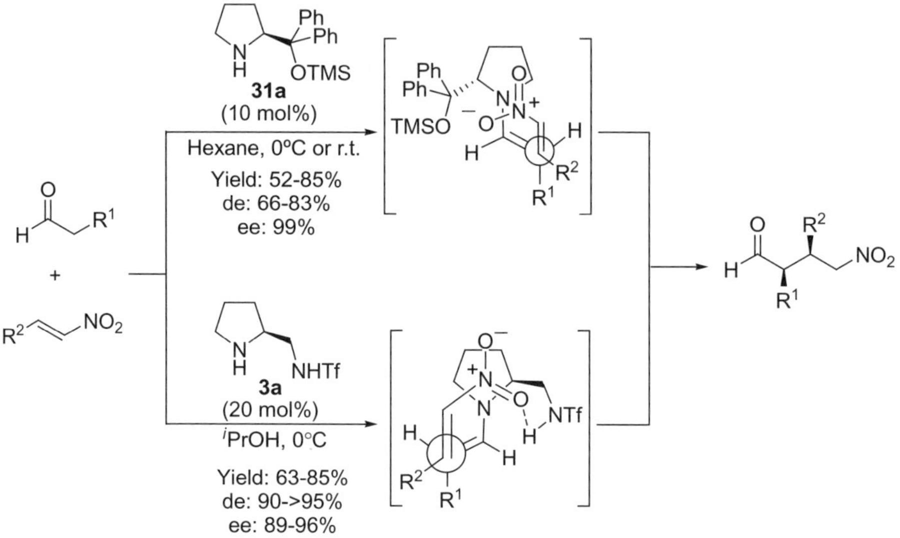

Scheme 2.12   Enantioselective Michael addition of aldehydes to nitroalkenes catalyzed by **31a** and **3a**.

bulk of the two substituents of the carbonyl moiety for the case of aldehyde
reagents simplifies enormously the enamine geometry/conformation control
problem usually associated with this reaction when ketones were employed.

Intensive research in this field has led to the design of a wide number of
different types of catalysts, most of them consisting of chiral pyrrolidine
incorporating differently functionalized side chains and which allow carrying
out the Michael reaction of simple aldehydes to nitroalkenes in a very efficient
way. In this context, perhaps *O*-TMS protected diphenylprolinol **31a**[43] and
pyrrolidine sulfonamide **3a**[13a,44] have been found to be the most efficient catalysts for this transformation providing excellent results in terms of yields and
stereoselectivities, and for a wide scope of aldehydes and nitroalkenes (Scheme
2.12). The key for the success of catalyst **31a** relied on the effect exerted by the
very bulky substituent at the pyrrolidine ring, which results in a very efficient
geometry control of the enamine intermediate together with an excellent ability
to discriminate between its enantiotopic faces *via* steric shielding. On the other
hand, the high stereoselectivity achieved by **3a** is interpreted in terms of the
formation of a rigid hydrogen-bonded transition state by interaction between
the highly acidic NHTf moiety and the nitro group.

Very intensive research in this field has led to the design of a wide number of
different chiral secondary amine catalysts for carrying out this transformation
in an efficient way (Figure 2.6). In this context, diamines such as **14a** and **14d**
have performed well in this reaction, also allowing, in the case of **14a**, the use of
α,α-disubstituted aldehydes, which led to the generation of quaternary stereocenters, although in moderate stereoselectivities.[45] Alexakis has studied in
detail the use of 2,2′-bipyrroline derivative **15a** and its bimorpholine analogue

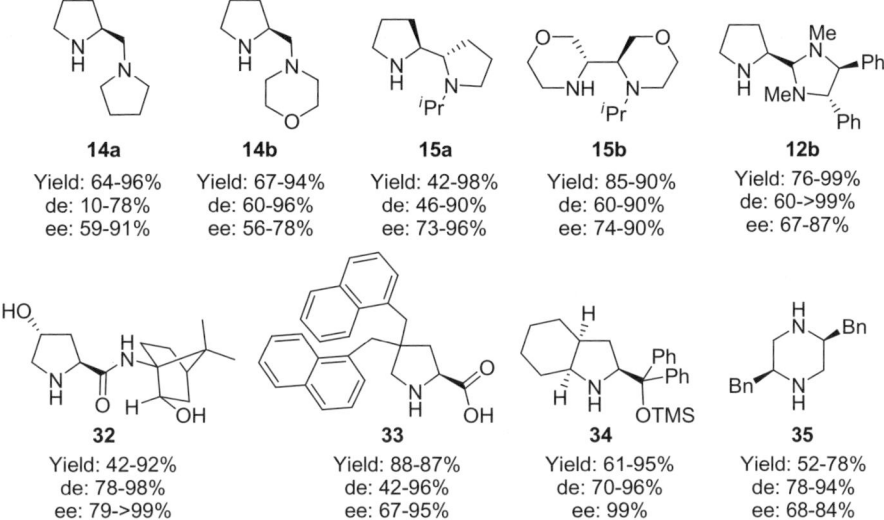

14a
Yield: 64-96%
de: 10-78%
ee: 59-91%

14b
Yield: 67-94%
de: 60-96%
ee: 56-78%

15a
Yield: 42-98%
de: 46-90%
ee: 73-96%

15b
Yield: 85-90%
de: 60-90%
ee: 74-90%

12b
Yield: 76-99%
de: 60->99%
ee: 67-87%

32
Yield: 42-92%
de: 78-98%
ee: 79->99%

33
Yield: 88-87%
de: 42-96%
ee: 67-95%

34
Yield: 61-95%
de: 70-96%
ee: 99%

35
Yield: 52-78%
de: 78-94%
ee: 68-84%

Figure 2.6 Some chiral pyrrolidine-based amines employed as catalysts in the enantioselective Michael addition of aldehydes to nitroalkenes.

15b,[46] leading to very interesting results in this reaction and also showing that microwave activation could be a useful tool for achieving better conversion in much lower reaction times without loss of selectivity, which, consequently, allowed the use of lower catalyst loadings.[46c] Prolinal-derived diaminal **12b**,[47] pyrrolidine-camphor derivative **32**,[48] 4,4-disubstituted proline **33**,[49] diphenylperhydroindol silyl ether **34**[50] and chiral piperazine **35**[51] can also play the role of organocatalysts in this transformation although, in general, it has to be said that the results are very variable depending on the structure of the aldehyde donor and the nitroalkene acceptor.

On the other hand, several primary amines have also been successfully employed as catalysts in the reaction using aldehydes as Michael donors. Important representative examples are shown in Figure 2.7, like bifunctional sulfamide **36**[52] and primary amine-thioureas **37a**[53] and **23b**,[54] the latter being developed by Jacobsen and very similar to that used by the same group in the reaction with ketones as donors (see Scheme 2.7) and which performs exceptionally well in the Michael reaction of α,α-disubstituted aldehydes to both aromatic and aliphatic nitroolefins.

There is an interesting variant of this reaction which involves the use of *tert*-butyldimethylsilyloxyacetaldehyde as Michael donors and chiral primary amine thiourea bifunctional catalyst **37b** (Scheme 2.13).[55] In this case, the diastereoselectivity of the reaction changed from the usually observed *syn* relative stereochemistry at the final Michael adduct to the formation of the *anti* diastereoisomer as the major product. This change in diastereoselectivity was explained in terms of the generation of a *Z*-enamine intermediate assisted by the formation of an intramolecular hydrogen bond between the secondary

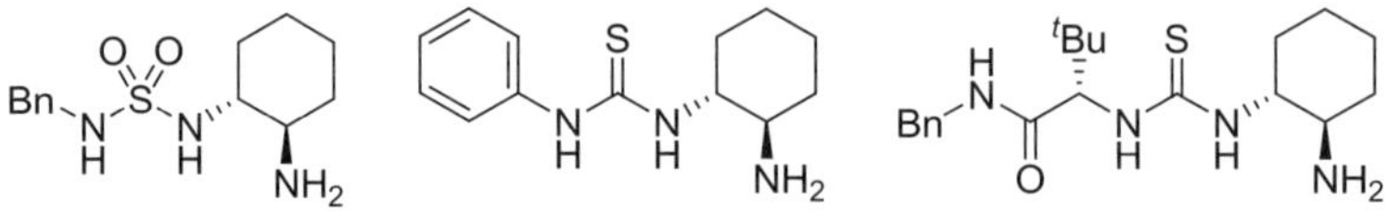

36	37a	23b
Yield: 41-99%	Yield: 27-98%	Yield: 78-94%
de: 50-75%	de: 50-70%	de: 46->99%
ee: 78-99%	ee: 41-98%	ee: 96-99%

Figure 2.7 Some chiral primary amines employed as catalysts in the enantioselective Michael addition of aldehydes to nitroalkenes.

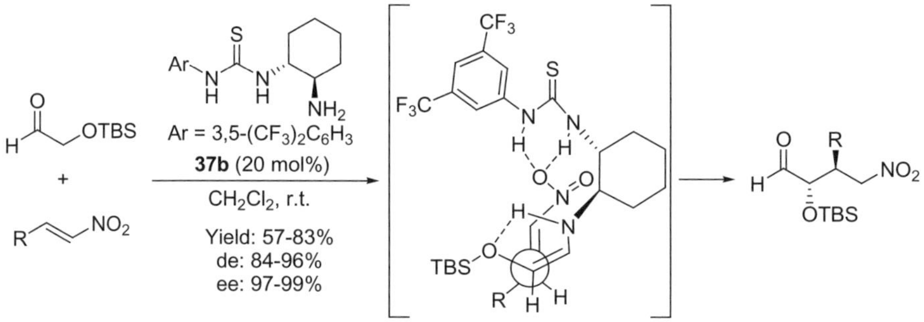

Scheme 2.13 Enantioselective Michael addition of *O*-TBS-acetaldehyde to nitroalkenes catalyzed by **37b**.

enamine group and the β-alkoxy substituent. Application of the Seebach–Golynski model accounts for the observed *anti* relative stereochemistry.

Nevertheless, all these above reported methodologies suffer from two important drawbacks. First of all, a large excess of donor source has to be employed in almost all examples reported in order to reach full conversion and hence high yields of the conjugate addition product. In fact, the use of a ten-fold excess of aldehyde with respect to the nitroalkane is a rather typical situation in many of the already-mentioned methodologies. Second, these catalysts are very effective on reactions with nitrostyrenes but when β-alkyl substituted nitroalkenes are employed, moderate yields or very low stereo-selectivities are typically obtained with a few exceptions as is the case, for example, of catalysts **31a** and **15a**. A solution to this problem has been provided with the preparation of catalysts expressly designed for this transformation and aimed to overcome these limitations. For example, Palomo has arrived at excellent results in the Michael reaction of aldehydes to a wide range of both aryl and alkyl substituted nitroalkenes, using equimolar amounts of aldehyde donor and nitroalkene acceptor and reaching excellent yields and stereo-selectivities and employing relatively low catalyst loadings (5–10%) with a modified 3-hydroxyprolinamide derivative **38**.[56] The catalyst design consisted

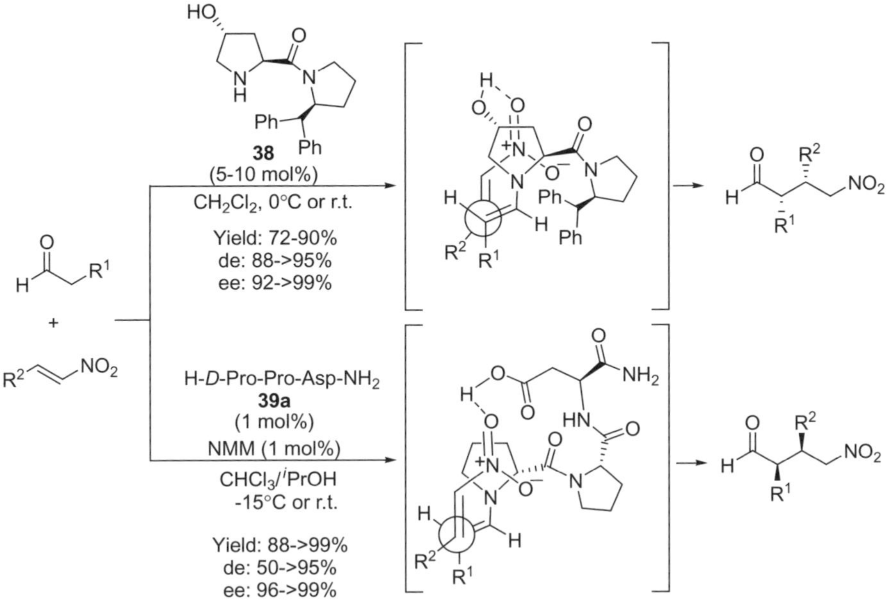

Scheme 2.14 Enantioselective Michael addition of aldehydes to nitroalkenes catalyzed by **38** and **39a**.

of the basic 2-substituted pyrrolidine skeleton, in which a bulky substituent was introduced at this position in order to exert an effective control on the enamine geometry and, in addition, a 4-OH substituent was placed at the pyrrolidine ring able to interact with the nitroalkane *via* hydrogen bonding and therefore favoring the formation of a well-defined transition state in the Michael addition reaction (Scheme 2.14). A similar design has been followed in the development of novel tripeptide **39a** as a very efficient catalyst for the Michael reaction of aldehydes to nitroalkenes with a remarkably wide substrate scope and tolerating both aryl- and alkyl-substituted nitroalkenes as electrophiles, although the aldehyde donor reagent has to be employed in a three-fold excess with respect to the nitroalkane (Scheme 2.14).[57] In this case, the tripeptide consisted of a terminal COOH group belonging to an asparagine α-amino acid, which is able to engage in a hydrogen bond with the nitro moiety and which was connected with a proline residue (the one responsible for the activation of the nucleophile by enamine formation) through another proline α-amino acid, the latter inducing a well-defined turn conformation which assisted the formation of a rigid transition state in which both reagents, the aldehyde donor (linked to the catalyst *via* enamine formation) and the Michael acceptor, were located close to each other. The high catalytic activity of this designed tripeptide was demonstrated by the unusually low catalyst loading required in the reaction (1 mol%) compared to other reported methodologies, which are typically in the range of 10–30 mol%.

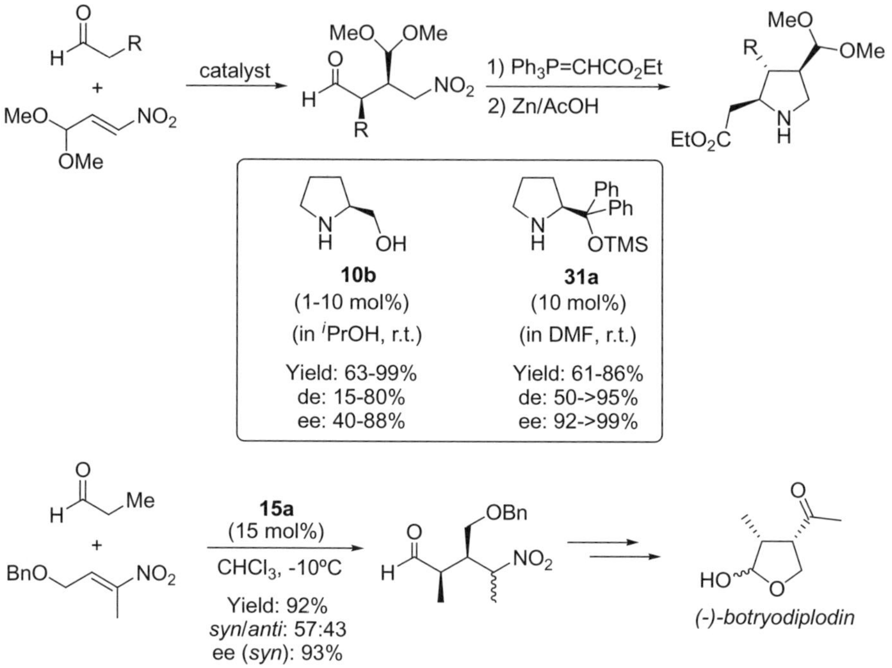

Scheme 2.15 Enantioselective Michael addition of aldehydes with functionalized nitroalkenes and an application to the total synthesis of densely functionalized homoprolines and natural product (−)-botryodiplodin.

However, despite these impressive advances, the use of functionalized nitroalkenes as acceptors still remains relatively undeveloped. A detailed study on the Michael reaction of aldehydes and β-nitroacrolein dimethyl acetal has been reported, concluding that simple prolinol **10b** was the most efficient catalyst for this particular transformation, allowing even the use of a 1% catalyst loading and the use of equimolar amounts of aldehyde and nitroalkane (Scheme 2.15).[58a] An improved version was also reported consisting of the use of *O*-TMS diphenylprolinol **31a** as catalyst which afforded much better diastereo- and enantioselectivities than prolinol.[58b] The synthetic applicability of the obtained Michael adducts was demonstrated with the development of a very simple and direct protocol for their transformation into densely functionalized homoproline derivatives. The experimental procedure involved a Wittig olefination followed by reaction with Zn dust, which induced a cascade process consisting of the chemoselective reduction of the nitro group followed by a fully diastereoselective intramolecular aza-Michael reaction. In a different report, a functionalized nitroalkene has been used by Alexakis as key substrate in the asymmetric synthesis of (−)-botryodiplodin, a natural product isolated from *botryodiplodia theobromae* compound, which displays interesting antibiotic and antileukemic activity.[59] The synthesis relied mainly on the high efficiency of the iPBP catalyst **15a** to catalyze the Michael reaction of aldehydes with nitroalkenes especially in the particular case in which propanal was used as Michael donor reagent, which usually led to a very fast and clean reaction. The Michael addition product was obtained in good yield but as a mixture of *syn/ anti* diastereoisomers in 93% and 74% ee respectively, although these were easily separated by chromatography. The major *syn* diastereoisomer, which had the required absolute configuration to reach the target natural product, was easily transformed into (–)-botryodiplodin using a very simple and straightforward procedure.

Another remarkable case is found in the use of nitrodienes as functionalized nitroalkenes (Scheme 2.16).[60] In this case, the Michael addition of aldehydes to this α,β,γ,δ-unsaturated system catalyzed by *O*-TMS-diphenylprolinol **31a** proceeded with exclusive 1,4-regioselectivity and delivered the corresponding Michael adducts incorporating an alkene side chain in good yields and excellent enantioselectivities, although mixtures of diastereoisomers in ratios varying from 96:4 to 70:30 were obtained. Importantly, the use of nitroalkenes incorporating a triple bond was also successfully studied, observing also in this case the formation of a single regioisomer and maintaining the excellent results with regard to chemical efficiency and stereoselectivity.

A particularly interesting situation which deserves a special analysis is found when the most simple possible aldehyde donors or nitroalkane acceptors, *i.e.*, acetaldehyde or nitroethylene, are employed in this reaction. The main problem associated with the use of these compounds is their high tendency to decompose and/or oligomerize together with the difficulty related to their manipulation and storage.[61] Moreover, the high reactivity of acetaldehyde to engage in self-aldol reactions proceeding *via* enamine intermediates, and therefore assisted by the presence of the secondary amine catalyst, is also a big problem that has to be overcome. In addition, the lack of substitution leads to a much higher

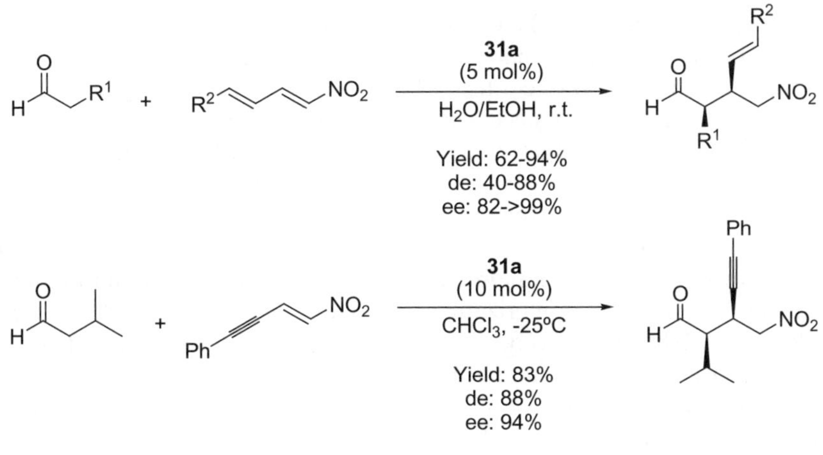

Scheme 2.16 Enantioselective Michael addition of aldehydes to extended poly-unsaturated nitroalkenes.

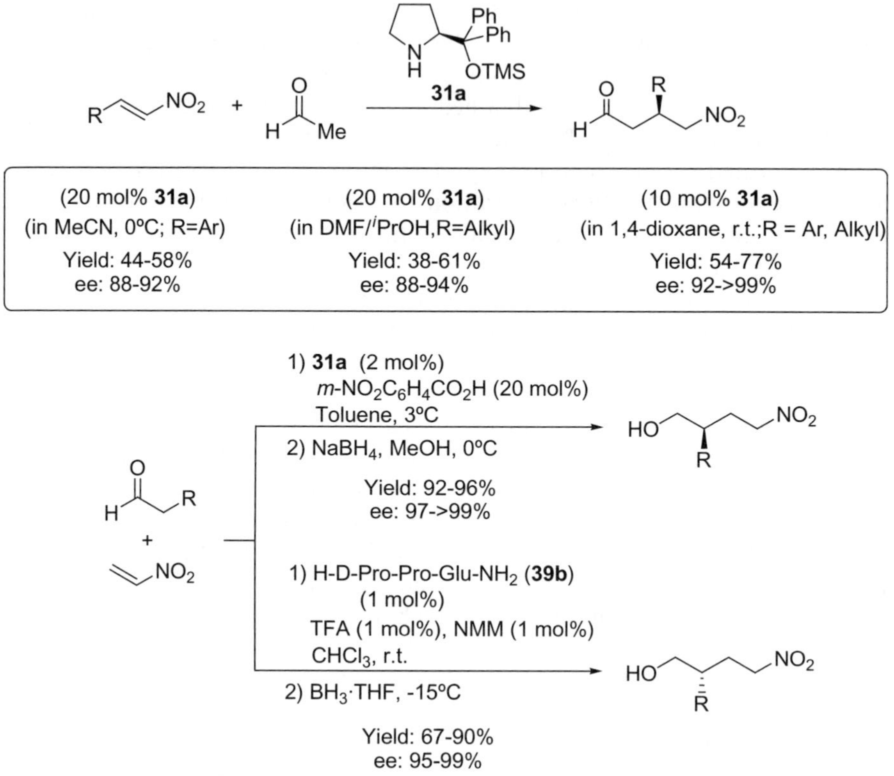

Scheme 2.17 Enantioselective Michael addition of acetaldehyde to β-substituted nitroalkenes and Michael addition of aldehydes to nitroethene.

conformational freedom at the transition state during the C–C bond formation step, which turns into a key parameter to be controlled in order to reach high enantioselectivities. In this context, the addition of acetaldehyde to a wide range of nitroalkenes has been carried out independently by two different research groups, both using *O*-TMS diphenylprolinol **31a** as catalyst (Scheme 2.17).[62] Experimental conditions change from one report to the other and while in one case MeCN (when nitrostyrenes were employed as Michael acceptors) or a DMF/*i*PrOH mixture (for β-alkyl substituted nitroolefins) was employed as the reaction solvent and the main efforts were focused on the manipulation of acetaldehyde to avoid oligomerization (a five-fold excess of a solution of freshly distilled acetaldehyde had to be employed and it was very slowly added to the reaction mixture *via* siring pump), in the other case the simple use of a ten-fold excess of acetaldehyde and 1,4-dioxane as reaction solvent was reported to be enough to reach the final Michael adducts in acceptable yields, although the use of a sealed tube as reaction vessel is also mentioned. On the other hand, the Michael addition of a set of different aldehydes to nitroethylene has also been

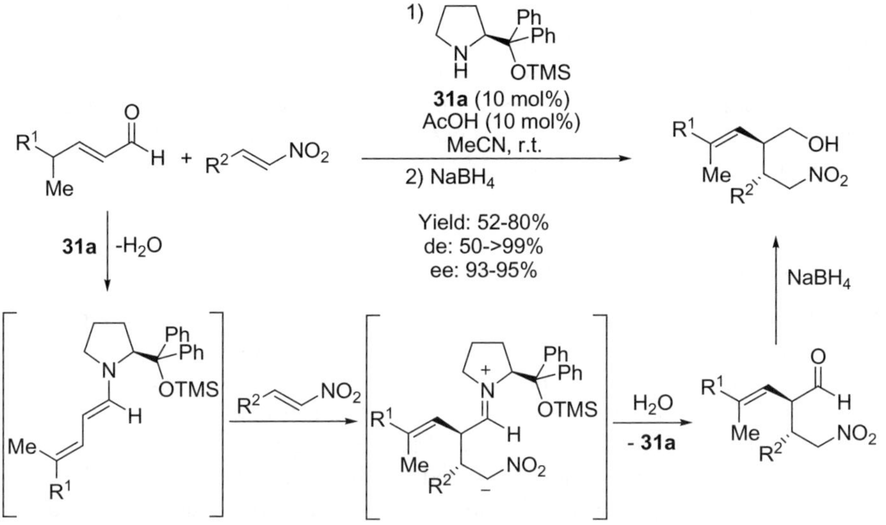

Scheme 2.18 Enantioselective Michael addition of enolizable α,β-unsaturated aldehydes to nitroalkenes *via* dienamine catalysis.

reported to proceed with excellent results, again using **31a** as catalyst and in the presence of *m*-nitrobenzoic acid,[63a] which plays the role of a Brønsted acid co-catalyst which assists the formation of the enamine intermediate. Alternatively, a modified version of tripeptide **39a** (**39b**) has also been reported to be an extremely active catalyst for this transformation, allowing the use of up to 1 mol% catalyst loading under the optimized reaction conditions, and reaching the target compounds in good yields and excellent enantioselectivities.[63b] In this case, the final amino acid of the tripeptide had to be changed from asparagine (in **39a**) to glutamic acid, in which the incorporation of the additional methylene group was observed to have a very positive influence on the catalytic activity. In this case, the use of *N*-methylmorpholine/TFA as co-catalyst is also required in the reaction, which probably contributes to the formation of the enamine intermediate in the same sense as *m*-nitrobenzoic acid did for the **31a**-catalyzed reaction. Dilution also seems to be a key parameter to be controlled if high yields of the final compounds are desired.

Finally, it has also to be mentioned that enolizable α,β-unsaturated aldehydes have also been employed as Michael donors in this context (Scheme 2.18). This reaction proceeds *via* formation of a dienamine nucleophilic intermediate,[64] which undergoes regioselective α-addition leading to the formation of the corresponding Michael adduct containing an α-substituted β,γ-unsaturated aldehyde moiety.[65] The conditions had to be carefully optimized and required the use of a γ,γ-disubstituted α,β-unsaturated aldehyde reagent and involved the use of catalyst **31a** in the presence of AcOH as Brønsted acid co-catalyst and acetonitrile as solvent. *In situ* reduction of the Michael adducts was also required, which possibly precluded the double-bond isomerization which would provide the corresponding more stable conjugated enal. Yields of the final products were good to moderate and good diastereoselectivities and excellent enantioselectivities were obtained for a variety of enals and nitroalkene substrates, including the use of both β-aryl and β-alkyl substituted nitroalkenes.

2.2.3   Reactions under Non-conventional Methodologies

Another issue that has led to intensive research in recent years is the development of more environmentally benign methodologies, especially regarding the solvent employed (with particular interest in the possibility of using water)[66] and the chance of recycling the catalyst. In this context, it has been proven that the O-TMS-diphenylprolinol **31a**-catalyzed Michael reaction of aliphatic aldehydes to different nitroalkenes could be carried out in water with a similar degree of efficiency as that displayed for the same catalyst when this reaction was carried out in an organic solvent.[67a] A modified 2-aminomethylpyrrolidine such as **14e** (Figure 2.8) in the presence of TFA as co-catalyst has also been successfully employed in the Michael reaction of cyclohexanone with different aromatic nitroalkenes in water.[67b] Alternatively, some catalysts have been developed that are specifically designed to be active in water media, which very often implies the introduction of long hydrophobic alkyl side chains in the catalyst structure. The aggregation of the organocatalyst with the hydrophobic reagents participating in the Michael reaction reduces the contact between these reagents and bulk water as the reaction proceeds, therefore behaving as they were dissolved in an organic solvent. Applying this concept, Barbas has brilliantly shown that brine and even seawater can be suitable media for carrying out the organocatalytic Michael reaction of ketones and aldehydes to nitroalkenes, using N,N-bis (n-decyl)-substituted diamine catalyst **14b** (Figure 2.8).[68] In this paper, the use of

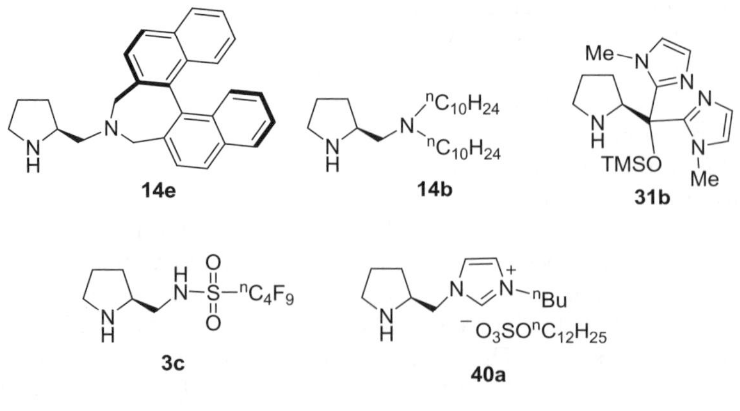

Figure 2.8   Some active catalysts in the Michael addition of aldehydes or ketones to nitroolefins in aqueous media.

electrolyte-rich solutions such as brine played a crucial role in the reaction, increasing the yield of the Michael addition product by avoiding the polymerization of the nitroalkene, one of the side reactions usually found in these transformations. A modified *O*-TMS-protected diarylprolinol derivative **31b** in which the two aryl groups present at the original catalyst were replaced by two hydrophilic methylimidazol-2-yl substituents has also demonstrated its abilities to catalyze this reaction in water.[69] Alternatively, Wang has explored the use of fluorous pyrrolidine **3c** as a potential catalyst able to promote this reaction in water. In this case, the presence of perfluorinated alkyl chains on its structure allowed recycling *via* solid–liquid extraction using a fluorous silica gel and it could be reused in the same reaction up to seven consecutive times with no important loss of performance.[70] Surfactant-type organocatalyst **40a** has also been found to efficiently catalyze the Michael addition of cyclohexanone to different nitroalkenes in water with excellent results.[71]

The search for recyclable chiral amine catalysts active in this transformation has also registered intense effort in recent years. A first approach has been focused in the field of functionalized chiral ionic liquids, which, owing to their wide possibilities for functionalization, present a powerful approach for the preparation of new organic catalysts which can be easily recovered and recycled.[72] For example, compound **40b** was employed in the enantioselective Michael addition of cyclohexanone, acetone and even isobutyraldehyde to a wide variety of nitroalkenes reaching outstanding levels of chemical efficiency and stereoselectivity.[73] However, the most attractive feature was the fact that the catalyst could be easily recovered from the crude reaction mixture by simple precipitation with diethyl ether and it could be reused in the same reaction up to four additional times with similar efficiency, although the reactions were found to be somewhat slower after the second run (Scheme 2.19). Pyrrolidine-sulfonamide **41a**-based functionalized ionic liquid has also been employed in a similar context with excellent results, also allowing recycling and reusing up to five times without loss of chemical efficiency or selectivity.[74] An additional remarkable feature of the latter catalyst **41a** is the possibility to employ the Michael donor in much lower amounts (up to three- to five-fold excess) compared to the usually very large excess of this reagent used in many other examples reported. Other pyrrolidine-based functionalized ionic liquids have also been developed and applied in the same context with different results.[75] In addition, in some cases, reactions using this kind of catalysts are also carried out in ionic liquids such as 1-*n*-butyl-3-methylimidazolium hexafluorofosfate (BMImPF6) as reaction solvent, which, as happened with catalyst **42**,[76] led to improved results compared to the use of standard organic solvents, showing that a synergistic effect between the functionalized ionic liquid catalyst and the solvent is operating. In addition, in these cases, the biphasic character of the ionic liquid/organic solvent mixture also allowed the isolation of the Michael addition products by extraction and the recovery of the BMImPF6/catalyst mixture which could be directly reused in subsequent reactions without further treatment. A recent report has also appeared describing the use of simple L-proline as catalyst in the Michael reaction of aldehydes with nitrostyrene

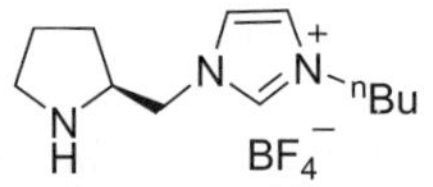

Scheme 2.19 Enantioselective Michael addition of cyclohexanone to nitroalkenes using recyclable functionalized ionic liquids as catalysts.

For catalyst **40b** (15 mol%) + TFA (5 mol%), neat, r.t. (R=p-ClC$_6$H$_4$):

	Time	Yield (%)	de (%)	ee(%)
run 1:	12h	92	96	95
run 2:	10h	100	98	99
run 3:	10h	92	92	94
run 4:	24h	93	95	93

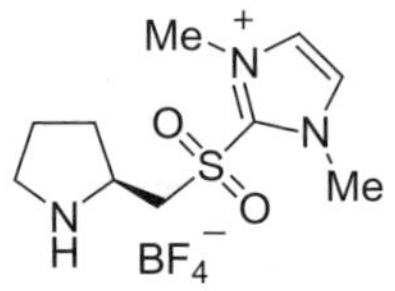

For catalyst **41a** (10 mol%), ^{i}PrOH, r.t. (R=Ph):

	Time	Yield (%)	de (%)	ee(%)
run 1:	12h	92	90	90
run 2:	16h	90	90	90
run 3:	18h	93	88	90
run 4:	24h	86	90	89
run 5:	48h	80	94	88

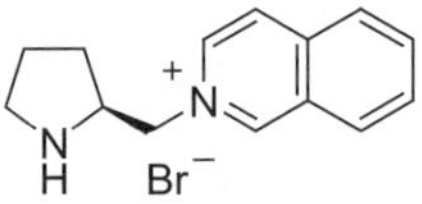

For catalyst **42** (10 mol%), BMImBF$_4$, r.t. (R=Ph):

	Time	Yield (%)	de (%)	ee(%)
run 1:	20h	94	88	99
run 2:	24h	94	88	99
run 3:	24h	92	88	99
run 4:	24h	90	88	97
run 5:	24h	90	88	96

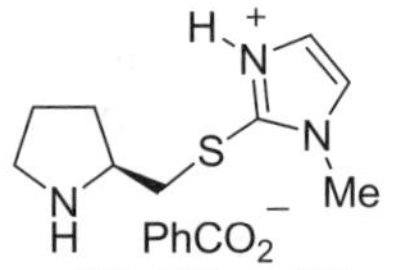

For catalyst **41b** (20 mol%), PEG-800, r.t. (R=Ph):

	Time	Yield (%)	de (%)	ee(%)
run 1:	12h	95	94	97
run 2:	12h	95	92	97
run 3:	12h	93	92	97
run 4:	18h	90	86	96
run 5:	18h	88	88	97
run 6:	24h	83	88	94
run 7:	24h	76	96	96
run 8:	24h	75	92	96

using ionic liquids as solvents.[77] Another possibility of catalyst recycling has been the use of poly(ethylene glycol)s as reaction solvent in which it is claimed that the organocatalyst **41b** forms a host–guest complex with the poly(ethylene glycol) solvent, as electrospray mass ionization spectrometry indicated, leading to a very effective catalytic system for the Michael addition of cyclohexanone to nitrostyrene. Recyclability of the catalytic poly(ethylene glycol)/**41b** mixture was also proven to be possible up to eight consecutive times.[78]

Chiral secondary amines attached to solid supports have also been used as a methodological approach to the development of recyclable catalysts.[79] In these cases, click 1,3-dipolar cycloaddition reactions between a modified solid support incorporating an alkyne moiety and a substituted pyrrolidine

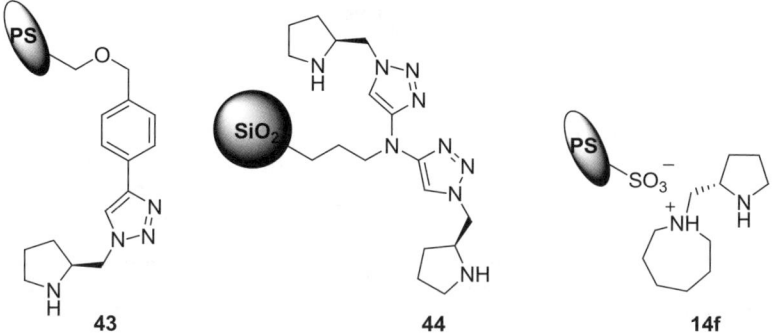

Figure 2.9 Some solid-supported catalysts for the Michael addition of aldehydes or ketones to nitroolefins.

incorporating an azide group have constituted the typical approach employed to attach the catalytically active unit to the solid support. Representative examples of this chemistry are shown in Figure 2.9 and include the polystyrene-supported pyrrolidine **43**,[80] and the silica-supported catalyst **44**,[81] the former being remarkably efficient with a broad scope of Michael donors and acceptors and also allowing the reaction to be carried out in water. Another interesting approach has also been devised for the attachment of the amine catalyst to the solid support *via* non-covalent interactions. In this case, an ammonium sulfonate species was formed when a modified polystyrene/sulfonic acid resin was mixed with a dialkylaminomethylpyrrolidine **14f**; this was applied for the immobilization of the catalyst to the resin and this was tested in the Michael addition of cyclohexanone to nitrostyrene with good results.[82] The main advantage of this approach is the possibility to recover the diamine catalyst **14f**, which also allows the combinatorial use of different diamine catalysts in this transformation. Finally, it has to be mentioned that recoverable dendritic organocatalysts have also been tested in this reaction with good results.[83]

2.3 Enones and Enals as Michael Acceptors

The use of α,β-unsaturated aldehydes and ketones as Michael acceptors in the reaction with aldehydes and/or ketones under enamine activation has been studied by several research groups, all of them reporting that the lower reactivity of the intermediate enamines toward 1,4-addition to enones or enals compared to nitroalkenes led to a much slower reaction and also to difficulties in reaching full conversion. In addition, when these reagents are employed as electrophiles, the participation of an activated form of the Michael acceptor as an iminium ion (see Chapter 3) is a matter of controversy even though the formation of a chiral enamine intermediate by itself could be enough for explaining the observed stereoselectivities and might also be enough for the

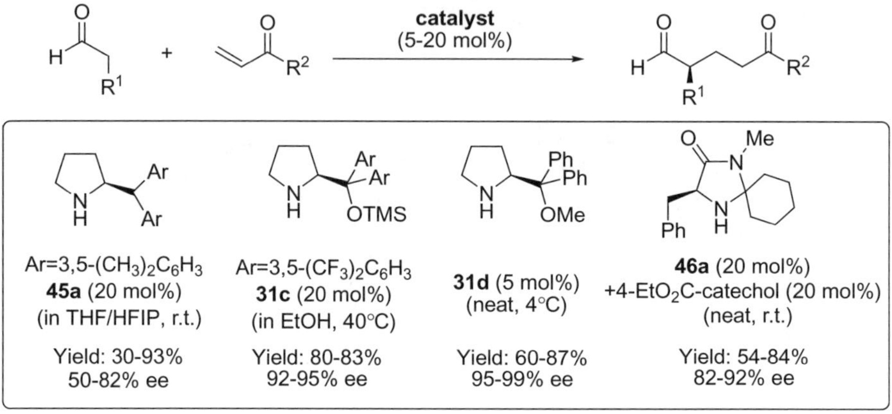

Scheme 2.20 Enantioselective intermolecular Michael reaction between aldehydes and alkyl vinyl ketones.

reaction to proceed without the further activation of the Michael acceptor counterpart.[84]

In this context, the Michael reaction between aldehydes and alkyl vinyl ketones has been carried out using **45a** as catalyst, affording the Michael adducts with good yields and moderate enantioselectivities (Scheme 2.20).[85] Nevertheless, the authors reported a slight nonlinear effect in the same reaction, which pointed toward two molecules of the catalysts being involved in the reaction mechanism and therefore a possible iminium-type activation of the electrophile might occur. Further research led to an improved protocol involving *O*-silyldiarylprolinol **31c** as catalyst and also observing that the reaction had to be run at a relatively high temperature (40 °C) in order to reach the final products in good yields and in an acceptable reaction time.[86] On the other hand, Gellman and coworkers demonstrated that *O*-methyldiphenylprolinol **31d** and imidazolidinone **46a** were also able to promote the same reaction (Scheme 2.20),[87] although they proposed that, in this case, the reaction proceeded exclusively *via* enamine activation of the aldehyde. The reaction also required the presence of a catechol derivative as an additive, explaining the role of this protic co-catalyst in terms of activation of the enone acceptor by hydrogen-bond donation, which also pointed toward the absence of iminium-type activation of the enone. The exclusive formation of an enamine intermediate in this case was also supported by carrying out control experiments using a previously generated enamine as nucleophile under the same reaction conditions.

An interesting variation of this reaction is shown in Scheme 2.21 and consists of a Michael reaction between aldehydes and quinones, which results in a procedure for carrying out the formal α-arylation of aldehydes.[88] In this case, the activation of the Michael donor as the corresponding enamine takes place followed by the conjugate addition to the highly electrophilic quinone reagent

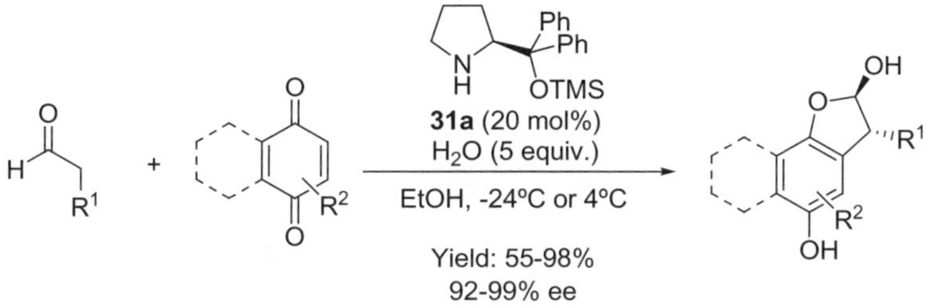

Scheme 2.21 Enantioselective intermolecular Michael reaction between aldehydes and quinones.

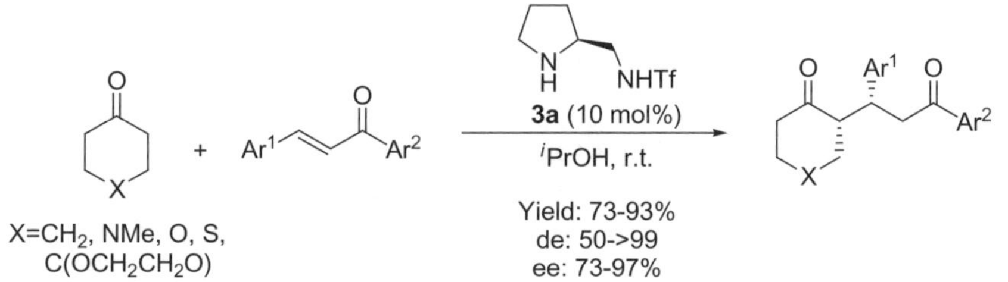

Scheme 2.22 Enantioselective Michael reaction between cyclic ketones and 1,4-diaryl enones catalyzed by **3a**.

and the final product is isolated as the corresponding hemiacetal form after releasing the catalyst and intermolecular 1,2-addition of the phenol moiety to the formyl group. The formation of this hemiacetal derivative might also play the role of a driving force for the reaction to proceed with full conversion.

A particularly difficult situation arises when combining in the same reaction the use of these rather unreactive acceptors such as enones with the incorporation of ketones as Michael donors in which the formation of the intermediate enamine by condensation with the amine catalyst is much more difficult. For this reason, the organocatalytic Michael addition of ketones to enones still remains rather unexplored. An example has been outlined in Scheme 2.22, in which it has been shown that pyrrolidine-sulfonamide **3a** could catalyze the Michael reaction between cyclic ketones and enones with remarkably good results, although the reaction scope was exclusively studied for the case of cyclic six-membered ring ketones as nucleophiles and 1,4-diaryl substituted enones as electrophiles.[89] In this system the authors also pointed toward a mechanism involving exclusively enamine-type activation of the nucleophile, with no contribution of any intermediate iminium species which could eventually activate the electrophile. Surprisingly, the use of primary amines as catalysts in this transformation has not been already considered.

On the other hand, an intramolecular version of this reaction has also been developed using imidazolidinone **46b** as catalyst (Scheme 2.23).[90] This

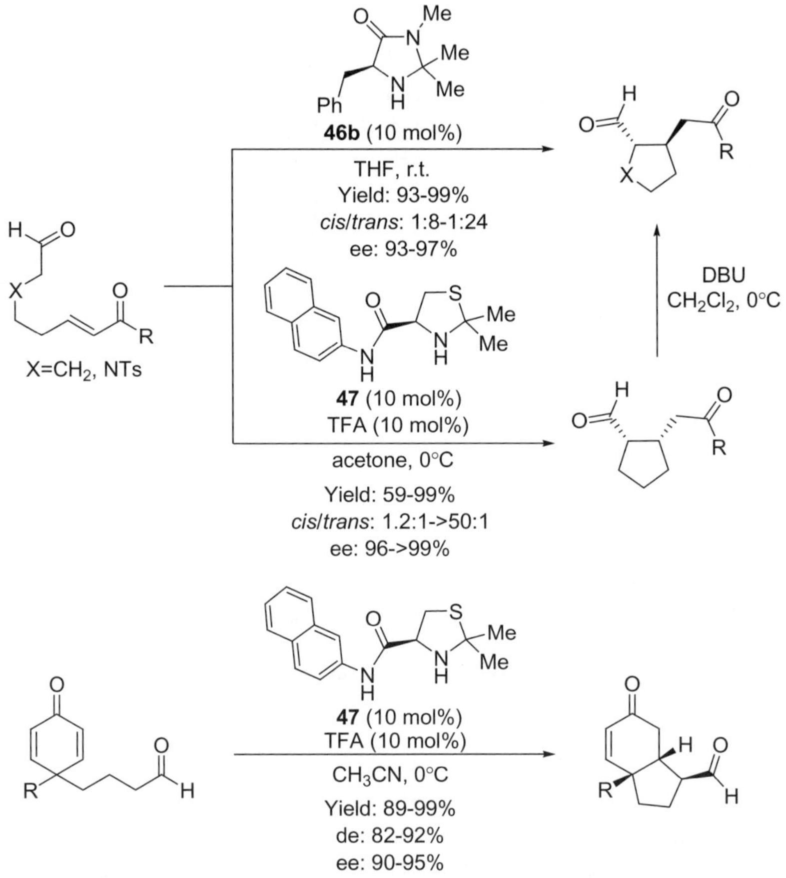

Scheme 2.23 Enantioselective intramolecular Michael reaction for the synthesis of cyclopentanes.

methodology allows the preparation of disubstituted cyclopentanes and five-membered ring heterocycles like pyrrolidines in a very efficient way and with a rather broad range of application regarding the substitution at the internal enone electrophile. In this case the authors were not able to clearly rule out the possible contribution of iminium-type intermediates, although the fact that the reaction using a substrate incorporating an α,β-unsaturated aldehyde as Michael acceptor on its structure furnished the final compound in remarkably lower enantioselectivity also suggested that the participation of activated enamine species should be the crucial element for this reaction to occur satisfactorily. Interestingly, modified thiazolidine catalyst **47**, derived from cysteine, was also employed in the same kind of transformation, and led to the corresponding cyclic compounds in excellent yield and enantioselectivity and with opposite relative configuration (*cis*) of the products obtained in the parent

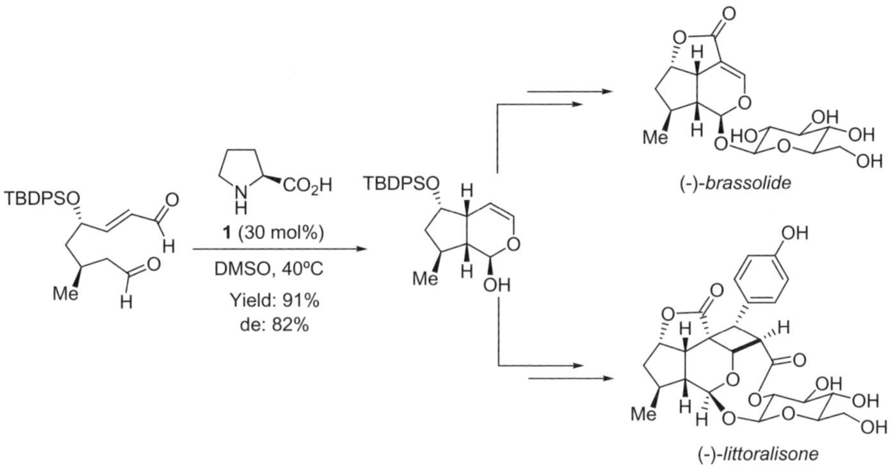

Scheme 2.24 Enantioselective intramolecular Michael reaction for the synthesis of brassolide and littoralisone.

reaction catalyzed by **46b** (Scheme 2.23).[91] Base-catalyzed isomerization allowed the quantitative transformation of these *cis* isomers into the corresponding thermodynamically more stable *trans* compounds. In addition, functionalized substrates incorporating a 2,5-cyclohexadien-1-one moiety as internal Michael acceptor have been employed as outstanding substrates in this transformation, furnishing in a single step and with excellent yields, diastereo- and enantioselectivities the bicyclo[4.3.0]nonene skeleton, which is present in a wide number of important families of natural products. Remarkably, the reaction also proceeded with the desymmetrization of the starting material, which led to the simultaneous generation of three stereogenic centers, including a quaternary one.

Alternatively, an intramolecular Michael reaction using an enal as acceptor has been developed by MacMillan as a key step toward the total synthesis of *brassolide and littoralisone*,[92] two natural products isolated from extracts of *verbena littoralis*, a plant used in traditional folk remedies which displays important biological activities. The transformation is shown in Scheme 2.24 and consisted of a proline-catalyzed reaction on an already chiral substrate furnishing a final bicyclic product in excellent yield and diastereoselectivity after intramolecular hemiacetal formation. Proof that the reaction proceeded under catalyst control was given by using the two enantiomers of proline, each of them furnishing a different diastereoisomer and therefore demonstrating that the reaction was not occurring under substrate control. Interestingly, the reaction provided the thermodynamically less stable diastereoisomer, observing that prolonged reaction times, or the use of apolar solvents, led to the preferential formation of the more stable *trans* diastereoisomer, which did not have the correct configuration required for the synthesis of the target products.

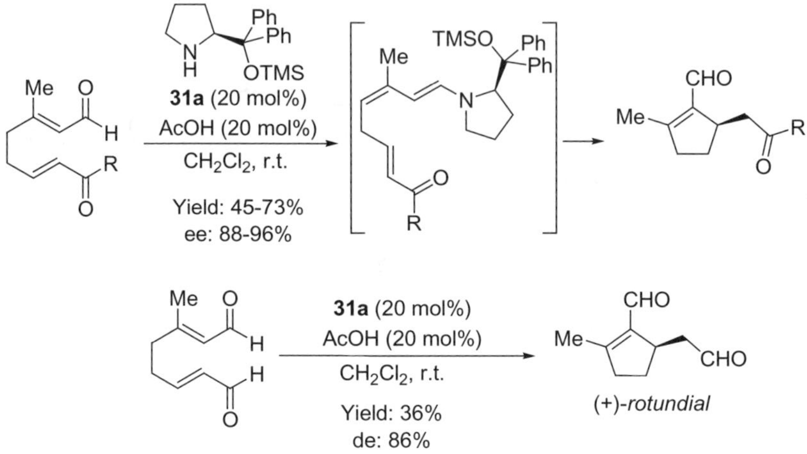

Scheme 2.25   Enantioselective intramolecular Rauhut–Currier reaction catalyzed by **31a**.

Using a highly polar solvent such as DMSO, the reaction proceeded cleanly to the kinetic *cis* product, which was employed as suitable starting material for both brassolide and littoralisone.

Finally, there is another relevant example that deserves to be mentioned, which entails a different activation profile of the Michael donor. An intramolecular Rauhut–Currier-type reaction has been developed employing *O*-TMS diphenylprolinol **31a** as catalyst for the enatioselective synthesis of chiral cyclopentenes, which involves the use of an α,β-unsaturated aldehyde as Michael donor and an enone as Michael acceptor (Scheme 2.25).[93] In this case, the activation of the enal as donor happens *via* the formation of a dienamine intermediate, which subsequently undergoes intramolecular Michael reaction with the enone. Although yields range from moderate to good, enantioselectivities were found to be very high in most cases and perhaps one of the most remarkable features of the methodology is associated to the good performance of the reaction using a β,β-disubstituted α,β-unsaturated aldehydes as Michael donors, which are supposed to be rather challenging substrates for the formation of an enamine-type intermediate. The fact that a dienamine is formed in this case overcomes this limitation and even facilitates the process. Other functionalities were also tested at the internal Michael acceptor such as a nitroalkene or another α,β-unsaturated aldehyde but with somewhat poorer results, highlighting the later case, in which the reaction provided a direct access to natural product, (+)-rotundial.

2.4   Doubly Activated Olefins as Michael Acceptors

The need for a highly reactive electrophile has also extended the field of application of the enamine activation manifold to the use of other highly electron-deficient olefins incorporating two electron-withdrawing groups at different positions.

For example, the addition to vinyl sulfones has attracted a lot of attention mainly because of the synthetic possibilities that the sulfone group affords in terms of its wide possibilities to undergo a wide variety of transformations. iPBP **15a** and prolinal-derived aminal **12b** catalysts have proved their usefulness in the Michael addition of aldehydes to this particular class of Michael acceptors, showing that good yields of the desired Michael adducts could be obtained under the optimized conditions, although with moderate enantioselectivities (Scheme 2.26).[51a,94] One of the main problems of this particular reaction is the occurrence of the base-catalyzed retro-addition reaction, which leads to the formation of undesired side products arising from the dimerization

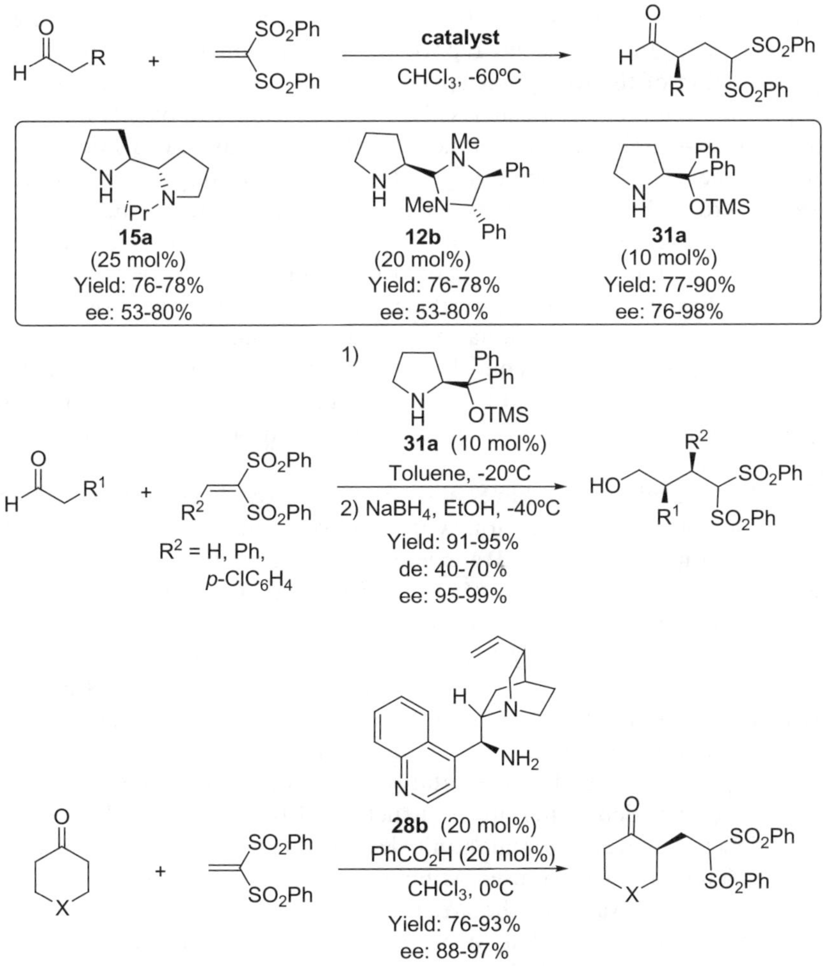

Scheme 2.26 Enantioselective Michael reaction of aldehydes with vinyl(bis)sulfones.

of the vinylsulfone reagent. The use of the less basic and sterically more demanding *O*-TMS diphenylprolinol catalyst **31a** provided a good solution to this problem and also furnished a higher enantioselectivity for a wide range of different enolizable aldehydes used as Michael donors. Interestingly, the absolute configuration of the products obtained by both the **15a**- and the **31a**-catalyzed reaction were found to be the same, which pointed toward the *N*-isopropylpyrrolidin-2-yl substituent in catalyst **15a** exerting its stereochemical influence exclusively by steric shielding of one of the diastereotopic faces of the intermediate enamine nucleophile. The need for the presence of the two sulfone groups was also demonstrated, showing that vinylsulfones with a single activating group were unable to react under many reaction conditions tried. Other different reports[94c,95] also showed the outstanding performances of catalyst **31a** in the same reaction (Scheme 2.26) and also allowed the use of β-substituted vinyl sulfones, leading to the formation of two contiguous stereogenic centers, although the final products were obtained as mixtures of diastereoisomers. In this second report, the final products were isolated after *in situ* reduction of the formyl group. The addition of ketones to vinylsulfones using primary amine catalyst **28b** has also been reported with outstanding results when cyclic six-membered ring ketones were employed as Michael donors.[96]

On the other hand, the use of vinylsulfones incorporating a second different activating group like a nitrile or an ethoxycarbonyl moiety has also been explored with success,[95] although the configurational instability of the stereogenic center containing the two activating groups led to the formation of mixtures of diastereoisomers. Nevertheless, the presence of these two different nitrile/ester and sulfone functional groups at the final adducts could also be employed as a source of chemical diversity and therefore the intermediate reduced Michael adducts could be converted into a variety of different potentially useful chiral building blocks after several chemoselective transformations (Scheme 2.27).

Vinyl bis(phosphonates) have also been employed as suitable Michael acceptors in the Michael reaction with aldehydes under enamine activation, with *O*-TMS diphenylprolinol **31a** emerging as a very efficient catalyst for this transformation (Scheme 2.28).[94b,97] In addition, vinylphosphonates incorporating an additional activating group of a different nature have also been employed, changing to catalyst **31c** and using very similar reaction conditions.[98] In this case, the reactivity of the phosphonate moiety has also been subsequently exploited for the preparation of α-methylene-δ-lactones (Scheme 2.27). The reaction using ketones as Michael donor has also been reported to be possible a proline-based diamine catalyst, although with variable yields and enantioselectivities depending on the structure of the ketone employed.[99]

Alkylidenemalonates and malononitriles constitute another class of doubly activated olefins that can be used as highly electrophilic Michael acceptors in this reaction. For example, the Michael addition of aldehydes with these compounds has been reported to proceed with very good yields and enantioselectivities using *O*-TMS diphenylprolinol **31a** as catalyst (Scheme 2.29).[100] On the other hand, the Michael addition of ketones to alkylidenemalonates has

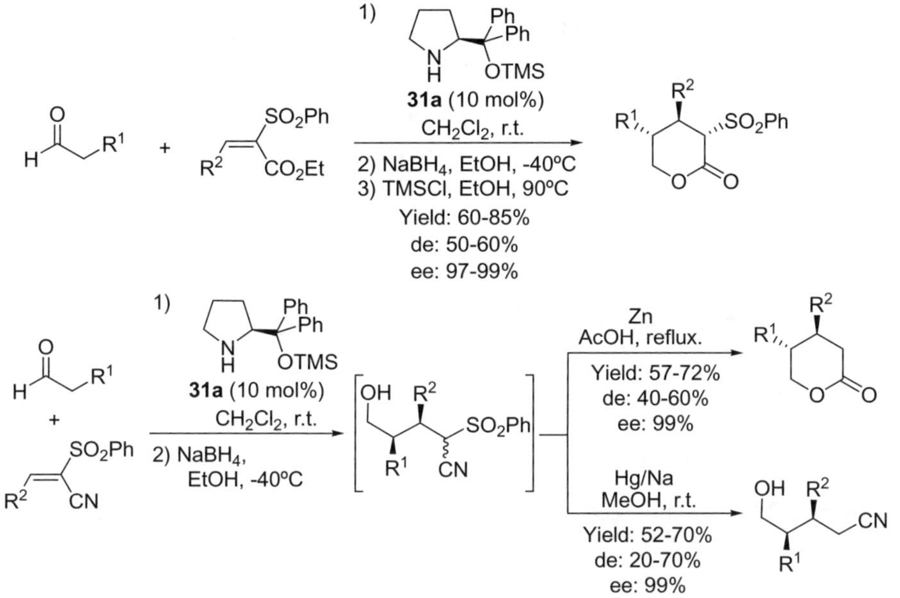

Scheme 2.27 Enantioselective Michael reaction of aldehydes with vinylsulfones incorporating a second different activating group.

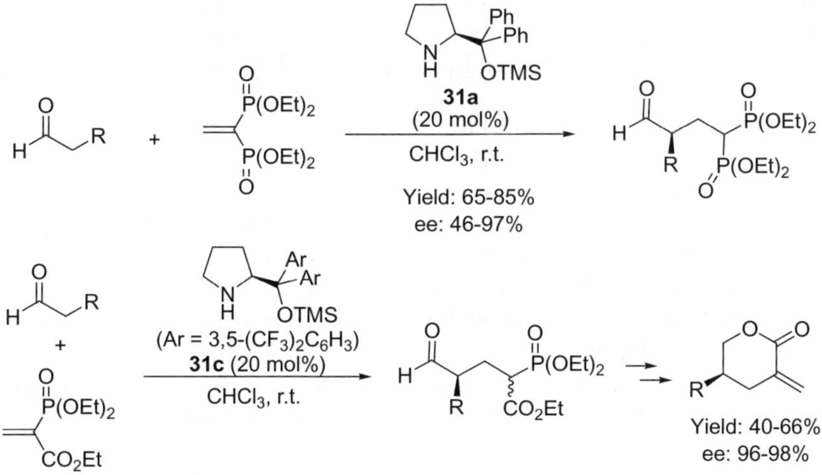

Scheme 2.28 Enantioselective Michael reaction using vinyl phosphonates as Michael acceptors.

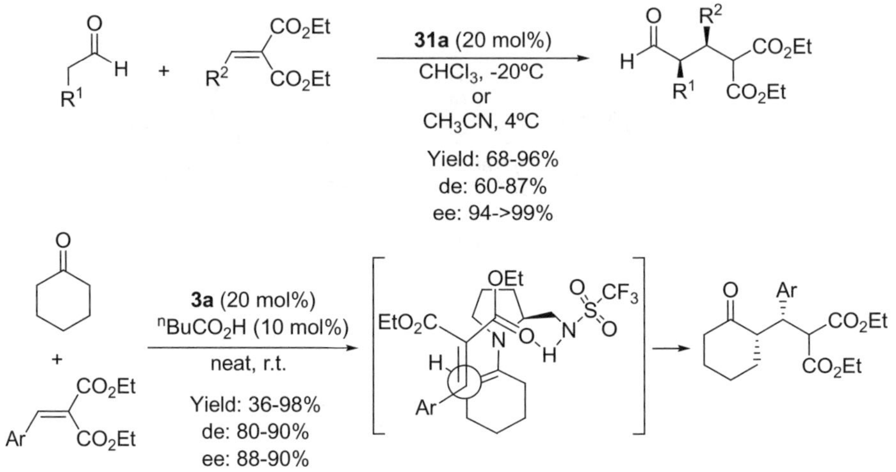

Scheme 2.29   Enantioselective Michael reaction using alkylidenemalonates as Michael acceptors.

also been carried out using **14a**,[24,101] **3a**[102] and **22b**[103] as catalysts, although, in many cases, the final adducts were obtained in variable yields, showing a high dependence with the nature of the substituents at both the ketone donor and the alkylidenemalonate acceptor. The best results were obtained for the **3a**-catalyzed addition of cyclohexanone to β-aryl substituted alkylidenemalonates. A transition state for the reaction is proposed in which the cooperative influence of the pyrrolidine moiety (for the formation of the enamine) and the sulfonamide group (which interacts with the malonate unit *via* hydrogen-bonding) is invoked to explain the stereochemical outcome of the reaction. The reaction between ketones and malononitriles has also been reported using a primary amine organocatalyst and proceeding with good yields and moderate enantioselectivities.[104]

The behavior of α-keto-α,β-unsaturated esters is slightly different. The Michael reaction between aldehydes and these kinds of doubly activated olefins proceeded smoothly in the presence of **31a** as catalyst but the conjugate addition products underwent spontaneous intramolecular hemiacetalization *via* the corresponding enol form of the α-ketoester moiety. Consequently, this reactivity was exploited for setting up a very efficient and simple protocol for the enantioselective synthesis of highly substituted dihydropyrones by carrying out the *in situ* oxidation of these Michael adducts (Scheme 2.30).[105] The final heterocyclic derivatives were obtained in excellent overall yield and outstanding enantioselectivity. Moreover, the reaction could also be carried out in water as solvent.

Finally, it has to be mentioned that not only *gem*-disubstituted diactivated olefins but also α,β-disubstituted Michael acceptors have been employed in this transformation. For example, the Michael addition of a series of aldehydes to

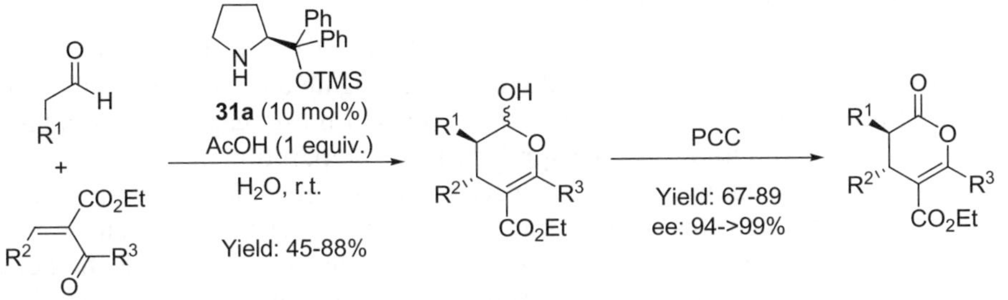

Scheme 2.30 Enantioselective synthesis of dihydropyrones.

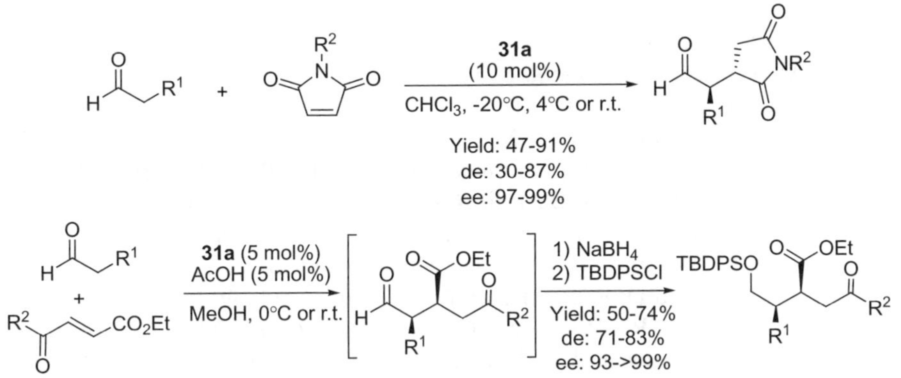

Scheme 2.31 Enantioselective Michael reaction using maleimides and γ-keto-α,β-unsaturated esters as Michael acceptors.

maleimides and to γ-keto-α,β-unsaturated esters and thioesters has been reported to occur smoothly with the assistance of **31a** as the most efficient catalyst (Scheme 2.31).[106] In the maleimide case the final adducts were obtained in moderate to good yields, good diastereoselectivity and excellent enantioselectivity while in the reactions with γ-keto-α,β-unsaturated esters, the final adducts had to be isolated after the *in situ* reduction of the formyl group followed by protection, with an acceptable yield for an overall three-step sequence. Remarkably, in this case the reaction was shown to proceed with excellent regioselectivity, which was governed by the better ability of the ketone group for the activation of the olefin moiety toward conjugate addition compared to that of the ester group. In addition, several synthetic protocols were devised for the chemical derivatization of the Michael adducts, which led to a very efficient and simple procedure for the preparation of different interesting chiral building blocks.

2.5 Concluding Remarks

The enamine activation concept has opened the way for the direct use of aldehydes and ketones as Michael donors in conjugate addition reactions with no need for previous modification in the form of a more reactive nucleophilic reagent. This methodology can be applied for the preparation of many different interesting chiral building blocks in a very simple and reliable way, with special emphasis in the high operational simplicity of these reactions, which do not need anhydrous solvents or an inert atmosphere. On the other hand, the need for a highly electrophilic Michael acceptor is a clear limitation of the methodology and the extension to other more common α,β-unsaturated carbonyl compounds such as α,β-unsaturated esters or amides is still undeveloped.

References and Notes

1. For several reviews exclusively focused on enamine activation see (a) S. Mukherjee, J. W. Yang, S. Hoffmann and B. List, *Chem. Rev.*, 2007, **107**, 5471; (b) S. Sulzer-Mossé and A. Alexakis, *Chem. Commun.*, 2007, 3123; (c) T. Kano and K. Maruoka, *Chem. Commun.*, 2008, 5465. For general reviews covering aminocatalysis see; (d) P. Melchiorre, M. Marigo, A. Carlone and G. Bartoli, *Angew. Chem. Int. Ed.*, 2008, **47**, 6138; (e) B. List, *Chem. Commun.*, 2006, 819; (f) B. List, *Synlett*, 2001, 1675.

2. (a) Z. G. Hajos and D. R. Parrish, *German Patent* DE 2102623, 1971; (b) Z. G. Hajos and D. R. Parrish, *J. Org. Chem.*, 1974, **39**, 1615; (c) U. Eder, G. Sauer and R. Wiechert, *German Patent* DE 2014757, 1971; (d) U. Eder, G. Sauer and R. Wiechert, *Angew. Chem. Int. Ed. Engl.*, 1971, **10**, 496.

3. B. List, R. A. Lerner and C. F. BarbasIII, *J. Am. Chem. Soc.*, 2000, **122**, 2395.

4. For an account on the evolution of the research of Barbas group from aldolase antibodies to proline see C. F. Barbas III, *Angew. Chem. Int. Ed.*, 2008, **47**, 42.

5. For reviews covering organocatalytic Michael reactions see(a) J. L. Vicario, D. Badia and L. Carrillo, *Synthesis*, 2007, 2065;(b) D. Almaçi, D. A. Alonso and C. Nájera, *Tetrahedron: Asymmetry*, 2007, **18**, 299; (c) S. B. Tsogoeva, *Eur. J. Org. Chem.*, 2007, 1701.

6. For some reviews see(a) C. Palomo and A. Mielgo, *Angew. Chem. Int. Ed.*, 2006, **45**, 7876;(b) A. Mielgo and C. Palomo, *Chem. Asian J.*, 2008, **3**, 922;(c) A. Lattanzi, *Chem. Commun.*, 2009, 1452.

7. D. Seebach and J. Golinski, *Helv. Chim. Acta*, 1981, **64**, 1413.

8. For some reviews see(a) Y.-C. Chen, *Synlett*, 2008, 1919;(b) L.-W. Xu, J. Luo and Y. Lu, *Chem. Commun.*, 2009, 1807;(c) L.-W. Xu and Y. Lu, *Org. Biomol. Chem.*, 2008, **6**, 2047.

9. (a) B. List, P. Pojarliev and H. J. Martin, *Org. Lett.*, 2001, **3**, 2423; (b) D. Enders and A. Seki, *Synlett*, 2002, 26. Almost simultaneously Barbas III reported the reaction between aldehydes and nitroalkenes (see

ref. 45(a)) and between ketones and alkylidenemalonates; (c) J. M. Betancort, K. Sakthivel, R. Thayumanavan and C. F. BarbasIII, *Tetrahedron Lett.*, 2001, **42**, 4441.

10. A. J. A. Cobb, D. A. Longbottom, D. M. Shaw and S. L. Ley, *Chem. Commun.*, 2004, 1808.
11. C. E. T. Mitchell, A. J. A. Cobb and S. V. Ley, *Synlett*, 2005, 611.
12. M. Arnó, R. J. Zaragozá and L. R. Domingo, *Tetrahedron: Asymmetry*, 2007, **18**, 157.
13. (a) J. Wang, H. Li, B. Lou, L. Zu, H. Guo and W. Wang, *Chem. Eur. J.*, 2006, **12**, 4231. See also; (b) D. Enders and S. Chow, *Eur. J. Org. Chem.*, 2006, 4578.
14. B. Ni, Q. Zhang and A. D. Headley, *Tetrahedron: Asymmetry*, 2007, **18**, 1443.
15. C. Wang, C. Yu, C. Liu and Y. Peng, *Tetrahedron Lett.*, 2009, **50**, 2363.
16. S. V. Pansare and R. L. Kirby, *Tetrahedron*, 2009, **65**, 4557.
17. (a) Y.-L. Cao, H.-H. Lu, Y.-Y. Lai, L.-Q. Lu and W.-J. Xiao, *Synthesis*, 2006, 3795; (b) Y.-J. Cao, M.-C. Ye, X.-L. Sun and Y. Tang, *Org. Lett.*, 2006, **8**, 2901. For other related systems see; (c) A. P. Carley, S. Dixon and J. D. Kilburn, *Synthesis*, 2009, 2509; (d) A. Lu, P. Gao, Y. Wu, Y. Wang, Z. Zhou and C. Tang, *Org. Biomol. Chem.*, 2009, **7**, 3141.
18. (a) D. Almasi, D. A. Alonso and C. Nájera, *Tetrahedron: Asymmetry*, 2006, **17**, 2064; (b) D. Almasi, D. A. Alonso, E. Gómez-Bengoa, Y. Nagel and C. Nájera, *Eur. J. Org. Chem.*, 2007, 2328.
19. M. Freund, S. Schenker and S. B. Tsogoeva, *Org. Biomol. Chem.*, 2009, **7**, 4279.
20. (a) B. Tan, X. Zeng, Y. Lu, P. J. Chua and G. Zhong, *Org. Lett.*, 2009, **11**, 1927. For the use of a similar sulfinyl derivative see; (b) X. Zeng and G. Zhong, *Synthesis*, 2009, 1545.
21. (a) F. Liu, S. Wang, N. Wang and Y. Peng, *Synlett*, 2007, 2415; (b) S.-W. Wang, J. Chen, G.-H. Chen and Y.-G. Peng, *Synlett*, 2009, 1457.
22. T. Mandal and C.-G. Zhao, *Tetrahedron Lett.*, 2007, **48**, 5803.
23. D. Díez, A. B. Antón, P. García, N. M. Garrido, I. S. Marcos, P. Basabe and J. G. Urones, *Tetrahedron: Asymmetry*, 2008, **19**, 2088.
24. (a) J. M. Betancort, K. Sakthivel, R. Thayumanavan, F. Tanaka and C. F. Barbas III, *Synthesis*, 2004, 1509; (b) S. V. Pansare and K. Pandya, *J. Am. Chem. Soc.*, 2006, **128**, 9624.
25. A. Alexakis and O. Andrey, *Org. Lett.*, 2002, **4**, 3611.
26. T. Ishii, S. Fujioka, Y. Sekiguchi and H. Kotsuki, *J. Am. Chem. Soc.*, 2004, **126**, 9558.
27. (a) D.-Q. Xu, L.-P. Wang, S.-P. Luo, Y.-F. Wang, S. Zhang and Z.-Y. Xu, *Eur. J. Org. Chem.*, 2008, 1049; (b) D.-Q. Xu, H.-D. Yue, S.-P. Luo, A.-B. Xia, S. Zhang and Z.-Y. Xu, *Org. Biomol. Chem.*, 2008, **6**, 2054.
28. (a) S. Luo, H. Xu, X. Mi, J. Li, X. Zheng and J.-P. Cheng, *J. Org. Chem.*, 2006, **71**, 9244; (b) S. Chandrasekhar, B. Tiwari, B. B. Parida and C. R. Reddy, *Tetrahedron: Asymmetry*, 2008, **19**, 495; (c) S. Chandrasekhar,

K. Millikarjun, G. Pavankumarreddy, K. V. Rao and B. Jagadeesh, *Chem. Commun.*, 2009, 4985.

29. (a) M.-K. Zhu, L.-F. Cun, A.-Q. Mi, Y.-Z. Jiang and L.-Z. Gong, *Tetrahedron: Asymmetry*, 2006, **17**, 491; (b) H. Chen, Y. Wang, S. Wei and J. Sun, *Tetrahedron: Asymmetry*, 2007, **18**, 1308.
30. (a) O. Andrey, A. Alexakis and G. Bernardinelli, *Org. Lett.*, 2003, **5**, 2559; (b) S. Belot, S. Sulzer-Mossé, S. Kehrli and A. Alexakis, *Chem. Commun.*, 2008, 4694.
31. (a) Y. Xu and A. Córdova, *Chem. Commun.*, 2006, 460; (b) Y. Xu, W. Zou, H. Sundén, I. Ibrahem and A. Córdova, *Adv. Synth. Catal.*, 2006, **348**, 418.
32. Y. Xiong, Y. Wen, F. Wang, B. Gao, X. Liu, X. Huang and X. Feng, *Adv. Synth. Catal.*, 2007, **349**, 2156.
33. Z. Yang, J. Liu, X. Liu, Z. Wang, X. Feng, Z. Su and C. Hu, *Adv. Synth. Catal.*, 2008, **350**, 2001.
34. H. Huang and E. N. Jacobsen, *J. Am. Chem. Soc.*, 2006, **128**, 7170.
35. (a) S. B. Tsogoeva and S. Wei, *Chem. Commun.*, 2006, 1451; (b) D. A. Yalalov, S. B. Tsogoeva and S. Schmatz, *Adv. Synth. Catal.*, 2006, **348**, 826.
36. (a) K. Liu, H.-F. Cui, J. Nie, K.-Y. Dong, X.-J. Li and J.-A. Ma, *Org. Lett.*, 2007, **9**, 923. For a related study using the same catalyst see; (b) Q. Gu, X.-T. Guo and X.-Y. Wu, *Tetrahedron*, 2009, **65**, 5265.
37. C. G. Kokotos and G. Kokotos, *Adv. Synth. Catal.*, 2009, **351**, 1355.
38. X. Jiang, Y. Zhang, A. S. C. Chan and R. Wang, *Org. Lett.*, 2009, **11**, 153.
39. (a) F. Xue, S. Zhang, W. Duan and W. Wang, *Adv. Synth. Catal.*, 2008, **350**, 2194; (b) R. Rasappan and O. Reiser, *Eur. J. Org. Chem.*, 2009, 1305.
40. S. H. McCooey and S. J. Connon, *Org. Lett.*, 2007, **9**, 599.
41. M. L. Clarke and J. A. Fuentes, *Angew. Chem. Int. Ed.*, 2007, **46**, 930.
42. T. Mandal and C.-G. Zhao, *Angew. Chem. Int. Ed.*, 2008, **47**, 7714.
43. (a) Y. Hayashi, H. Gotoh, T. Hayashi and M. Shoji, *Angew. Chem. Int. Ed.*, 2005, **44**, 4212. For a recent application see; (b) D. Bonne, L. Salat, J.-P. Dulcère and J. Rodríguez, *Org. Lett.*, 2008, **10**, 5409.
44. W. Wang, J. Wang and H. Li, *Angew. Chem. Int. Ed.*, 2005, **44**, 1369.
45. (a) J. M. Betancort and C. F. Barbas III, *Org. Lett.*, 2001, **3**, 3737; (b) N. Mase, R. Thayumanavan, F. Tanaka and C. F. Barbas III, *Org. Lett.*, 2004, **6**, 2527.
46. (a) O. Andrey, A. Alexakis, A. Tomassini and G. Bernardinelli, *Adv. Synth. Catal.*, 2004, **346**, 1147; (b) S. Mosse, M. Laars, K. Kriis, T. Kanger and A. Alexakis, *Org. Lett.*, 2006, **8**, 2559; (c) S. Mosse and A. Alexakis, *Org. Lett.*, 2006, **8**, 3577. See also M. Laars, K. Ausmees, M. Uudsemaa, T. Tamm, T. Kanger and M. Lopp, *J. Org. Chem.*, 2009, **74**, 372.
47. A. Quintard, C. Bournaud and A. Alexakis, *Chem. Eur. J.*, 2008, **14**, 7504.

48. (a) R. J. Reddy, H.-H. Kuan, T.-Y. Chou and K. Chen, *Chem. Eur. J.*, 2009, **15**, 9294. See also; (b) C. Chang, S.-H. Li, R. J. Reddy and K. Chen, *Adv. Synth. Catal.*, 2009, **351**, 1273.

49. L. Gu and G. Zhao, *Adv. Synth. Catal.*, 2007, **349**, 1629.

50. R.-S. Luo, J. Weng, H.-B. Ai, G. Lu and A. S. C. Chan, *Adv. Synth. Catal.*, 2009, **351**, 2449.

51. M. T. Barros and A. M. F. Phillips, *Eur. J. Org. Chem.*, 2007, 178.

52. X. Zhang, S. Liu, X. Li, M. Yan and A. S. C. Chan, *Chem. Commun.*, 2009, 833.

53. X. Zhang, S. Liu, J. Lao, G. Du, M. Yan and A. S. C. Chan, *Tetrahedron: Asymmetry*, 2009, **20**, 1451.

54. M. P. Lalonde, Y. Chen and E. N. Jacobsen, *Angew. Chem. Int. Ed.*, 2006, **45**, 6366.

55. H. Uehara and C. F. Barbas III, *Angew. Chem. Int. Ed.*, 2009, **48**, 9848.

56. C. Palomo, S. Vera, A. Mielgo and E. Gómez-Bengoa, *Angew. Chem. Int. Ed.*, 2006, **45**, 5984.

57. (a) M. Wiesner, J. D. Revell and H. Wennemers, *Angew. Chem. Int. Ed.*, 2008, **47**, 1871; (b) M. Wiesner, M. Neuburger and H. Wennemers, *Chem. Eur. J.*, 2009, **15**, 10103.

58. (a) E. Reyes, J. L. Vicario, D. Badia and L. Carrillo, *Org. Lett.*, 2006, **8**, 6135; (b) N. Ruiz, E. Reyes, J. L. Vicario, D. Badia, L. Carrillo and U. Uria, *Chem. Eur. J.*, 2008, **14**, 9357.

59. O. Andrey, A. Vidonne and A. Alexakis, *Tetrahedron Lett.*, 2003, **44**, 7901.

60. (a) S. Belot, A. Massaro, A. Tenti, A. Mordini and A. Alexakis, *Org. Lett.*, 2008, **10**, 4557. For a related reaction using ketones as Michael donors see; (b) T. He, J.-Y. Qiang, H.-L. Song and X.-Y. Wu, *Synlett*, 2009, 3195.

61. For a highlight on organocatalytic reactions with acetaldehyde see B. Alcaide and P. Almendros, *Angew. Chem. Int. Ed.*, 2008, **47**, 4632.

62. (a) P. García-García, A. Ladépêche, R. Halder and B. List, *Angew. Chem. Int. Ed.*, 2008, **47**, 4719; (b) Y. Hayashi, T. Itoh, M. Ohkubo and H. Ishikawa, *Angew. Chem. Int. Ed.*, 2008, **47**, 4722.

63. (a) Y. Chi, L. Guo, N. A. Kopf and S. H. Gellman, *J. Am. Chem. Soc.*, 2008, **130**, 5608; (b) M. Wiesner, J. D. Revell, S. Tonazzi and H. Wennemers, *J. Am. Chem. Soc.*, 2008, **130**, 5610.

64. S. Bertelsen, M. Marigo, S. Brandes, P. Diner and K. A. Jørgensen, *J. Am. Chem. Soc.*, 2006, **128**, 12973.

65. B. Han, Y.-C. Xiao, Z.-Q. He and Y.-C. Chen, *Org. Lett.*, 2009, **11**, 4660.

66. (a) M. Raj and V. K. Singh, *Chem. Commun.*, 2009, 6687; (b) J. Para-dowska, M. Stodulski and J. Mlynarsi, *Angew. Chem. Int. Ed.*, 2009, **48**, 4288. For a discussion of the role of water in organocatalytic reactions see; (c) D. G. Blackmond, A. Armstrong, V. Coombe and A. Wells, *Angew. Chem. Int. Ed.*, 2007, **46**, 3798.

67. (a) S. Zhu, S. Yu and D. Ma, *Angew. Chem. Int. Ed.*, 2008, **47**, 545; (b) V. Singh and V. K. Singh, *Org. Lett.*, 2007, **9**, 1117.

68. N. Mase, K. Watanabe, H. Yoda, K. Takabe, F. Tanaka and C. F. Barbas III, *J. Am. Chem. Soc.*, 2006, **128**, 4966.
69. J. Wu, B. Ni and A. D. Headley, *Org. Lett.*, 2009, **11**, 3354.
70. (a) L. Zu, J. Wang, H. Li and W. Wang, *Org. Lett.*, 2006, **8**, 3077. For a related fluorous catalyst see; (b) L. Zu, H. Li, J. Wang, X. Yu and W. Wang, *Tetrahedron Lett.*, 2006, **47**, 5131.
71. S. Luo, X. Mi, S. Liu, H. Xu and J.-P. Cheng, *Chem. Commun.*, 2006, 3687.
72. For some reviews see(a) S. Toma, M. Meciarova and R. Sebesta, *Eur. J. Org. Chem.*, 2009, 321;(b) P. Domínguez de Maria, *Angew. Chem. Int. Ed.*, 2008, **47**, 6960;(c) M. Gruttadauria, F. Giacalone and R. Noto, *Chem. Soc. Rev.*, 2008, **37**, 1666;(d) S. Luo, L. Zhang and J.-P. Cheng, *Chem. Asian J.*, 2009, **4**, 1184.
73. S. Luo, X. Mi, L. Zhang, S. Liu, H. Xu and J.-P. Cheng, *Angew. Chem. Int. Ed.*, 2006, **45**, 3093.
74. B. Ni, Q. Zhang, K. Dhungana and A. D. Headley, *Org. Lett.*, 2009, **11**, 1037.
75. Z. Yacob, J. Shah, J. Leistner and J. Liebscher, *Synlett*, 2008, 2342.
76. D.-Q. Xu, B.-T. Wang, S.-P. Luo, H.-D. Yue, L.-P. Wang and Z.-Y. Xu, *Tetrahedron: Asymmetry*, 2007, **18**, 1788.
77. M. Meciarova, S. Toma and R. Sebesta, *Tetrahedron: Asymmetry*, 2009, **20**, 2403.
78. D. Q. Xu, S. P. Luo, Y. F. Wang, A. B. Xia, H. D. Yue, L. P. Wang and Z. Y. Xu, *Chem. Commun.*, 2007, 4393.
79. For a review see M. Gruttadauria, F. Giacalone and R. Noto, *Chem. Soc. Rev.*, 2008, **37**, 1666.
80. (a) E. Alza, X. C. Cambeiro, C. Jimeno and M. A. Pericàs, *Org. Lett.*, 2007, **9**, 3717; (b) E. Alza and M. A. Pericás, *Adv. Synth. Catal.*, 2009, **351**, 3051. For other PS-supported catalysts employed in this reaction see; (c) Y. Chuan, G. Chen and Y. Peng, *Tetrahedron Lett.*, 2009, **50**, 3054; (d) L. Tuchman-Sukron and M. Portnoy, *Adv. Synth. Catal.*, 2009, **351**, 541.
81. (a) Y.-B. Zhao, L.-W. Zhang, L.-Y. Wu, X. Zhong, R. Li and J.-T. Ma, *Tetrahedron: Asymmetry*, 2008, **19**, 1352. See also; (b) P. Li, L. Wang, Y. Zhang and G. Wang, *Tetrahedron*, 2008, **64**, 7366.
82. S. Luo, J. Li, L. Zhang, H. Xu and J.-P. Cheng, *Chem. Eur. J.*, 2008, **14**, 1273.
83. G. Lv, R. Jin, W. Mai and L. Gao, *Tetrahedron: Asymmetry*, 2008, **19**, 2568.
84. M. P. Patil and R. B. Sunoj, *Chem. Asian J.*, 2009, **4**, 714.
85. P. Melchiorre and K. A. Jørgensen, *J. Org. Chem.*, 2003, **68**, 4151.
86. J. Franzén, M. Marigo, D. Fielenbach, T. C. Wabnitz, A. Kjaersgaard and K. A. Jørgensen, *J. Am. Chem. Soc.*, 2005, **127**, 18296.
87. (a) Y. Chi and S. H. Gellman, *Org. Lett.*, 2005, **7**, 4253; (b) T. J. Peelen, Y. Chi and S. H. Gellman, *J. Am. Chem. Soc.*, 2005, **127**, 11598.

88. J. Alemán, S. Cabrera, E. Maerten, J. Overgaard and K. A. Jørgensen, *Angew. Chem. Int. Ed.*, 2007, **46**, 5520.

89. (a) W. Wang, H. Li, L. Zu and W. Wang, *Adv. Synth. Catal.*, 2006, **348**, 425. For a similar system see; (b) D.-Z. Xu, S. Shi, Y. Liu and Y. Wang, *Tetrahedron*, 2009, **65**, 9344.

90. (a) M. T. Hechavarria Fonseca and B. List, *Angew. Chem. Int. Ed.*, 2004, **43**, 3958For a related work using a proline-derived anilide as catalyst see; (b) M. Kikuchi, T. Inagaki and H. Nishiyama, *Synlett*, 2007, 1075.

91. Y. Hayashi, H. Gotoh, T. Tamura, H. Yamaguchi, R. Masui and M. Shoji, *J. Am. Chem. Soc.*, 2005, **127**, 16028.

92. I. K. Mangion and D. W. C. MacMillan, *J. Am. Chem. Soc.*, 2005, **127**, 3696.

93. E. Marqués-López, R. P. Herrera, T. Marks, W. C. Jacobs, D. Könning, R. M. De Figueiredo and M. Christmann, *Org. Lett.*, 2009, **11**, 4116.

94. (a) S. Mossé and A. Alexakis, *Org. Lett.*, 2005, **7**, 4361; (b) S. Sulzer-Mossé, A. Alexakis, J. Mareda, G. Bollot, G. Bernardinelli and Y. Filinchuk, *Chem. Eur. J.*, 2009, **15**, 3204; (c) A. Quintard and A. Alexakis, *Chem. Eur. J.*, 2009, **15**, 11109. See also; (d) Q. Zhu and Y. Lu, *Org. Lett.*, 2008, **10**, 4803.

95. A. Landa, M. Maestro, C. Masdeu, A. Puente, S. Vera, M. Oiarbide and C. Palomo, *Chem. Eur. J.*, 2009, **15**, 1562.

96. Q. Zhu, L. Cheng and Y. Lu, *Chem. Commun.*, 2008, 6315.

97. S. Sulzer-Mossé, M. Tissot and A. Alexakis, *Org. Lett.*, 2007, **9**, 3749.

98. L. Albrecht, B. Richter, H. Krawczyk and K. A. Jørgensen, *J. Org. Chem.*, 2008, **73**, 8337.

99. M. T. Barros and A. M. F. Phillips, *Eur. J. Org. Chem.*, 2008, 2525.

100. (a) G.-L. Zhao, J. Vesely, J. Sun, K. E. Christensen, C. Bonneau and A. Córdova, *Adv. Synth. Catal.*, 2008, **350**, 657.

101. R. Chowdhury and S. K. Ghosh, *Org. Lett.*, 2009, **11**, 3270.

102. C.-L. Cao, X.-L. Sun, J.-L. Zhou and Y. Tang, *J. Org. Chem.*, 2007, **72**, 4073.

103. J. Liu, Z. Yang, X. Liu, Z. Wang, Y. Liu, S. Bai, L. Lin and X. Feng, *Org. Biomol. Chem.*, 2009, **7**, 4120.

104. L. Yue, W. Du, Y.-K. Liu and Y.-C. Chen, *Tetrahedron Lett.*, 2008, **49**, 3881.

105. J. Wang, F. Yu, X. Zhang and D. Ma, *Org. Lett.*, 2008, **10**, 2561.

106. (a) G.-L. Zhao, Y. Xu, H. Sundén, L. Eriksson, M. Sayah and A. Córdova, *Chem. Commun.*, 2007, 734; (b) J. Wang, A. Ma and D. Ma, *Org. Lett.*, 2008, **10**, 5425; (c) S. Zhu, Y. Wang and D. Ma, *Adv. Synth. Catal.*, 2009, **351**, 2563.

CHAPTER 3

Enantioselective Conjugate Addition Reactions via *Iminium Activation*

3.1 Introduction

A second possibility for a primary or secondary amine to catalyze a conjugate addition consists on the reversible formation of an iminium ion upon condensation with an appropriate Michael acceptor such as an α,β-unsaturated aldehyde or ketone.[1] The formation of the iminium ion leads to the lowering of the LUMO of the Michael acceptor, therefore becoming activated for the conjugate addition reaction, which now takes place much faster than the possible background uncatalyzed reaction between the nucleophile and the α,β-unsaturated aldehyde or ketone. After the conjugate addition step takes place, an enamine intermediate is generated, which, upon hydrolysis furnishes the product and releases the amine catalyst for its participation on a subsequent catalytic cycle (Scheme 3.1). Importantly, it has to be noted that the participation of an enamine-type intermediate species in the catalytic cycle of the conjugate addition reactions proceeding *via* iminium activation opens the way for a domino process to occur, provided that a suitable electrophile is included in the reaction able to react with this enamine intermediate either inter- or intramolecularly. These domino processes initiated by Michael addition reactions will be specifically covered in Chapter 7.

Interestingly, some very early examples of iminium-mediated conjugate additions can be found in the chemical literature. For example, Langenbeck and coworkers in 1937 reported the conjugate addition of water to crotonaldehyde catalyzed by piperidine or sarcosine[2] and Woodward, in 1981,

RSC Catalysis Series No. 5
Organocatalytic Enantioselective Conjugate Addition Reactions: A Powerful Tool for the Stereocontrolled Synthesis of Complex Molecules
By Jose L. Vicario, Dolores Badía, Luisa Carrillo and Efraim Reyes
© J. L. Vicario, D. Badía, L. Carrillo and E. Reyes 2010
Published by the Royal Society of Chemistry, www.rsc.org

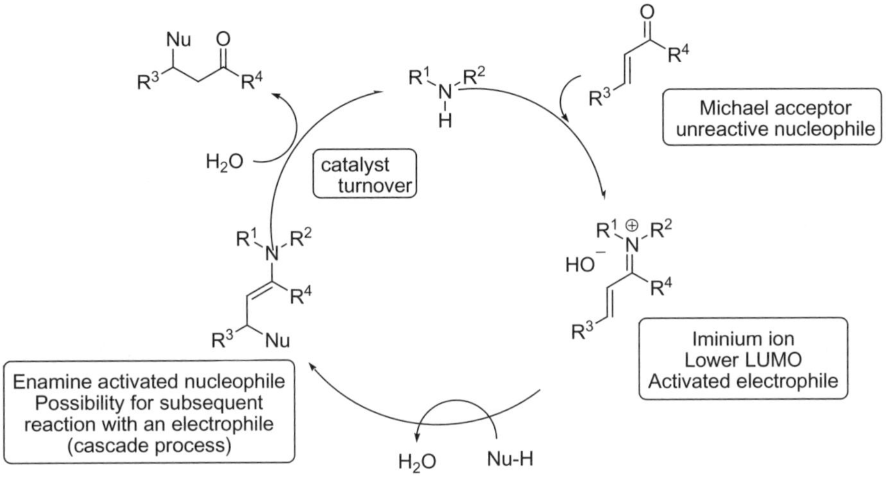

Scheme 3.1 Catalytic cycle for the Michael reaction proceeding *via* iminium activation.

reported a D-proline-catalyzed cascade process consisting of a retro-sulfa-Michael/sulfa-Michael/intramolecular aldol reaction involving iminium intermediates.[3] Between 1991 and 1993, Yamaguchi reported the use of alkaline prolinate salts in the asymmetric Michael reaction using malonates and nitroalkanes as nucleophiles.[4] In all these cases, the participation of an intermediate iminium species was proposed.

Nevertheless, the rational development and application of this concept has to be found in the seminal report on amine-catalyzed enantioselective Diels–Alder cycloadditions reported in 2000 by MacMillan (Scheme 3.2).[5] In this paper, MacMillan outlined that the formation of an iminium ion after condensation between the α,β-unsaturated aldehyde dienophile with the amine catalyst should result in a significant lowering the energy of its LUMO in a similar way as a Lewis acid does when complexing with an electron-deficient alkene, thus becoming activated for the cycloaddition reaction. On the other hand, it was also outlined that the dynamic processes involving complexation/decomplexation, which allow catalyst turnover in the Lewis acid-mediated transformations, would also find a counterpart in the reversibility of the reaction operating in the formation of iminium ions from amines and aldehydes or ketones. Once this concept was clearly established and understood from the mechanistic point of view, this chemistry was extended to other transformations involving α,β-unsaturated aldehydes and/or ketones, which included the Michael reaction, which since then has become a very active field of research.

As it happened in enamine activation reactions, the most remarkable feature of this strategy is the high atom efficiency of the process and also the operational simplicity of the experimental protocol, because of the compatibility of the reagents and intermediates participating in the reaction toward the presence

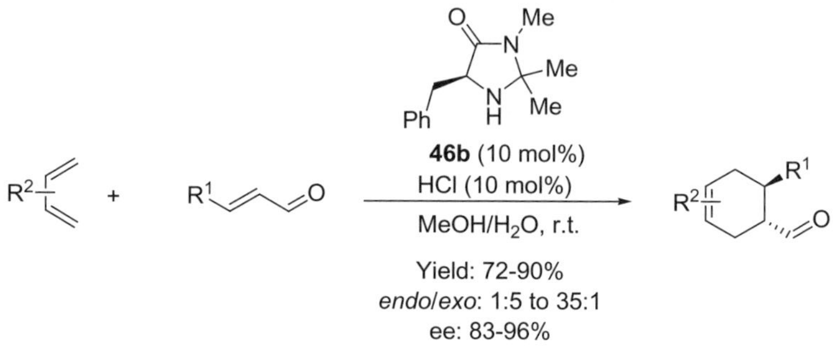

Scheme 3.2   The enantioselective Diels–Alder reaction developed by MacMillan.

of oxygen or moisture. This feature allows carrying out the reaction without the need of dried solvents or an inert atmosphere. Also the possibility of using α,β-unsaturated aldehydes as Michael acceptors is a relevant feature because these are not the kind of reagents that can be easily employed when working with transition metals. On the other hand, the covalent nature of the interaction between the catalyst and the substrate and the need for reversible reactions for the attachment and the removal of the catalyst cause an issue of the catalytic activity and the low TOF numbers usually observed. This issue is even more problematic when the formation of an iminium ion by condensation with an α,β-unsaturated aldehyde or ketone is required if we compare with the enamine activation case, in which the condensation reaction between a simple enolizable aldehyde or ketone and the catalysts is usually much faster. In addition, the ambident character of the Michael acceptor causes the issue of the 1,4- *vs.* 1,2-regioselectivity problem, the former being an important side reaction which would eventually lead to catalyst consumption (through competitive aza-Michael reaction). For all these reasons, the rate of the dynamic processes involved very often determines the viability of the catalytic reaction. One way to circumvent these problems has been the incorporation of additives, which increase the rate of these processes, and, therefore, it is very often found that Brønsted acid co-catalysts are included in the reaction design in order to help in the formation of the iminium ion intermediate. In other cases, water has to be incorporated in variable amounts in order to assist the final hydrolysis step.

3.1.1   Factors Influencing the Stereocontrol in Michael Additions Proceeding *via* Iminium Activation

The catalytic cycle operating for a generic Michael-type reaction of a nucleophile (Nu-H) to an α,β-unsaturated aldehyde or ketone proceeding *via* iminium activation has been indicated in Scheme 3.1. As it can be seen in this proposed mechanism, all the steps involving iminium formation and hydrolysis are supposed to be in dynamic equilibrium, therefore concluding that the conjugate

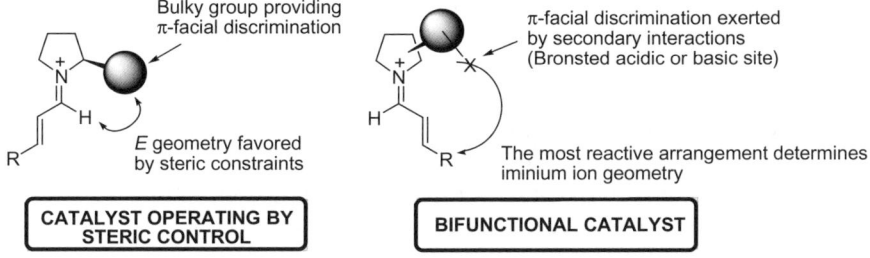

Figure 3.1 Factors to be considered when designing a chiral secondary amine catalyst to be used in Michael reactions under iminium activation.

addition step should appear as the relevant process with regard to stereochemical control. As a consequence of this, it is postulated that it should be this point at which the catalyst must exert its stereochemical influence by controlling the approach of the nucleophile by one of the two diastereotopic faces of the iminium intermediate. As happened in the enamine activation case, this stereodiscrimination can be carried out either by steric bias or by introducing a stereodirecting element (Figure 3.1). However, the reversible nature of some conjugate addition reactions has to be considered, for example, in the cases of the aza-Michael, the sulfa-Michael or other hetero-Michael reactions because other steps of the catalytic cycle may show up as the rate-determining step of the overall catalytic process. The geometry control on the iminium ion is also a key parameter to be controlled when designing an efficient catalyst for this transformation because mixtures of Z/E isomers would turn into low enantioselectivity on the reaction. This is not a problem when bifunctional catalysts are employed because the stereodirecting element incorporated at the catalyst determines the more reactive arrangement of the iminium ion during the conjugate addition step. However, this is a crucial factor in the design of a catalyst exerting the stereodifferentiation by steric bias because the volume of the substituent at the pyrrolidine ring has to be large enough not only for inducing the required high level of π-facial discrimination but also for determining a preferred geometry for the iminium ion. This issue is not a big problem when α,β-unsaturated aldehydes are used as Michael acceptors, the less sterically encumbered E iminium ion the being favored isomer. Finally, the *s-cis/s-trans* conformational flexibility of the α,β-unsaturated system does not represent a big problem in these reactions because of the high energy barrier required for the iminium ion intermediate to adopt the *s-cis* conformation.

The different reactivity of aldehydes and ketones toward condensation with amines is also a differentiating element when using enals or enones as Michael donors under iminium activation. As in the enamine activation case, working with α,β-unsaturated aldehydes usually leads to faster reactions or better conversions but the same reaction with enones in many cases turns out to be a very slow or even non-existent reaction. Stereochemical control is also more problematic when α,β-unsaturated ketones are employed because the presence of two more similar substituents at the carbonyl group can lead to the formation of mixtures of *Z/E* iminium isomers and therefore to poor enantioselectivities. As mentioned in the previous chapter, the use of primary amines as aminocatalysts provides an efficient solution to both the reactivity and the stereoselectivity problems because, on one hand, the condensation reaction between the enone and a primary amine is sterically more feasible than with a secondary amine and, on the other, the higher differences in steric bulk between the coplanar substituents across the conjugate system allow an effective geometry control on the iminium intermediate.[6]

Finally, the nature of the counteranion associated with the iminium cation is an additional element that has to be considered in this reaction design. This anion can play the role of a base, assisting the deprotonation of the nucleophile source (see Scheme 3.1) or in other cases remains as a simple spectator in the catalytic cycle. However, the tight interaction existing between the iminium cation and the corresponding counteranion can be used to incorporate additional chiral information in the reaction design. In fact, this novel approach (also called asymmetric counterion-directed catalysis, ACDC) has appeared as a very effective tool for carrying out enantioselective conjugate additions with excellent levels of stereoselection. In these cases, an achiral amine catalyst is employed for the formation of the iminium ion and a chiral Brønsted acid is incorporated as an additive in substoichiometric amount (typically a chiral phosphoric acid or an α-amino acid).

3.2   Michael Reaction

The conjugate addition of stabilized carbon nucleophiles to α,β-unsaturated carbonyl compounds or related derivatives (the Michael reaction) under iminium activation requires the use of a pro-nucleophile reagent incorporating a rather acidic C–H hydrogen, because the activation of the Michael donor by deprotonation with strong bases (for the generation of an enolate nucleophile) is not compatible with the reacting species participating in the catalytic cycle, in particular the molecules of water generated after the formation of the iminium ion. For this reason, this methodology has been applied to conjugate addition reactions using 1,3-dicarbonyl compounds and nitroalkanes as Michael donors, which can undergo deprotonation under the very mild reaction conditions employed. In fact, the acidity of the carbon pro-nucleophile is a parameter that very often requires a fine-tuning for a successful reaction. In this context, a recent study has pointed out that there must exist a pK_a barrier for nucleophile activation in conjugate additions of carbon nucleophiles proceeding *via* iminium activation, which should lie between the pK_a values of 16 and 17.[7] Nevertheless, some examples have also been developed for the application of several neutral nucleophilic reagents in this transformation, which do not require a previous deprotonation step like silyl enol ethers or enamides, among others, although it has to be said that important limitations are normally found regarding the tolerance of the reaction to structural variations in these reagents.

It has to be pointed out that simple enolizable aldehydes and ketones, which are not acidic enough compounds to be directly used as pro-nucleophiles in this context, can nevertheless be employed as Michael donors in the reaction with enals or enones, which have been previously activated as the corresponding iminium ion, but their use requires prior activation *via* enamine activation. In these cases, it is usually proposed that the amine catalyst is involved in a dual activation profile interacting with both the Michael donor and the acceptor, although the enamine activation of the pro-nucleophile is mandatory for the reaction to occur, the activation of the acceptor being of less relevance in most cases.[8] For these reasons, this chemistry has been covered in Chapter 2.

3.2.1 α,β-Unsaturated Aldehydes as Michael Acceptors

As already mentioned before, α,β-unsaturated aldehydes are very appropriate Michael acceptors to be employed in conjugate addition reactions under iminium activation by a chiral secondary amine. In this context, among the many different chiral amine catalysts developed for this particular transformation, with no doubt, *O*-trialkylsilyl-α,α-diaryl prolinol derivatives **31** (see Scheme 3.3) constitute a privileged family of catalysts which have demonstrated an outstanding performance in this reaction.[9] The first reports dealing with the use of this type of catalysts (**31a** and **31c**) were developed independently by the groups of Jørgensen and Hayashi, in the context of several reactions proceeding *via* enamine activation. The concept behind the catalyst design, when applied to the activation of the Michael acceptor by iminium ion formation, is essentially based on the presence of a very bulky diaryltrimethylsilyloxy substituent, which allows an excellent geometry control in the formation of the intermediate iminium ion and also exerts a very efficient sterical bias on the stereotopic faces of the Michael acceptor. These catalysts are nowadays commercially available but they can be easily prepared in multigram quantities starting from proline in only four steps.

In this context, 1,3-dicarbonyl compounds and nitroalkanes constitute excellent candidates to be used as Michael donors in this reaction because of the ability of the two carbonyl moieties or the nitro group to stabilize an adjacent negative charge, allowing the *in situ* generation of the corresponding enolate or nitronate respectively under mild conditions. In particular, the rich reactivity profile displayed by the nitro group makes the conjugate addition of nitroalkanes to enals a very powerful method in organic synthesis for the stereoselective preparation of many different chiral building blocks. Several reports illustrate the use of catalyst **31a** for the reaction between nitromethane and different alkyl- and aryl-substituted α,β-unsaturated aldehydes (Scheme 3.3), obtaining excellent yields and enantioselectivities.[10] It has to be pointed out that, in all cases, an acid co-catalyst had to be introduced in the reaction scheme, the former being incorporated in order to assist the formation of the iminium intermediate, or, alternatively, the addition of a base like LiOAc was found to be needed in order to help in the formation of the more reactive

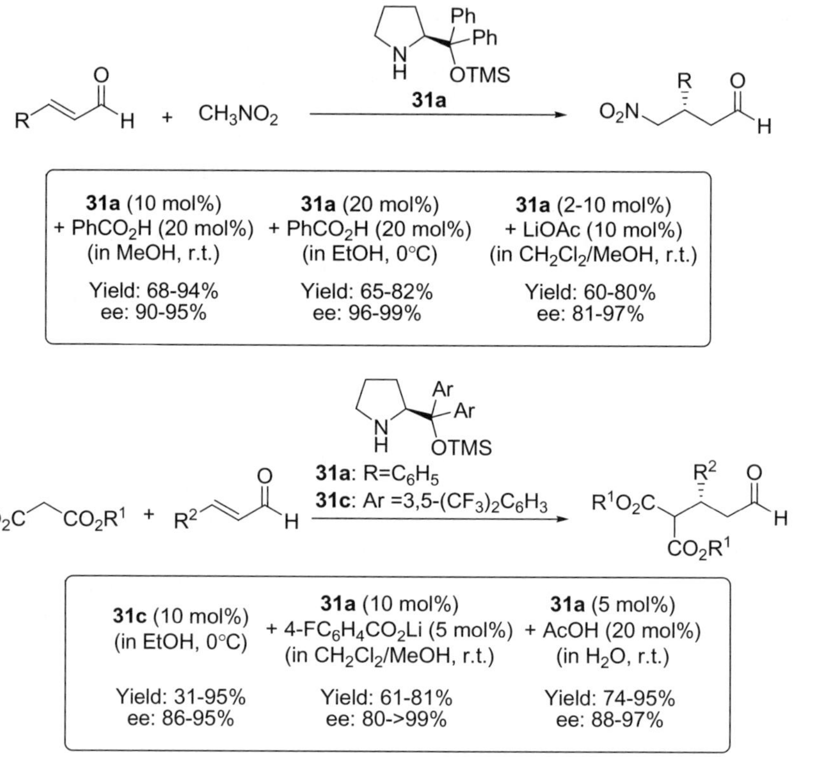

Scheme 3.3 Enantioselective **31a** and **31c**-catalyzed Michael reaction of nitromethane to α,β-unsaturated aldehydes.

nitronate nucleophile. An important and still unsolved limitation associated to this reaction is the fact that when nitroalkanes other than nitromethane that generate an additional stereogenic center in the reaction such as nitroethane are employed, the final products are typically obtained as mixtures of diastereoisomers, which is associated to this reaction. A very similar situation is found on the Michael reaction of malonates with enals (Scheme 3.3),[11] with several examples reporting the use of **31a** and modified diarylprolinol derivative **31c** as catalysts and obtaining the final conjugate addition products with excellent yields and enantioselectivities, also for a wide variety of differently substituted substrates. In this case, catalyst **31c** could be used directly without the incorporation of any other additive provided that the reaction is carried out in EtOH as solvent or, alternatively, requiring either the incorporation of a base co-catalyst or the use of an aqueous acidic medium for the reaction to proceed with good yields. In some cases, the nature of the alkoxy substituents at the malonate reagents had an important influence on the reaction, usually observing that the introduction of bulky groups such as ^{i}Pr led to an important decrease in the conversion.

The applicability of these methodologies has also been demonstrated with the synthesis of several important chiral drugs such as ($-$)-*baclofen*, a GABA receptor agonist used for the treatment of spinal cord injury-induced spasms, ($+$)-*pregabalin*, an anticonvulsant drug and the antidepressants ($-$)-*paroxetine* and ($+$)-*femoxetine*. In all cases, highly efficient synthetic procedures were developed to reach these target molecules in only a few steps.

There is also an important report detailing the preparation of a new family of O-TMS-α,α-dialkylprolinol catalysts specifically designed for carrying out this reaction using water as the only solvent (Scheme 3.4).[12] The principles of the catalyst design are based on previous reports in which the introduction of hydrophobic long alkyl side chains in the structure of several proline-based compounds resulted in an increased catalytic activity in several enamine-mediated reactions in water. These side chains presumably contribute to the formation of a hydrophobic pocket assisting the incorporation of the organic reagents into the catalytic site. Alternatively, the formation of micellar structures in water media has also been proposed to be the reason for the excellent performance of these kinds of catalysts. In this case, linear alkyl side chains of different lengths were introduced in the place of the two aryl groups present in the basic structure of diarylprolinols **31a** and **31c**. After some control experiments, O-triphenylsilyl protected dihexyl derivative **31e** was identified as the most efficient catalyst with the related dimethyl, dipropyl, dinonyl and didodecyl derivatives being less active in the model reaction. The use of this catalyst in water as the only solvent resulted in a high yielding and very enantioselective procedure for the Michael addition of nitromethane to a wide variety of α,β-unsaturated aldehydes and also for the same reaction using malonates as

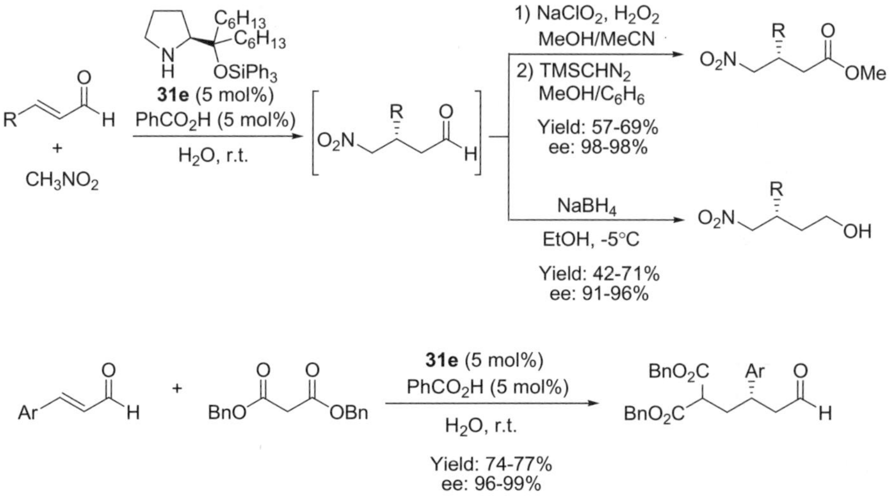

Scheme 3.4   Enantioselective Michael reaction of nitromethane and dibenzylmalonate to α,β-unsaturated aldehydes in water.

nucleophiles although, in the latter case, the study was limited to the use of two examples of aromatic enals.

A very interesting crossed Michael reaction between nitroalkenes and α,β-unsaturated aldehydes has also been reported in which the cooperative use of two different catalytic species able to activate independently the nucleophile and the electrophile (dual activation strategy) has been explored (Scheme 3.5).[13] In this case, diphenylprolinol **48a** was employed as the catalyst engaged in the activation of the enal by iminium ion formation and also one equivalent of trimethylphosphite was included as promoter of the reaction, which was needed for the activation of the nitroalkane Michael donor. The reaction therefore proceeded through a first step in which the conjugate addition of the phosphite to the nitroalkene delivered a nitronate nucleophile intermediate, which next underwent conjugate addition to the iminium ion intermediate obtained after the condensation of **48a** with the α,β-unsaturated aldehyde substrate. A subsequent β-elimination released the $(MeO)_3P$ reagent and a final hydrolysis step released the product and the catalyst ready to participate in the following catalytic cycle. Using these conditions a wide range of different β-methyl nitrostyrenes and α,β-unsaturated aldehydes reacted with each other furnishing the corresponding Michael adducts in excellent yield and enantioselectivities, although again as mixtures of diastereoisomers in 1:1 to 1:2.5 ratio, favoring the formation of the *anti* isomer. This lack of diastereoselectivity, once again,

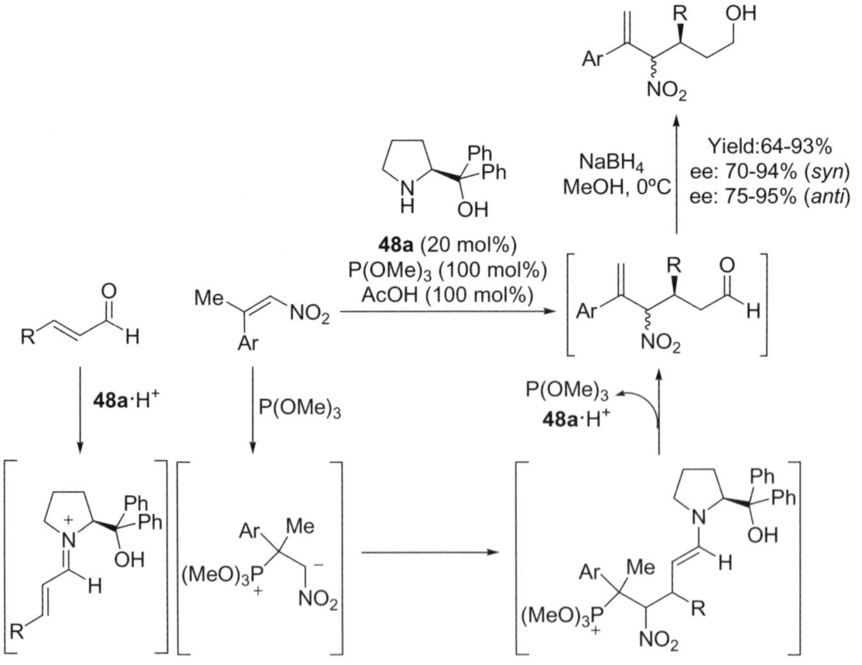

Scheme 3.5 **48a**-catalyzed enantioselective crossed Michael reaction between nitroalkenes and α,β-unsaturated aldehydes.

illustrates the difficulties associated with the stereochemical control of the reaction when substituted nitroalkanes that generate an additional stereogenic center at the α-position with regard to the nitro group are employed.

Bis(phenylsulfonyl)methane has also been employed as an acidic carbon pronucleophile related to malonates and 1,3-diketones with success in the Michael reaction with α,β-unsaturated aldehydes using **31c** as catalyst (Scheme 3.6).[14] The reaction showed a remarkable substrate scope when alkyl-substituted enals were employed but failed when cinnamaldehyde was tested as Michael acceptor. Alternatively, a more acidic cyclic *gem*-bissulfone has been used as Michael donor, keeping the high yields and enantioselectivities observed for the reaction and also allowing to expand the scope of the reaction to several aromatic enals.[15] In all cases, the chemistry of the sulfonyl group was employed to generate a methyl group after metal-mediated desulfuration or, alternatively,

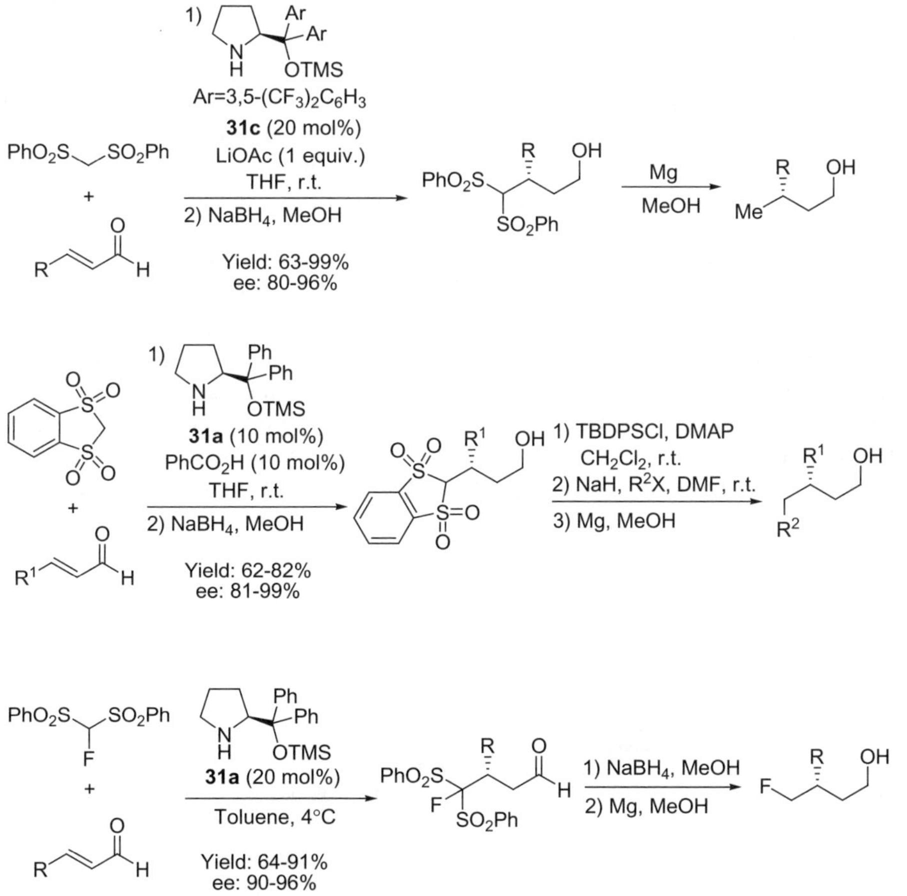

Scheme 3.6 Enantioselective Michael reaction of bis(phenylsulfonyl)methanes to α,β-unsaturated aldehydes.

procedures have been developed for the formal conjugate addition of an alkyl anion by introducing the alkyl chain by simple alkylation of the bissulfone moiety followed by desulfonylation. Fluorobis(phenylsulfonyl)methane has also been employed in this context, using the same catalyst and both alkyl- and aryl-substituted α,β-unsaturated aldehydes could be employed as Michael donors furnishing high yields and enantioselectivities.[16] The obtained adducts were also subjected to metal-mediated desulfuration resulting in an overall conjugate addition of a fluoromethyl anion.

The chemistry of the sulfonyl group has also been brilliantly exploited to achieve the formal conjugate addition of alkenyl and alkynyl anions to α,β-unsaturated aldehydes under iminium activation (Scheme 3.7).[17] The Michael addition of β-(5-phenyltetrazol-1-yl)sulfonyl ketones under catalysis by **31c** delivered cleanly the expected conjugate addition products which were subsequently subjected, without purification, to a sequence of reactions consisting of protection of the formyl moiety as an acetal, followed by base-promoted Smiles rearrangement and final acid hydrolysis. In this way, the corresponding β-alkynyl-substituted aldehydes were obtained in a one-pot operation in good yields and excellent enantioselectivities. Alternatively, the one-pot sequence consisting in the **31c**-catalyzed addition of β-(5-phenyltetrazol-1-yl)sulfonyl ketones to enals followed by NaBH$_4$ reduction resulted in the formation of an alkene moiety after the Smiles rearrangement, which proceeded with almost complete *E*-selectivity. The same chemistry has been expanded to the use of enones as Michael acceptors.[18]

The particular situation arising when cyclic 1,3-dicarbonyl compounds are employed as Michael donors in this reaction has been studied in detail by the groups of Rueping and Jørgensen (Scheme 3.8), observing that, after the conjugate addition reaction, the Michael adduct typically underwent a subsequent intramolecular hemiacetalization reaction reaching to a more stable form of the tautomer of the 1,3-dicarbonyl moiety. This special behavior gives rise to the formation of compounds with dihydropyran structure in a single step. In the reaction of 1,3-cyclohexanedione with enals,[19] the catalyst

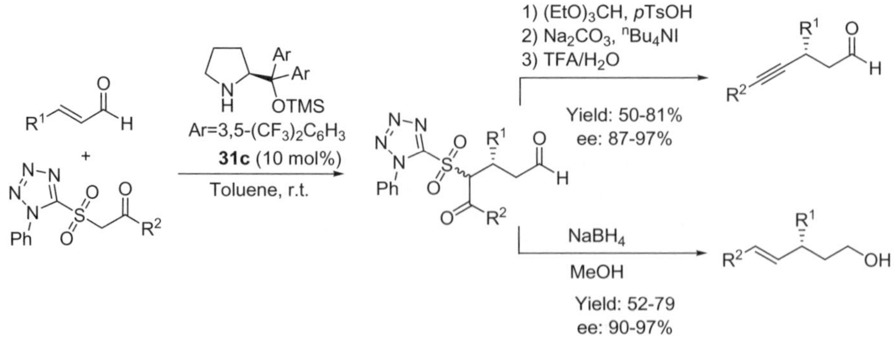

Scheme 3.7   Enantioselective Michael reaction of bis(sulfonyl)methanes to α,β-unsaturated aldehydes.

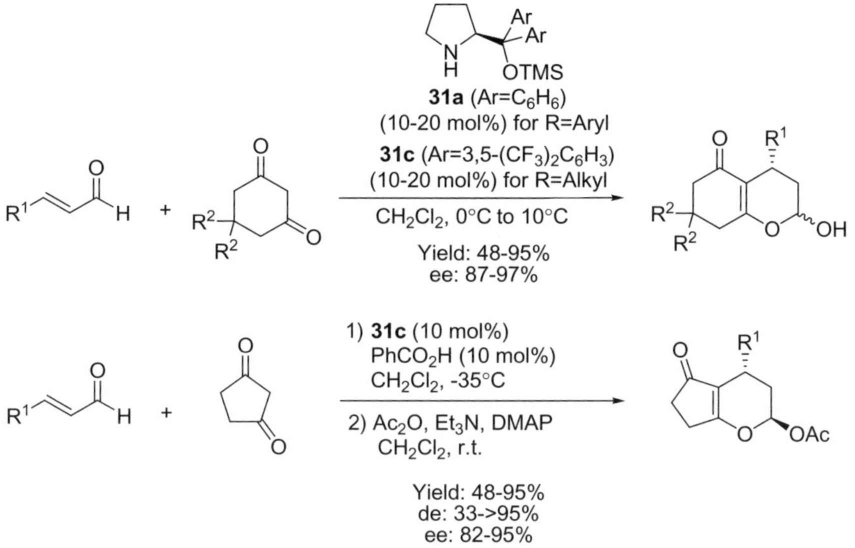

Scheme 3.8 Enantioselective Michael reaction of cyclic 1,3-diketones to α,β-unsaturated aldehydes.

employed had to be changed depending on the nature of the β-substituent of the Michael acceptor, observing that the best results were obtained using **31c** for alkyl-substituted enals, while the simple *O*-TMS diphenylprolinol **31a** showed up as an excellent promoter of the reaction involving aromatic enals. On the other hand, the reaction with the smaller 1,3-cyclopentanedione Michael donor was found to proceed satisfactorily with **31c** as catalyst for a wide variety of aliphatic, aromatic and even functionalized α,β-unsaturated aldehydes.[20] In this case, the final adducts were isolated as the corresponding acetylated derivatives. The higher acidity of the cyclic 1,3-diketones compared to the acidity of malonates is probably the main reason because this reaction proceeds much faster and with better conversions than the corresponding Michael addition of malonates to the same substrates. In addition, the stabilization of the final Michael addition as the cyclic hemiacetal form also contributes to the acceleration of the reaction by reducing the concentration of the Michael adduct in solution and therefore pushing the equilibriums participating in the process toward the formation of the final products. A very clever application of this reaction design to other different cyclic 1,3-dicarbonyl compounds or analogous derivatives has allowed the preparation of many different oxygen heterocycles such as pyranonaphthoquinones, β-lapachones, benzopyranes, chromenones, quinolinones or pyranones among others.[21]

The principle of vinylogy can also be applied to search for other C–H acidic compounds as suitable candidates to undergo conjugate addition under iminium activation.[22] This is the case of the vinylogous Michael addition of α,α-dicyanoolefins to α,β-unsaturated aldehydes catalyzed by diphenylprolinol

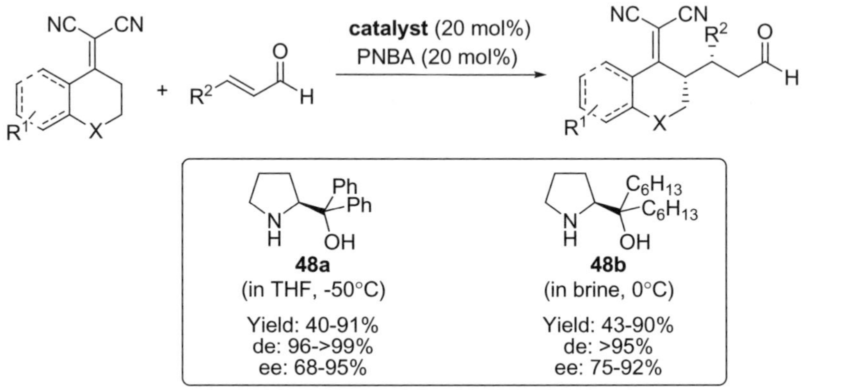

Scheme 3.9 Vinylogous Michael reaction of dicyanoacrylates with α,β-unsaturated aldehydes.

48a (Scheme 3.9). The corresponding γ-addition products were obtained with excellent yields and stereoselectivities and showing a remarkable substrate scope with regard to the substitution pattern at the enal and at the benzo-fused aromatic ring present at the structure of the pro-nucleophile.[23] As in other cases, the inclusion of a Brønsted acid co-catalyst such as *para*-nitrobenzoic acid (PNBA) was required to facilitate the condensation step between the catalyst and the Michael acceptor. Further research has led to the use of modified prolinols containing hydrophobic alkyl side chains as catalysts for carrying out exactly the same reaction using brine as reaction solvent. In this case, the design of the catalyst was carried out following a similar concept as that applied to the Michael reaction of nitroalkanes and malonates to enals in water which was depicted in Scheme 3.4.

The enhanced acidity of cyclic oxazolones compared with the corresponding open-chain *N*-alkoxycarbonyl α-amino acid analogues also allowed their use as nucleophiles in the Michael reaction with α,β-unsaturated aldehydes, demonstrating the high utility of these heterocycles as masked forms of α-amino acid-derived enolates. In this context, several racemic 4-alkyl- and 4-aryl-substituted oxazolones have been employed in the **31c**-catalyzed conjugate addition to enals, obtaining the expected adducts in excellent yields and with an outstanding level of stereocontrol (Scheme 3.10).[24] The presence of the C-4-stereogenic center at the oxazolone moiety, which led to the generation of two new stereogenic centers at the final adduct, had not any influence on the stereochemical outcome of the reaction, because it had to proceed *via* the corresponding achiral enolate. On the other hand, the C-2 substituent exerted a big influence in both the yield and the diastereoselectivity of the reaction, observing that installing a bulky benzhydryl group at this position resulted in a dramatic increase in the diastereoselectivity of the reaction, leading to the almost exclusive formation of a single diastereoisomer in very high enantiomeric

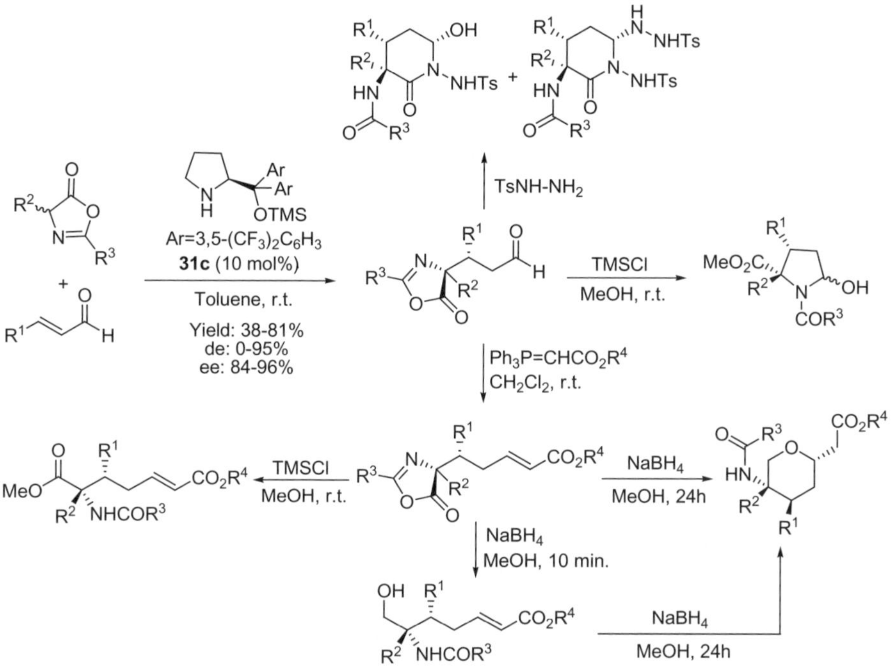

Scheme 3.10 Michael reaction of 4-substituted oxazolones with α,β-unsaturated aldehydes catalyzed by **31c** and further transformations carried out on the obtained adducts.

excess. An intensive study on the synthetic utility of the obtained adducts was also carried out and therefore several protocols were designed for the selective elaboration of the different functionalities present in these molecules. In particular, the obtained Michael adducts could be converted into α,α-disubstituted α-amino esters, polysubstituted prolines, δ-lactams, β-amino alcohols and tetrahydropyrans using very simple procedures which proceeded in high yields and maintaining the stereochemical integrity of the starting materials.

In all the examples presented previously, the high acidity of the pronucleophile is derived either from the incorporation of an extremely efficient electron-withdrawing group such as the nitro group or, alternatively, by the presence of two activating groups as, for example, in the malonate case. Important efforts have been carried out directed to the identification of a suitable approach, which allows the use of molecules incorporating a single activation group as Michael donors, in particular focused on the use of simple ester nucleophiles. In this context, the use of trifluoroethyl thioesters as nucleophiles in iminium-mediated Michael reactions has been surveyed, based on the hypotheses that thioesters are more acidic compounds than the corresponding esters, and also that the electronic influence exerted by the highly electron-withdrawing nature of the trifluoroethyl substituent might increase even more

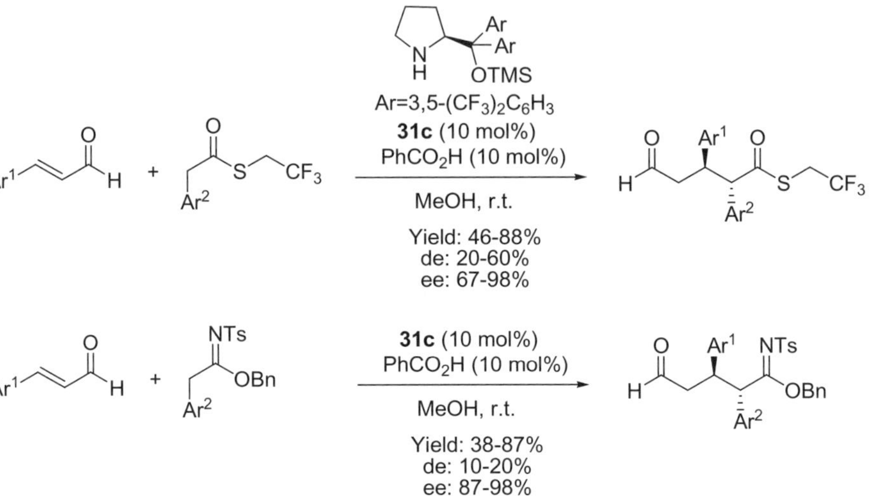

Scheme 3.11   **31c**-catalyzed Michael reaction of simple ester surrogates to aromatic α,β-unsaturated aldehydes.

the acidity of these α-C–H protons. These hypotheses were confirmed with the finding that catalyst **31c** was able to promote the Michael reaction of different arylacetic acid-derived trifluoroethyl thioesters and α,β-unsaturated aldehydes in a remarkably efficient way (Scheme 3.11).[7] The reaction was limited to α-aryl substituted thioesters which appear over the pK_a barrier needed for their activation and furnished enantiomeric excesses lying between 67 and 96% ee, although typically mixtures of *syn*/*anti* diastereoisomers were obtained with a slight preference for the formation of the *anti* isomer. The scope of Michael acceptor was also restricted to aromatic enals, as the reaction using croton-aldehyde as electrophile furnished the final compound with lower enantio-selectivity. Following the same line, *N*-tosylimidates, which show a similar acidity to trifluoroethyl thioesters, have also been tested as ester surrogates in the same context with similar results (Scheme 3.11).[25] In this case, excellent enantioselectivities of the corresponding Michael addition products were obtained, although the yield of the reaction was found to be usually moderate and 1:1 mixtures of diastereoisomers were obtained in all cases. As it happened in the trifluoroethyl thioester case, the reaction was exclusively limited to the use of nucleophiles derived from phenylacetic acid and to aromatic α,β-unsaturated aldehydes as Michael acceptors.

Oxindoles are a particular class of cyclic amides which are acidic enough to play the role of pro-nucleophiles in conjugate additions under iminium acti-vation. In particular, the reaction between 3-alkyl substituted oxindoles and α,β-unsaturated aldehydes leading to the formation of a quaternary stereocenter at the heterocyclic unit has been studied by several authors (Scheme 3.12). The use of *O*-trialkylsilyldiarylprolinols like **31a** as catalysts led

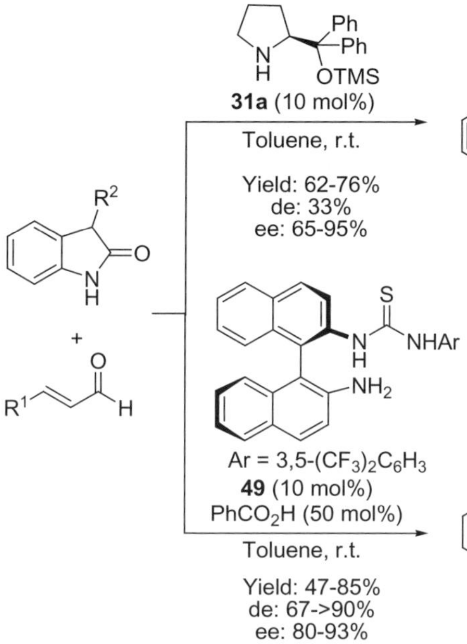

Scheme 3.12 Michael reaction of oxindoles to α,β-unsaturated aldehydes.

to moderate to good yields of the corresponding adducts, which were obtained with excellent enantioselectivities but as a mixture of diastereoisomers in 2:1 ratio.[26] Alternatively, bifunctional primary amine-thiourea catalyst **49** was found to be a good promoter for this reaction,[27] obtaining moderate to good yields of the final products for a variety of differently substituted oxindoles and α,β-unsaturated aldehydes with enantioselectivities in the range of 80–93% ee. Moreover, the reaction proceeded with a remarkably high degree of diastereoselection, allowing the installation of two contiguous stereogenic centers, one of them a quaternary one, in a very fast and efficient way.

Cyclopentadiene has also been employed as acidic C-nucleophile in conjugate additions proceeding *via* iminium ion formation in what can also be considered as an ene reaction (Scheme 3.13).[28] In this case, *O*-TBS diphenyl prolinol **31f** was found as the most effective catalyst for this transformation, which also needed the incorporation of *p*-nitrophenol as Brønsted acid cocatalyst assisting the formation of the iminium ion. The Michael adducts were obtained as mixtures of regioisomers arising after an isomerization process occurring on the initially formed 5-cyclopentadienyl adducts which subsequently furnished the more stable 1- and 2-substituted cyclopentadienes. Nevertheless, the combined yield of the reaction was high and excellent enantioselectivities were reached for all the cases studied. On the other hand, the substrate scope was restricted to the use of aromatic α,β-unsaturated aldehydes as Michael acceptors, with no data provided regarding the use of β-alkyl

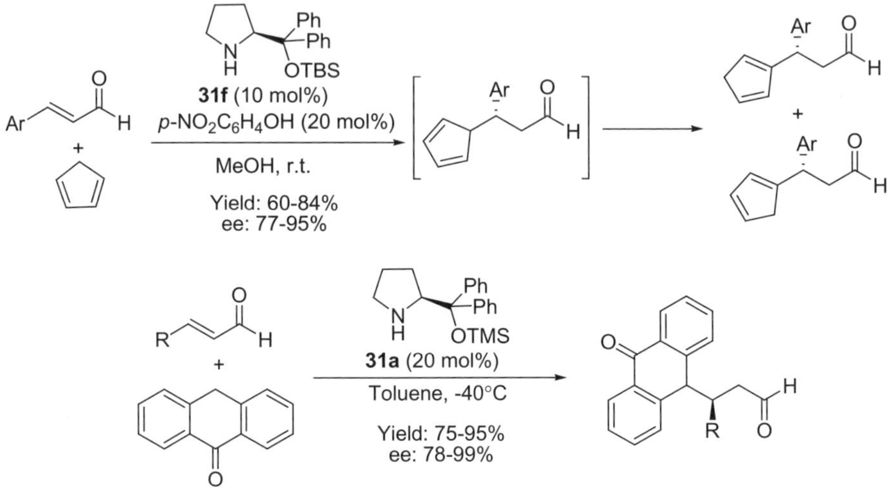

Scheme 3.13 Michael reaction of cyclopentadiene or anthrone with α,β-unsaturated aldehydes.

substituted enals. Anthracenones have also been employed as C–H acidic pro-nucleophiles in a similar context, furnishing the corresponding conjugate addition products in good yields and enantioselectivities for a variety of both β-aryl and β-alkyl substituted enals, using in this case *O*-TMS-diphenylprolinol **31a** as the most efficient catalyst.[29]

An alternative approach for carrying out Michael additions under iminium activation overcoming the requisite use of C–H acidic pro-nucleophiles is the formation of an activated form of the starting carbonyl compound under conditions compatible with the presence of water and the amine catalyst. As the formation of metal enolates is not allowed, enamides and silyl enol ethers can show up as useful candidates to be used as Michael donors in this process, provided that the intermediate iminium ion is reactive enough to initiate the conjugate addition process without the assistance of a Lewis acid, which is also not compatible with the presence of water in the reaction medium. The con-jugate addition of enamides to α,β-unsaturated aldehydes has been reported to proceed very efficiently using *O*-silylated diarylprolinols **31f** and **31a** as cata-lysts respectively (Scheme 3.14).[30] This reaction has been proposed to happen as an aza-ene reaction (concerted mechanism) although, alternatively, the reaction can also be considered as an stepwise process starting with the con-jugate addition of the enamine nucleophile to the iminium ion, followed by hydrolysis, imine/enamine interconversion and final hemiaminal formation. The nature of the *N*-acyl group played an important role in the reaction, showing that the use of *N*-Boc enamines as nucleophiles needed rather high temperatures (70 °C) in order to reach full conversion, although they main-tained excellent levels of enantioselectivity. On the contrary, the reaction

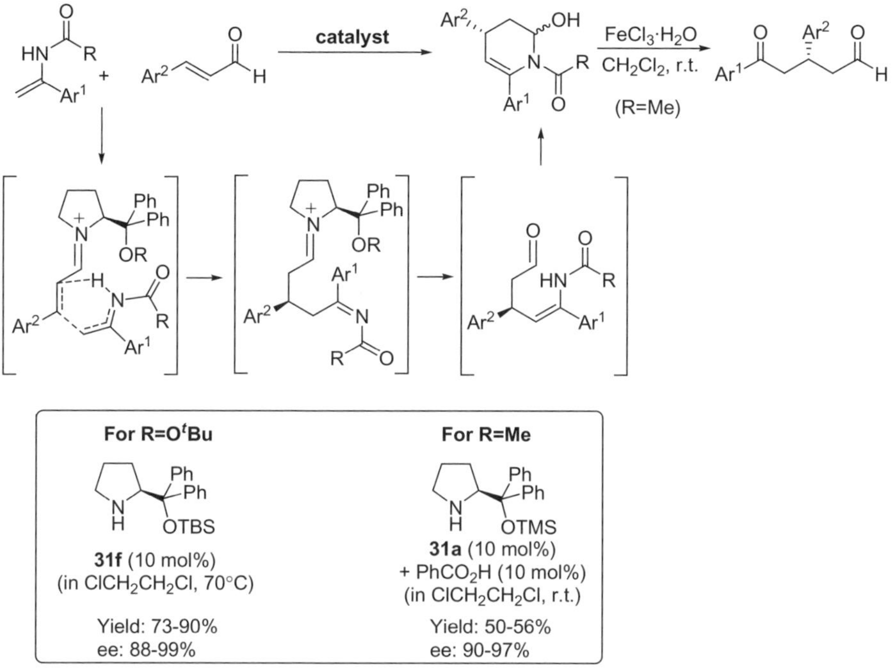

Scheme 3.14 Enantioselective Michael addition of enamides to α,β-unsaturated aldehydes.

involving *N*-acetylenamines could be carried out at r.t. with the aid of a Brønsted acid co-catalyst. In the former case, the final hemiaminal adducts were obtained as mixtures of α/β isomers in excellent yields and enantioselectivities for a wide variety of α,β-unsaturated aldehydes, although restricted to aromatic substituents at the β-position. In the latter, the authors carried out the *in situ* hydrolysis of the hemiaminal adduct isolating the corresponding 5-oxoaldehydes in overall yields around 50% but with excellent enantioselectivity.

Alternatively, the iminium-activation strategy has also been applied to the Mukaiyama–Michael reaction, which involves the use of silyl enol ethers as nucleophiles. In this context, imidazolidinone **50a** was identified as an excellent chiral catalyst for the enantioselective conjugate addition of silyloxyfuran to α,β-unsaturated aldehydes, providing a direct and efficient route to the γ-butenolide architecture (Scheme 3.15).[31] This is a clear example of the chemical complementarity between organocatalysis and transition-metal catalysis, with the latter usually furnishing the 1,2-addition product (Mukaiyama aldol) while the former proceeds *via* 1,4-addition when ambident electrophiles such as α,β-unsaturated aldehydes are employed. This reaction needed the incorporation of 2,4-dinitrobenzoic acid (DNBA) as a Brønsted acid co-catalyst assisting the formation of the intermediate iminium ion, and also two equivalents of water had to be included as additive for the reaction to proceed to completion, which

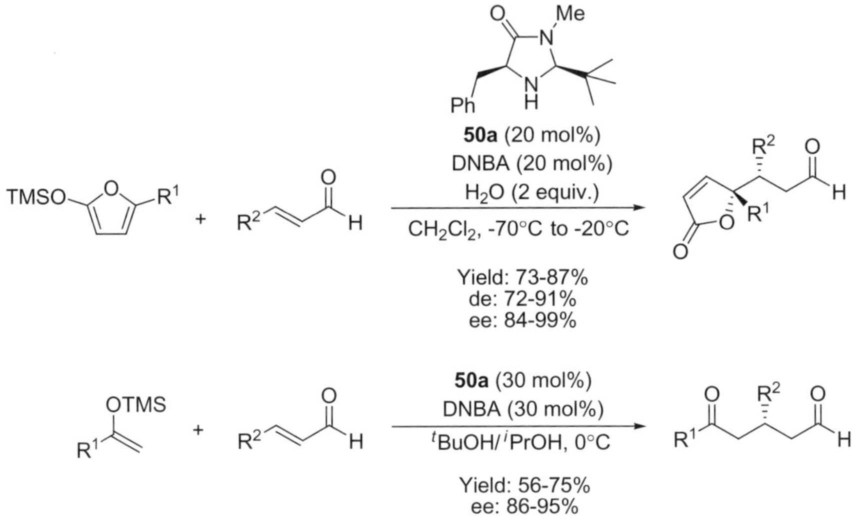

Scheme 3.15 Mukaiyama–Michael reactions with α,β-unsaturated aldehydes.

were needed to overcome the loss of water from the catalytic cycle, thought to happen because of the generation of the trimethylsilyl cation, which prevents catalyst turnover. Under the optimized conditions, the reaction proceeded with good yields and with excellent levels of stereocontrol for a series of α,β-unsaturated aldehydes and different silyloxyfuranes. The outstanding performance of this imidazolidinone catalyst is related to its ability to control the iminium geometry together with the very effective steric shielding achieved for one of the enantiotopic faces of the Michael acceptor. In addition, the applicability of this reaction in synthesis has been demonstrated with some examples of total syntheses of biologically active compounds.[31,32] Later on, it has been reported that the same catalyst **50a** was also able to catalyze the Mukaiyama–Michael reaction of acyclic silyl enol ethers with α,β-unsaturated aldehydes, obtaining the corresponding Michael adducts in excellent yields and enantioselectivities.[33] As happened in the MacMillan case, 2,4-dinitrobenzoic acid had to be included as additive and also scavenging of the trimethylsilyl cation formed as the reaction proceeded forward had to be addressed, which in this case could be carried out by using a *t*BuOH/*i*PrOH mixture as reaction solvent.

3.2.2 α,β-Unsaturated Ketones as Michael Acceptors

It has already been mentioned that one of the first examples in which the catalytic use of chiral amines was suggested for the activation of Michael acceptors in Michael addition reactions were the early reports by Yamaguchi, which focused on the conjugate addition of malonates and nitroalkanes to enones. In these early reports, rubidium prolinate salt **51** was identified as the most efficient catalyst for this reaction, although enantioselectivities remained moderate in almost all cases.[34] The Michael addition of dibenzyl malonate to cyclic enones and to benzylideneacetone using proline-based ammonium salt **52** reported by Kawara and Taguchi in 1994 (Scheme 3.16) provided similar results in terms of yields and enantioselectivities.[35] Importantly, the absolute configuration of the Michael adducts obtained in the **52**-catalyzed reaction resulted in being the opposite to that found for the same rubidium prolinate-catalyzed reaction. Therefore, and taking into account the very similar structure of both catalysts, a tentative model for the stereochemical outcome of the reaction was provided involving the ability of the ammonium substituent to direct the attack of the malonate enolate *via* electrostatic interactions. Consequently, in the **51**-mediated reaction, the carboxylate substituent should only exert the stereochemical control by exclusive steric effects.

Once in the modern organocatalysis era and with the mechanistic rationale for the iminium activation concept in hand, many different and more efficient methodologies have been developed for this particular reaction. For example, and still focused on the use of secondary amines as catalysts, imidazolidine **53a**[36] and proline-tetrazole **2a**[37] catalysts have been developed for the conjugate addition of malonates to acyclic enones (Scheme 3.17). For the **53a**-catalyzed reaction, this proceeded well in terms of yields and enantioselectivities for a wide range of differently substituted arylideneacetones and for cyclohexenone but yields tend to decrease when more bulky substituents were placed around the carbonyl moiety. Importantly, the enantioselectivity of the reaction was very dependent upon the nature of the malonate reagent, observing that dibenzyl malonate and diethylmalonate furnished the best results. The most

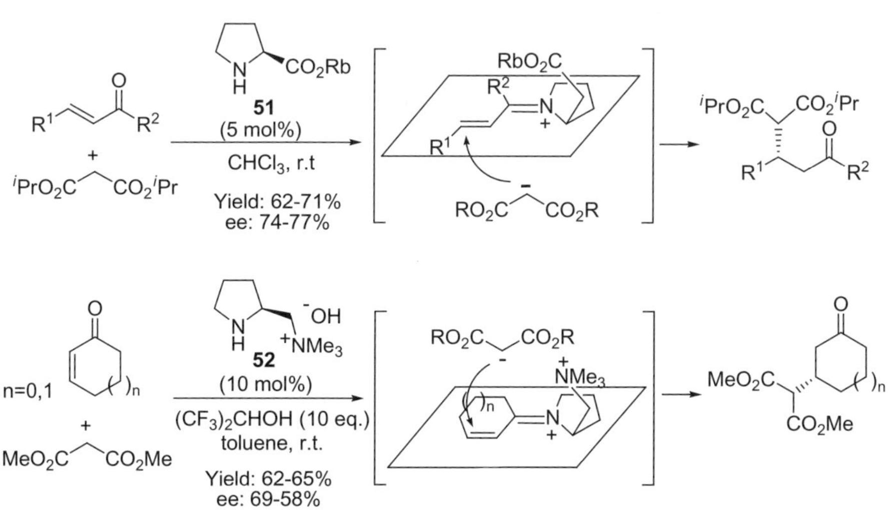

Scheme 3.16 The Michael reaction of malonates and enones catalyzed by rubidium prolinate and proline-derived tetraalkylammonium salt **52**.

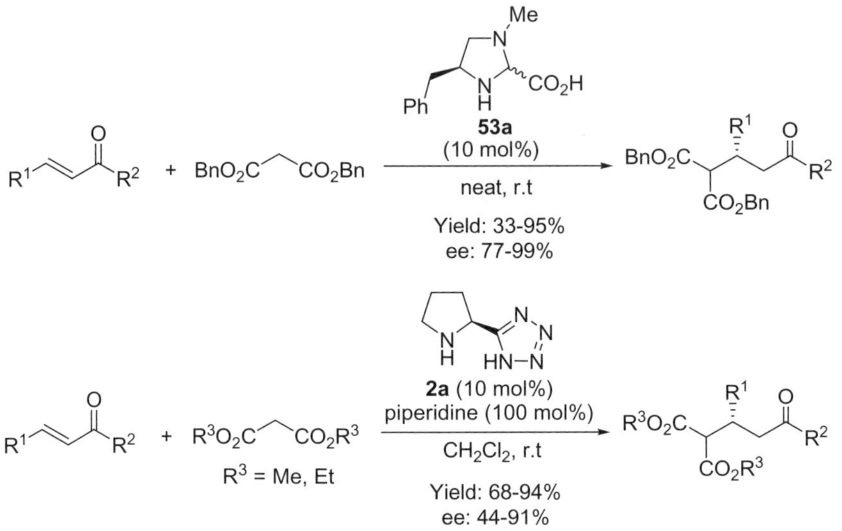

Scheme 3.17   The Michael addition of malonates to enones catalyzed by **53a** and **2a**.

important drawbacks associated to this protocol were the need for a large excess of the malonate nucleophile, which was usually employed as the reaction solvent, and also the long reaction times required. On the other hand, proline tetrazole catalyst **2a** was expressly employed in an effort to overcome the need for such large amounts of nucleophile and also with a focus on the preferable use of the more accessible and cheap diethyl and dimethyl malonates as Michael donors. In this context, the **2a**-catalyzed reaction of diethylmalonate to several enones proceeded in a very efficient way, providing excellent yields and enantioselectivities and using only a slight excess of Michael donor reagent. However, this methodology showed a more narrow substrate scope with regard to the range of enones amenable to be employed because, while excellent results were obtained for cyclohexenone and arylideneacetones, the performance of this catalyst in the reaction with aliphatic enones was significantly poorer. In addition, it has to be mentioned that, in this case, the reaction needed for the incorporation of one equivalent of piperidine as additive, which is thought to assist the reaction by helping in the deprotonation of the malonate pro-nucleophile, increasing the concentration of the corresponding enolate in solution. Other modified pyrrolidine catalysts containing functional side chains able to interact with the malonate reagent have also been developed recently.[38]

As happened when α,β-unsaturated aldehydes were employed as Michael acceptors, the conjugate addition of cyclic 1,3-dicarbonyl compounds to enones has received particular attention, in this case mainly because warfarin, an anticoagulant generic drug, can be directly prepared by using this reaction (see Scheme 3.18). In this context, modified imidazolidine catalyst **53b** has shown its outstanding performance in the Michael reaction of a wide range of different cyclic 1,3-dicarbonyl compounds with acyclic enones, showing a remarkably wide substrate scope regarding the substitution pattern at the Michael acceptor.[39] Even though enantioselectivities remained around 80% ee in most cases, usually more than 99% ee could be obtained after a single recrystallization process. One of the examples presented represents the direct synthesis of (*R*)-*warfarin* and this reaction could be scaled up to one-kilogram scale without any decrease in yield or enantioselectivity. The real catalytic species participating in this reaction is not completely clear. Initially, it was proposed that the formation of a cyclic aminal as the intermediate undergoing the conjugate addition reaction explained much better the observed high enantioselectivity, ruling out the formation of the iminium ion because calculations suggested a poor shielding of the alkene if this kind of intermediate would participate in the reaction. However, in a later report, the same reaction was studied employing diphenylethylenediamine **54** (a primary amine) as catalyst,[40] providing similar results to those furnished by imidazolidine **53b** (Scheme 3.18). Consequently, an alternative intermediate was proposed involving the direct formation of a bis-imine intermediate by condensation of two equivalents of the enone with **54**, which should be present in the reaction medium because of the inherent instability of catalyst **53b** toward the presence of water. It was also proposed that the hydrolysis of the catalyst would also generate glyoxylic acid which might operate as a Brønsted acid co-catalyst in the formation of an activated bis-iminium ion by protonation, which is in accordance to the observed increase of enantioselectivity for the **54**-catalyzed reaction in the presence of AcOH as additive.

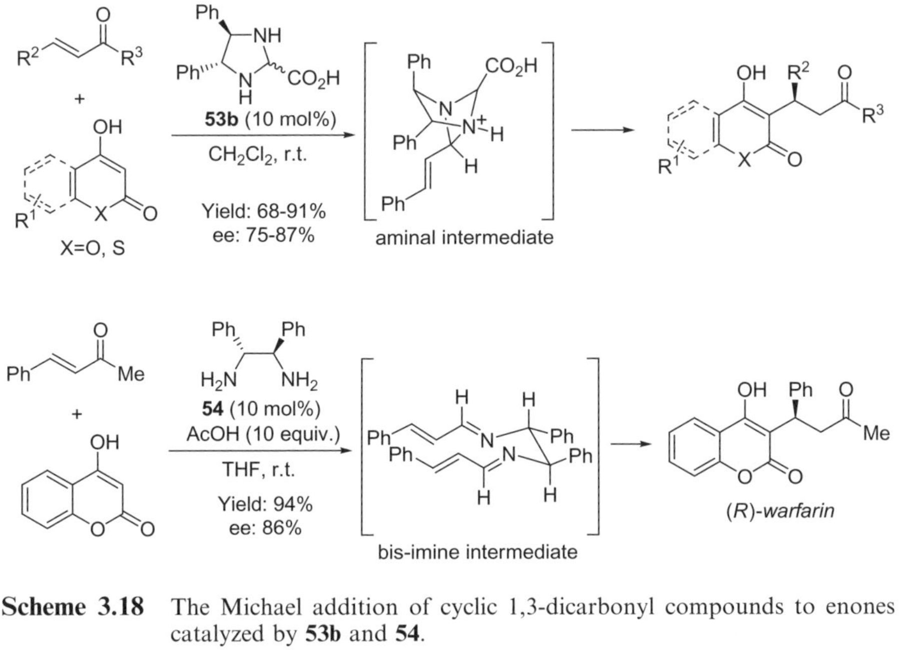

Scheme 3.18 The Michael addition of cyclic 1,3-dicarbonyl compounds to enones catalyzed by **53b** and **54**.

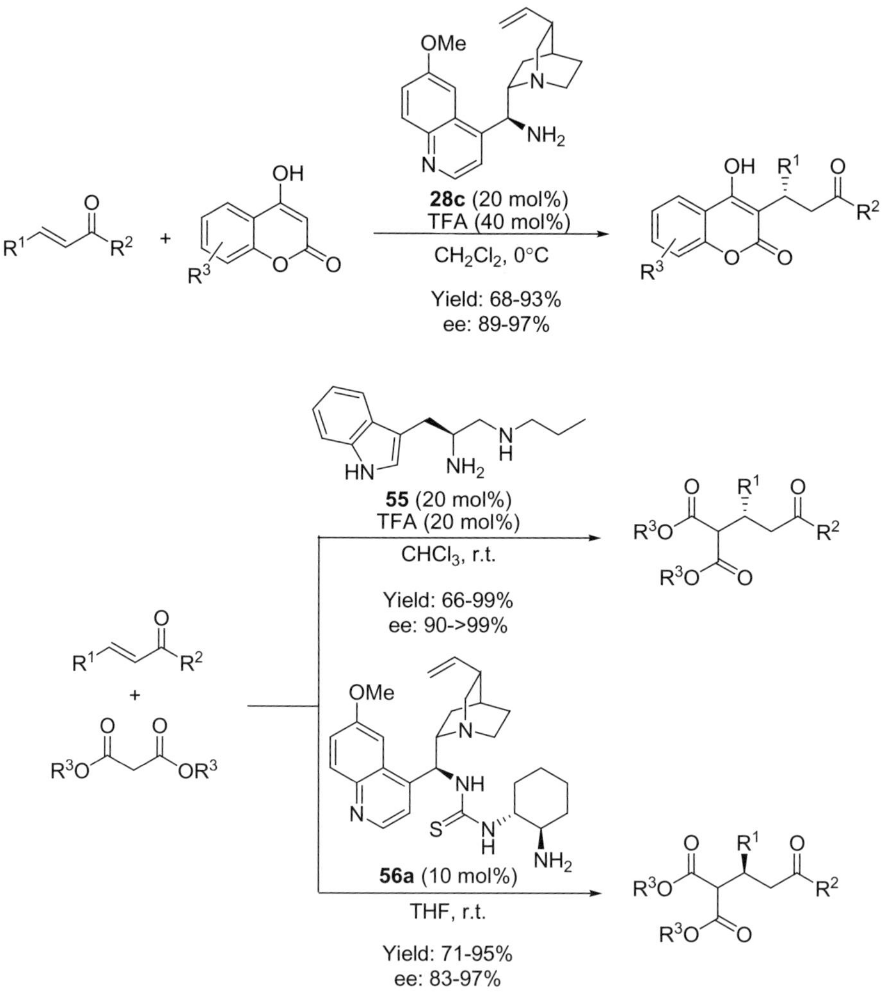

Scheme 3.19   The Michael addition of 1,3-dicarbonyl compounds to enones catalyzed by bifunctional primary amine catalysts.

The higher activity of primary amines in the reaction involving enones as Michael acceptors has also been extended to the use of different bifunctional catalysts (Scheme 3.19), which usually contain a primary amine functionality connected to a basic site by means of a chiral scaffold, as is the case in the use of **28c**[41] and **55**.[42] These diamine catalysts have been found to be excellent promoters of the Michael reaction of enones with cyclic 1,3-dicarbonyl compounds and malonates respectively, the tertiary amine basic site present at the catalyst structure being responsible for assisting in the deprotonation of the Michael donor in order to increase the concentration of the nucleophile species. In a different approach, bifunctional thiourea-primary amine catalyst **56a** has also

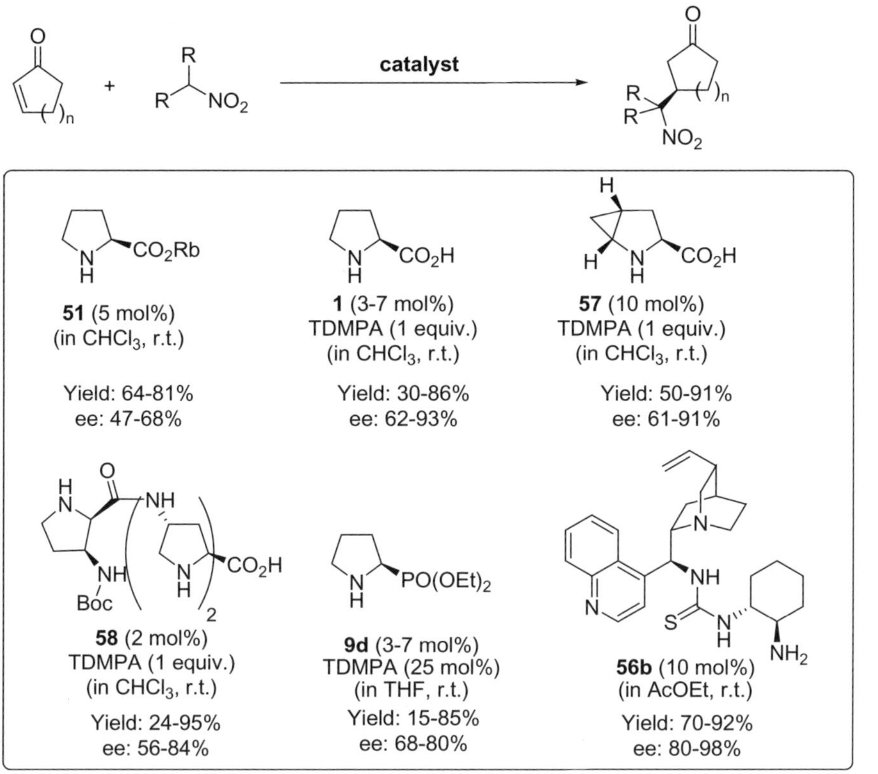

Scheme 3.20 Enantioselective Michael addition of nitroalkanes to cyclic enones catalyzed by chiral secondary amines.

been successfully employed in the Michael reaction of malonates to cyclic and acyclic enones. In this case, the principle behind the design of the catalyst consisted of the incorporation of a fragment containing a thiourea moiety capable of interacting with the malonate *via* hydrogen bonding in order to help in the preorganization of the reactants before the reaction takes place.[43]

Regarding the use of nitroalkanes as Michael donors, the majority of the reports have focused on the use of cyclic enones as Michael acceptors (Scheme 3.20). In this context, even before the modern concept of organocatalysis had been introduced, a very early attempt can be found in the literature involving the use of rubidium prolinate **51** as catalyst for the Michael addition of nitroalkanes to cyclic and acyclic enones proceeding *via* the formation of an iminium intermediate.[44] Although good yields were obtained, the enantioselectivities were only moderate in the best case. Afterwards, the proline-catalyzed Michael reaction of nitroalkanes to cycloalkenones was also studied, showing an improved performance than that furnished by rubidium prolinate, although the reaction scope was still restricted to the use of cyclic enones.[45] An improved version of the reaction was reported later on involving the use of *trans*-4,5-methanoproline **57** as catalyst,[46] which was followed with the use and development of other new secondary amine catalysts to be used in this reaction, all of them based on the proline motif, such as tripeptide **58**[47] and α-amino-phosphonate **9d**.[48] Remarkably, in all these examples, excepting that in which rubidium prolinate salt **51** was involved, a basic additive such as *trans*-2,5-dimethylpiperazine (TDMPA) had to be incorporated in the reaction either as a co-catalyst or as a stoichiometric additive. This was proposed to participate in the reaction by assisting the deprotonation of the nitroalkane donor in order to increase the concentration of the reactive nitronate nucleophile. Bifunctional primary amine-thiourea catalyst **56b** has also been employed in this reaction furnishing, in general, higher yields and enantioselectivities than the other secondary amine catalysts and also showing a remarkable substrate scope regarding the Michael donor, allowing the use of both cyclic and acyclic enones.[49] The bifunctional nature of the catalysts also allowed carrying out the reaction without the need of any external base additive. It should also be underlined that, in all cases when nitroalkanes with different substituents at the α-carbon were employed, the final products were typically obtained as mixtures of diastereoisomers in ratios ranging from 1:1 to 1:5, which is still an unsolved issue associated to this reaction, in a clear parallelism to the behavior of the same reaction in which α,β-unsaturated aldehydes were used as Michael acceptors.

On the other hand, imidazolidines **53a** and **53c** have also been successfully employed in this reaction, showing a similar behavior to that observed in the same reaction with malonate nucleophiles and with **53c** appearing as the most efficient catalyst (Scheme 3.21).[50] The use of differently substituted nitroalkenes was also evaluated but diastereoselectivities still remained rather low, obtaining

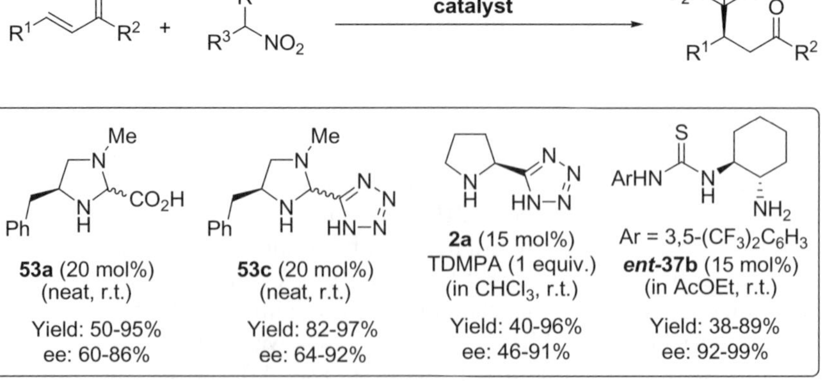

Scheme 3.21 Enantioselective Michael addition of nitroalkanes to acyclic enones catalyzed by imidazolidines **53a** and **53c**, proline tetrazole **2a** and amine-thiourea *ent*-**37b**.

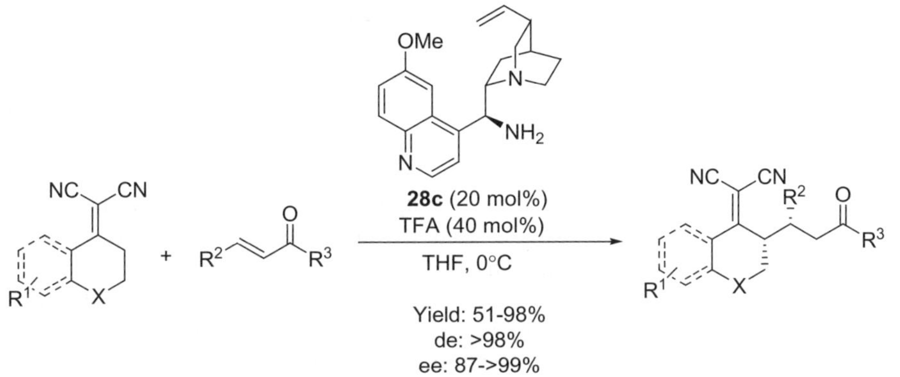

Scheme 3.22 Enantioselective vinylogous Michael addition of dicyanoacrylates to enones.

diastereomeric ratios between 2:1 and 1:1. Remarkably, even though moderate yields were also obtained in several cases, it is pointed out that the reaction was extremely clean, with the absence of any by-product, the only reason for the moderate yields obtained in some cases being a matter of conversion. This allowed carrying out the reaction on a kilogram scale with recovery and reuse of the catalyst without any decrease in catalytic activity or enantioselectivity. On the other hand, proline tetrazole analogue **2a** afforded similar results to those furnished by **53c**, that is, good to moderate yields (probably due to the lack of conversion) and enantioselectivities up to 91% ee, although it performed remarkably well in the Michael addition of nitroalkanes to cyclohexenone.[51] Also in this case, the use of substituted nitroalkanes led to mixtures of diastereoisomers in approximately 1:1 ratio. Bifunctional primary amine-thiourea **ent-37b** has also been employed with success in this reaction.[52]

Finally, the vinylogous Michael addition of α,α-dicyanoolefins with α,β-unsaturated ketones under iminium activation by a chiral primary amine catalyst has also been reported using enones as Michael acceptors involving the use of a chiral primary amine catalyst such as **28c** which was required for a more effective formation of the intermediate iminium ion (Scheme 3.22).[53] In this case, the inclusion of a Brønsted acid co-catalyst was reported to accelerate the reaction by facilitating the condensation step between the catalyst and the Michael acceptor.

3.3 Conjugate Friedel–Crafts Alkylation Reaction

The conjugate Friedel–Crafts alkylation is a powerful strategy for the chemical modification of electron-rich aromatic substrates allowing the building up of complex structures, which is very often the synthetic tool of choice when preparing a highly substituted heterocyclic compound. Developments in asymmetric versions of this reaction have been dominated by metal catalysis using suitable Michael acceptors which should be able to interact with a Lewis acid. However, the use of simple enones and, especially, α,β-unsaturated aldehydes as Michael acceptors has not found applicability in this reaction because of their tendency to undergo 1,2-addition rather than conjugate addition in the presence of Lewis acids. The appearing of the iminium activation concept has represented a breakthrough in this area and nowadays these elusive Michael acceptors can be used as substrates in many examples of conjugate Friedel–Crafts alkylations.

MacMillan was the first in applying this concept to the conjugate Friedel–Crafts reaction using chiral imidazolidinones as chiral secondary amine catalysts. The first attempt was carried out using pyrroles as nucleophiles, with catalyst **46b** furnishing outstanding results (Scheme 3.23).[54] However, the authors found that applying the same conditions to heteroaromatics of significantly lower reactivity such as indoles led to lower yields and enantioselectivities, which encouraged the authors to carry out a thoughtful study on the reaction directed toward a better understanding of the factors playing in

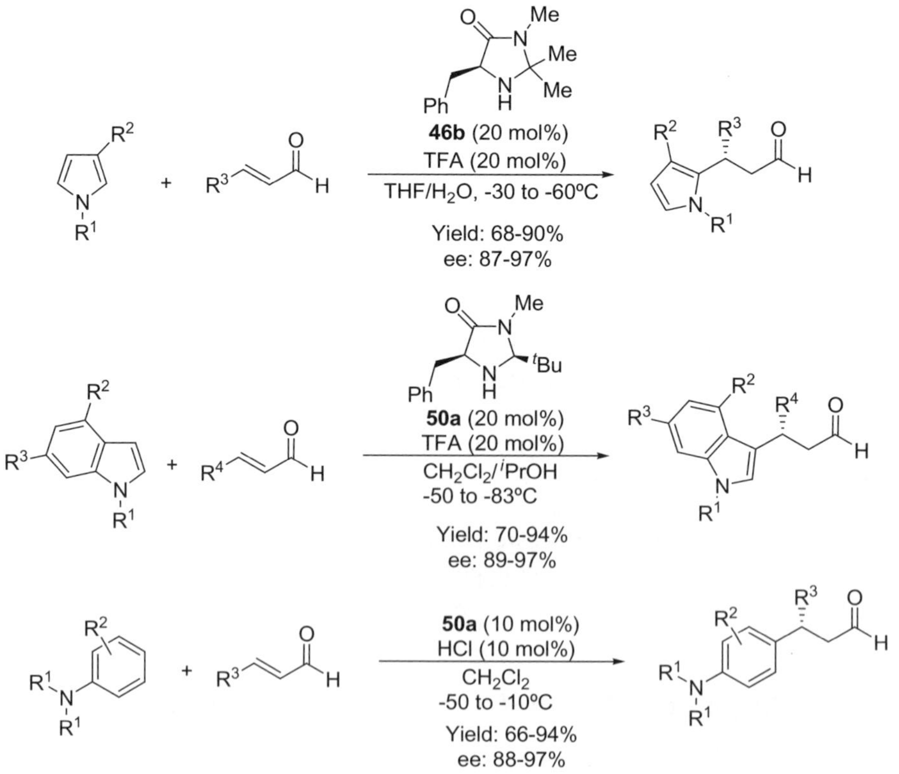

Scheme 3.23   Enantioselective conjugate Friedel–Crafts alkylations using imidazolidinone catalysts.

this transformation and also to overcome the limitations found.[55] As a consequence, based on computational design, a crucial catalyst modification was carried out based on the hypothesis that the presence of two methyl groups at the catalyst structure in **46b** led to lower availability of the nitrogen lone pair, which hampered the formation of the iminium intermediate. Therefore, one of the methyl groups was substituted by a smaller hydrogen atom and, in addition, a *tert*-butyl group was introduced in place of the other methyl group (the one *cis* to the benzyl moiety), which was supposed to increase the capacity of the catalyst to exert steric shielding of one of the enantiotopic faces of the α,β-unsaturated aldehyde. This new design, also referred to as second-generation MacMillan catalyst, proved to be extremely effective in this transformation and also in the addition of electron-rich benzenes (Scheme 3.23).[56] The exact explanation for the different behavior regarding stereocontrol provided by these two catalysts has been provided by computational methods.[57] Calculations at high theory level indicated that the minimum energy geometry of the intermediate iminium ion in the transition state for catalyst **46b** placed the phenyl ring of the benzyl moiety away from the reactive site, as a consequence of a stabilizing C–H···Me π-interaction between one of the methyl groups at C-2 of the imidazolidinone ring and this phenyl ring. This conformation is essentially avoided in the case of catalyst **50a** because of the additional steric hindrance provided by the ^{t}Bu group, which makes the system adopt a conformation in which the phenyl ring of the Bn moiety and the reactive α,β-unsaturated moiety are placed much closer to each other than with **46b**. Several modified analogues of these imidazolidinone catalysts have also demonstrated their ability in this reaction,[58] and also other catalysts not based in the imidazolidinone skeleton have also been developed for the conjugate Friedel–Crafts alkylation of indoles and 4,7-dihydroindoles with α,β-unsaturated aldehydes.[59] Moreover, different intramolecular variants have also been developed for the formation of fused aromatic and heteroaromatic compounds in a very efficient and stereoselective way.[60]

The reaction using enones as Michael acceptors needed the development of the more active primary amine catalysts. In fact, an initial attempt, a Mac-Millan-type catalyst, was tested in the reaction of indole with 5-methyl-3-buten-2-one, furnishing a moderate yield and a low enantioselectivity (28% ee).[61] Changing to primary amine catalysts like cinchonine derivative **59a** showed that this could play the role of a very effective promoter of the reaction between indoles and a wide range of different acyclic enones, obtaining good to moderate yields and achieving enantioselectivities between 50 and 85% ee under the best experimental conditions (Scheme 3.24). In this reaction, the incorporation of a strong Brønsted acid co-catalyst such as TfOH and also a rather high catalyst loading (30 mol%) were required.[62] A similar catalyst system formed by **28a** in combination with a chiral amino acid as Brønsted acid co-catalyst was employed by Melchiorre and coworkers in the same reaction, reaching improved yields of the addition products with a higher level of enantiocontrol and also slightly reducing the catalyst loading.[63] This strategy in which a significant improvement in the performance of an amine catalyst is

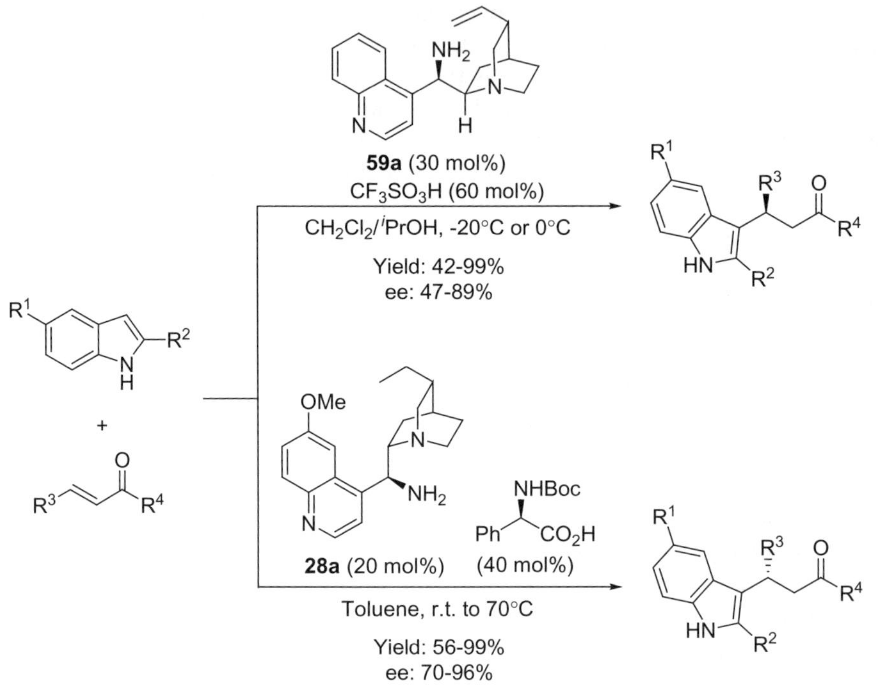

Scheme 3.24   Enantioselective conjugate Friedel–Crafts alkylations using enones as Michael acceptors.

improved by using a combination of an amine catalyst with a chiral acid co-catalyst, also known as asymmetric counterion directed catalysis, had already been developed by List in the context of the conjugate hydrogen transfer reactions (see Section 3.4).

A problem associated to this Friedel–Crafts approach for the introduction of functionalized lateral side chains on aromatic rings is the lack of flexibility in the regioselectivity of the reaction, which is exclusively determined by the substitution pattern or the nature of the aromatic ring and also the need for an electron-rich substrate as Michael donor. A new reaction design has been developed in this context consisting of the use of aryl trifluoroborates as activated π-nucleophiles in the conjugate addition to α,β-unsaturated aldehydes under iminium activation (Scheme 3.25). This new reaction design is inspired on the Petasis reaction, in which boronic acids are employed as effective π-activation groups that enable unreactive arenes to undergo 1,2-addition to iminium ions. This new approach allows a complete control on the regio-selectivity of the reaction by introducing the borate group at the desired position of the arene or heteroarene nucleophile. Under the optimized conditions, excellent yields and enantioselectivities were obtained for several differently substituted electron deficient aryltrifluoroborates reacting with a variety

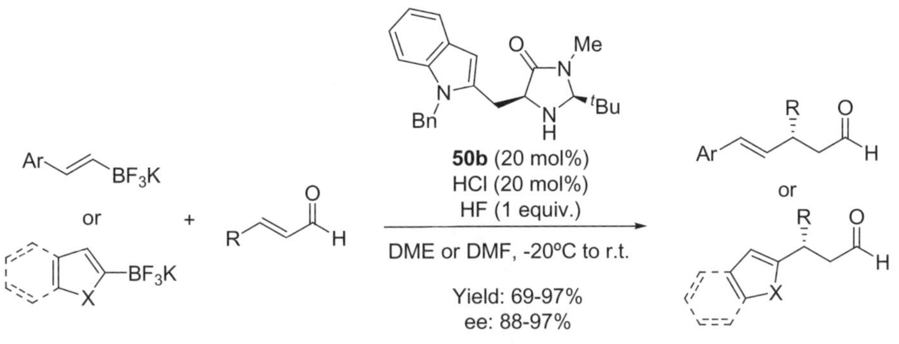

Scheme 3.25 Asymmetric **50b**-catalyzed conjugate addition of organotrifluoroborate salts to α,β-unsaturated aldehydes.

of α,β-unsaturated aldehydes.[64] Although some limitations were found on the tolerance toward variation at the nucleophile (the simple phenyltrifluoroborate salts were found to be unreactive under the reaction conditions), the reaction showed a remarkably broad substrate scope with regard to the α,β-unsaturated aldehyde reagent employed. It was also found that the addition of one equivalent of hydrofluoric acid was crucial for the reaction to proceed to completion, this additive being required for the quenching of the boron trifluoride, which is formed as the reaction proceeds forward. The methodology could also be extended to the use of vinylborates as Michael donors with success, remarkably expanding the versatility and synthetic applicability of this methodology.

3.4 Conjugate Hydrogen Transfer Reaction

The possibility of carrying out enantioselective conjugate reductions applying the iminium activation concept started with the introduction of Hantzsch esters as organic compounds which are able to transfer simultaneously a hydride and a proton to the Michael acceptor.[65] This approach mimics the way in which reductions are accomplished in biological processes, with the Hantzsch ester behaving as hydride-reduction cofactors like NADH or $FADH_2$ do in the related enzyme-mediated reaction. A non-stereoselective version was carried out in an initial approach by reducing α,β-unsaturated aldehydes with a commercially available Hantzsch ester in the presence of dibenzylamine as catalyst,[66] and soon afterwards the asymmetric version of this reaction was reported using imidazolidinone **50a** as catalyst and a bulkier Hantzsch ester as hydride-donor reagent (Scheme 3.26).[67] The reaction proceeded with very high efficiency, providing excellent yields and enantioselectivities for several β-disubstituted α,β-unsaturated aldehydes, although in all cases incorporating an aromatic substituent at the β-position. Interestingly, the double-bond geometry of the Michael acceptor was not shown to have any influence on the stereo-chemical outcome of the reaction, observing that enals of different configuration furnished the same enantiomer after reduction. This allowed the use of mixtures of *Z* and *E* enals as substrates in this reaction, therefore avoiding the need for a highly diastereomerically pure Michael acceptor in order to reach good enantioselectivity in the reduction. A similar reduction protocol was simultaneously and independently reported,[68] showing that imidazolidinone **50c** containing a single stereogenic center could play the role of an outstanding catalyst for this transformation, also using a different Hantzsch ester as hydride donor (Scheme 3.26). This reaction was applicable even to enals containing both aliphatic and aromatic substituents at the β-position and even substrates containing bulky substituents underwent reduction with high efficiency. They also found the same enantioconvergency in this reaction, which allowed the use of *Z/E* mixtures of the starting Michael acceptors to furnish equally high

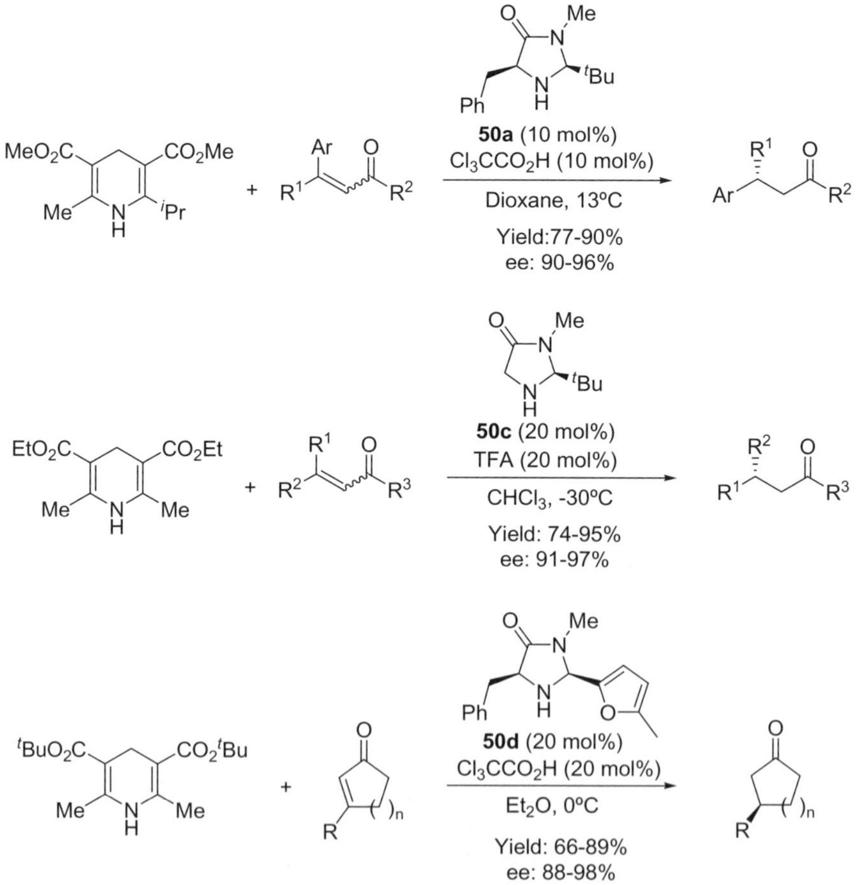

Scheme 3.26 Enantioselective transfer hydrogenation of α,β-unsaturated enones.

enantioselectivities. This behavior was explained in terms of a rapid Z/E isomerization of the starting materials catalyzed by the imidazolidinone catalyst during the formation of the intermediate iminium ion. Cyclic enones can also be reduced in a very efficient way using in this case imidazolidinone **50d** as catalyst (Scheme 3.26),[69] although in this case the structure of the Hantzsch ester had to be optimized in order to achieve the best enantioselectivity. There is still no report relating the use of primary amines in the transfer hydrogenation of enones.

This reaction provided a perfect basis for the development of the asymmetric counteranion directed catalysis concept by List.[70] This model relies upon the close association which occurs between ion pairs in polar reactions proceeding *via* charged intermediates which are conducted in non-polar organic solvents. Consequently, it was hypothesized that the counteranion associated to the intermediate iminium cation could be placed near the position in which the conjugate addition would take place with the possibility to exert stereochemical influence in the reaction. This suggested the possibility of carrying out enantioselective conjugate addition reactions *via* iminium activation by using an achiral ammonium salt catalyst incorporating a chiral counteranion able to translate its stereochemical information to the iminium intermediate because of the proximity between each other. This hypothesis was confirmed when it was

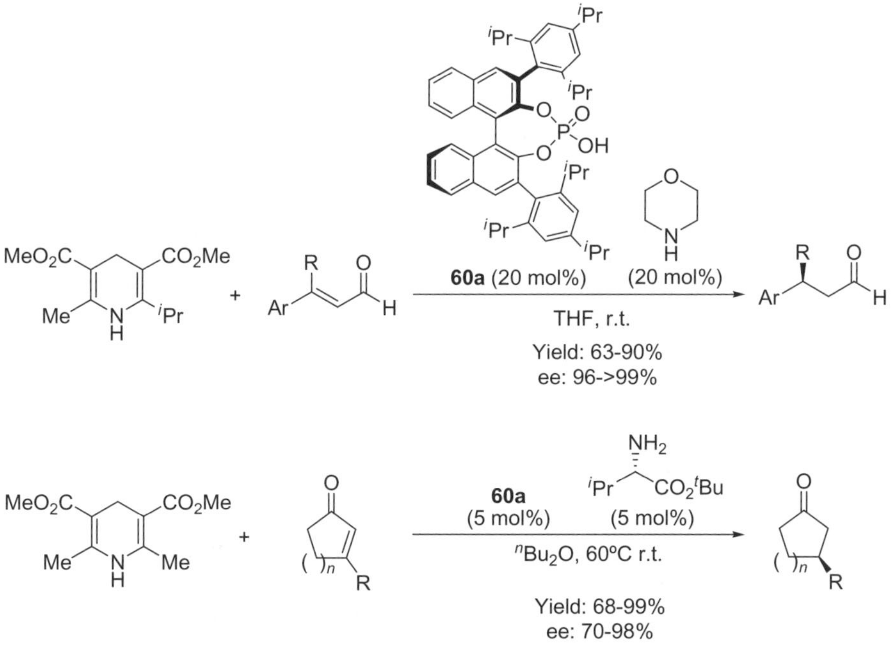

Scheme 3.27 Enantioselective transfer hydrogenation α,β-unsaturated aldehydes and enones by applying the asymmetric counterion directed catalysis concept.

found that chiral phosphoric acid **60a** (TRIP) together with morpholine were able to catalyze the conjugate reduction of α,β-unsaturated aldehydes in a very efficient way (Scheme 3.27). This concept was also applied to the enantiose-lective transfer hydrogenation of enones but, in this case, a primary amine was employed as the iminium-forming reagent because of the already mentioned better ability of primary amines for the formation of iminium ions with enones compared to the secondary amine counterparts.[71] In this particular case ami-noesters have been employed as amine catalysts in this reaction design together with the chiral phosphoric acid TRIP (**60a**). The *tert*-butyl ester of valine was identified as the most appropriate primary amine catalyst for this transfor-mation in terms of both the yield and the enantioselectivity, although, as both constituents of the catalytic salts were chiral, an interesting double stereo-differentiation phenomenon was observed, which led to the identification of the corresponding *matched* and *mismatched* combination of reagents. These opti-mized conditions were extended to a wide range of enones providing excellent results, especially for cyclic enones, and obtaining slightly lower enantioselec-tivities with acyclic enones. In addition, a remarkably low catalyst loading (5 mol%) was necessary for achieving such excellent results.

3.5   Conjugate Addition of Heteroatom-based Nucleophiles

The iminium activation concept has been successfully extended to the conjugate addition of heteroatom-centered nucleophiles to α,β-unsaturated aldehydes and enones. In general, the main problem associated with the use of sulfur-, oxygen-, nitrogen- or phosphorous-based Michael donors is related to the reversibility of the conjugate addition process, which leads to low conversions or low configurational stability of the final compounds. For this reason, most of the methods reported in this context have incorporated an additional elec-trophile in the reaction design in order to quench the hetero-Michael addition product in a typical cascade process, which overrides the reversibility of the reaction. All these cascade reactions will be presented in Chapter 7, which is especially dedicated to cascade reactions initiated by organocatalytic conjugate addition processes.

3.5.1   Oxa-Michael Reactions

The conjugate addition of oxygen heterocycles to α,β-unsaturated aldehydes or ketones under iminium activation is one of the most challenging reactions among all organocatalytic Michael-type reactions.[72] In addition to the already mentioned low configurational stability of the oxa-Michael adducts, the low nucleophilicity exhibited by the most common oxygen nucleophiles like water, alcohols or carboxylates and their tendency to undergo 1,2-addition rather than the desired conjugate addition process makes this transformation especially difficult.

A clear indication of the challenging nature of this reaction can be seen in the fact that an example of an organocatalytic oxa-Michael reaction can be found as early as in 1937, when the conjugate addition of water to crotonaldehyde was reported to be possible under iminium catalysis, although in a non-stereo-selective version,[2] and it was not until 2007 when the first enantioselective amine-catalyzed oxa-Michael addition to enals was reported.[73] In this case, the problem of the low reactivity of the oxygen nucleophile and the reversibility of the reaction was circumvented by choosing aromatic oximes as Michael donors, with *O*-TMS diarylprolinol **31c** showing up as a very efficient catalyst for this transformation (Scheme 3.28). The reaction was remarkably fast (usually 1 hour to reach full conversion) and also required the *in situ* reduction of the conjugate addition products to the corresponding primary alcohols for pur-ification because of their tendency for racemization. The reaction showed to have a rather broad substrate scope for alkyl-substituted enals but failed in the case of β-aryl substituted substrates. An important advantage of the metho-dology was found on the easy cleavage of the *O*–*N* bond, which allowed a very simple procedure for the asymmetric synthesis of 1,3-diols. The same kind of nucleophiles were employed afterwards in the reaction using enones as Michael acceptors,[74] in this case requiring the combination of a primary amine derived from hydroquinine as catalyst and a chiral Brønsted acid co-catalyst (*N*-Boc-phenylglycine), in a similar approach (asymmetric counterion-directed cata-lysis) to that shown before for the conjugate Friedel–Crafts reaction of indoles with enones (see Scheme 3.28). As it happened with the reaction with enals, the scope of the reaction was limited to aliphatic enones.

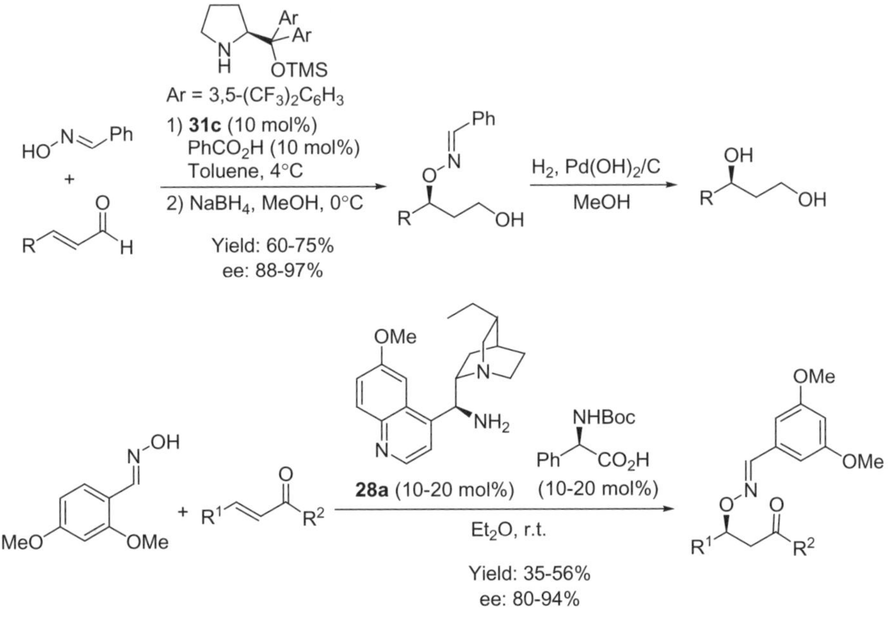

Scheme 3.28 The conjugate addition of oximes to enals and enones.

Some attempts to use aliphatic alcohols as nucleophiles in the enantioselective conjugate addition to enals have been carried out, but with no success with regard to the stereochemical control (Scheme 3.29).[75] For example, the conjugate addition of methanol to several aliphatic enals using axially chiral amines derived from **BINOL** as catalysts proceeded with enantioselectivities around 50% ee and also a **31c**-catalyzed intramolecular reaction leading to the formation of the tetrahydropyrane skeleton has been reported to proceed with 57% ee.

Hydrogen peroxide has also been employed as nucleophile in this context with different results (Scheme 3.30).[76] While the reaction using aldehydes as Michael acceptors usually furnishes the corresponding α,β-epoxy aldehydes *via* cascade oxa-Michael addition followed by intramolecular electrophilic attack of the intermediate enamine to the peroxide moiety (see Section 7.3.7), the same reaction using enones as electrophiles can be blocked at the first oxa-Michael addition step because the β-hydroperoxide aldehyde formed after the 1,4-addition can undergo spontaneous intramolecular addition to the carbonyl group, furnishing a 1,2-dioxolan-3-ol derivative as final reaction product. The formation of this cyclic derivative operates as the driving force for the reaction, avoiding the retro-addition process and therefore accounts for the high conversion observed and avoids racemization of the final product. In this context, the primary amine **28c**-catalyzed reaction of several enones with H_2O_2

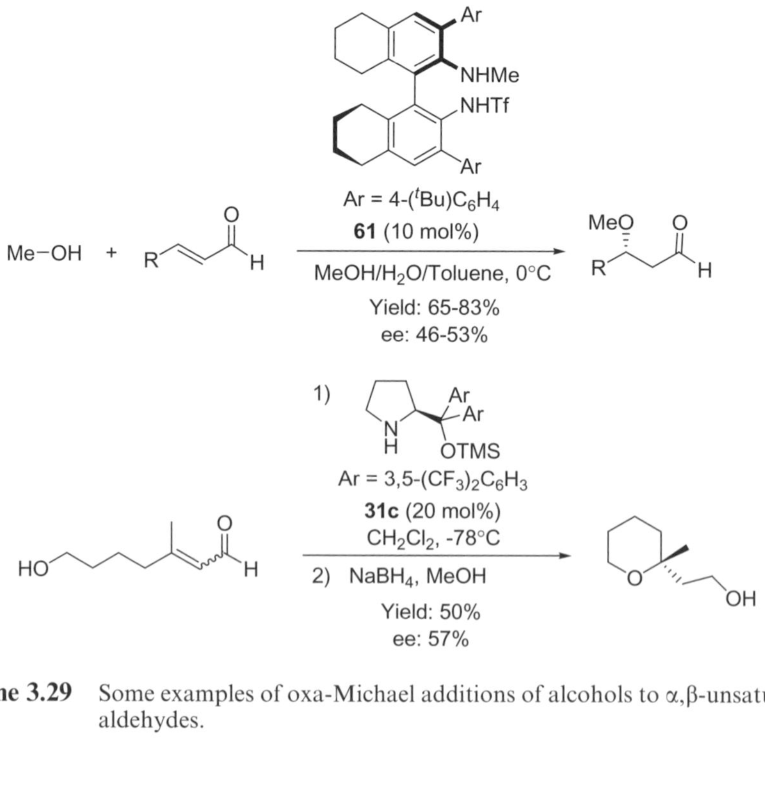

Scheme 3.29   Some examples of oxa-Michael additions of alcohols to α,β-unsaturated aldehydes.

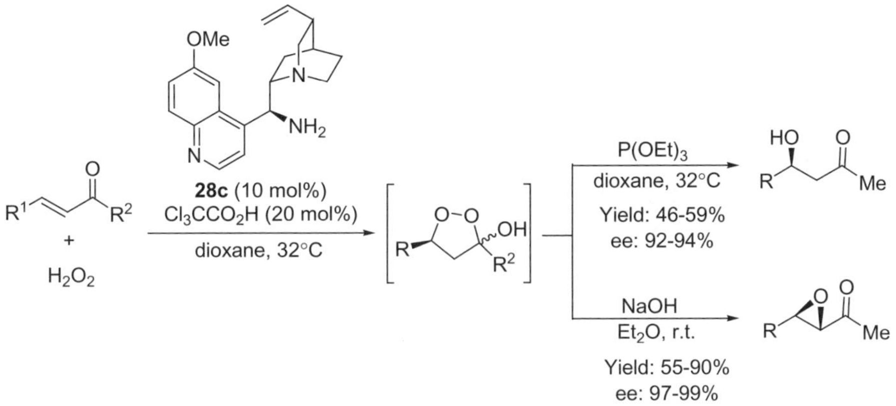

Scheme 3.30 **28c**-catalyzed enantioselective hydroperoxydation of enones and subsequent transformations.

proceeded with moderate yields but with excellent enantioselectivities for several β-alkyl substituted α,β-unsaturated ketones, with no example provided related to the use of β-aryl substituted enones. An interesting feature of this methodology is the possibility of converting the 1,2-dioxolane adducts into β-hydroxy ketones by reduction with P(OEt)$_3$ and also the easy access to epoxides by carrying out a base-promoted sequential hemiacetal cleavage/intramolecular substitution process.

The reaction between alkyl hydroperoxides and enones has also been carried out using the same catalyst **28c** and showing a similar behavior (Scheme 3.31).[77] While the reaction usually leads to the formation of an epoxide as final product, under the optimized conditions and, in particular, by the correct election of the alkyl hydroperoxide nucleophile, the reaction can be directed to the formation of the β-peroxy-substituted ketone product in excellent yields and enantioselectivities. Reduction of these adducts allowed the preparation of the corresponding β-hydroxy ketone, therefore showing that the methodology is suitable for carrying out the formal enantioselective conjugate addition of OH to enones.

3.5.2 Sulfa-Michael Reactions[78]

The Michael-type addition of thiols to α,β-unsaturated aldehydes has been reported to proceed very efficiently using *O*-TMS protected diarylprolinol **31c** as catalyst (Scheme 3.32). The low configurational stability of the adducts at r.t. led the authors to modify the reaction conditions, which finally involved carrying out the reaction at –24 °C and reducing the adducts *in situ*, furnishing the corresponding γ-thio alcohols in excellent yields and enantioselectivities.[79] Both alkyl- and aryl-substituted enals could be used as suitable Michael

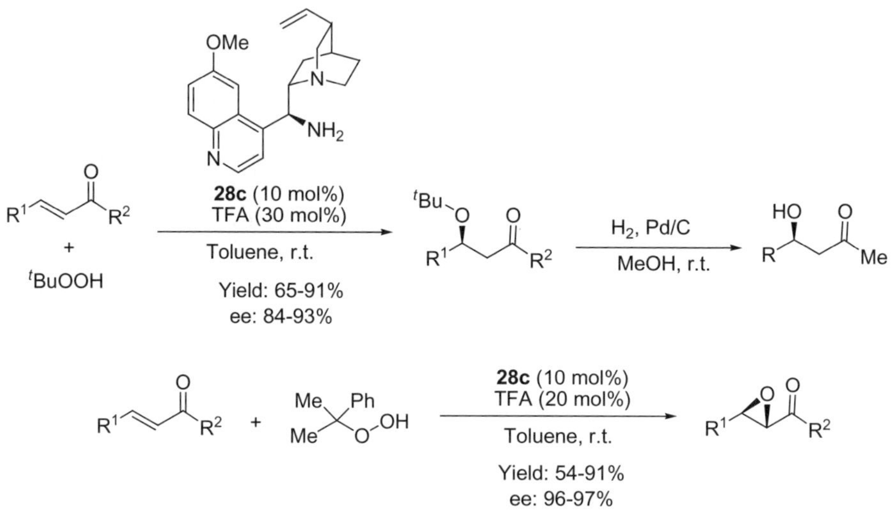

Scheme 3.31   **28c**-catalyzed enantioselective β-peroxydation of enones.

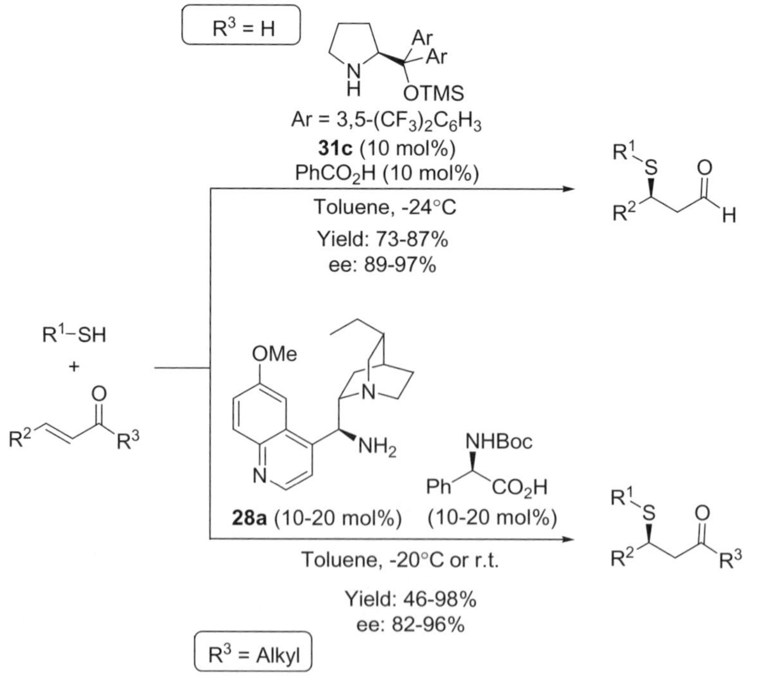

Scheme 3.32   Enantioselective conjugate addition of thiols to α,β-unsaturated aldehydes and ketones.

acceptors in this reaction and furthermore different thiols like *tert*-butyl mercaptan, benzyl mercaptan or even functionalized thiols were evaluated resulting to be in all cases equally effective nucleophiles to be used in this reaction. The same reaction involving α,β-unsaturated ketones as Michael acceptors has also been reported,[80] also using the asymmetric counterion directed catalysis approach like that shown in Scheme 3.28 for the oxa-Michael addition of oximes to enones. In this case, the addition of both *tert*-butyl- and benzyl mercaptan to several enones was found to proceed with moderate to good yields and enantioselectivities. On the other hand, a cyclic enone such as cyclohexenone was also evaluated as Michael acceptor with excellent results.

3.5.3 Aza-Michael Reactions

The conjugate addition of nitrogen nucleophiles to carbonyl compounds represents one of the most attractive procedures for the asymmetric synthesis of β-amino carbonyl compounds, which are key constituents of many biologically active compounds and also useful building blocks in total synthesis.[81] The development of enantioselective organocatalytic methods has also been the focus of intensive research, especially in the last few years.[82] However, aza-Michael reaction using the iminium activation concept represents one of the most challenging transformations in this field, mainly because of the additional chemoselectivity issues that have to be addressed in this particular case. The fact that in an amine-catalyzed aza-Michael reaction both the catalyst and the nucleophile are amine species means that the role of both reagents participating in the reaction must be clearly established. For instance, the chiral amine chosen as catalyst must not undergo conjugate addition reaction, which would eventually lead to catalyst consumption and, on the other hand, the amine reagent selected as nucleophile must not participate in iminium ion formation, which would lead to the formation of a racemic product.

In this context, the key for the success of an aza-Michael reaction under iminium activation relies mainly on the correct design of the nitrogen nucleophile to be employed. For example, *N-tert*-butyldimethylsilyloxycarbamates show up as a very efficient Michael donor for the reaction with α,β-unsaturated aldehydes using chiral imidazolidinone *ent*-50a as catalyst (Scheme 3.33).[83] These reagents show an enhanced nucleophilicity due to the so-called α-effect, exerted by the presence of the trialkylsilyloxy moiety directly attached to the nitrogen atom, while the introduction of the carbamate functionality results in the formation of a non-basic *N*-protected β-amino carbonyl compound as reaction product, thus avoiding the reversibility of the reaction. The closely related *N*-methoxycarbamates can also be employed as nucleophiles in the same reaction, in this case using *O*-trimethylsilyldiphenylprolinol **31a** as catalyst, although achieving somewhat lower yields.[84] The addition of *N*-benzyloxy carbamates to enones has also been carried out with the help of a bifunctional catalyst containing a primary amine moiety able to condense easily with the Michael acceptor and incorporating a basic site able to interact with the *N*-

nucleophile, which also assists its deprotonation (Scheme 3.33).[85] In this way, a very well-organized transition state is formed, which results in very high level of stereoinduction. The reaction proceeded with excellent yields and enantio-selectivities in most cases, although the structure of the catalyst had to be optimized depending on the substitution pattern at the Michael acceptor. Catalyst **28b** furnished excellent results for β-alkyl-substituted enones but the use of enones with aromatic substituents at the β-position required the use of the more active derivative **28c**.

Another way to overcome the reversibility problem of the aza-Michael reaction is the use of a bifunctional nucleophile such as simple *N*-hydroxy

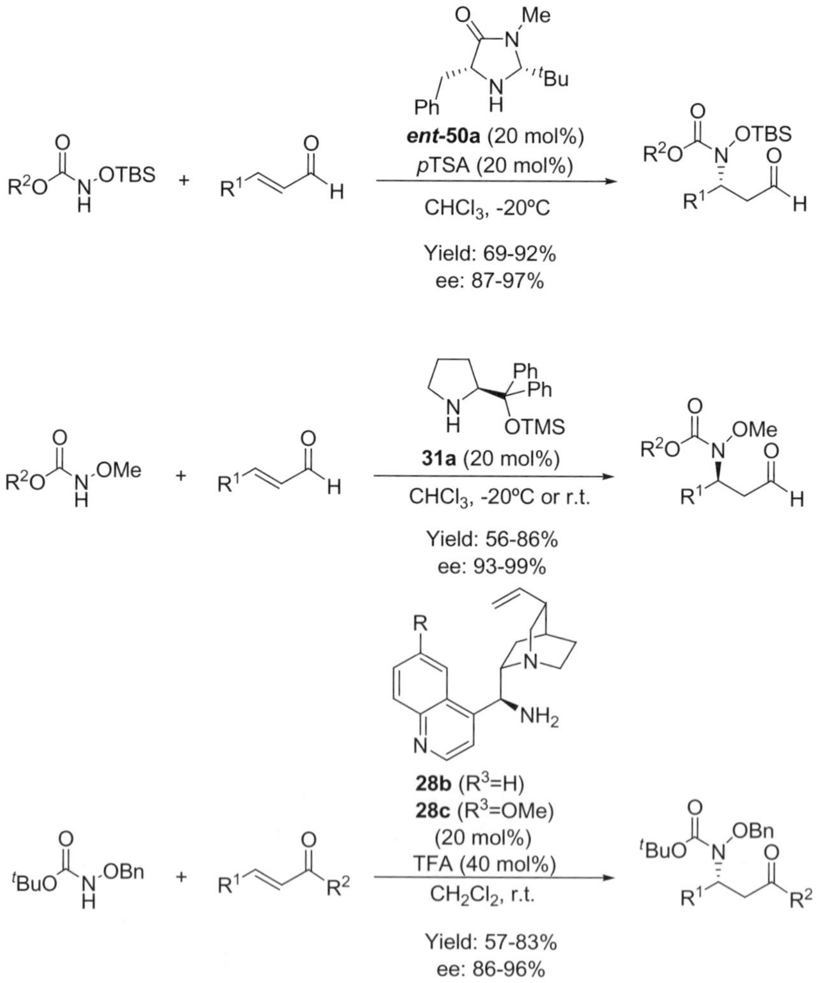

Scheme 3.33 Enantioselective aza-Michael reactions using *N*-trialk-ylsilyloxycarbamates and *N*-alkoxycarbamates as nucleophiles.

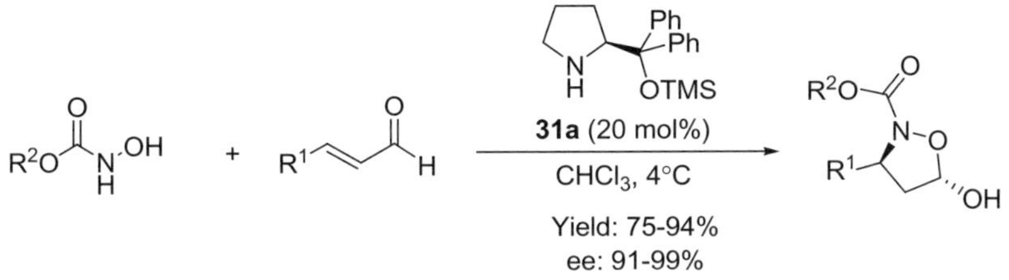

Scheme 3.34 Enantioselective aza-Michael reactions between *N*-hydroxycarbamates and α,β-unsaturated aldehydes.

carbamates, which are able to undergo a subsequent hemiacetal formation with the formyl moiety therefore pushing the equilibria toward the formation of the final products (Scheme 3.34). These kinds of nucleophiles also show an enhanced nucleophilicity by the presence of the OH group directly linked to the nitrogen atom. In this case, the conjugate addition of these reagents to α,β-unsaturated aldehydes could be easily carried out using **31a** as catalyst, and furnishing the final conjugate addition products as the corresponding 5-hydroxyoxazolines with high yields and enantioselectivities, and as single diastereoisomers.[86]

The use of nitrogen heterocycles as suitable *N*-nucleophiles in this reaction design has also been covered in the recent literature. For example, 1,2,4-triazole, 1,2,3-triazole, benzotriazole and 5-phenytetrazole add smoothly to α,β-unsaturated aldehydes in the presence of **31c** as catalyst furnishing the final adducts with a remarkable degree of stereocontrol.[87] The best results were obtained in the reaction with 1,2,4-triazole (Scheme 3.35), which proceeded with complete regioselectivity and high enantioselectivity at room temperature, mainly due to the fact that the aza-Michael adducts were found to be configurationally stable at r.t. and did not racemize easily. In the reactions with the other heterocycles, the temperature of the reaction had to be carefully controlled because of the tendency shown by the adducts to racemize and also the conjugate addition products had to be reduced *in situ* for isolation and purification. In addition, while the reaction with 5-phenyltetrazole occurred with complete regioselectivity, the use of 1,2,3-triazoles led to the formation of mixtures of regioisomers, which could be separated by chromatography and isolated as highly enantio-enriched compounds. In an independent and almost simultaneous work, imidazolidinone **50a** was also identified as a very efficient catalyst for the aza-Michael reaction of 5-phenyltetrazole with α,β-unsaturated aldehydes (Scheme 3.35).[88] In this case the reversibility of the reaction was avoided by carrying it out at very low temperatures (−78 or −88 °C), which resulted in very long reaction times for reaching synthetically useful yields and, as happened in the **31c**-catalyzed reaction, the final compounds had to be reduced *in situ* for isolation and purification. The reaction proceeded with complete regioselectivity, affording exclusively *N*2–alkyl-substituted tetrazole derivatives and good yields and enantioselectivities were obtained for a wide

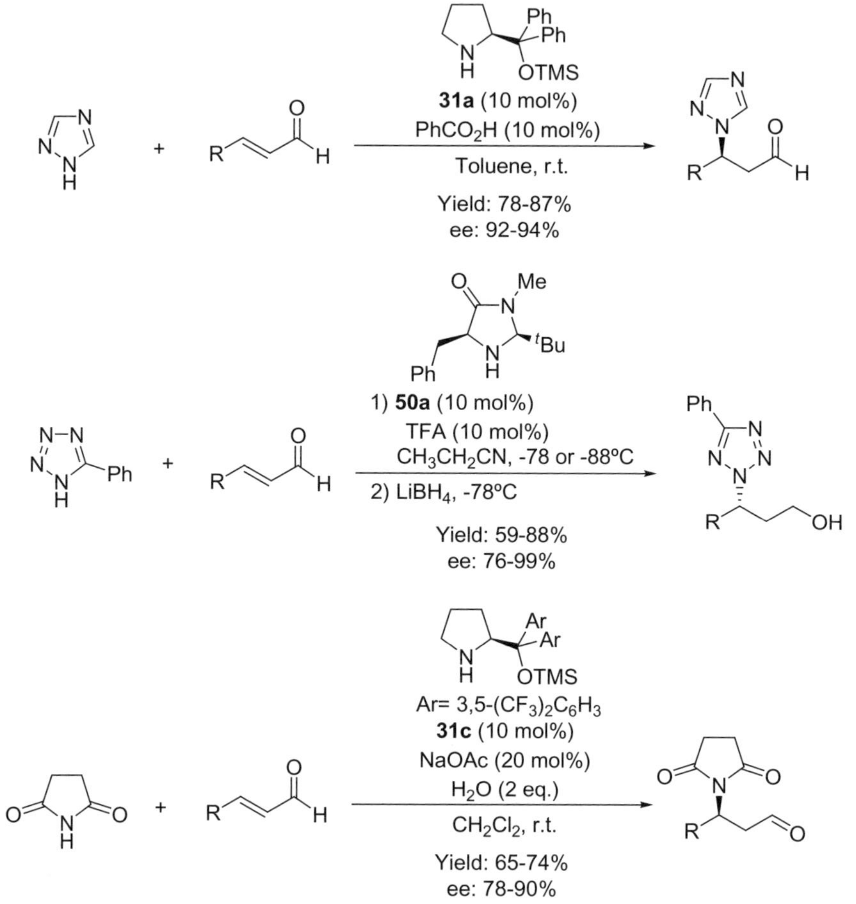

Scheme 3.35 Enantioselective aza-Michael reactions using nitrogen heterocycles as nucleophiles.

range of different α,β-unsaturated aldehydes. Alternatively, the use of succinimide as nucleophile has also been evaluated, which provided slightly lower enantioselectivities than 1,2,4-triazole in the reaction catalyzed by **31c** and using α,β-unsaturated aldehydes as Michael acceptors (Scheme 3.35).[89] In addition, NaOAc had to be included in this case as a basic additive in order to increase the nucleophilicity of succinimide by promoting its deprotonation. Several reports have also appeared investigating the conjugate addition of pyrazoles to enals catalyzed by *O*-trialkylsilyldiarylprolinols[90] and the reaction of benzotriazol, 9-phenyltetrazol and 2-pyrazolin-5-ones with enones catalyzed by primary amines derived from quinine.[91]

Intramolecular versions of this reaction have also been studied, specially directed toward the preparation of several natural products (Scheme 3.36). The

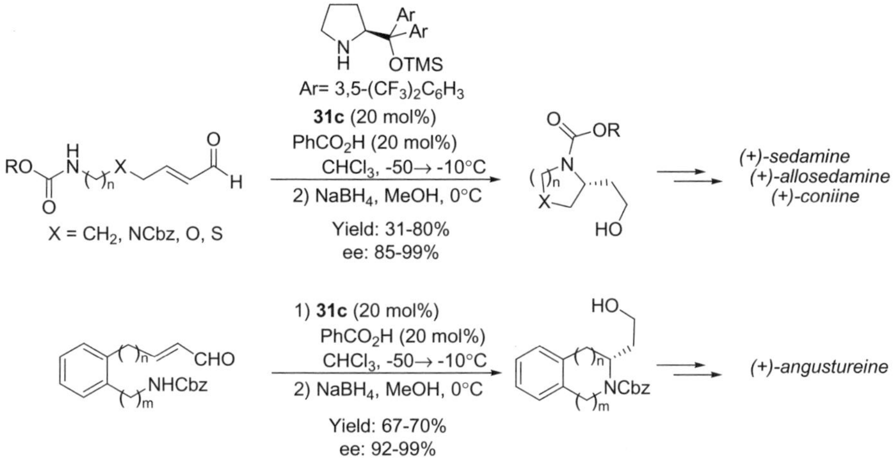

Scheme 3.36 Enantioselective intramolecular aza-Michael reactions.

reaction design includes a functionalized α,β-unsaturated aldehyde containing a functionalized side chain at the β-position, which also incorporated a suitable protected amino group located at a convenient position, thus allowing the formation of a heterocyclic structure after the intramolecular aza-Michael reaction takes place. As happened in the previous examples, the reversibility of the reaction resulted in low configurational stability of the obtained adducts and therefore reduction of the formyl group had to be carried out *in situ* for isolation of the final compounds. Catalyst **31c** was employed as the most effective promoter of the reaction using substrates with tethers of different lengths between the enal and the amino moieties and also evaluating the incorporation of intercalating heteroatoms for the preparation of different heterocyclic structures. This reaction was applied to the asymmetric synthesis of several piperidine alkaloids such as (+)-*sedamine*, (+)-*allosedamine* and (+)-*coniine*.[92] In addition, a similar intramolecular reaction in which an aromatic ring was intercalated into the tether was used in the asymmetric preparation of tetrahydroquinolines, tetrahydroisoquinolines and indolines, including the biologically active alkaloid (+)-*angustureine* (Scheme 3.36).[93]

3.5.4 Phospha-Michael Reactions[94]

The phospha-Michael reaction has been the last hetero-Michael reaction to be developed under iminium activation. In addition to the selectivity issues that have to be addressed, the identification of a suitable phosphorous nucleophile has been the most difficult task to overcome when developing the reaction because of the high tendency of phosphines toward oxidation in the presence of air. The first example was developed independently by Melchiorre[95] and

Córdova[96] in 2007 and involved the use of diphenylphosphine as nucleophile and *O*-trimethylsilyl diarylprolinols **31a** or **31c** as catalysts in the presence of different Brønsted acid co-catalysts (Scheme 3.37). In both reports, the phospha-Michael adducts had to be reduced *in situ* and the phosphine moiety had to be transformed to the corresponding phosphine-borane complex for better isolation and purification of the final products. Other phosphorous nucleophiles have been tested in this reaction with different success. For example, Melchiorre reported that di-*tert*-butylphosphine was inert under the optimized reaction conditions, diphenylphosphine-borane complex furnished a racemic product and diphenylphosphine delivered only the 1,2-addition side product, while Córdova also tested a wider variety of other different phosphorous compounds, concluding that only trivalent reagents were active as nucleophiles undergoing conjugate addition under iminium activation, but diphenylphosphine still being the most appropriate one. The nature of the Brønsted acid co-catalyst was also decisive for achieving good stereocontrol, with the slightly more acidic *p*-nitrobenzoic acid and *p*-fluorobenzoic acid as the additives furnishing the best enantioselectivities. The reaction appeared to proceed much

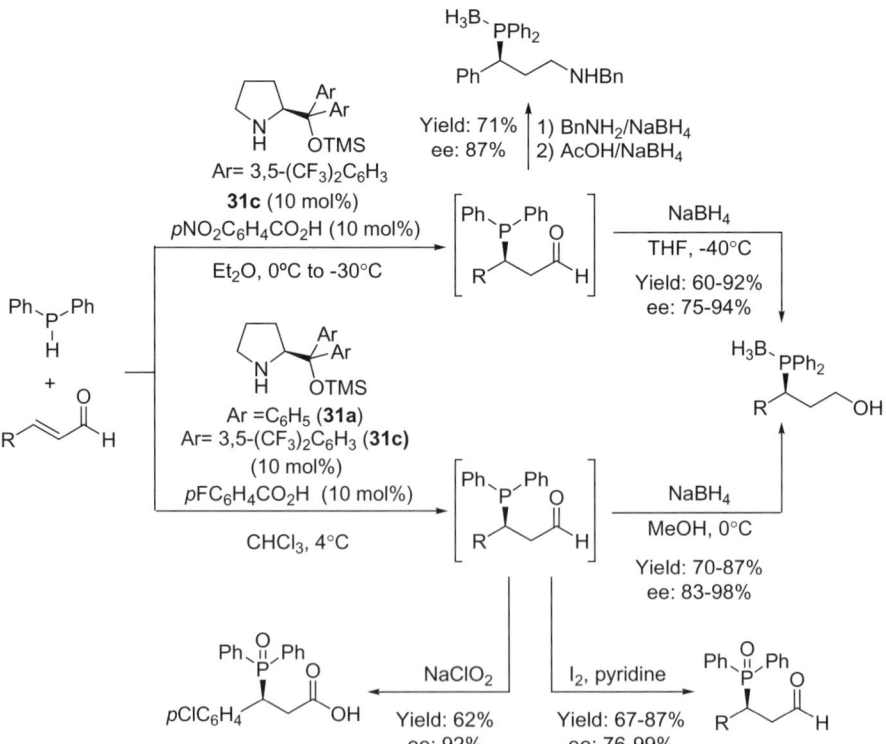

Scheme 3.37 Enantioselective phospha-Michael reactions between diphenylphosphine and α,β-unsaturated aldehydes.

easier with aromatic enals, while alkyl-substituted α,β-unsaturated aldehydes were more problematic substrates, especially when the β-alkyl substituent was a small group such as methyl or ethyl. In such cases, experimental conditions such as the dilution appeared as crucial details for achieving good results. The authors also surveyed the reactivity of the obtained β-phosphino aldehyde adducts by carrying out several modifications like the *in situ* reductive amination to furnish γ-phosphino amines or the *one-pot* phospha-Michael reaction/oxidation sequence furnishing β-phosphine oxide carboxylic acids or β-phosphine oxide aldehydes depending on the oxidant employed.

On the other hand, improved conditions were found for carrying out the conjugate addition of P(O*i*Pr)₃ to α,β-unsaturated aldehydes using **31c** as catalyst (Scheme 3.38), delivering dialkylphosphonate-type conjugate adducts as the final compounds.[97] In this case, a different mechanistic pathway has been postulated involving an intermolecular nucleophilic substitution at the phosphorous atom after the phospha-Michael addition step in order to deliver the final phosphonate moiety. In this case, it was proposed that this second S$_N$ reaction was the rate-determining step of the process and, consequently, an external nucleophile such as sodium iodide was incorporated as an additive which might catalyze this reaction, leading to an improvement on the conversion and the enantioselectivity of the reaction. Under these improved conditions, a wide variety of alkyl- and aryl-substituted α,β-unsaturated aldehydes underwent conjugate addition reaction with P(O*i*Pr)₃ delivering the corresponding β-phosphonate aldehydes in moderate yields and good enantioselectivities.

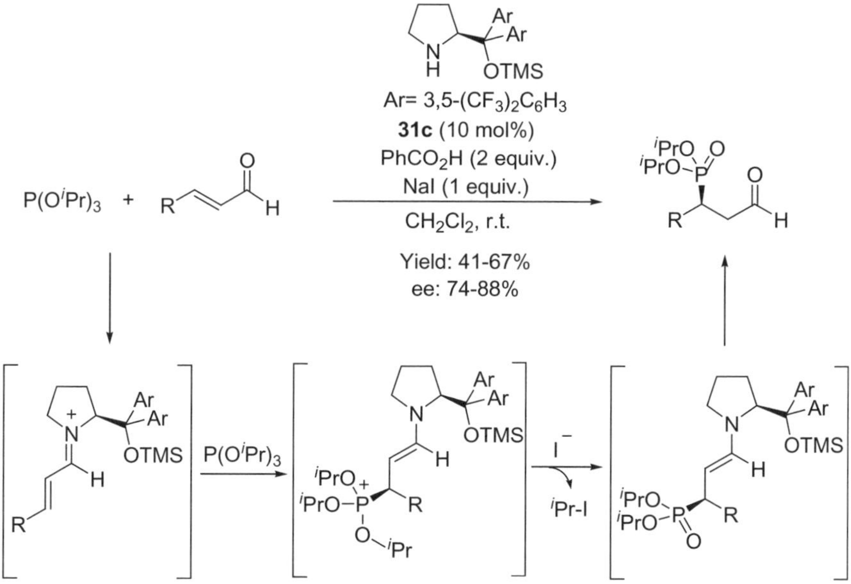

Scheme 3.38 Enantioselective conjugate addition of P(O*i*Pr)₃ to α,β-unsaturated aldehydes.

3.6 Concluding Remarks

The iminium activation concept has shown up as a very powerful approach for carrying out enantioselective conjugate additions of many different nucleophiles to α,β-unsaturated aldehydes and ketones under operationally simple procedures and very often reaching excellent yields and stereoselectivities. Moreover, the possibility of using enals as Michael acceptors is also another remarkable feature of this methodology when thinking of its applicability as a general tool in the total synthesis of complex molecules. In this context, in many cases, this methodology represents a true advantage with respect to the related metal-mediated methodologies. However, limitations are still found, especially when choosing the Michael donor reagent for which it has been shown that a careful tuning of its acidity and nucleophilicity is very often needed to achieve good conversions and selectivities. Another important drawback that still remains unsolved is the very high catalyst loadings required in most of the reactions developed to date, which is an important problem when scaling up these transformations for industrial production. In this context, the development of recyclable catalysts is becoming a growing field of research in this area. Finally, there are some conjugate addition reactions which still need further development, as is the case of some hetero-Michael reactions in which the inherent reversibility of the reaction needs an expressly designed nucleophile or, alternatively, the use of additional synthetic steps in order to isolate the conjugate addition compounds as a configurationally stable modified derivative.

References and Notes

1. For a specific review covering the iminium activation concept, see A. Erkkilä, I. Majander and P. M. Pihko, *Chem. Rev.*, 2007, **107**, 5416.
2. W. Langebeck and R. Sauerbier, *Chem. Ber.*, 1937, **70**, 1540.
3. R. B. Woodward, E. Logusch, K. P. Nambiar, K. Sakan, D. E. Ward, B.-W. Au-Yeung, P. Balaram, L. J. Browne, P. J. Card, C. H. Chen, R. B. Chenevert, A. Fliri, K. Frobel, H.-J. Gais, D. G. Garratt, K. Hayakawa, W. Heggie, D. P. Hesson, D. Hoppe, I. Hoppe, J. A. Hyatt, D. Ikeda, P. A. Jacobi, K. S. Kim, Y. Kobuke, K. Kojima, K. Krowicki, V. J. Lee, T. Lautert, S. Malchenko, J. Martens, R. S. Matthews, B. S. Ong, J. B. Press, T. V. Rajan Babu, G. Rousseau, H. M. Sauter, M. Suzuki, K. Tatsuta, L. M. Tolbert, E. A. Truesdale, I. Uchida, Y. Ueda, T. Uyehara, A. T. Vasella, W. C. Vladuchick, P. A. Wade, R. M. Williams and H. N.-C. Wong, *J. Am. Chem. Soc.*, 1981, **103**, 3210.
4. (a) M. Yamaguchi, N. Yokota and T. Minami, *J. Chem. Soc. Chem. Commun.*, 1991, 1088; (b) M. Yamaguchi, T. Shiraishi and M. Hirama, *Angew. Chem. Int. Ed. Engl.*, 1993, **32**, 1176. See also; (c) M. Yamaguchi, T. Shiraishi and M. Hirama, *J. Org. Chem.*, 1996, **61**, 3520.
5. K. A. Ahrendt, C. J. Borths and D. W. C. MacMillan, *J. Am. Chem. Soc.*, 2000, **122**, 4243.

6. For a specific account on iminium activation using primary amines see G. Bartoli and P. Melchiorre, *Synlett*, 2008, 1759.
7. D. A. Alonso, S. Kitagaki, N. Utsumi and C. F. Barbas III, *Angew. Chem. Int. Ed.*, 2008, **47**, 4588.
8. M. P. Patil and R. B. Sunoj, *Chem. Asian J.*, 2009, **4**, 714.
9. (a) C. Palomo and A. Mielgo, *Angew. Chem. Int. Ed.*, 2006, **45**, 7876; (b) A. Mielgo and C. Palomo, *Chem. Asian J.*, 2008, **3**, 922; (c) A. Lattanzi, *Chem. Commun.*, 2009, 1452.
10. (a) H. Gotoh, H. Ishikawa and Y. Hayashi, *Org. Lett.*, 2007, **9**, 5307; (b) L. Zu, H. Xie, H. Li, J. Wang and W. Wang, *Adv. Synth. Catal.*, 2007, **349**, 2660; (c) Y. Wang, P. Li, X. Liang, T. Y. Zhang and J. Ye, *Chem. Commun.*, 2008, 1232.
11. (a) S. Brandau, A. Landa, J. Franzen, M. Marigo and K. A. Jørgensen, *Angew. Chem. Int. Ed.*, 2006, **45**, 4305; (b) A. Ma, S. Zhu and D. Ma, *Tetrahedron Lett.*, 2008, **49**, 3075; (c) Y. Wang, P. Li, X. Liang and J. Ye, *Adv. Synth. Catal.*, 2008, **350**, 1383; (d) X. Companyó, M. Hejnová, M. Kamlar, J. Vesely, A. Moyano and R. Rios, *Tetrahedron Lett.*, 2009, **50**, 5021. For a modified version of this catalyst with an ionic liquid moiety used in this reaction see; (e) O. V. Maltsev, A. S. Kucherenko and S. G. Zlotin, *Eur. J. Org. Chem.*, 2009, 5134.
12. C. Palomo, A. Landa, A. Mielgo, M. Oiarbide, A. Puente and S. Vera, *Angew. Chem. Int. Ed.*, 2007, **46**, 8431.
13. C. Zhong, Y. Chen, J. L. Petersen, N. G. Akhmedov and X. Shi, *Angew. Chem. Int. Ed.*, 2009, **48**, 1279.
14. (a) J. L. Garcia Ruano, V. Marcos and J. Alemán, *Chem. Commun.*, 2009, 4435. For a later report see; (b) A.-N. Alba, X. Companyó, A. Moyano and R. Rios, *Chem. Eur. J.*, 2009, **15**, 11095.
15. A. Landa, A. Puente, J. I. Santos, S. Vera, M. Oiarbide and C. Palomo, *Chem. Eur. J.*, 2009, **15**, 11954.
16. (a) A.-N. Alba, X. Companyó, A. Moyano and R. Rios, *Chem. Eur. J.*, 2009, **15**, 7035; (b) F. Ullah, G.-L. Zhao, L. Deiana, M. Zhu, P. Dziedzic, I. Ibrahem, P. Hammar, J. Sun and A. Córdova, *Chem. Eur. J.*, 2009, **15**, 10013; (c) S. Zhang, Y. Zhang, Y. Ji, H. Li and W. Wang, *Chem. Commun.*, 2009, 4886. For the use of this nucleophile with enones see; (d) H. W. Moon, M. J. Cho and D. Y. Kim, *Tetrahedron Lett.*, 2009, **50**, 4896.
17. M. Nielsen, C. B. Jacobsen, M. W. Paixao, N. Holub and K. A. Jørgensen, *J. Am. Chem. Soc.*, 2009, **131**, 10581.
18. M. W. Paixao, N. Holub, C. Vila, M. Nielsen and K. A. Jørgensen, *Angew. Chem. Int. Ed.*, 2009, **48**, 7338.
19. M. Rueping, E. Sugiono and E. Merino, *Chem. Eur. J.*, 2008, **14**, 6329.
20. P. T. Franke, B. Richter and K. A. Jørgensen, *Chem. Eur. J.*, 2008, **14**, 6317.
21. (a) M. Rueping, E. Sugiono and E. Merino, *Angew. Chem. Int. Ed.*, 2008, **47**, 3046; (b) M. Rueping, E. Merino and E. Sugiono, *Adv. Synth. Catal.*, 2008, **350**, 2127.
22. For a review: H.-L. Cui and Y. C. Chen, *Chem. Commun.*, 2009, 4479.

23. (a) J.-W. Xie, L. Yue, D. Xue, X.-L. Ma, Y.-C. Chen, Y. Wu, J. Zhu and J.-G. Deng, *Chem. Commun.*, 2006, 1563; (b) J. Lu, F. Liu and T.-P. Loh, *Adv. Synth. Catal.*, 2008, **350**, 1781.
24. (a) S. Cabrera, E. Reyes, J. Alemán, A. Milelli, S. Kobbelgaard and K. A. Jørgensen, *J. Am. Chem. Soc.*, 2008, **130**, 12031. For a later report see; (b) Y. Hayashi, K. Obi, Y. Ohta, D. Okamura and H. Ishikawa, *Chem. Asian J.*, 2009, **4**, 246.
25. A. Massa, N. Utsumi and C. F. Barbas III, *Tetrahedron Lett.*, 2009, **50**, 145.
26. N. Bravo, I. Mon, X. Companyó, A.-N. Alba, A. Moyano and R. Rios, *Tetrahedron Lett.*, 2009, **50**, 6624.
27. P. Galzerano, G. Bencivenni, F. Pesciaioli, A. Mazzanti, B. Giannichi, L. Sambri, G. Bartoli and P. Melchiorre, *Chem. Eur. J.*, 2009, **15**, 7846.
28. H. Gotoh, R. Matsui, H. Ogino, M. Shoji and Y. Hayashi, *Angew. Chem. Int. Ed.*, 2006, **45**, 6853.
29. A.-N. Alba, N. Bravo, A. Moyano and R. Rios, *Tetrahedron Lett.*, 2009, **50**, 3067.
30. (a) Y. Hayashi, H. Gotoh, R. Matsui and H. Ishikawa, *Angew. Chem. Int. Ed.*, 2008, **47**, 4012; (b) L. Zu, H. Xie, H. Li, J. Wang, X. Yu and W. Wang, *Chem. Eur. J.*, 2008, **14**, 6333.
31. S. P. Brown, N. C. Goodwin and D. W. C. MacMillan, *J. Am. Chem. Soc.*, 2003, **125**, 1192.
32. J. Robichaud and F. Tremblay, *Org. Lett.*, 2006, **8**, 597.
33. W. Wang and J. Wang, *Org. Lett.*, 2005, **7**, 1637.
34. (a) M. Yamaguchi, N. Yokota and T. Minami, *J. Chem. Soc. Chem. Commun.*, 1991, 1088; (b) M. Yamaguchi, T. Shiraishi and M. Hirama, *Angew. Chem. Int. Ed. Engl.*, 1993, **32**, 1176. See also; (c) M. Yamaguchi, T. Shiraishi and M. Hirama, *J. Org. Chem.*, 1996, **61**, 3520.
35. A. Kawara and T. Taguchi, *Tetrahedron Lett.*, 1994, **35**, 8805.
36. N. Halland, P. S. Aburel and K. A. Jørgensen, *Angew. Chem. Int. Ed.*, 2003, **42**, 661.
37. (a) K. R. Knudsen, C. E. T. Mitchell and S. V. Ley, *Chem. Commun.*, 2006, 66; (b) V. Wacholowski, K. R. Knudsen, C. E. T. Mitchell and S. V. Ley, *Chem. Eur. J.*, 2008, **14**, 6155.
38. (a) S. V. Pansare and R. Lingampally, *Org. Biomol. Chem.*, 2009, **7**, 319; (b) E. Riguet, *Tetrahedron Lett.*, 2009, **50**, 4283.
39. N. Halland, T. Hansen and K. A. Jørgensen, *Angew. Chem. Int. Ed.*, 2003, **42**, 4955.
40. H. Kim, C. Yen, P. Preston and J. Chin, *Org. Lett.*, 2006, **8**, 5239.
41. J.-W. Xie, L. Yue, W. Chen, W. Du, J. Zhu, J.-G. Deng and Y.-C. Chen, *Org. Lett.*, 2007, **9**, 413.
42. (a) Y.-Q. Yang and G. Zhao, *Chem. Eur. J.*, 2008, **14**, 10888. For other examples of bifunctional catalysts employed in the asymmetric synthesis of warfarin see; (b) Z. Dong, L. Wang, X. Chen, X. Liu, L. Lin and X. Feng, *Eur. J. Org. Chem.*, 2009, 5192; (c) T. E. Kristensen, K. Vestli, F. K. Hansen and T. Hansen, *Eur. J. Org. Chem.*, 2009, 5185.

43. P. Li, S. Wen, F. Yu, Q. Liu, W. Li, Y. Wang, X. Liang and J. Ye, *Org. Lett.*, 2009, **11**, 753.
44. M. Yamaguchi, T. Shiraishi, Y. Igarashi and M. Hirama, *Tetrahedron Lett.*, 1994, **35**, 8233.
45. S. Hanessian and V. Pham, *Org. Lett.*, 2000, **2**, 2975.
46. S. Hanessian, Z. Shao and J. S. Warrier, *Org. Lett.*, 2006, **8**, 4787.
47. S. B. Tsogoeva, S. B. Jagtap, Z. A. Ardemasova and V. N. Kalikhevich, *Eur. J. Org. Chem.*, 2004, 4014.
48. M. Malmgren, J. Granander and M. Amedjkouh, *Tetrahedron: Asymmetry*, 2008, **19**, 1934.
49. P. Li, Y. Wang, X. Liang and J. Ye, *Chem. Commun.*, 2008, 3302.
50. (a) N. Halland, R. G. Hazell and K. A. Jørgensen, *J. Org. Chem.*, 2002, **67**, 8331; (b) A. Prieto, N. Halland and K. A. Jørgensen, *Org. Lett.*, 2005, **7**, 3897.
51. C. E. T. Mitchell, S. E. Brenner and S. V. Ley, *Chem. Commun.*, 2005, 5346.
52. K. Mei, M. Jin, S. Zhang, P. Li, W. Liu, X. Chen, F. Xue, W. Duan and W. Wang, *Org. Lett.*, 2009, **11**, 2864.
53. J.-W. Xie, W. Chen, R. Li, M. Zeng, W. Du, L. Yue, Y.-C. Chen, Y. Wu, J. Zhu and J.-G. Deng, *Angew. Chem. Int. Ed.*, 2007, **46**, 389.
54. A. Paras and D. W. C. MacMillan, *J. Am. Chem. Soc.*, 2001, **123**, 4370.
55. J. F. Austin and D. W. C. MacMillan, *J. Am. Chem. Soc.*, 2002, **124**, 1172.
56. N. A. Paras and D. W. C. MacMillan, *J. Am. Chem. Soc.*, 2002, **124**, 7894.
57. (a) R. Gordillo, J. Carter and K. N. Houk, *Adv. Synth. Catal.*, 2004, **346**, 1175. See also; (b) J. B. Brazier, G. Evans, T. J. K. Gibbs, S. J. Coles, M. B. Hursthouse, J. A. Platts and N. C. O. Tomkinson, *Org. Lett.*, 2009, **11**, 133.
58. (a) H. D. King, Z. Meng, D. Denhart, R. Mattson, R. Kimura, D. Wu, Q. Gao and J. E. Macor, *Org. Lett.*, 2005, **7**, 3437; (b) Y. Zhang, L. Zhao, S. S. Lee and J. Y. Ying, *Adv. Synth. Catal.*, 2006, **348**, 2027; (c) L. Hong, L. Wang, C. Chen, B. Zhang and R. Wang, *Adv. Synth. Catal.*, 2009, **351**, 772.
59. (a) T. Tian, B.-J. Pei, Q.-H. Li, H. He, L.-Y. Chen, X. Zhou, W.-H. Chan and A. W. M. Lee, *Synlett*, 2009, 2115; (b) Z.-J. Wang, J.-G. Yang, J. Jin, X. Lv and W. Bao, *Synthesis*, 2009, 3994; (c) L. Hong, W. Sun, C. Liu, L. Wang, K. Wong and R. Wang, *Chem. Eur. J.*, 2009, **15**, 11105; (d) L. Hong, C. Liu, W. Sun, L. Wang, K. Wong and R. Wang, *Org. Lett.*, 2009, **11**, 2177.
60. (a) C.-F. Li, H. Liu, J. Liao, Y.-J. Cao, X.-P. Liu and W.-J. Xiao, *Org. Lett.*, 2007, **9**, 1847; (b) H.-H. Lu, H. Liu, W. Wu, X.-F. Wang, L.-Q. Lu and W.-J. Xiao, *Chem. Eur. J.*, 2009, **15**, 2742.
61. D.-P. Li, Y.-C. Guo, Y. Ding and W.-J. Xiao, *Chem. Commun.*, 2006, 799.
62. W. Chen, W. Du, L. Yue, R. Li, Y. Wu, L.-S. Ding and Y.-C. Chen, *Org. Biomol. Chem.*, 2007, **5**, 816.
63. G. Bartoli, M. Bosco, A. Carlone, F. Pesciaioli, L. Sambri and P. Melchiorre, *Org. Lett.*, 2007, **9**, 1403.

64. S. Lee and D. W. C. MacMillan, *J. Am. Chem. Soc.*, 2007, **129**, 15438.
65. For some reviews see (a) H. Adolfsson, *Angew. Chem. Int. Ed.*, 2005, **44**, 3340; (b) S. G. Ouellet, A. M. Walji and D. W. C. MacMillan, *Acc. Chem. Res.*, 2007, **40**, 1327; (c) S.-L. You, *Chem. Asian J.*, 2007, **2**, 820.
66. J. W. Yang, M. T. Hechavarria-Fonseca and B. List, *Angew. Chem. Int. Ed.*, 2004, **43**, 6660.
67. J. W. Yang, M. T. Hechavarria-Fonseca, N. Vignola and B. List, *Angew. Chem. Int. Ed.*, 2005, **44**, 108.
68. (a) S. G. Ouellet, J. B. Tuttle and D. W. C. MacMillan, *J. Am. Chem. Soc.*, 2005, **127**, 32. For an example directed to the application of this methodology in total synthesis see; (b) T. J. Hoffman, J. D. Dash, J. H. Rigby, S. Arseniyadis and J. Cossy, *Org. Lett.*, 2009, **11**, 2756.
69. (a) J. B. Tuttle, S. G. Ouellet and D. W. C. MacMillan, *J. Am. Chem. Soc.*, 2006, **128**, 12662. For a computational study on this reaction see; (b) O. Gutierrez, R. G. Iafe and K. N. Houk, *Org. Lett.*, 2009, **11**, 4298.
70. S. Mayer and B. List, *Angew. Chem. Int. Ed.*, 2006, **45**, 4193.
71. N. J. A. Martin and B. List, *J. Am. Chem. Soc.*, 2006, **128**, 13368.
72. For a general review on oxa-Michael reactions see C. F. Nising and S. Bräse, *Chem. Soc. Rev.*, 2008, **37**, 1218.
73. (a) S. Bertelsen, P. Dinér, R. L. Johansen and K. A. Jørgensen, *J. Am. Chem. Soc.*, 2007, **129**, 1536; (b) N. R. Andersen, S. G. Hansen, S. Bertelsen and K. A. Jørgensen, *Adv. Synth. Catal.*, 2009, **351**, 3193. For a related report using acetone oximes as Michael donors see; (c) A. Pohjakallio and P. Pihko, *Chem. Eur. J.*, 2009, **15**, 3960.
74. A. Carlone, G. Bartoli, M. Bosco, F. Pesciaioli, P. Ricci, L. Sambri and P. Melchiorre, *Eur. J. Org. Chem.*, 2007, 5492.
75. (a) D. Díez, M. G. Nuñez, A. Benéitez, R. F. Moro, I. S. Marcos, P. Basabe, H. Broughton and J. G. Urones, *Synlett*, 2009, 390; (b) T. Kano, Y. Tanaka and K. Maruoka, *Tetrahedron*, 2007, **63**, 8658.
76. C. M. Reisinger, X. Wang and B. List, *Angew. Chem. Int. Ed.*, 2008, **47**, 8112.
77. X. Lu, Y. Liu, B. Sun, B. Cindric and L. Deng, *J. Am. Chem. Soc.*, 2008, **130**, 8134.
78. For a general review on sulfa-Michael reactions see D. Enders, K. Lüttgen and A. A. Narine, *Synthesis*, 2007, 959.
79. M. Marigo, T. Schulte, J. Franzén and K. A. Jørgensen, *J. Am. Chem. Soc.*, 2005, **127**, 15710.
80. P. Ricci, A. Carlone, G. Bartoli, M. Bosco, L. Sambri and P. Melchiorre, *Adv. Synth. Catal.*, 2008, **350**, 49.
81. For some reviews on the asymmetric aza-Michael reaction see (a) J. L. Vicario, D. Badía, L. Carrillo, J. Etxebarria, E. Reyes and N. Ruiz, *Org. Prep. Proc. Int.*, 2005, **37**, 513; (b) L.-W. Xu and C.-G. Xia, *Eur. J. Org. Chem.*, 2005, 633.
82. D. Enders, C. Wang and J. X. Liebich, *Chem. Eur. J.*, 2009, **15**, 11058.
83. Y. K. Chen, M. Yoshida and D. W. C. MacMillan, *J. Am. Chem. Soc.*, 2006, **128**, 9328.

84. (a) J. Vesely, I. Ibrahem, R. Rios, G.-L. Zhao, Y. Xu and A. Córdova, *Tetrahedron Lett.*, 2007, **48**, 2193. For a related example using a chiral sulfonyl hydrazine as organocatalyst see; (b) L.-Y. Chen, H. He, B.-J. Pei, W.-H. Chan and A. W. M. Lee, *Synthesis*, 2009, 1573.
85. X. Lu and L. Deng, *Angew. Chem. Int. Ed.*, 2008, **47**, 7710.
86. I. Ibrahem, R. Rios, J. Vesely, G.-L. Zhao and A. Córdova, *Chem. Commun.*, 2007, 849.
87. P. Dinér, M. Nielsen, M. Marigo and K. A. Jørgensen, *Angew. Chem. Int. Ed.*, 2007, **46**, 1983.
88. U. Uria, J. L. Vicario, D. Badia and L. Carrillo, *Chem. Commun.*, 2007, 2509.
89. H. Jiang, J. B. Nielsen, M. Nielsen and K. A. Jørgensen, *Chem. Eur. J.*, 2007, **13**, 9068.
90. Q. Lin, D. Meloni, Y. Pan, M. Xia, J. Rodgers, S. Shepard, M. Li, L. Galya, B. Metcalf, T.-Y. Yue, P. Liu and J. Zhou, *Org. Lett.*, 2009, **11**, 1999.
91. (a) G. Luo, S. Zhang, W. Duan and W. Wang, *Synthesis*, 2009, 1564; (b) S. Gogoi, C.-G. Zhao and D. Ding, *Org. Lett.*, 2009, **11**, 2249.
92. (a) S. Fustero, D. Jiménez, J. Moscardó, S. Catalán and C. del Pozo, *Org. Lett.*, 2007, **9**, 5283. For a later report involving similar substrates and products see; (b) E. C. Carlson, L. K. Rathbone, H. Yang, N. D. Collett and R. G. Carter, *J. Org. Chem.*, 2008, **73**, 5155.
93. S. Fustero, J. Moscardó, D. Jiménez, M. D. Pérez-Carrión, M. Sánchez-Roselló and C. del Pozo, *Chem. Eur. J.*, 2008, **14**, 9868.
94. For a leading review on the asymmetric phospha-Michael reaction see D. Enders, A. Saint-Dizier, M.-I. Lannou and A. Lenzen, *Eur. J. Org. Chem.*, 2005, 29.
95. A. Carlone, G. Bartoli, M. Bosco, L. Sambri and P. Melchiorre, *Angew. Chem. Int. Ed.*, 2007, **46**, 4504.
96. (a) I. Ibrahem, R. Rios, J. Vesely, P. Hammar, L. Eriksson, F. Himo and A. Córdova, *Angew. Chem. Int. Ed.*, 2007, **46**, 4507; (b) I. Ibrahem, P. Hammar, J. Vesely, R. Rios, L. Eriksson and A. Córdova, *Adv. Synth. Catal.*, 2008, **350**, 1875.
97. E. Maerten, S. Cabrera, A. Kjaersgaard and K. A. Jørgensen, *J. Org. Chem.*, 2007, **72**, 8893.

CHAPTER 4

Enantioselective Conjugate Addition Reactions via Hydrogen-bonding Activation

4.1 Introduction

Hydrogen bonding interactions play a key role in many biochemical reactions which are crucial to the processes of life. In fact, the ability of biological receptors to recognize selectively a given molecule or the mechanism of action of many enzymes when catalyzing a wide variety of biochemical transformations is mostly governed by the formation of one or more hydrogen bonds. In particular, molecules capable of acting as hydrogen-bond donors are able to activate electrophiles such as carbonyl or azomethine compounds by the formation of a hydrogen-bonded structure, which releases electronic density from the electrophile resulting in a significant decrease in energy of its LUMO and therefore becoming activated toward the reaction with a nucleophile present in the reaction medium. This mechanism of action closely resembles the one operating in the activation of electrophiles by metal-centered Lewis acid catalysts, in which the same LUMO-lowering effect happens as a consequence of the formation of a dative bond between the metal and a Lewis basic site at the electrophile. In this context, the conjugate addition reaction has also played a central role during the development of new reactivity using the activation by chiral Brønsted acids.[1] In this particular case, the activation of the acceptor *via* hydrogen bonding releases electronic density from the olefin and, at the same time, provides the required sterical bias which allows stereochemical control of the reaction.

RSC Catalysis Series No. 5
Organocatalytic Enantioselective Conjugate Addition Reactions: A Powerful Tool for the Stereocontrolled Synthesis of Complex Molecules
By Jose L. Vicario, Dolores Badía, Luisa Carrillo and Efraim Reyes
© J. L. Vicario, D. Badía, L. Carrillo and E. Reyes 2010
Published by the Royal Society of Chemistry, www.rsc.org

However, despite this close mechanistic analogy between Brønsted acid and Lewis acid catalysis, the use of small hydrogen donor molecules to promote organic reactions has remained as a rather undeveloped approach. Brønsted acid catalysis of a wide set of reactions such as transesterification, acetalization, Fischer esterification, hydrolysis of esters, amides or nitriles among others is well documented in the literature since the very early foundations of organic synthesis. However, most of these reactions involve the use of strong acid catalysts, which precludes the application of this concept to asymmetric catalysis, in which selective activation of the different functionalities is required in order to reach conformationally rigid intermediates and well-defined transition states. Nevertheless, the development of new compounds able to engage selectively in hydrogen bonding interactions with a given functionality has opened the possibility to carry out the fine-tuning of the reactivity of the species participating in a given reaction; as a consequence, a huge number of new methodologies have been reported which allow carrying out many different reactions which were formerly exclusively available under metal-centered Lewis acid catalysis.[2]

The weak nature of the hydrogen bond leads to a weak enthalpic binding between the Brønsted acid catalyst and the Michael acceptor, which turns into inferior turnover frequencies, especially compared with the parent situation found in Lewis acid catalysis. The use of catalysts incorporating multiple hydrogen donors opens up as a useful strategy in order to overcome this unfavorable situation, which results in an increase of these binding affinities with the substrate and also provides a much better organization in the transition state, which also leads to improved stereoselectivities. For this reason, most of the catalysts developed in this field are multifunctional structures incorporating multiple elements with abilities to form hydrogen bonds (as either H-donors or H-acceptors) favoring a well-defined orientation of the reagents participating in the reaction. This feature also produces better chemoselectivity control and a wider functional group tolerance, which also allows working with very complex molecules and therefore opens the way for the application of these methodologies in total synthesis, where highly functionalized molecules have to be manipulated with complete selectivity.

Nevertheless, the foundations of this field can be found many decades ago, starting approximately when chemists started to investigate the mechanism of action of several enzymes for which activation of the reagents by hydrogen bonding was found to play a crucial role. Independently, several small organic molecules containing H-donors were also found to be able to catalyze some relevant organic transformations. For example, in 1985, Hine identified 1,8-biphenylienediol **62a** as an effective promoter of the ring-opening reaction of phenyl glycidyl ether with diethylamine, also pointing out that a cooperative effect of both hydroxy groups could also be involved in the activation of the electrophile (Scheme 4.1).[3] Further research led to the identification of dinitroderivative **62b** as a suitable catalyst for the Diels–Alder reaction between cyclopentadiene and several α,β-unsaturated aldehydes and ketones (Scheme 4.1),[4] in which double hydrogen-bond donation was also proposed to

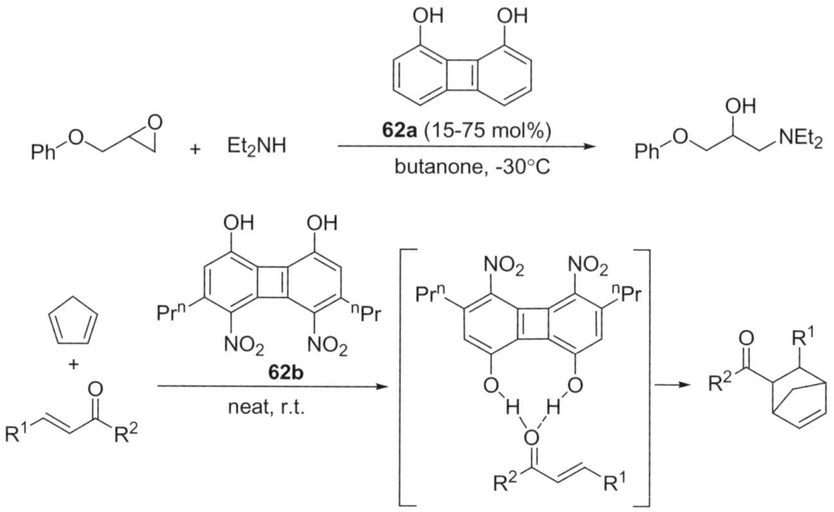

Scheme 4.1   Pioneering examples of catalytic reactions using small organic molecules as H-donors.

account for the activation of the dienophile. This proposal was also in accordance with a theoretical study by W. L. Jorgensen, who also explained the rate-accelerating effect of Diels–Alder reactions and Claisen rearrangements in water with respect to non-protic solvents by means of the same double hydrogen-bonding interaction with the substrate.[5] This double hydrogen-bonding interaction implied that the activation of the electrophile/dipolarophile should occur *via* a conformationally restricted intermediate, which also established the base for the design of new catalysts with improved abilities and also showed the way for the development of chiral catalysts able to promote the reaction in a stereocontrolled fashion, although the development of the first enantioselective example still had to wait for a few years.

Parallel reports by other authors working on solid state chemistry indicated that H-donors were able to interact selectively with different functional groups forming stable Brønsted acid/base pairs. For example, studies by Etter and coworkers showed that *N,N*'-diarylureas co-crystallized with compounds containing a wide variety of Lewis basic functional groups like, for example, nitroarenes, forming stable complexes in which each of the hydrogen donors from the urea interacted with each of the two oxygen atoms of the nitro group.[6] This provided the basis for the development of a new family of urea-based catalysts operating *via* selective double hydrogen-bonding with the substrate, which was first reported by Curran who found that the incorporation of conveniently substituted diarylureas such as **63** as additives in different proportions were able to enhance the yield and diastereoselectivity of the allylation of α-sulfinyl radicals with allyltributylstannane (Scheme 4.2).[7] Later on, the same group reported the **63**-catalyzed Claisen rearrangement, observing a noticeable

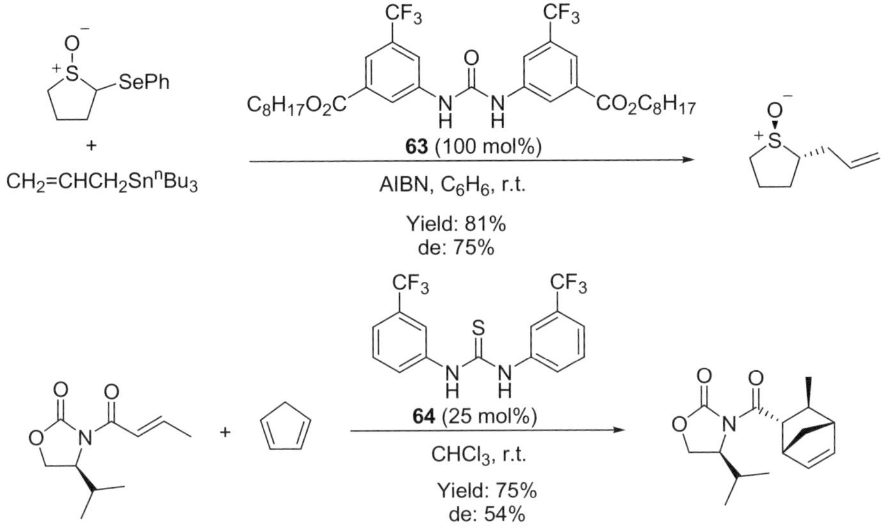

Scheme 4.2   Some pioneering examples of urea- and thiourea-mediated reactions.

rate acceleration effect, although with rather high catalyst loadings.[8] In the same line, Schreiner introduced the use of thioureas such as **64** as catalysts for diastereoselective Diels–Alder reactions (Scheme 4.2).[9] The higher N–H acidity of thioureas compared with ureas, which was also enhanced by the introduction of electron-withdrawing CF_3 groups at both aryl substituents, led to a more active catalyst. Further research by other authors has also contributed to the development of this basic concept in many reactions such as 1,2-additions to azomethine compounds,[10] the Baylis–Hillman reaction[11] and Friedel–Crafts alkylation,[12] among others. In all these cases, achiral ureas and thioureas were employed in order to accelerate the reaction by activating one reagent (the electrophile or dienophile) *via* hydrogen-bonding interactions.

The first steps toward catalytic enantioselective reactions using H-bonding catalysis should be attributed to Wynberg in the late 1970s, in the context of the quinine-catalyzed Michael reaction between several C- and S-nucleophiles and enones, outlining the importance of electrophile activation by H-bonding interaction with the free OH present at the catalyst.[13] Some years later, the enantioselective Strecker reaction using chiral thiourea **65a** as catalyst reported by Jacobsen in 1998 (Scheme 4.3)[14] represented a real breakthrough in the field, demonstrating the enormous power of the H-bonding activation concept and opening a new horizon of possible applications of this methodology as a general and useful tool in asymmetric synthesis. Importantly, computational studies supported by a series of experiments allowed proposal of a mechanism showing that the thiourea catalyst was able to interact with the imine electrophile by formation of a double H-bond network, which resulted in a rigid transition state that accounted for the high enantioselectivity observed.

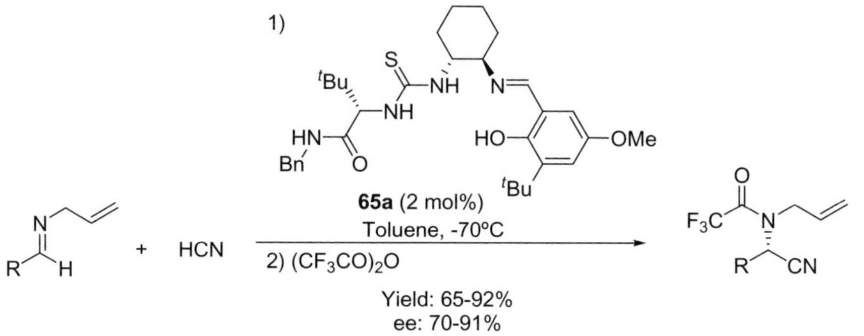

Scheme 4.3   The thiourea-catalyzed Strecker reaction reported by Jacobsen.

Moreover, calculations also suggested that the final product only participated in a single hydrogen bonding with the catalyst, resulting in a weaker interaction and therefore providing an explanation for the catalyst turnover. This study set the basis for the subsequent development of a wide family of chiral thioureas.

Nevertheless, it has also to be pointed out that several other previous examples of organocatalytic enantioselective transformations exist in which hydrogen bonding also plays an important role in controlling the stereochemistry of the process, although in these cases these H-bonding interactions can be considered as secondary interactions which assist the formation of a rigid intermediate, other types of activation mechanisms being involved in the activation of the reagents (Scheme 4.4). The proline-mediated aldol reaction presented in Chapter 2 is a representative example of this situation, where activation of the electrophile *via* enamine formation should be considered as the key aspect related to the catalytic activity of proline, although organization of the transition state *via* hydrogen bonding is also present and important to account for the excellent stereoselectivities obtained. This situation is also found in the enantioselective hydrocyanation of aldehydes using cyclic dipeptide **66** as catalyst reported by Inoue in 1981 (Scheme 4.4),[15] in which the catalysts act as a chiral base activating the nucleophile by deprotonation and also in the enolate alkylation under PTC conditions reported by a research group at Merck in 1984.[16] In all these cases, H-bond formation contributes to the preorganization of the reagents in the transition state.

4.1.1   Some Relevant Aspects Influencing the Stereocontrol in Michael Additions Proceeding *via* Hydrogen-bonding Activation

As previously pointed out, the mechanism operating in the activation of a Michael acceptor with a chiral catalyst *via* hydrogen bonding involves the release of electronic density from the olefin and, at the same time, the backbone

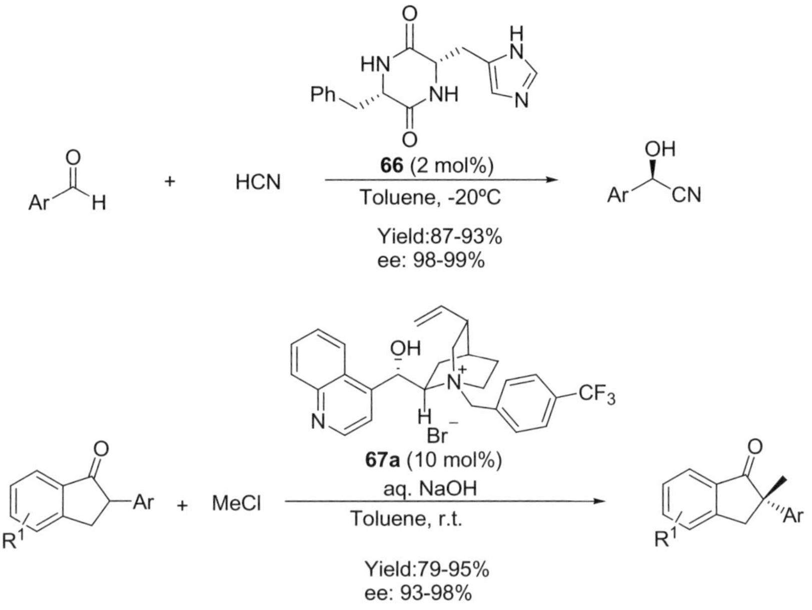

Scheme 4.4 Some early reports on enantioselective transformation in which H-bonding contributes to stereocontrol.

structure of the catalyst should provide the required steric bias which should allow stereochemical control of the reaction. Therefore, the main principle to be attended in the design of a suitable catalyst for this transformation is related to its ability to engage in hydrogen bonding with the basic sites at the electron-withdrawing group that activates the olefin, preferably by forming a conformationally rigid network using multiple H-donor sites, which also contributes to a well-defined orientation of the reagents in the transition state.

As can be seen in the previous section, thioureas have emerged as one of the most efficient classes of catalysts working under H-bonding activation, with a superior ability as catalysts to that provided by the parent ureas. The higher NH acidity of thioureas also complements their poorer abilities as hydrogen-bond acceptors through the sulfur atom, which precludes self-association of the catalyst. The overall principle behind the design of most thiourea catalysts employed to date involves a modular structure in which the substitution of one of the nitrogen atoms is used to modulate the electronic density (and hence the acidity) of the thiourea moiety and the substituent at the other nitrogen incorporates the chiral information at the carbon backbone. In this context, most authors have followed a general design for the development of efficient chiral thiourea catalysts according to the principles shown in Figure 4.1. First, an aryl ring containing electron-withdrawing groups (very often the 3,5-$(CF_3)_2C_6H_3$ group) is incorporated at one of the nitrogen atoms, which releases

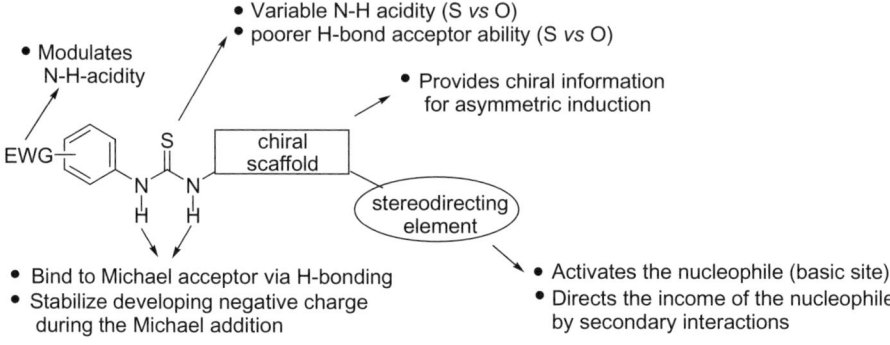

Figure 4.1 General principles applied to the design of chiral thiourea catalysts.

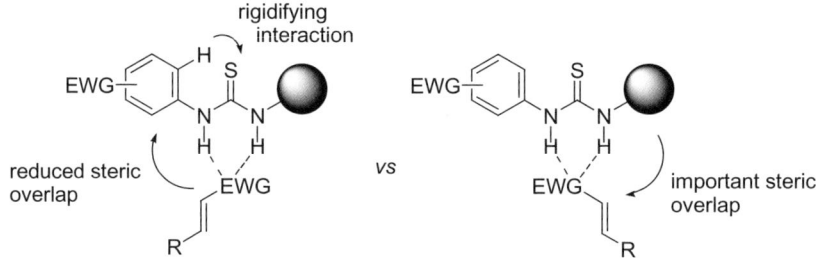

Figure 4.2 Conformational issues to be considered regarding the thiourea-Michael acceptor binding geometry.

electron density from the thiourea moiety and increases the acidity of both NH groups. Second, a substituent incorporating chiral information is usually introduced at the second nitrogen atom, and it is very often found that a second functionality is introduced at this chiral scaffold which interacts with the nucleophile directing the stereochemical outcome of the reaction by a secondary interaction. This can be either a basic site, which also assists the reaction by activation of the nucleophile by deprotonation, or an additional H-donor, which engages in another H-bonding interaction with basic sites present at the nucleophile structure.

Another issue to be controlled is related to the conformational freedom associated to the Michael acceptor and the orientation that it takes when engaging in H-bonding with the catalyst. In this case, two possible scenarios can be anticipated for an α,β-unsaturated carbonyl compound or related derivative to interact with the two H-donors of the thiourea catalyst (Figure 4.2). Assuming an efficient steric shielding/stereodirecting effect by the chiral information present at the catalyst, each of these two binding geometries leads to the exposition of a different stereotopic face of the Michael acceptor and, therefore, to a non-stereoselective process if this situation is not controlled. This geometry control on the binding mode of the acceptor to the catalyst is usually controlled by the inherent nature of the two very different substituents introduced at the thiourea N atoms. The aryl substituent represents a very small effective volume, especially compared with the usually large chiral *N*-alkyl substituent placed at the other nitrogen atom, which causes the Michael acceptor to place the β-substituents far away from this bulkier substituent in order to override steric overlap. Another situation which also assists this high degree of conformational control is related to the electronic properties of the *N*-aryl substituted thiourea moiety, for which it has been proposed that a significant rotational barrier around the C_{aryl}–N bond occurs as a result of an attractive interaction between the *ortho* hydrogen atoms (which are also polarized by the electron-withdrawing substituent introduced at the aromatic ring) and the sulfur heteroatom. This rigidifying interaction restricts the conformational movement associated to the Ar–N bond rotation and reduces the steric hindrance derived from the presence of this aromatic substituent, magnifying the difference between the two substituents at the *N*-atoms of the thiourea.

Compounds other than thioureas have also been developed as active catalysts in Michael-type reactions. These include not only the closely related guanidine analogues[17] but also other compounds which are only capable of engaging in a single H-bonding interaction with the substrate. This is, for example, the case for chiral phosphoric acids (Figure 4.3)[18] in which bulky substituents are placed at the *ortho* positions of the binaphthyl system in order to exert an effective conformational control of the activated Michael acceptor by steric effects, which also provides the required shielding of one of its stereotopic faces. However, in many cases the P=O moiety is also claimed to participate in the reaction, acting as a second point for interaction with the nucleophile, allowing a better organization of the intermediates and a more effective stereochemical control in the approximation of the reagents during the reaction.

4.2 Conjugate Addition of Stabilized Carbon Nucleophiles

As also happened in the iminium activation case, the Michael reaction has been the most studied conjugate addition reaction in this context, also serving as a

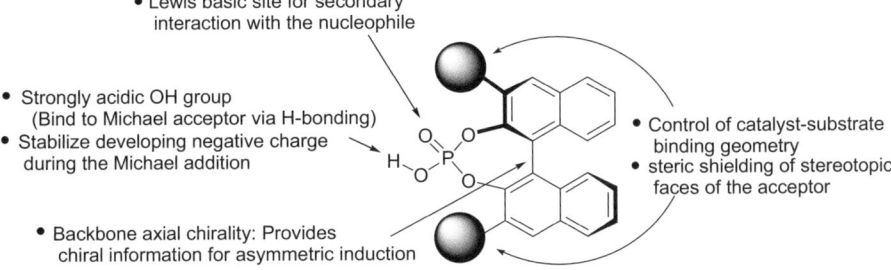

Figure 4.3   Relevant aspects of the design of chiral phosphoric acids as H-donor catalysts.

test ground for the development of new families of catalysts. Also in this case, the use for compounds incorporating a rather acidic C–H hydrogen as Michael donor is mandatory, because the activation of the Michael donor by deprotonation with strong bases (for the generation of an enolate nucleophile) is not compatible with the reacting species participating in the catalytic cycle, in particular with the acidic protons of the catalysts which have to engage in the activation *via* hydrogen bonding. For this reason, the conjugate addition of malonate-type compounds to α,β-unsaturated carbonyl compounds has been the reaction of choice for many research groups because of the high acidity of the starting pro-nucleophile, which guarantees an easy deprotonation in the reaction medium. It has to be pointed out that enolizable aldehydes and ketones have also been employed as Michael donors in the reaction with Michael acceptors which are also activated by H-bond formation, but these require their previous activation as the corresponding enamine, needing a bifunctional catalyst containing both a primary/secondary amine and an H-donor site. As we have considered that the activation of the nucleophile *via* enamine formation is the key feature for the viability of these processes, the activation of the acceptor being of less relevance in most cases, this chemistry has been covered in Chapter 2.

4.2.1 Nitroalkenes as Michael Acceptors

The use of nitroalkenes as Michael acceptors and chiral thioureas as catalysts turns into a particularly synergistic situation for asymmetric catalysis. In this case, the thiourea moiety finds a direct way for interacting with the nitro group due to the effective formation of two hydrogen bonds between the NH groups and the highly electron-rich oxygen atoms of the nitro moiety. This thiourea-nitroalkene interaction is rather strong, occurring even in the presence of water molecules in the reaction medium and also guarantees a high chemoselective reaction if other basic groups or potential Michael acceptor moieties are present at the substrate structure. The first example of one of these catalysts specifically designed for the Michael reaction of 1,3-dicarbonyl compounds with nitroalkenes is the bifunctional thiourea **68a** developed by Takemoto (Scheme 4.5).[19] This incorporates the usual *N*-aryl substituted thiourea moiety together with an additional dimethylamino basic site placed at a convenient position on the chiral backbone. This dimethylamino group is proposed to activate the acidic malonate reagent by assisting the formation of the corresponding enolate by deprotonation and it should also play a key role in the stereochemical control of the reaction, by directing the approach of the nucleophile toward the Michael acceptor. The reaction between different malonates and nitrostyrenes proceeded with excellent yields and enantioselectivities, although β-alkyl substituted nitroalkenes were found to be more challenging substrates furnishing a slightly inferior degree of stereocontrol in the reaction. Proof of the synthetic utility of the methodology has also been given by addressing the enantioselective synthesis of (*R*)-(–)-*baclofen*

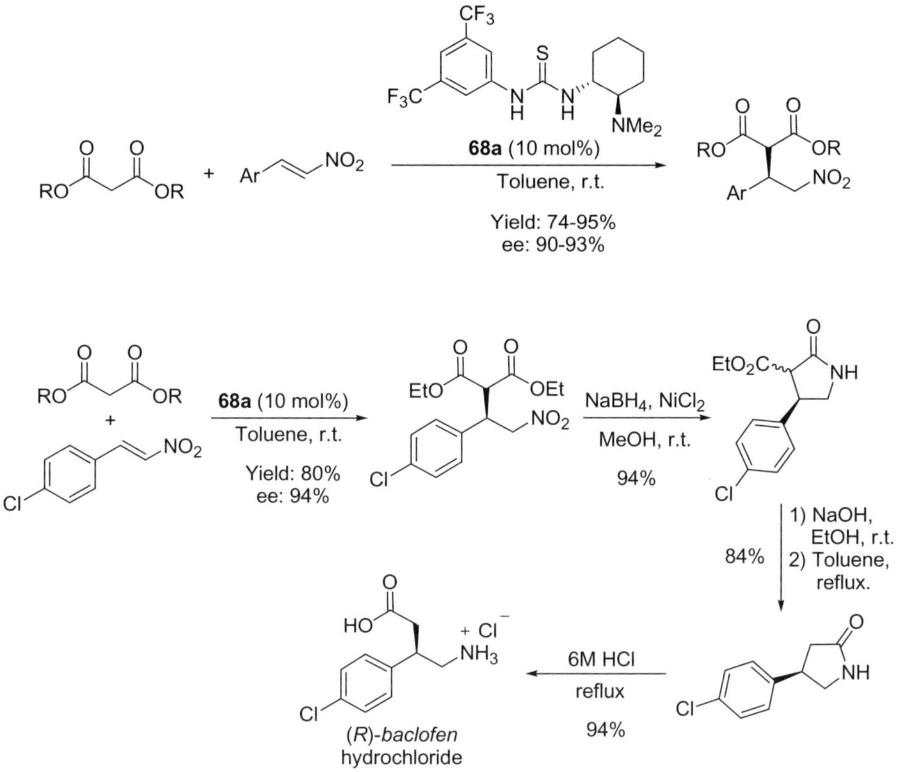

Scheme 4.5 **68a**-catalyzed enantioselective Michael addition of malonates to nitroalkenes and application to the total synthesis of (*R*)-(−)-*baclofen*.

antispasmoic drug, with 38% overall yield in six steps starting from 4-chlorobenzaldehyde.

Remarkably, the reaction also tolerated the use of differently substituted α-alkyl-β-ketoesters as Michael donors, leading to the formation of compounds containing two contiguous stereogenic centers, one of them a quaternary one, in good diastereo- and enantioselectivities (Scheme 4.6).[19] Masked cyclic 1,3-diketones such as 2-hydroxy-1,4-naphthoquinols have also been successfully applied in this context also showing the extraordinary performance of catalyst **68a** in terms of both yields and enantioselectivities.[20] There is also one example of a vinylogous Michael reaction between α,α-disubstituted dicyanoacrylates and nitroalkenes for which Takemoto's catalyst has also been identified as the most efficient promoter of the reaction.[21] Moreover a solid-supported version of this catalyst has also been developed by Takemoto himself and tested in the reaction of diethylmalonate with nitrostyrene with success.[22]

Different experimental and theoretical studies have been carried out in order to elucidate the exact reaction pathway which could account for the observed results and could also provide a reliable basis for the development of other

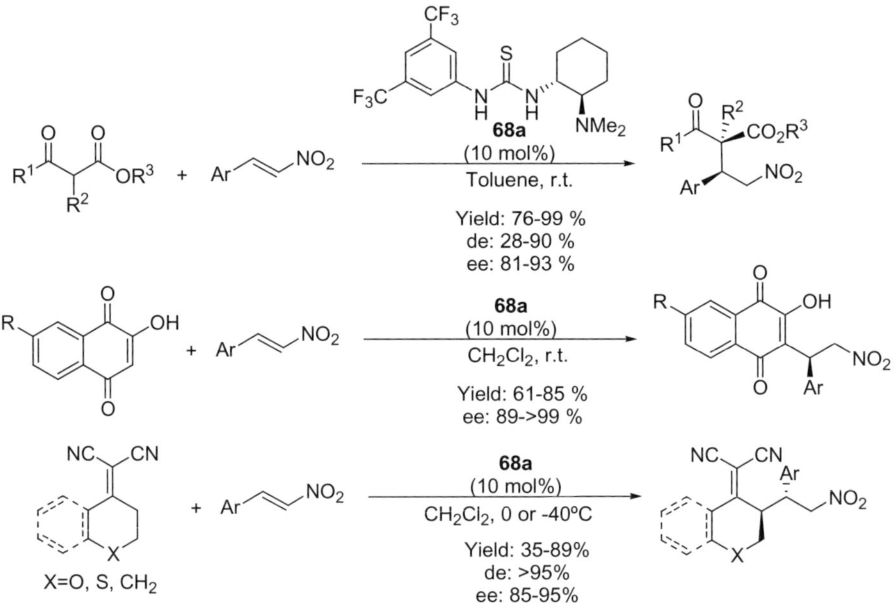

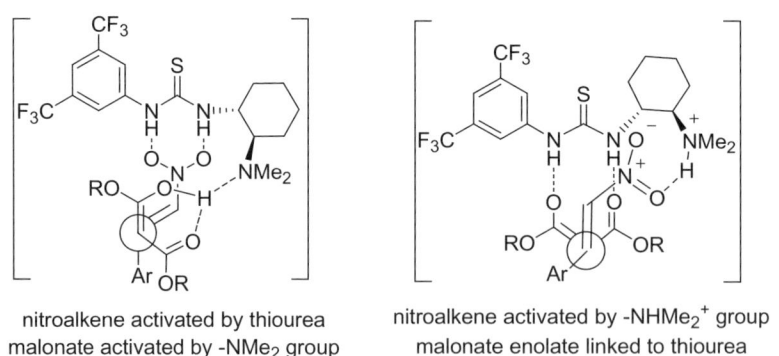

Scheme 4.6   **68a**-catalyzed enantioselective Michael addition of 1,3-dicarbonyl compounds to nitrostyrenes.

nitroalkene activated by thiourea
malonate activated by -NMe$_2$ group

nitroalkene activated by -NHMe$_2^+$ group
malonate enolate linked to thiourea

Figure 4.4   The two proposed models explaining the **68a**-catalyzed Michael reaction of malonates with nitrostyrenes.

more efficient and active catalysts (Figure 4.4). Kinetic studies by Takemoto demonstrated that the reaction was first order in the catalyst, the malonate and the nitrostyrene reagents,[19a] pointing toward the formation of a trimolecular complex in the transition state in which the nitroalkane would remain activated by the thiourea moiety by double H-bonding interaction and the dimethyl-amino moiety would be responsible for the activation of the nucleophile by deprotonation, as it was previously anticipated. The absence of nonlinear effects when using **68a** with variable enantiomeric purity also discarded the possibility of association between catalyst molecules. Computational studies carried out by Soós and Papai[23] also supported the cooperative contribution of the thiourea and the dimethylamino moieties in the activation of both the nitroalkene acceptor and the malonate donor, although an alternative possibility was outlined with a calculated lower activation energy for the corresponding transition state. In this proposal, activation of the nitroalkane would take place through the protonated dimethylammonium moiety (formed after the previous first deprotonation step of the Michael donor) by a single hydrogen-bonding interaction and the enolate derived of the 1,3-dicarbonyl compound would bind the thiourea moiety *via* double H-bonding network. Both proposed pathways provide a satisfactory explanation for the stereochemical outcome of the reaction and for the high enantioselectivities obtained.

Chiral ureas and thioureas derived from cinchona alkaloids represent another type of structurally related bifunctional catalysts which have been successfully employed in this transformation. In this case, the presence of the basic quinuclidine nucleus near to the (thio)urea moiety in a well-defined chiral environment results in a very successful architecture for the design of strongly active and very efficient catalysts which have been employed in the conjugate addition of malonates to nitroalkenes using remarkably low catalyst loadings, which in some cases have reached values near 0.5 mol%. It is important to point out that the absolute configuration at C-9 of the cinchona substructure turned out to be crucial for reaching high levels of stereocontrol (Scheme 4.7)[24] observing that, while urea derivative **69** prepared from the natural source was observed to be a very unsuccessful catalyst for the reaction between

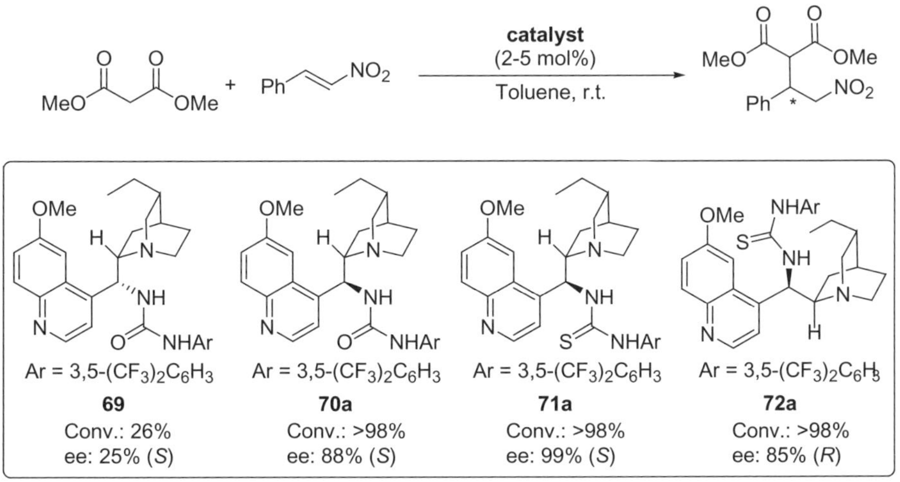

Scheme 4.7 Influence of cinchona catalysts' structure on the yield and enantioselectivity of the Michael reaction of dimethyl malonate with nitrostyrene.

nitrostyrene and dimethyl malonate, the 9-*epi* diastereoisomer **70a** appeared as a highly active catalyst in the same transformation. This is also an indication of the cooperativity in the mode of action of the urea moiety and the basic qui-nuclidine site, which needs to be positioned in the adequate spatial arrangement for the synergic activation of both the Michael donor and the acceptor. As can be seen in Scheme 4.7, changing to the more efficient H-donor thiourea ana-logue **71a** allowed to increase the enantioselectivity of the reaction and, inter-estingly, the use of pseudoenantiomeric quinidine-based compound **72a** led to a very efficient reaction but with an opposite sense of asymmetric induction compared to that observed for the quinine-based catalyst **71a**.

Therefore, under the optimized conditions, 9-*epi*-dihydroquinine-derived thiourea **71a** performed very well in the Michael reaction of diethyl malonate with nitroalkenes, furnishing the final compounds with enantioselectivities around 90% ee (Scheme 4.8).[24] Remarkably, this catalyst also seemed to tol-erate the use of β-alkyl substituted nitroalkenes as substrates, furnishing the expected final compounds with only a small decrease in enantioselectivity, although only a few examples were studied in that case. In an independent work, 9-*epi*-cinchonine-derived thiourea **72b** was also reported to be a very efficient catalyst for this reaction,[25] obtaining the final Michael adducts in

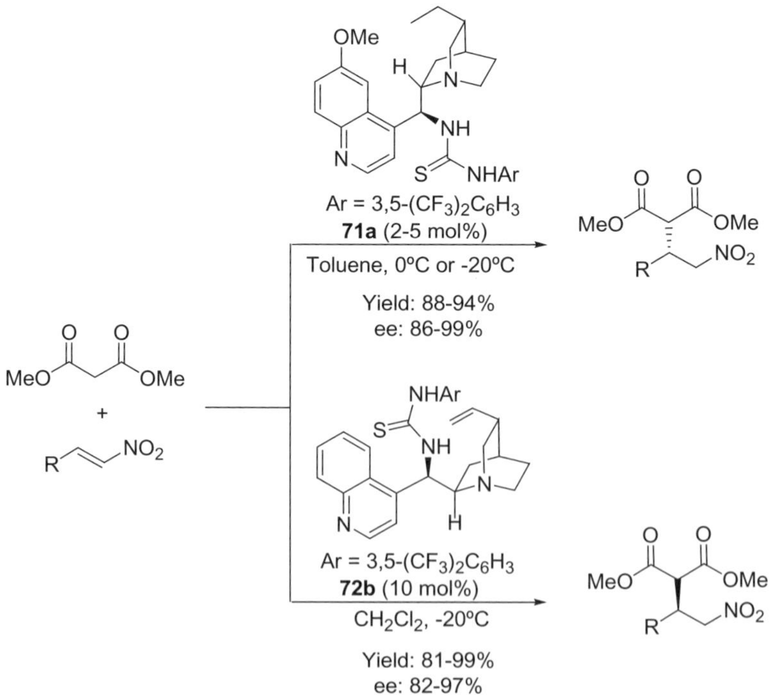

Scheme 4.8   Enantioselective Michael addition of malonates to nitroalkenes catalyzed by cinchona-thiourea compounds.

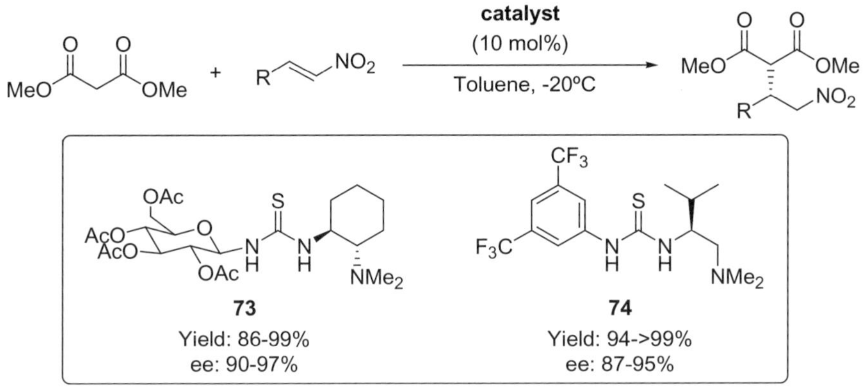

Scheme 4.9 Enantioselective Michael addition of malonates to nitroalkenes catalyzed by thiourea compounds.

excellent yields and enantioselectivities and also with the opposite configuration to the products obtained in the related **71a**-catalyzed reaction. This methodology, using either these catalysts or other closely related ones, has been employed by several authors for the enantioselective synthesis of different biologically active compounds,[26] showing the utility of this approach as a very efficient and general tool in synthesis.

Several other bifunctional thiourea catalysts incorporating a Brønsted basic site have been developed for the Michael addition of malonates to nitroalkenes (Scheme 4.9). For example, thiourea **73** derived from (*S*,*S*)-1,2-diaminocyclo-hexane and a peracetylated β-D-glucopiranoside was identified as a very efficient catalyst, also showing a remarkably good performance when aliphatic nitroolefins were employed.[27] The development of catalyst **74** by Pedrosa and coworkers represented an interesting alternative, proving to be able to promote this reaction in a very efficient manner, reaching excellent yields and enantioselectivities of the corresponding Michael adducts.[28] While Takemoto's catalyst **68a** and other related ones like **73** rely on the use of the rather expensive enantiomerically pure *trans*-1,2-diaminocyclohexane reagent as the basic chiral skeleton for the construction of the catalyst, the synthesis of **74** can be easily performed starting from L-valine. Moreover, such a modular structure allows the easy preparation of a library of catalysts by simply starting from different α-amino acids.

Acetylacetone has also been the subject of several studies when used as 1,3-dicarbonyl compound suitable to engage in a Michael reaction with nitroalkenes under H-bonding catalysis. In this context, Takemoto's catalyst **68a**[19b] and valine-derived thiourea **74**[28] have been tested in the reaction with acetylacetone with nitrostyrene furnishing good results, although no extensive study was carried out in order to evaluate the substrate scope with regard to the substitution at the nitroalkene. On the other hand, several functionalized thioureas have been expressly surveyed in this reaction, providing a detailed study

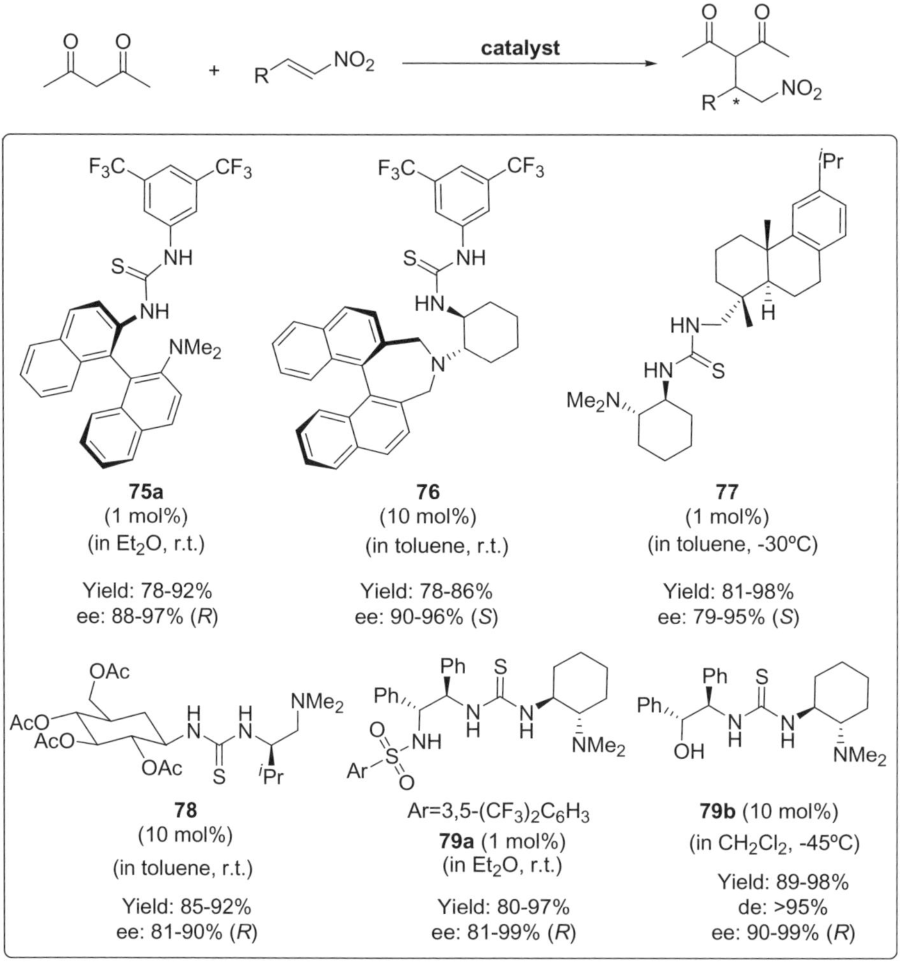

Scheme 4.10 Enantioselective Michael addition of acetylacetone to nitroalkenes catalyzed by multifunctional thioureas.

regarding the catalyst activities and the influence that the substitution pattern at the nitroalkene reagent has in the reaction (Scheme 4.10). For example, axially chiral binaphthyl-containing thioureas **75a**[29] and **76**,[30] dehydroabietic-derived compound **77**,[31] thiourea **78** incorporating both an amino acid and a carbohydrate substituents[32] and highly functionalized thiourea-amine-sulfonamide compound **79a**[33] have been successfully tested in this reaction with good results. All these catalysts proved to be efficient promoters for the reaction using nitrostyrene derivatives as Michael acceptors but, interestingly, catalyst **79a** also showed an acceptable performance in reactions with β-alkyl substituted nitroalkenes, which are more challenging substrates, as has been shown in many of the preceding examples. The incorporation of the additional

H-bonding site together with the usual thiourea-tertiary amine bifunctional architecture is claimed to provide an additional point of activation and also is proposed to assist a better organization of the reagents in the transition state, leading to a more active and stereoselective catalyst. These features have also been exploited in the reaction using **79b** as catalyst and other different 1,3-dicarbonyl compounds as Michael donors like cyclic α-substituted β-keto-esters,[34] leading to the simultaneous formation of two stereogenic centers. In this case, excellent yields and enantioselectivities were obtained and a remarkably high diastereoselection was also achieved.

Malonic acid half-thioesters can also be employed as suitable nucleophiles in the Michael reaction with nitroalkenes using this kind of activation. These particular malonate reagents can be used as an acetyl carbanion equivalent leading to the formation of γ-nitrothioesters due to a decarboxylation process occurring after or during the conjugate addition step. Interestingly, cinchona-urea **70b** was identified as the best catalyst for the reaction, while the corresponding thiourea showed up as a significantly less active promoter of the reaction between *p*-methoxyphenylthioacetylacetic acid and nitrostyrene (Scheme 4.11).[35] The results were strongly dependent upon the conditions employed and, after optimization, it was found that the reaction could be carried out with a wide variety of nitroalkenes, obtaining excellent yields of the final compounds using THF as solvent, although enantioselectivities remained at values around 55–67% ee. Alternatively, enantioselectivities up to 73–90% ee could be obtained using ethyl vinyl ether (EVE) as solvent, but in this case much lower yields were observed. Another remarkable feature of this methodology is the fact that the use of β-alkyl substituted nitroalkenes as Michael acceptors did not result in a particularly complicated situation, observing that these substrates behaved in a similar way to nitrostyrene derivatives.

The use of nitroalkanes as Michael donors produces a particularly difficult situation mainly because of the chemoselectivity issues that have to be

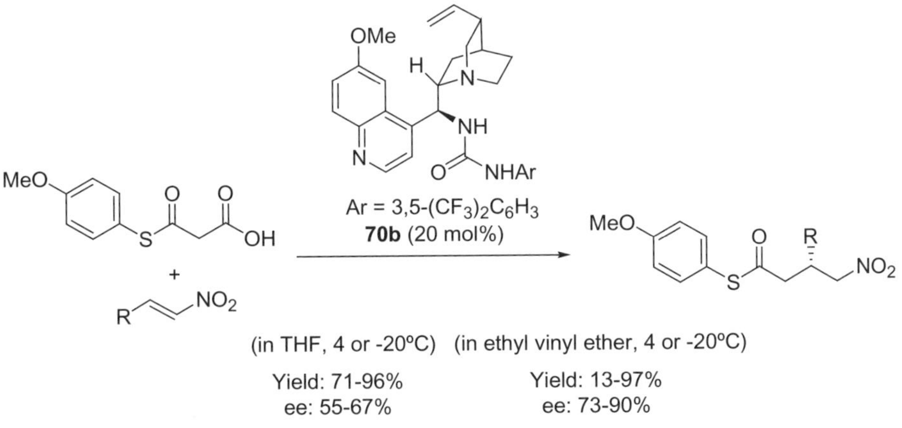

(in THF, 4 or -20°C)   (in ethyl vinyl ether, 4 or -20°C)

Yield: 71-96%   Yield: 13-97%
ee: 55-67%   ee: 73-90%

Scheme 4.11   Enantioselective Michael addition of malonic acid half-thioesters to nitroalkenes catalyzed by **70b**.

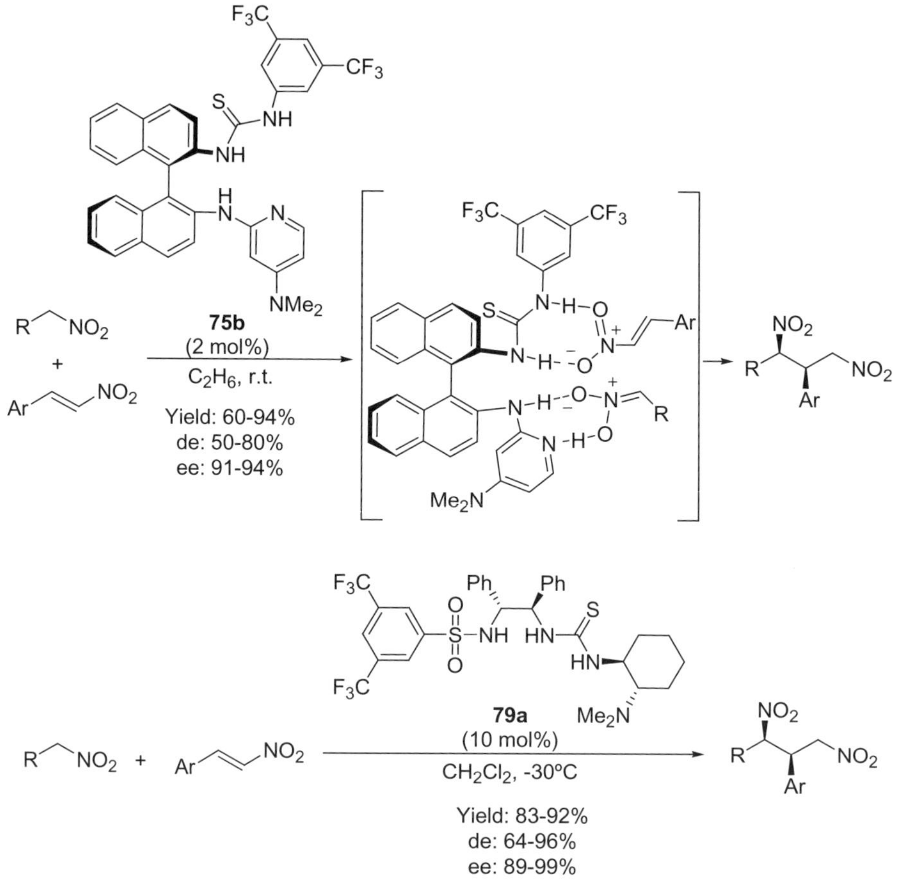

Scheme 4.12 Enantioselective Michael addition of nitroalkanes to nitroalkenes.

controlled, which are derived from the presence of two nitrocompounds in the reaction scheme. In this case, the catalyst has to differentiate clearly between both reagents, activating one of them *via* H-bonding with the thiourea moiety and the other one *via* secondary interaction with other functionality incorporated at the catalyst structure. An illustrative example can be found on Scheme 4.12, in which a new catalyst **75b** was designed containing an amino-pyridine moiety able to participate selectively in the activation of the nitroalkane Michael donor by assisting its deprotonation and also by stabilizing the formed nitronate *via* double H-bonding interactions. The amino-pyridine substructure was connected to the usual thiourea moiety by a binaphthyl-type axially chiral backbone. The reaction was studied on nitro-propane and nitrobutane as Michael donors and a variety of nitrostyrenes as acceptors, leading to the generation adducts with two contiguous stereogenic centers in good yields, diastereo- and enantioselectivities.[36] Remarkably, it required a rather low catalyst loading under the optimized conditions, although a large excess of nitroalkane reagent (up to 30-fold excess) had to be used to reach to full conversion. The reaction using β-alkyl-substituted nitroalkenes was reported to proceed with a much lower stereoselectivity, with both enantio- and diastereoselectivity being significantly affected, which is also an indication of the difficulties associated to the use of these nitroalkenes as Michael acceptors. Interestingly, there are no data reported regarding the use of nitromethane as Michael donor, which would furnish the final Michael adducts containing a single stereogenic center, therefore avoiding the need for dia- stereoselectivity control. Thiourea **79a** containing multiple H-donor sites has also been tested in this reaction with success,[37] providing similar results to those reported for catalyst **75b**, that is, good yields and stereoselectivities were obtained in the reaction between several nitroalkanes and nitrostyrenes but somewhat poorer results were observed in the reaction with an aliphatic nitroalkene substrate; no data were reported regarding the possibility to use nitromethane in the reaction, although it has to be pointed out that a chal- lenging, more hindered Michael donor such as 2-nitropropane was also able to undergo the reaction, affording the final product in good yield but moderate enantioselectivity. In this case, the optimized conditions involved the use of the nitroalkane and nitroalkene reagents in a 4:1 ratio.

The search toward the identification of other C–H acidic compounds able to participate as pro-nucleophiles in this reaction has also been intense. For example, racemic 4-substituted oxazolones, which were also employed as Michael donors in Michael reactions with α,β-unsaturated aldehydes under iminium activation (see Scheme 3.10 in Chapter 3), have also been employed in the reaction with nitroalkenes using cinchona-thiourea catalyst **71b**, leading to the formation of the corresponding Michael adducts in good yields and ste- reoselectivities (Scheme 4.13).[38] Interestingly, these oxazolones are precursors of a bidentate nucleophile as a result of the delocalization of the negative charge both at C-2 and at C-4 and, in this context, the regiochemistry of the addition was found to be strongly dependent on the nature of the substituents at these positions of the oxazolone reagent. While 2-aryl substituted com- pounds furnished cleanly the corresponding C-2 addition products, regio- chemistry changed when aryl substituents were placed at the 4-position and alkyl substituents at the 2-position, and only products arising from the C-4 addition were isolated. The obtained adducts were also identified as suitable precursors of a wide variety of highly functionalized chiral building blocks, in particular providing an easy and direct access to quaternary α,α-dialkyl-α- amino acids.

In a different work, the group of Dixon has also identified racemic 5-aryl-1,3- dioxolan-4-ones as suitable pro-nucleophiles in this reaction (Scheme 4.14), using in this case cinchonine-thiourea catalyst **72b** and obtaining the final adducts with moderate to good yields, excellent diastereoselectivities and enantioselectivities in the range of 60–89% ee.[39] The highly functionalized Michael adducts obtained also proved to be excellent precursors for a wide variety of interesting chiral building blocks by means of simple transformations

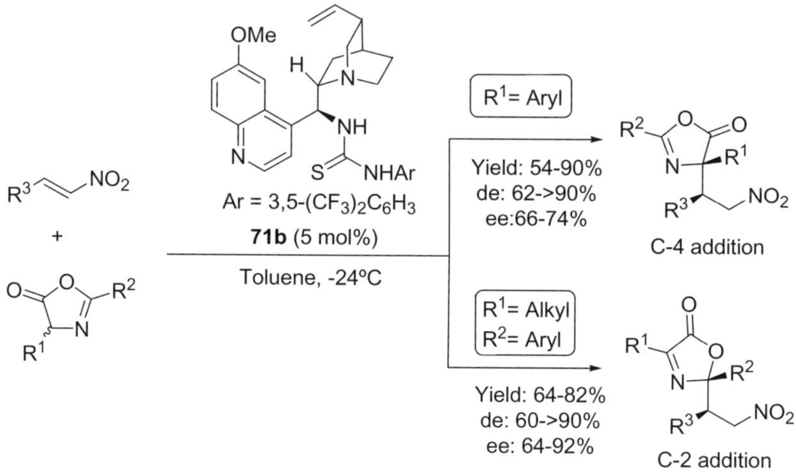

Scheme 4.13 Enantioselective Michael addition of racemic 4-substituted oxazolones to nitroalkenes.

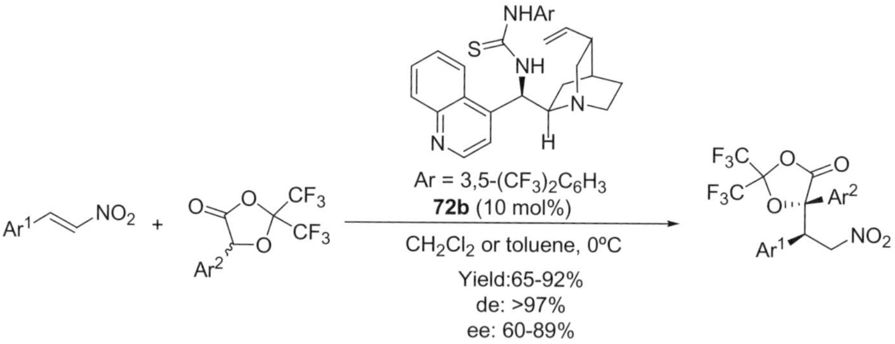

Scheme 4.14 Enantioselective Michael addition of racemic 5-aryl-1,3-dioxolan-4-ones to nitroalkenes.

which operated selectively at the different functionalities present on the molecule.

Racemic 3-alkyl oxindoles have also been found to be useful pro-nucleophiles in this context (Scheme 4.15).[40] For this transformation, a modified version of Takemoto's catalyst was identified as the most efficient promoter of the reaction, which was found to have a remarkably wide substrate scope, allowing many different substitution patterns both at the oxindole and at the nitroalkene reagent, and even tolerating well the use of β-alkyl substituted nitroalkenes as Michael acceptors. Yields, diastereo- and enantioselectivities were found to be excellent in almost all the cases studied. The reaction also accepted very well the use of the simple nitroethene, another very challenging

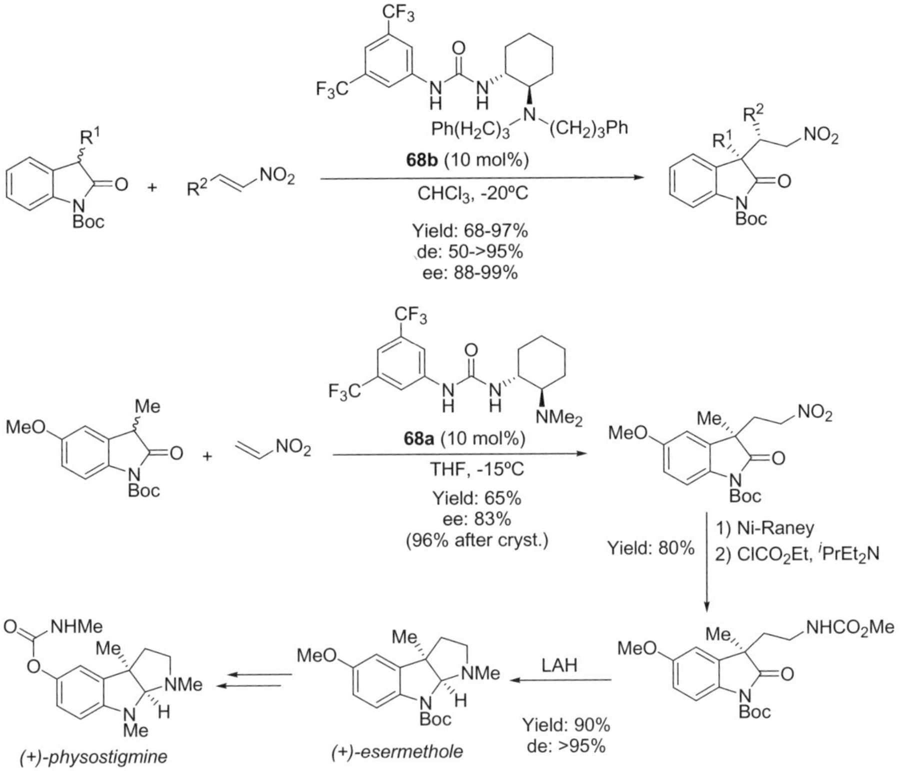

Scheme 4.15 Enantioselective Michael addition of racemic 3-alkyl oxindoles to nitroalkenes and application to the total synthesis of (+)-*physostigmine*.

nitroalkene reagent, using the original version of Takemoto's catalyst, which furnished the corresponding Michael adduct in 83% ee, which could be further improved to 96% ee after a single recrystallization process. This feature was exploited for the application of this methodology to the enantioselective formal total synthesis of natural product (+)-*physostigmine* via formation of (+)-*esermethole*, which is a known intermediate in the synthesis of the target compound.

Anthracenones are another class of C–H acidic compounds suitable to be employed in this reaction (Scheme 4.16) and, in fact, Takemoto's catalyst has been identified as the most efficient catalyst among a series of different thioureas tested, which also included a family of different cinchona alkaloid-derived candidates.[41] The reaction proceeded satisfactorily for a wide variety of aromatic nitroalkenes tested but poorer results were obtained in the case of the β-alkyl substituted Michael acceptors.

It also has to be remembered that enolizable aldehydes or ketones can also be used as Michael donors in this reaction using thiourea-containing catalysts for

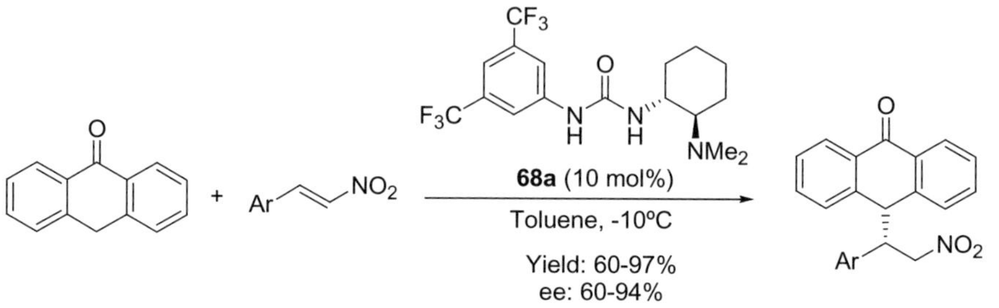

Scheme 4.16   Enantioselective **68a**-catalyzed Michael addition of anthracenones to nitrostyrenes.

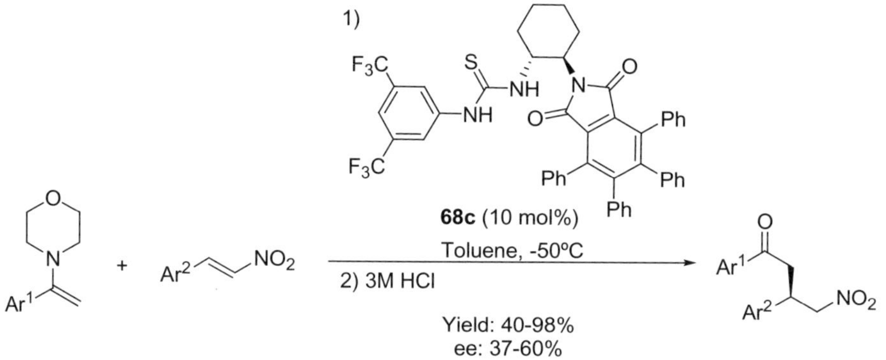

Scheme 4.17   Enantioselective Michael addition of enamines to nitroalkenes.

the activation of the nitroalkene electrophile, although, as pointed out previously (see Chapter 2, Section 2.2), it was considered that this thiourea-nitroalkene interaction played a secondary role compared with the need for the activation of the Michael donor in the form of the corresponding enamine. Nevertheless, there is also an example reporting the addition of pre-formed enamines to nitroalkenes catalyzed by simple thiourea catalysts. In this case, several enamines derived from aryl methyl ketones and morpholine were found to react with both aliphatic and aromatic nitroalkenes in the presence of catalyst **68c** furnishing the corresponding γ-nitroketones after hydrolytic work-up in good yields, although with moderate enantioselectivities (Scheme 4.17).[42]

Other different catalysts incorporating functionalities other than thioureas, which are also capable of interacting with the nitro group *via* hydrogen bonding have been surveyed in this reaction and, in this context, chiral guanidines represent another important group of compounds which have been employed as chiral catalysts in several enantioselective Michael reactions.[43] This is the case, for example, for the bifunctional guanidine **80**, which has been employed

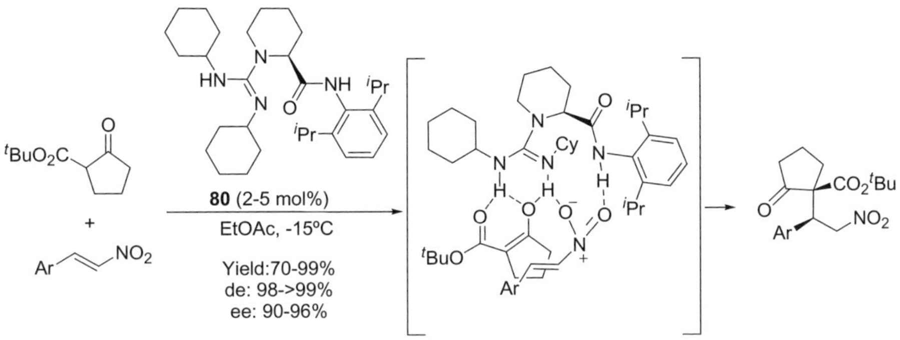

Scheme 4.18   Enantioselective **80**-catalyzed Michael addition of 2-*tert*-butoxy-carbonylcyclopentanone to nitrostyrenes.

as an outstanding catalyst for the Michael addition of cyclic β-ketoesters to nitrostyrene derivatives (Scheme 4.18).[44] Under the optimized conditions a remarkably wide range of aromatic nitroalkenes reacted with 2-*tert*-butoxycarbonyl cyclopentanone leading to the corresponding Michael adducts in excellent yields and enantiomeric excesses, and also as single diastereoisomers in most cases. This bifunctional catalyst is proposed to activate the pronucleophile by an effective deprotonation process assisted by the highly basic guanidine moiety. In addition, this guanidine group was also proposed to engage in multiple hydrogen bonding interactions with both the nitroalkene and the enolate nucleophile, the latter also establishing an additional H-bond with the lateral arylamide moiety, which also possesses another H-donor site. This multiple array of H-bonding interactions leads to a very well-ordered arrangement of the reagents in the transition state and accounts for the excellent diastereo- and enantioselectivities achieved. Moreover, **80** was also found to be a very active catalyst, and it was observed that all reactions were completed in around 32 h using a 2 mol% catalyst loading. This rate-accelerating ability is thought to be a consequence of the strongly basic guanidine moiety employed to activate the nucleophile.

Another very active guanidine type catalyst had already been reported before in the reaction of dimethylmalonate to nitroalkenes (Scheme 4.19).[45] Although, in this case, the axially chiral guanidine **81** did not incorporate any other secondary functionality for establishing multiple interactions with both the Michael donor and the acceptor, computational studies on other related guanidine-catalyzed Michael reactions (see Scheme 4.45) indicate that this compound also behaves as a bifunctional catalyst, in which the NH group is also involved in the activation of the nitroalkene by H-bonding,[46] the incoming trajectory of the electrophile being controlled by the axial chirality of the catalyst, in particular by the presence of the bulky 3,6-bis(3,5-di-*tert*-butylphenyl)phenyl groups placed at the 3 and 3′ positions of the binaphthyl backbone. In this sense, guanidine **81** was able to catalyze the

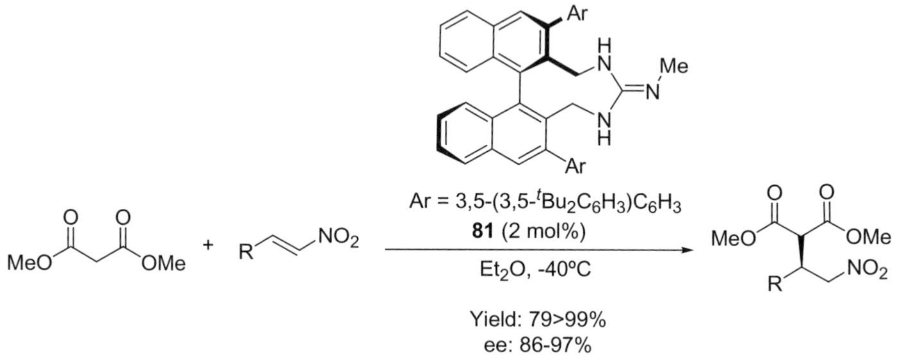

Scheme 4.19   Enantioselective Michael addition of dimethylmalonate to nitroalkenes catalyzed by **71**.

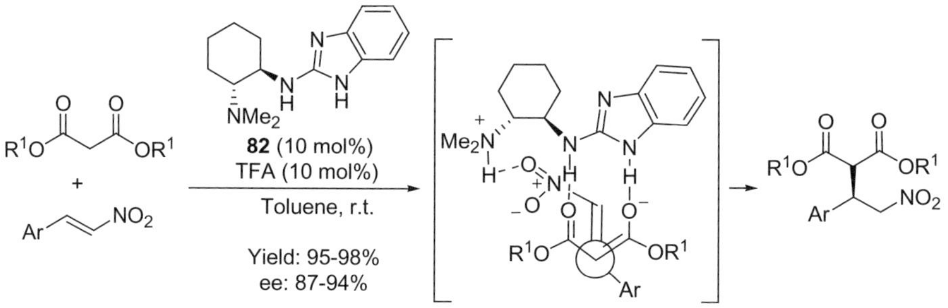

Scheme 4.20   Enantioselective Michael addition of malonates to nitroalkenes catalyzed by **82**.

reaction between dimethylmalonate and many different nitroalkenes, including β-alkyl substituted ones, in only 2–10 h and using a 2 mol% catalyst loading. It also has to be mentioned that other different chiral guanidines have been employed in several attempts to carry out conjugate additions of malonates or related compounds to nitroalkenes but furnishing low to moderate levels of enantioselection.[47]

There is also a modified version of Takemoto's catalyst, which incorporates a benzimidazole heterocycle as the H-bonding donor site in place of the thiourea moiety.[48] This catalyst **82** has been tested with success in several Michael-type reactions of 1,3-dicarbonyl compounds to nitroalkenes, in particular focused on the use of malonates as donors (Scheme 4.20), providing the corresponding adducts in excellent yields and enantioselectivities. β-Ketoesters have also been tested, although in this case the performance of the catalyst was found to be highly dependent on the structure of the β-ketoester employed. It has also to be pointed out that the reaction required the incorporation of a Brønsted acid co-catalyst such as TFA for achieving the best enantioselectivity, although the presence of this co-catalyst did not have any influence in the catalytic activity.

A very interesting feature of this catalyst can be found in its recyclability, showing that it could be recovered in *c.a.* 94% from the reaction mixture after extractive acid/base work-up with no loss of activity. The authors also carried out a computational study directed to the better understanding of the reaction pathway, observing that the favored mechanistic pathway did not arise from the usual double activation by the bifunctional catalyst in which the H-donor activates the nitroalkene and the tertiary amine group activates the pro-nucleophile by assisting its deprotonation. On the contrary, calculations indicated that the preferred pathway would arise from a first deprotonation step of the malonate pro-nucleophile exerted by the tertiary amino group, which should be followed by the Michael addition step *via* a transition state in which the malonate-derived enolate engaged in double H-bonding interaction with the benzimidazole H-donor and the ammonium salt moiety formed after the deprotonation step would direct the income of the nitroalkene reagent by a single H-bonding interaction. These calculations are also in agreement with those reported by Soós and Pápai during their studies, which focused on **68a**-catalyzed reaction (see Schemes 4.5 and 4.6 and Figure 4.4).

Some other molecules which participate in the activation of the electrophile by the donation of a single hydrogen bond have been tested in this reaction with different results. These include the simple cinchona alkaloids quinine and cinchonine used initially by Wynberg in some of the first pioneering examples of Michael additions to enones, already mentioned at the beginning of this chapter.[13] In this context, Deng has demonstrated that quinidine **83a** is a very poor promoter for the Michael reaction of dimethylmalonate with nitrostyrene, but the corresponding derivative **83b** containing a free OH at the quinoline moiety (6'-OH), also known as cupreidine,[49] afforded the corresponding Michael adducts in excellent yields and enantioselectivities (Scheme 4.21).[50] This observation, together with the fact that the reaction proceeded with much more efficiency in aprotic solvents, points toward the participation of this phenolic OH group in the activation of the electrophile *via* H-bonding interaction. A remarkable feature of this methodology is also the possibility of constructing both enantiomers of the same Michael adduct using the two pseudoenantiomeric cupreidine or cupreine alkaloids **83b** or **84a** (Scheme 4.20).

Further work led to the optimization of a methodology of even wider scope, regarding the possibility to use other 1,3-dicarbonyl compounds such as β-keto esters, 1,3-diketones, β-nitroesters and β-cyanoesters (Scheme 4.22).[51] Moreover, α-substituted pro-nucleophiles could also be used with success, leading to the formation of the final adducts with excellent diastereo- and enantioselectivities. In this case, the structure of the catalyst needed some optimization depending on the nature of the nucleophile employed. In particular, a careful modulation of the OR group placed at C-9 of the quinine skeleton was required, which pointed toward the possibility that conformational flexibility at this position could play an important role in the reaction.

In this context, the reaction of 2-fluoromalonates with nitrostyrenes has also been successfully carried out using **83d** as catalyst.[52] The direct addition of nitroalkanes to nitroalkenes has also been performed using the same modified

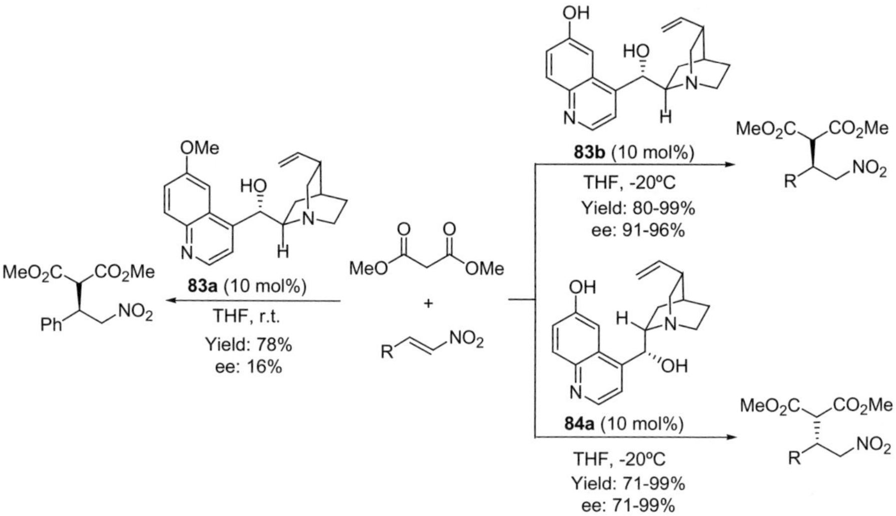

Scheme 4.21   Enantioselective Michael addition of dimethylmalonate to nitroalkenes quinidine and quinine derivatives.

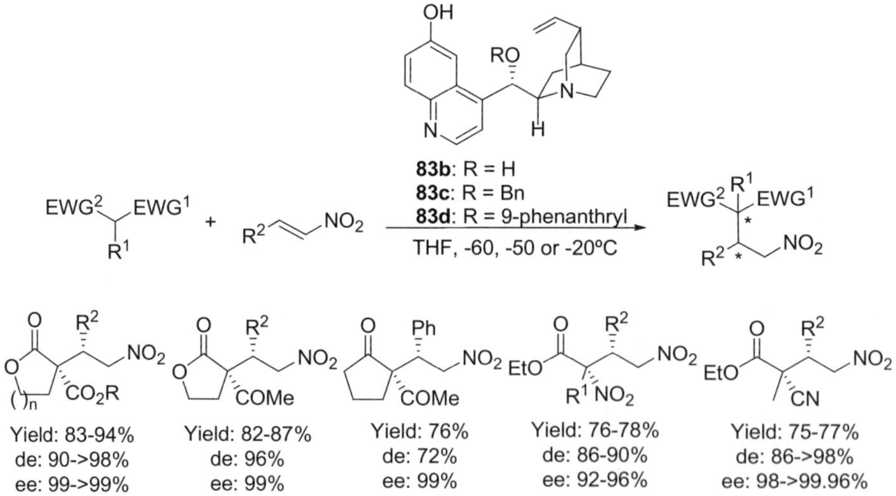

Scheme 4.22   Enantioselective Michael addition of 1,3-dicarbonyl compounds or related derivatives to nitroalkenes catalyzed by **83d**.

cinchona alkaloid **83d**, although, in this case, somewhat lower enantioselection was observed.[53]

A stereochemical model has also been proposed which accounts for the observed results (Scheme 4.23),[51] indicating that activation of the electrophile

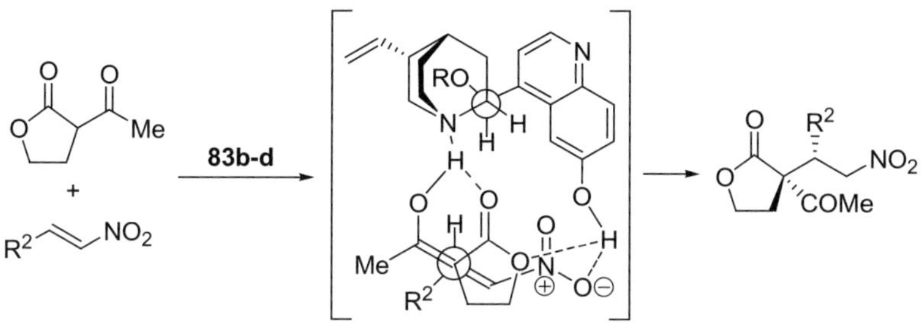

Scheme 4.23 Proposed stereochemical model for the **83b-d**-catalyzed Michael reaction of 1,3-dicarbonyl compounds to nitroalkenes.

should occur *via* H-bonding interaction with the phenolic OH group and the quinuclidine basic moiety would be involved in activation of the 1,3-dicarbonyl pro-nucleophile by deprotonation, which would also direct its attack from one of the enantiotopic faces of the nitroalkane. The authors also proposed that this phenolic OH group should also be engaged in a second simultaneous interaction with the ester moiety present at the β-ketoester pro-nucleophile, leading to a well-defined orientation of the enolate and therefore explaining also the high diastereoselectivity of the reaction when β-alkyl substituted 1,3-dicarbonyl compounds were employed. This second hypothesis was supported by the lower diastereoselectivity observed in the reaction with 1,3-diketones, where this additional H-bonding interaction is not possible.

Changing the secondary alcohol moiety to an amino or sulfonamido group and inverting the configuration of this stereogenic center has a striking influence in the way the catalyst performs. In fact, 9-*epi*-amino cinchona alkaloid **28c** has been found to be a very efficient catalyst for the Michael addition of 1,3-diaryl-1,3-propanediones to nitroalkenes.[54] Importantly, catalyst **28c** can not engage in H-bonding activation of the electrophile through the phenolic OH group, which points toward the implication of the primary amine group as the new H-donor site involved in the activation of the nitroalkene. This contrasts with the low enantioselectivity displayed by the analogous quinine **84b** (see Scheme 4.24), in which also the phenolic group is blocked as a methyl ether derivative and the secondary OH is supposed to be involved in H-bonding with the nitroalkene. An explanation for this behavior can be found in the opposite configuration of this C-9 stereogenic center, which facilitates the catalyst to adopt easily a correct conformation of lower energy, which delivers both activated reagents, the Michael donor and the acceptor, close to each other. An additional benefit associated to the use of this catalyst is related to the fact that the product was found to be very insoluble in Et_2O, while **28c** remained dissolved. This allowed recycling the catalyst, which after seven cycles still kept its excellent performance in the reaction.

In a different work, *N*-sulfonylamido 9-*epi* quinine catalyst **85** was also successfully employed in the Michael reaction of β-ketoesters to nitrostyrenes, also suggesting the same type of activation profile (Scheme 4.25).[55] In this case,

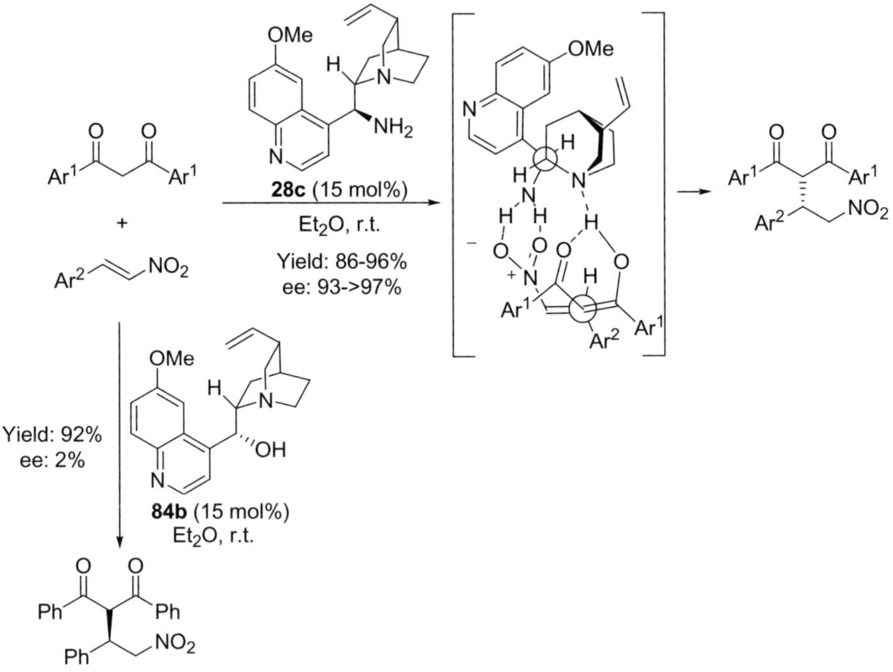

Scheme 4.24 Michael reaction of 1,3-dicarbonyl compounds to nitroalkenes catalyzed by **28c**.

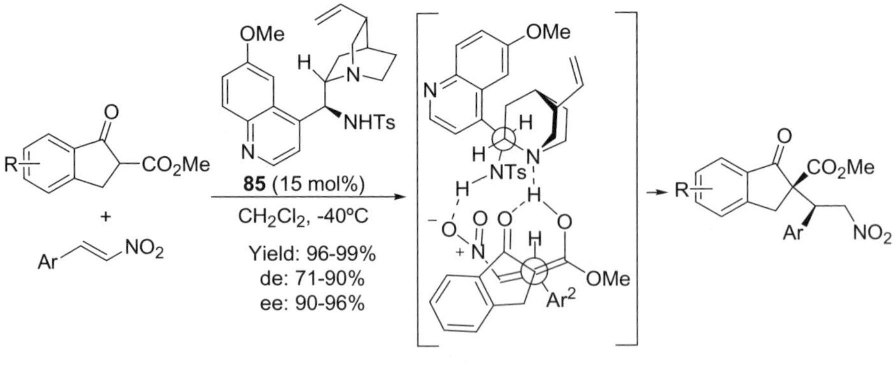

Scheme 4.25 Michael reaction of β-ketoesters to nitroalkenes catalyzed by **85**.

the quinuclidine nucleus would engage in activation of the nucleophile and the nitroalkene would be activated by the sulfonamido group by means of a single H-bonding interaction. Under the optimized conditions, a set of different Michael adducts were obtained with excellent yields, diastereo- and enantio-selectivities, although the reaction was only tested using nitrostyrenes as

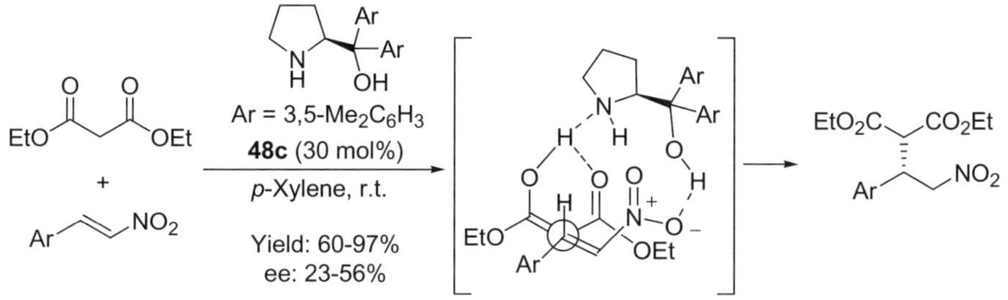

Scheme 4.26   **48c**-catalyzed Michael addition of diethylmalonate to nitrostyrenes.

Michael acceptors, with no example reported regarding the use of β-alkyl substituted nitroalkenes.

To end this section, it has also to be pointed out that very simple and small molecules like α,α-diarylprolinols have also been found to catalyze the Michael reaction of malonates with nitrostyrenes, although only moderate enantioselectivities could be achieved under the optimized conditions (Scheme 4.26).[56] In this case, it is proposed that the catalyst activates the nitroalkane by establishing a single H-bonding interaction *via* its OH group and the remaining secondary amine functionality would engage in activation of the nucleophile *via* deprotonation. This proposal is based on previous related works by the author in other conjugate addition reactions.[57]

4.2.2   Enones as Michael Acceptors

The use of enones as Michael acceptors in conjugate additions under H-bonding activation represents a more challenging situation compared to the use of nitroalkenes. On one hand, the lower Lewis basicity of the carbonyl group in enones compared to that of the nitro group leads to a weaker interaction with the Brønsted acid catalyst, which makes the activation of the electrophile more difficult. This might result in either a low conversion or a situation in which the competitive uncatalyzed non-stereoselective background reaction prevails. On the other hand, enones can only participate in H-bonding interactions with a catalyst using a single oxygen atom, in contrast with the two oxygen atoms involved in the activation of nitroalkenes, a key event operating in most of the examples presented previously. This turns into a more difficult control of the spatial arrangement of the reagents during the conjugate addition step, as required if a highly stereoselective reaction is desired.

Despite this, bifunctional thiourea-tertiary amine catalysts have also emerged as useful and very convenient compounds for the activation of enones in Michael reactions. The mentioned problems associated to the single position available for H-bonding interactions and to the lower Brønsted basicity of the carbonyl moiety are circumvented by the formation of a double H-bonding network in which both lone pairs at the oxygen atom participate with two

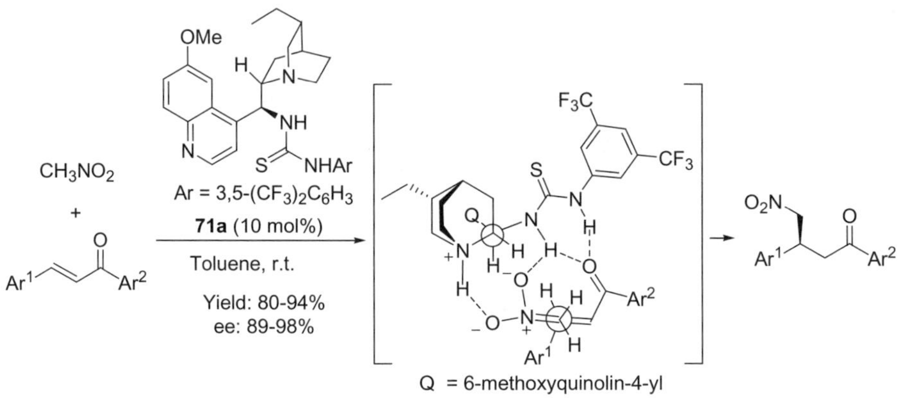

Scheme 4.27 71a-catalyzed Michael addition of nitromethane to chalcones.

H-donor sites of the catalyst. This leads not only to a very effective activation of the enone electrophile, but also to a well-defined transition state which leads to a highly stereoselective reaction. The first example reported in this context was the conjugate addition of nitromethane to chalcones developed by Soós using quinine-based thiourea catalyst **71a** (Scheme 4.27).[58] Under the optimized conditions, excellent yields and enantioselectivities were obtained for a variety of differently substituted chalcones. Interestingly, the authors demonstrated by computational methods that the most plausible mechanism participating in the reaction involved a two-step sequence with the participation of the catalyst in a previous stage by promoting the deprotonation of nitromethane by the tertiary amine moiety,[23a] as also happened in the **71a**-catalyzed Michael addition of malonates to nitroalkenes (see Scheme 4.8). In the conjugate addition step, the so-formed thiourea-ammonium salt would participate by activating the nucleophile *via* H-bonding interactions with the thiourea moiety and the chalcone acceptor would also engage in double H-bonding interaction with one of the thiourea H-donors and with the tertiary ammonium salt moiety.

This approach has also been applied to the use of malonates as C–H acidic carbon pro-nucleophiles, observing that the reaction performed excellently with a wide range of different chalcones and β-aryl substituted 2-butenones (Scheme 4.28).[59] Moreover, other 1,3-dicarbonyl compounds and related derivatives such as malononitriles, β-ketoesters, 2,4-pentanedione and ethyl nitroacetate have also been tested with success in the reaction using chalcone as Michael acceptor. In an independent work, the use of cyanoacetates was also surveyed with good results, although in this case mixtures of diastereoisomers were typically obtained.[60] In all these cases, the stereochemical outcome of the reaction was consistent with the model proposed by Soós.

Takemoto's catalyst has been employed in the Michael addition of α-substituted cyanoacetates to aryl vinyl ketones (Scheme 4.29).[61] The substitution

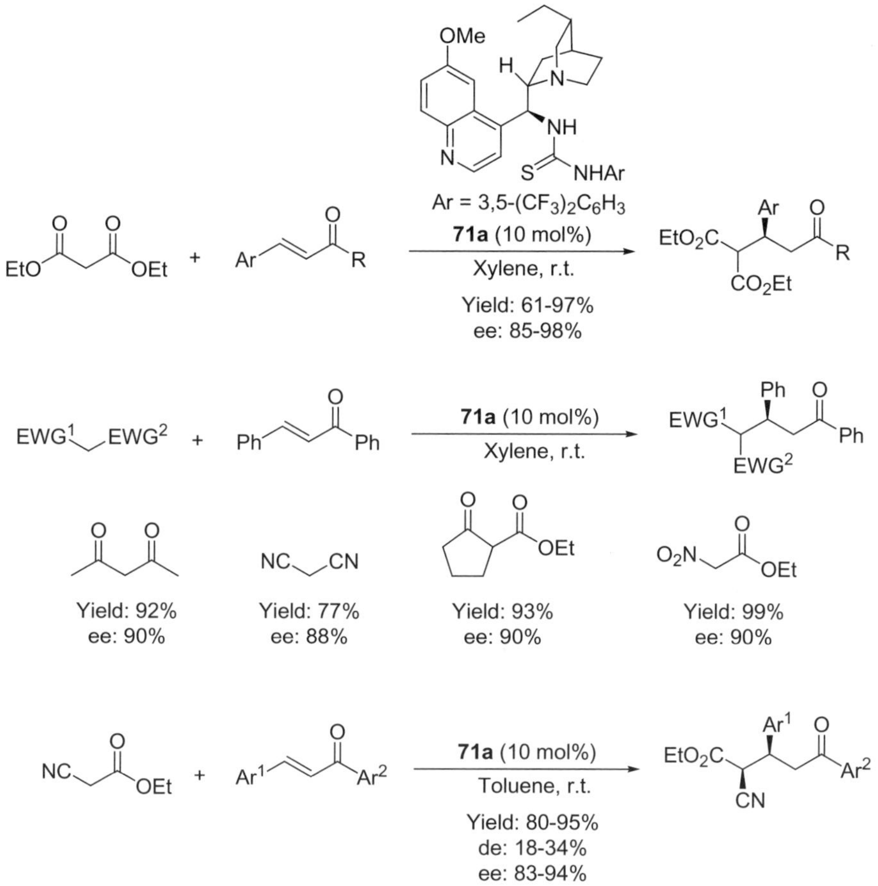

Scheme 4.28 **71a**-catalyzed Michael addition of nitromethane to chalcones.

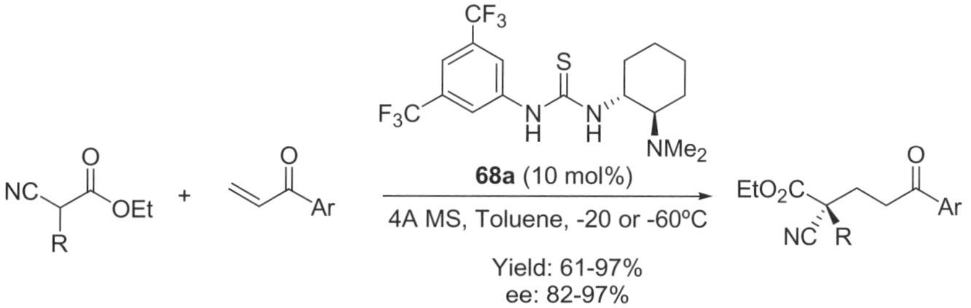

Scheme 4.29 The **68a**-catalyzed Michael addition of α-substituted cyanoacetates to aryl vinyl ketones.

pattern at the pro-nucleophile led to the formation of a quaternary stereocenter at the γ-position with respect to the ketone carbonyl, obtaining the final products in good yields and enantioselectivities and also showing a very broad substrate tolerance with regard to substitution at the Michael acceptor and at the α-substituent of the cyanoacetate reagent. The highly functionalized nature of the obtained adducts also allowed the preparation of a wide range of different interesting chiral building blocks by carrying out selective transformations on the different functional groups present at their structure.

Alternatively, simpler cinchona alkaloids like **84d** have also been employed as outstanding catalysts in the Michael addition of β-ketoesters to vinyl ketones (Scheme 4.30).[62] As happened for the same reaction using nitroalkenes as Michael acceptors (see Scheme 4.23), the basic quinuclidine nucleus is proposed to activate the pro-nucleophile by deprotonation and the free phenolic OH group would activate the enone by H-bonding interaction, also contributing to the formation of a rigid intermediate. The reaction also showed to be fairly general regarding the use of different cyclic and acyclic β-ketoesters of variable structures and also with regard to substitution at the vinyl ketone Michael acceptor. Excellent yields and enantioselectivities were obtained in all cases for the final conjugate addition products. In addition, cyclic enones were also successfully tested in the reaction leading to compounds containing two contiguous stereogenic centers as single diastereoisomers.

Interestingly, the addition of malononitriles to chalcones has been successfully carried out using natural quinine **84a** as catalyst (Scheme 4.31).[63] This contrasts with the poor performance of the same catalyst in the Michael addition of malonates to nitroalkenes (see Scheme 4.24), which was explained in terms of the difficulty in forming the intermediate in which both reagents, the

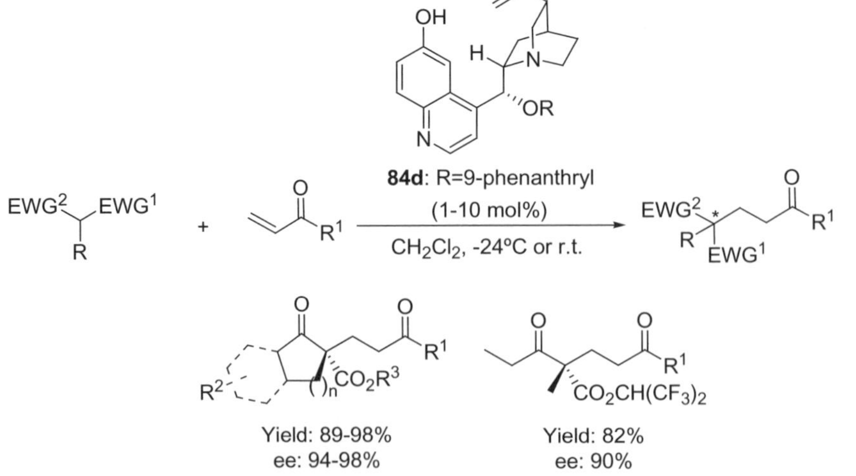

Scheme 4.30 Enantioselective Michael addition of β-ketoesters to vinyl ketones catalyzed by **83d**.

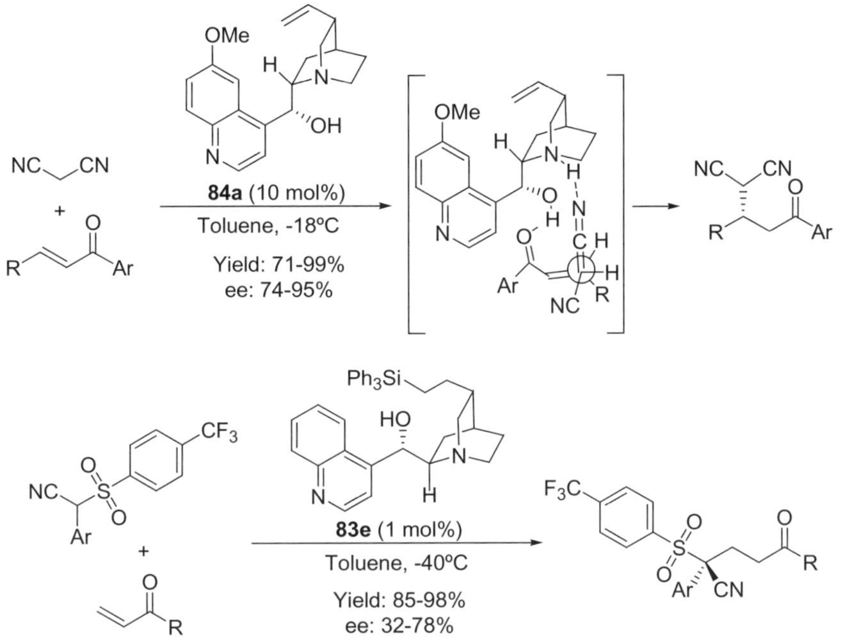

Scheme 4.31 Enantioselective Michael addition of malononitrile and cyanosulfones to enones.

Michael donor and the nitroalkane, were placed close to each other by the formation of an H-bonded network. In this case, the less sterically demanding linear structure of the malononitrile-derived enolate would probably provide an explanation for this behavior, facilitating the formation of the ternary complex involved in the highly stereoselective pathway. Some preliminary calculations also indicate that a favorable π-stacking interaction between the quinoline residue of the catalyst and the aryl group of the chalcone would also assist the formation of this intermediate and would also explain the drop on the enantioselectivity observed when alkenyl methyl ketones were used as substrates. In a similar way, the Michael addition of α-substituted cyanosulfones to vinyl ketones has also been reported to occur efficiently using modified cinchonine-derived compound **83e** as catalyst.[64] This methodology allowed the preparation of compounds containing a quaternary stereogenic center which also contain different functionalities suitable to be selectively modified and also tolerated well the use of cyclic enones as Michael donors, leading in this case to the simultaneous formation of two stereocenters. The structure of this catalyst also suggests a similar type of ternary complex involved in the reaction that accounts for the high stereoselectivities observed.

Quinine **84a** has also been found to be an excellent catalyst for the conjugate addition of β-ketoesters to quinones. In this case, the reaction proceeded in the usual way, but the obtained conjugate addition products, in which the quinone

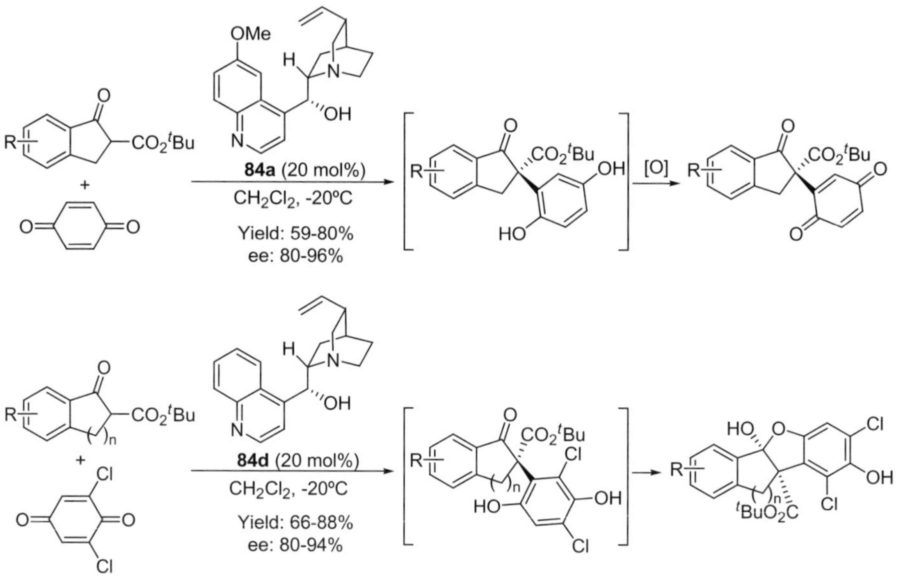

Scheme 4.32   Enantioselective Michael addition of β-ketoesters to quinones.

moiety resulted in being incorporated as the corresponding more stable 1,4-diphenol form, underwent *in situ* oxidation by the presence of either atmospheric oxygen or excess quinone reagent, delivering the final quinone-type adducts (see Scheme 4.32).[65] This oxidation process was not observed when a quinone incorporating electron-withdrawing substituent such as 2,6-dichloroquinone was employed, in which the aromatic diphenol moiety underwent intramolecular 1,2-addition to the β-ketoester residue, resulting in the stabilization of the Michael adduct as the related hemiacetal form. For this case, it was found that cinchonidine **84d** was a more efficient catalyst than quinine.

Guanidines have also been employed as catalysts in Michael reactions with enones and, for example, bicyclic guanidine **86** has been found to perform well in the Michael reaction of different 1,3-dicarbonyl compounds with cyclopentenones, providing the final Michael adducts with excellent yields and enantioselectivities (Scheme 4.33).[66] In this case, the authors found that the use of triethylamine as solvent resulted in a dramatic increase in the rate of the reaction, without affecting significantly its stereochemical outcome with respect to the same reaction in toluene, which is the usually chosen solvent when working with this type of catalyst.

The application of this catalyst was extended to the use of Michael acceptors containing two different electron-withdrawing groups at the two ends of the olefin moiety, such as 4-oxoimides (Scheme 4.34).[67] This results in a very regioselective reaction, due to the better efficiency of the ketone moiety to activate the Michael acceptor compared with the imide group. This effect could

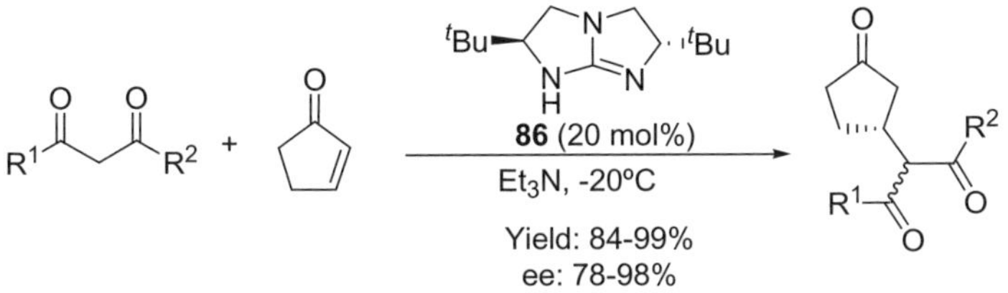

Scheme 4.33 Chiral guanidine-catalyzed enantioselective Michael addition of 1,3-dicarbonyl compounds to enones.

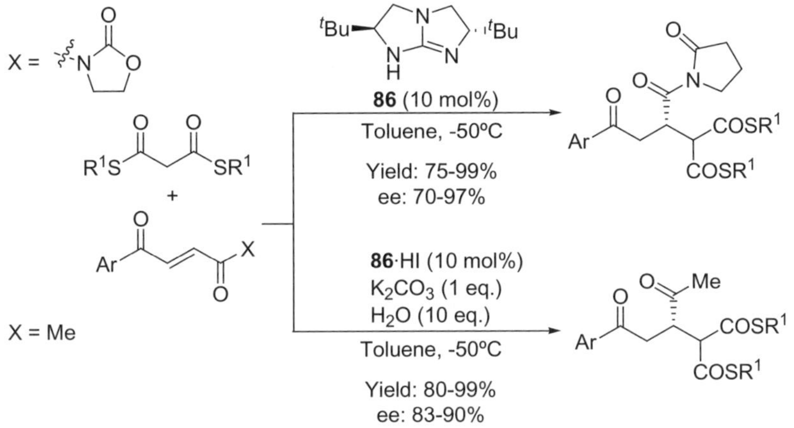

Scheme 4.34 **86**-catalyzed enantioselective Michael addition of 1,3-dicarbonyl compounds to enones.

also be extended to 1,4-diketone substrates incorporating an alkyl and an aryl ketone activating group and the better ability as electron-withdrawing group of the latter being the dominating effect with regard to regioselectivity.

There is also one example in which a chiral phosphoric acid has been employed as catalyst in the reaction. In particular, the addition of several cyclic β-ketoesters to methyl vinyl ketone was found to occur smoothly in the presence of several chiral phosphoric acids (Scheme 4.35).[68] As mentioned earlier, a key feature of the chiral phosphoric acid catalyst is the backbone binaphthyl axial chirality together with the incorporation of bulky substituents at the 2′ positions. In this case, **60b** was identified as an appropriate promoter of the reaction leading to the corresponding Michael adducts in excellent yields, although with moderate enantioselectivity. In addition, the authors succeeded in applying this reaction to a procedure to carry out a subsequent Robinson-type annulation.

Finally, it should be pointed out that this methodology has been extended to the use of other nucleophiles that are not based exclusively on stabilized carbanions like enolates or related species. In particular, alkenylboronic acids have

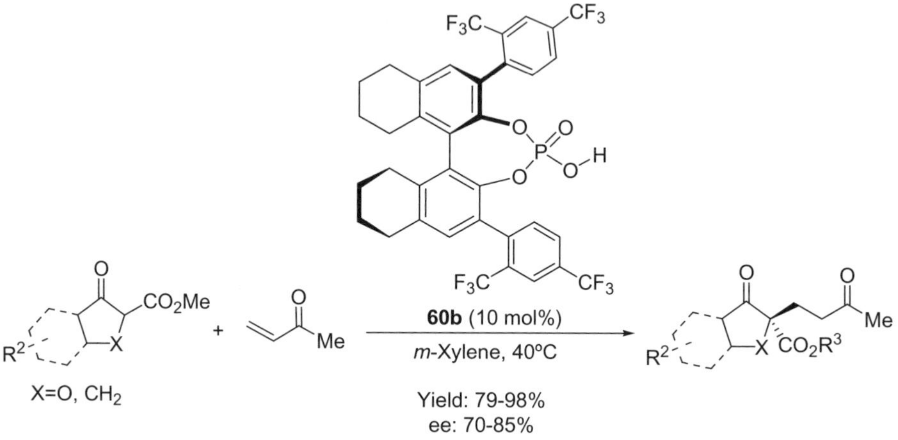

Scheme 4.35 Enantioselective Michael addition of β-ketoesters to vinyl ketones catalyzed by chiral phosphoric acid **60b**.

been found to be useful nucleophiles in the Michael reactions to enones under H-bonding activation, using chiral thiourea **87** as catalyst. The reaction required the use of a γ-hydroxylated α,β-unsaturated enone as electrophile, which is proposed to engage in double H-bonding with the boronic acid nucleophile, and the catalyst, directing the reaction in an intramolecular way, with the chiral backbone of the thiourea providing the required steric bias (Scheme 4.36).[69] There is also a previous report in which chiral BINOL **88** was found to catalyze the reaction of a wide variety of alkenylboronates to enones with excellent results,[70] but in this case the authors proposed that the activation of the nucleophile should take place by simple exchange of ligands at boron, leading to a chiral boronate intermediate but with no H-bonding interaction involved during the conjugate addition step. Scheme 4.36

4.2.3 α,β-Unsaturated Aldehydes as Michael Acceptors

The use of α,β-unsaturated aldehydes as electrophiles in this reaction results in an even more challenging situation compared to the same reaction involving enones as Michael acceptor, mainly due to the usual preference of enals to engage in the competitive 1,2-addition rather than undergoing the desired conjugate addition reaction. In spite of this, several reports have appeared showing that these compounds can be employed as very convenient Michael acceptors in reactions with C–H acidic pro-nucleophiles under H-bonding activation. For example, Deng has extended the conditions found for the conjugate addition of β-ketoesters to vinyl ketones depicted in Scheme 4.30, to the use of α,β-unsaturated aldehydes as Michael acceptors, also observing that a similar quinidine-based compound such as **83d** was the most effective catalyst

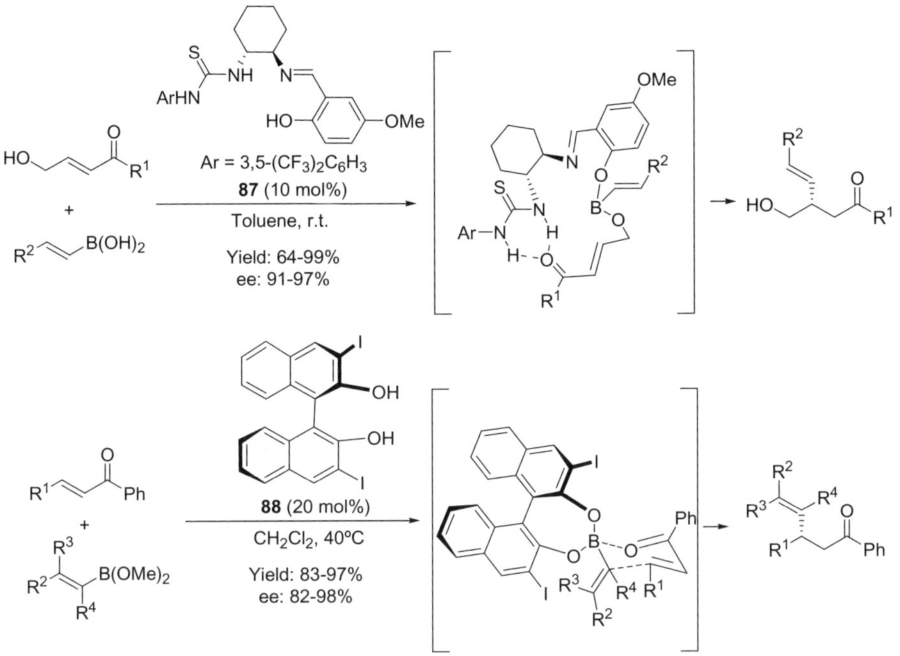

Scheme 4.36 Enantioselective Michael addition of alkenylboronic acids and alkenylboronates to enones.

for the reaction (Scheme 4.37).[71] Importantly, the use of a more challenging α,β-unsaturated aldehyde like acrolein as electrophile was also surveyed with great success in the reaction, obtaining excellent results regarding both chemical efficiency and enantiocontrol. Moreover, applicability of this methodology in the preparation of valuable compounds was also demonstrated with a short and efficient total synthesis of natural product (+)-*tanikolide*, using this reaction as a key step in the synthesis.

In the same paper, the Michael addition of α-aryl substituted α-cyanoesters to acrolein was also studied, also obtaining excellent yields and enantioselectivities for a variety of different Michael donors (Scheme 4.38). Remarkably, the structure of the catalyst needed additional optimization for achieving the best results, which were obtained by modifying the substituent at the secondary OH group of the central cinchonine framework from 9-phenanthryl to the more sterically demanding 6-chloro-2,5-diphenylpyrimidin-4-yl group. This new catalyst **83f** showed a very good performance in this reaction.

A vinylogous Michael reaction has also been reported using enolizable doubly activated alkylidenes as Michael donors and acrolein as Michael acceptor (Scheme 4.39).[72] In this case, the reaction furnished regioselectively the corresponding α-addition products, the unsaturation remaining in the final adduct at the β-position. After optimizing the reaction conditions, modified

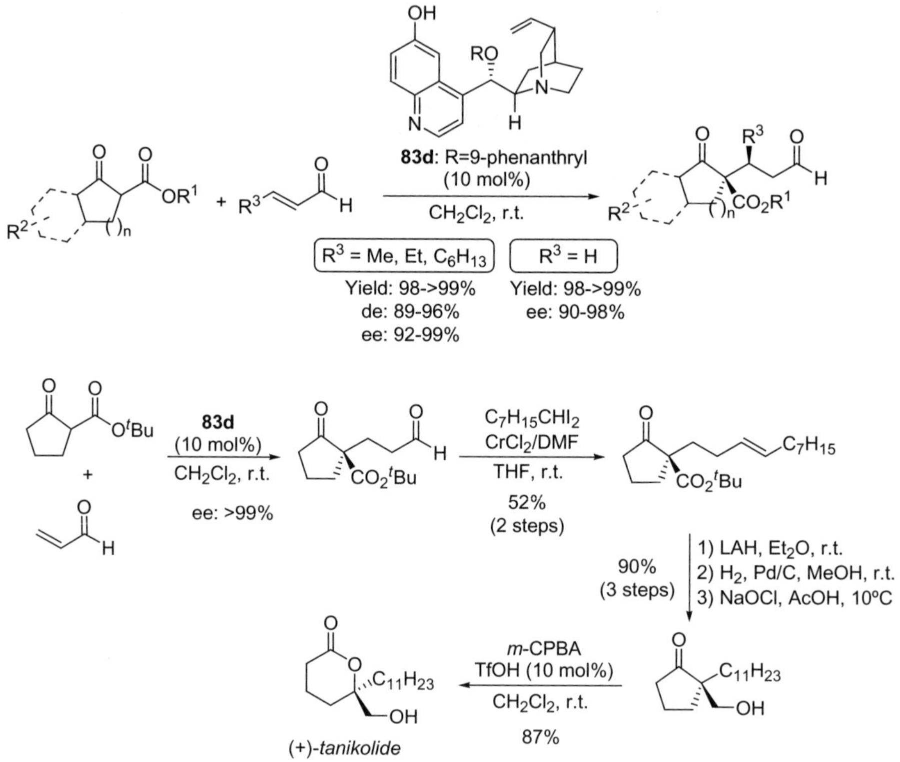

Scheme 4.37 Enantioselective Michael addition of α-alkyl β-ketoesters to α,β-unsaturated aldehydes and application to the total synthesis of (+)-*tanikolide*.

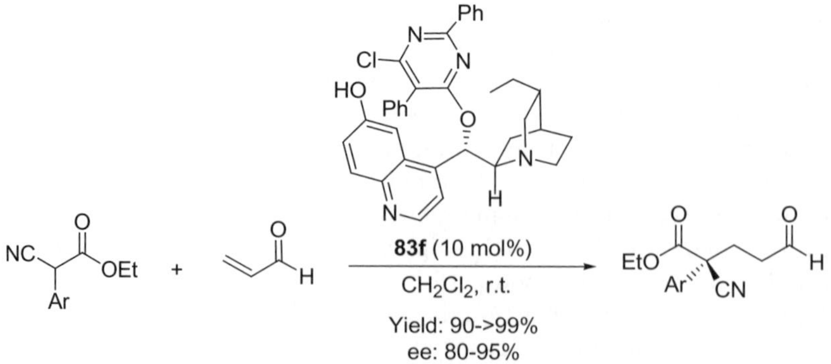

Scheme 4.38 Enantioselective Michael addition of α-alkyl α-cyanoesters to acrolein.

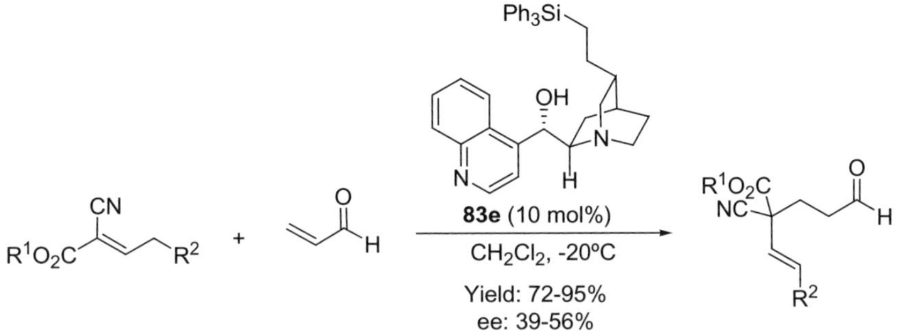

Scheme 4.39 Enantioselective vinylogous Michael addition of activated alkylidene reagents to acrolein.

cinchonine alkaloid **83e** was identified as the most efficient one furnishing the final compounds in good yields and moderate enantioselectivities.

4.2.4 Acrylic Acid Derivatives as Michael Acceptors

When acrylates or related compounds are employed as electrophiles in conjugate addition reactions, a conveniently substituted C–H pro-nucleophile has to be employed in order to generate a new stereogenic center. For this reason, most of the methodologies reported in this context involve the use of 2-substituted 1,3-dicarbonyl compounds or related derivatives as Michael donors.

Dixon has studied in depth the conjugate addition of β-ketoesters to many different acrylic acid derivatives, also using cinchona alkaloids of general structure **83** as bifunctional catalysts. In an initial attempt, simple ethyl acrylate and ethyl thioacrylate failed to react with a model pro-nucleophile (a β-ketoester) but the use of the corresponding naphthyl thioacrylate furnished the corresponding conjugate addition product in a fast and clean way using DABCO as catalyst. When focused on the stereoselective version, it was found that a wide variety of α-substituted β-ketoesters reacted in a very efficient way with aryl acrylates and aryl thioacrylates using **83d** as catalyst, furnishing the final compounds with excellent yields and enantioselectivities (Scheme 4.40).[73] Moreover, the reaction could also be extended to the use of *N*-acryloyl pyrrole with the same excellent results, which allowed the preparation of a variety of different compounds by carrying out chemoselective manipulations on the *N*-acylpyrrole moiety. It should also be pointed out that the quinidine-based catalyst used in this work furnished the final compounds with an opposite absolute configuration at the newly created stereogenic center to those obtained in the **84c**-catalyzed Michael addition of 1,3-dicarbonyl compounds to vinyl ketones already mentioned before (see Scheme 4.30), which is consistent with the pseudoenantiomeric nature of the catalysts **83d** and **84c**.

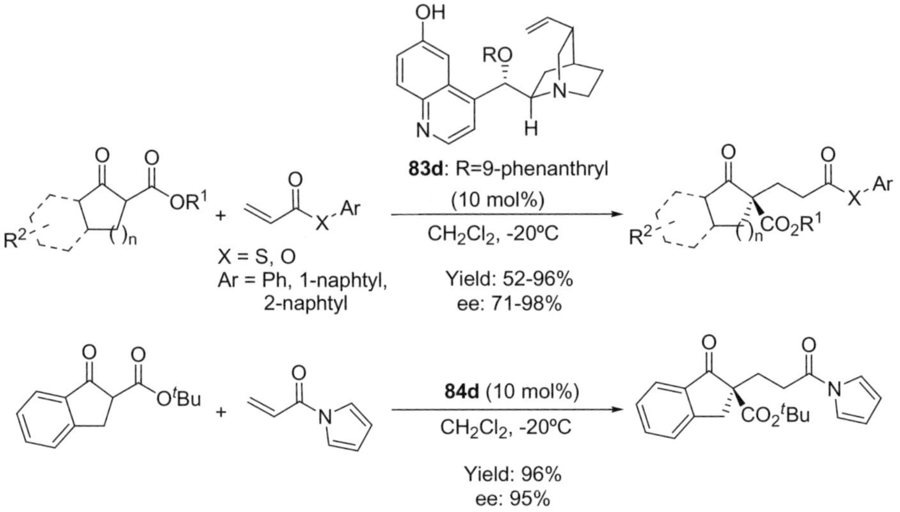

Scheme 4.40   Enantioselective Michael addition of α-alkyl β-ketoesters to acrylic acid derivatives.

An interesting report has described the use of α-chlorocyanoacrylate as Michael acceptor, leading to the formation of adducts containing two stereogenic centers in non-adjacent 1,3-positions. In this case, quinine-based catalyst **84c** was also found to perform well in the reaction, furnishing the final compounds in good yields and very high stereoselectivities (Scheme 4.41).[74] A model was also proposed to account for the high level of diastereoselection obtained, which was interpreted in terms of a catalyst-controlled intramolecular protonation of the intermediate formed after the conjugate addition step. Moreover, a very interesting situation was found when quinine-based thiourea catalyst **71b** was employed in the same reaction,[75] observing that in this case the reaction proceeded with reversal stereodiscrimination regarding the first conjugate addition step, while the subsequent protonation process furnished the final product with the same configuration at the tertiary α-stereogenic center as that furnished by catalyst **84c**. This behavior results in a very powerful reaction, allowing the preparation of any desired isomer of the final adducts by careful selection of the chiral catalyst employed.

Moreover, as is usually found in most of the asymmetric reactions catalyzed by cinchona alkaloid derivatives, the opposite enantiomer of each diastereomeric Michael adduct could also be obtained by simply changing the catalyst to the corresponding pseudoenantiomeric quinidine-based compounds of type **83** or **71**. Once again, a model was proposed to account for the observed results, involving a conformationally rigid intermediate in which both the pronucleophile and the electrophile were attached to the catalyst by the formation of multiple H-bonds, explaining the reversal of the diastereoselection by the epimeric nature of C-9 in catalyst **71b** with respect to **84c**. The stereochemical

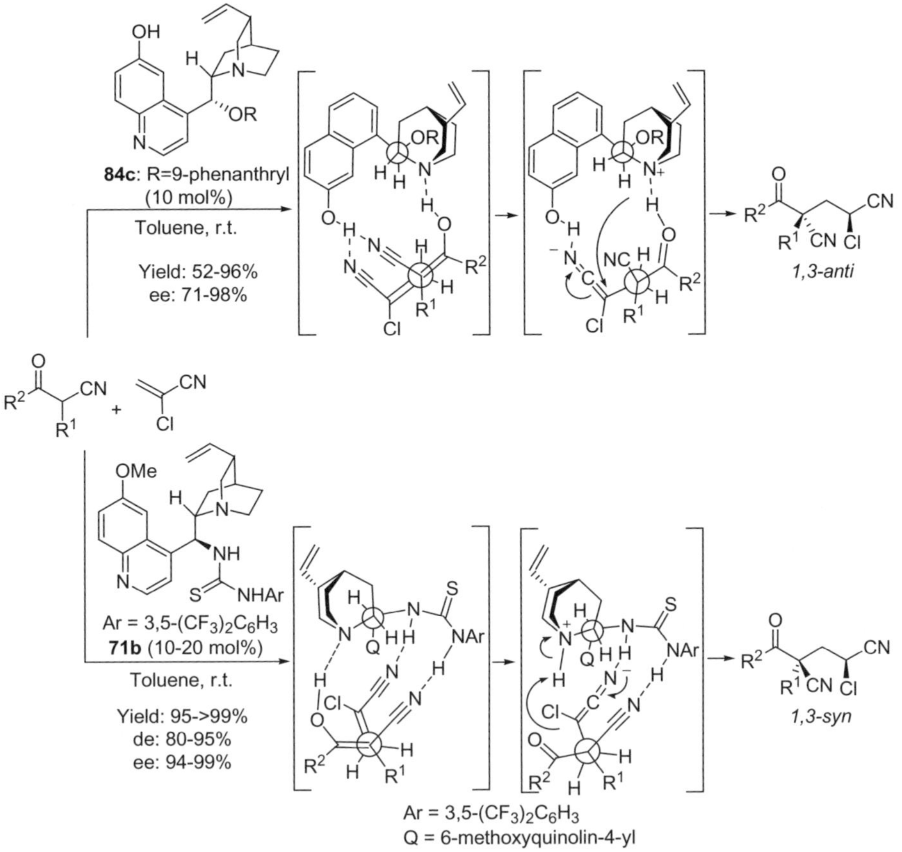

Scheme 4.41   Enantioselective Michael addition of α-alkyl α-cyanoketones to α-chlorocyanoacrylate catalyzed by **84c** or **71b**.

outcome of the protonation step was also explained in terms of catalyst-controlled intramolecular protonation of the intermediate generated after the conjugate addition process. Deng himself has also reported the **71b**-catalyzed reaction of α-alkyl-substituted α-cyanoesters and α-cyanoketones to acrylonitrile, leading to excellent yields and enantioselectivities.[69]

Finally, chiral guanidine **89a** has been used as a catalyst in the Michael reaction of a glycine imine with ethyl acrylate (Scheme 4.42).[76] In this case, the catalyst is proposed to be involved exclusively in the activation of the nucleophile by exerting its deprotonation, a crucial role being played by the presence of the primary OH-group at the catalyst structure interpreted in terms of H-bonding interaction with the ethoxycarbonyl moiety. However, the participation of the guanidine as a bifunctional catalyst, with the NH group involved in the activation of the acrylate Michael acceptor by the formation of a hydrogen bond can also be considered as a plausible explanation, in line with

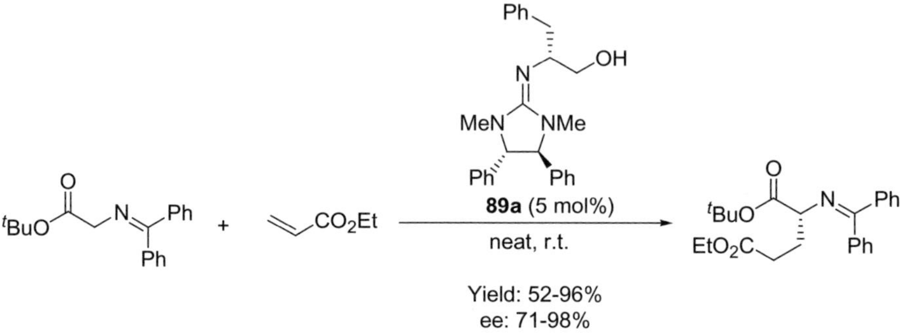

Scheme 4.42 **89a**-catalyzed enantioselective Michael addition of a glycine imine to *tert*-butyl acrylate.

the other proposals made in order to account for the stereochemical outcome in guanidine-mediated reactions.

4.2.5 Other Michael Acceptors

Many other activated olefins have been employed as acceptors in Michael reactions with stabilized carbon nucleophiles under H-bonding activation. A very illustrative example is that related to the use of conjugated imides as Michael acceptors in which the structural similarities between the imide and the nitro functionalities were exploited by Takemoto in the conjugate addition using a wide variety of C–H acidic pro-nucleophiles (Scheme 4.43).[77] In the first approach the conjugate addition of malononitrile to *N*-enoylpyrrolidin-2-one was found to occur in good yields and enantioselectivities using thiourea **68a** as catalyst. A stereochemical model was proposed involving activation of the electrophile by double H-bonding interaction between both carbonyl oxygen atoms of the imide moiety and the thiourea acidic hydrogen atoms. However, the reaction was found to be very dependent upon the concentration of reagents and, more importantly, rather long reaction times were required to reach to full conversion. A clever modification on the structure of the imide acceptor led to the identification of conjugated 2-methoxybenzimides as highly reactive electrophiles, allowing to carry out the reaction in much shorter reaction time and also with improved enantioselectivities. An explanation was provided to explain this enhanced electrophilicity which involved stabilization of one conformer of the acceptor by intramolecular H-bonding between the NH group of the imide and the methoxy substituent, decreasing the electron density at the nitrogen atom and also favoring a very convenient orientation of the substrate toward the formation of H-bonding interactions with the thiourea catalyst. The methodology proved to be very general regarding the use of differently substituted β-alkyl and β-aryl α,β-unsaturated imides and also related to the possibility of employing other C–H acidic pro-nucleophiles like

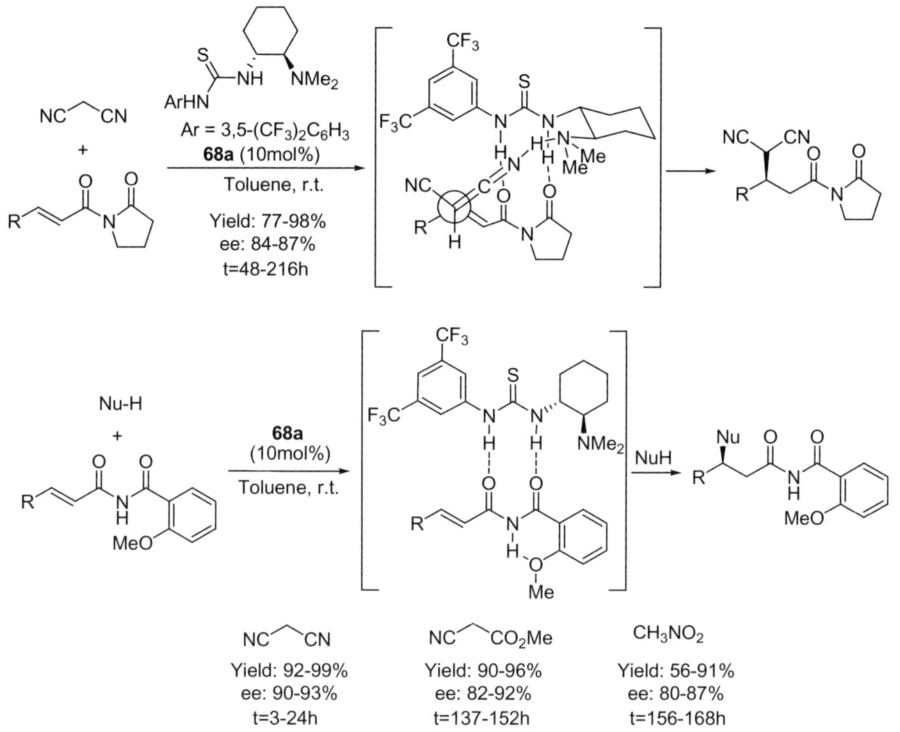

Scheme 4.43 Enantioselective Michael addition of malononitrile, methyl cyanoacetate and nitromethane to *N*-enoylimides.

methyl cyanoacetate or nitromethane although, in the latter case, enantio-selectivities were somewhat lower and longer reaction times were required.

The Michael addition of β-ketoesters and 1,3-diketones to maleimides has been reported to occur in a very efficient way using natural cinchona alkaloids such as quinine **84a** or the pseudoenantiomeric quinidine **83d** as catalysts (Scheme 4.44).[78] The reaction yielded a family of products containing two stereogenic centers, one of them a quaternary one, and, under the optimized conditions, it proceeded with excellent yields and diastereo- and enantioselec-tivities. Both enantiomers of the Michael adducts could be easily obtained using either quinine or quinidine. Computational studies[79] indicated that the bifunctional nature of the catalyst was crucial to achieve such high stereo-selectivities, showing that a ternary complex was formed as intermediate in which both reagents, the Michael donor and the Michael acceptor, were linked to the catalyst before the conjugate addition step took place. As is usually found with these kinds of catalysts, the quinuclidine moiety was proposed to engage in the activation of the nucleophile by exerting its deprotonation and the secondary OH group would proceed to activate the electrophile by an H-bonding interaction. In a different report, a vinylogous Michael reaction of

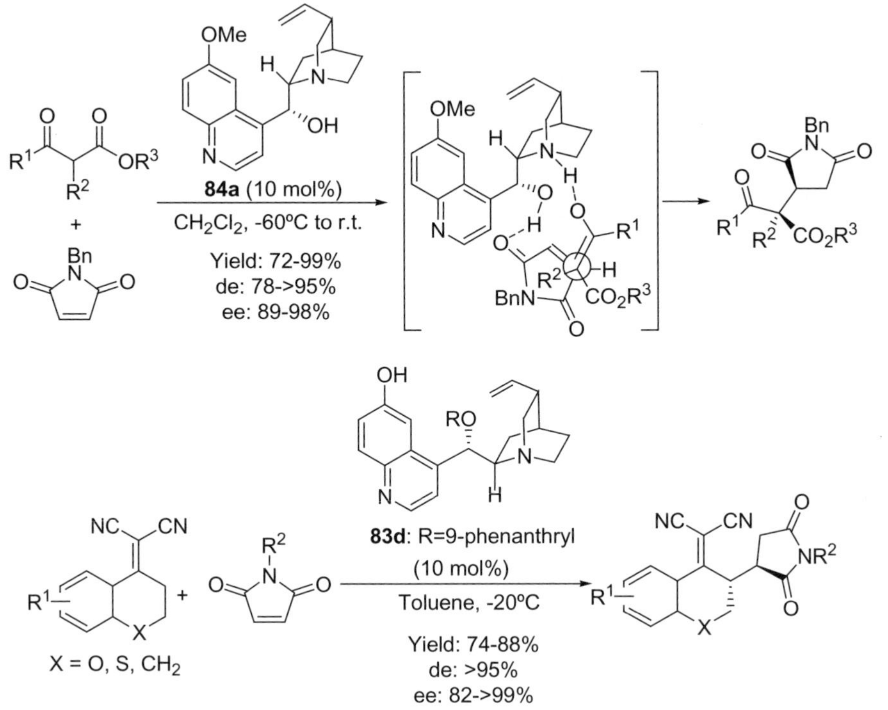

Scheme 4.44   Enantioselective Michael addition using maleimide as Michael acceptor.

α,α-dicyanoolefins to maleimides has been carried out using modified cinchona alkaloid **83d** as catalyst in which the activation of the Michael acceptor had to occur *via* the phenolic OH moiety.[80]

Alternatively, bicyclic guanidine **86** has been found to perform well in the Michael reaction of malonates, dithiomalonates, β-ketoesters and 1,3-diketones with maleimides, providing the final Michael adducts with excellent yields and enantioselectivities.[81] As an extension of this work, α-fluoro β-keto esters have also been successfully employed as Michael donors, leading to the formation of two contiguous stereogenic centers, one of them a fluorine-containing quaternary one (Scheme 4.45).[46] Under the optimized reaction conditions, a variety of 3-aryl-2-fluoro-3-ketoesters reacted with *N*-ethyl maleimide providing the conjugate addition products in excellent yield, as single distereoisomers and in enantioselectivities up to >99% in most cases. For this case, computational studies confirmed the participation of the guanidine as a bifunctional catalyst, by activating the nucleophile due to the basic character of the C=N moiety and interacting with the maleimide *via* H-bonding through the NH moiety.

Vinylsulfones have also been used as Michael acceptors suitable to be activated by H-bonding interactions with a cinchone-type catalyst. In particular,

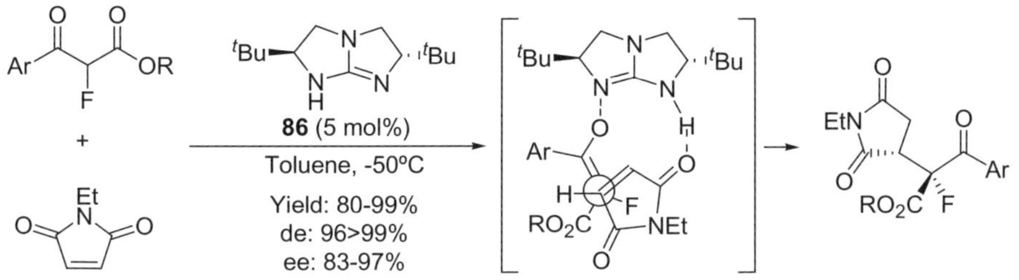

Scheme 4.45 The **86**-catalyzed Michael addition of α-fluoro-β-keto esters to maleimides.

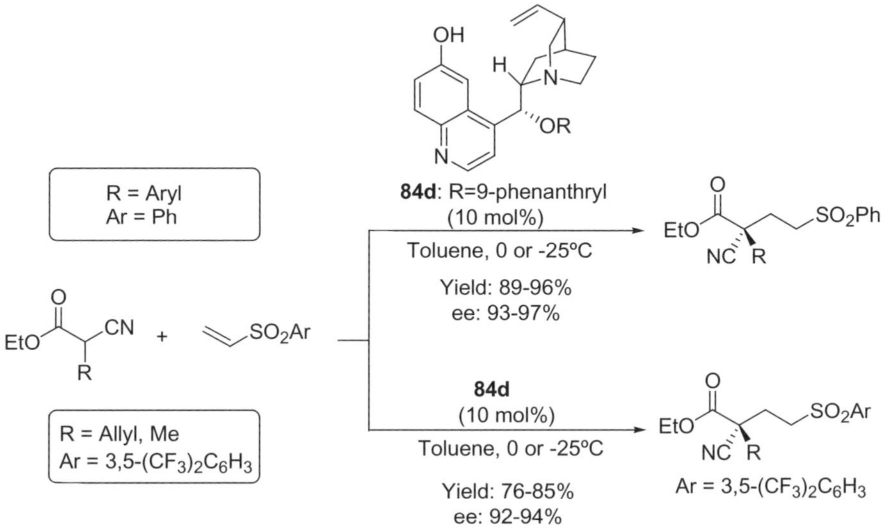

Scheme 4.46 Enantioselective Michael addition of α-alkyl α-cyanoketones to vinyl sulfones catalyzed by **84c**.

84d was found to be an outstanding catalyst for the reaction of α-aryl substituted cyanoacetates with vinyl sulfones, leading to the formation of highly functionalized chiral compounds containing a quaternary stereogenic center (Scheme 4.46).[82] Excellent yields and enantioselectivities were obtained for a wide range of different Michael donors tested. Interestingly, the use of β-alkyl substituted cyanoacetates required a more electron-withdrawing sulfone functionality in order to reach full conversion.

Thioureas have also been employed as catalysts in the same reaction and, in this context, the conjugate addition of α-substituted ethyl cyanoacetate to vinylsulfones, catalyzed by different thioureas, has been studied in detail, providing excellent results (Scheme 4.47).[83] In this paper, it was found that **90** was a very efficient catalyst for the reaction when the α-substituent of the cyanoacetates pro-nucleophile was an aromatic group but α-alkyl substituted

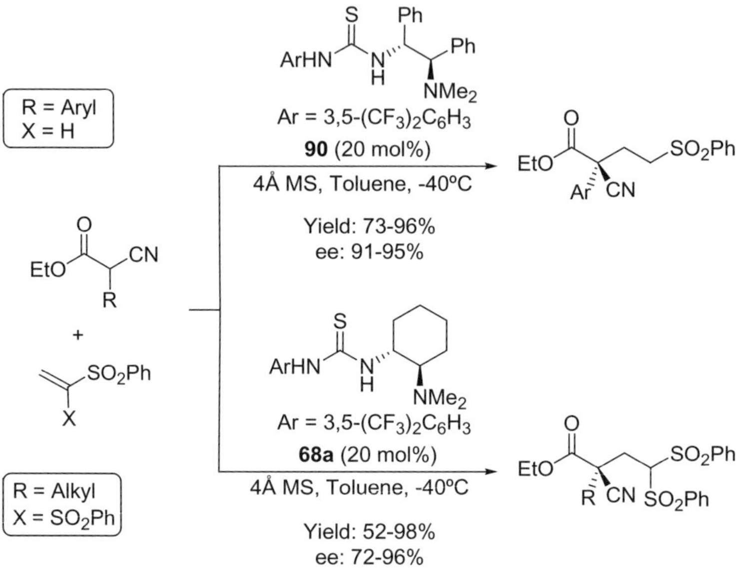

Scheme 4.47 Enantioselective Michael addition of α-substituted cyanoacetates to vinylsulfones.

derivatives failed to react under different conditions tested. This limitation was overcome with the use of a doubly activated bis(phenylsulfone) as Michael acceptor, which allowed to carry out the reaction with several α-alkyl substituted cyanoacetates due to its higher electrophilic character. In this case, the structure of the thiourea also had to be optimized again, observing that Takemoto's catalyst **68a** was the best one for the reaction, although enantioselectivities were found to be slightly lower than those obtained in the previous case.

Continuing with the use of thioureas as catalysts, the complementarity between the thiourea and the nitro functional groups has also been employed in order to achieve highly enantioselective conjugate additions using nitroalkanes as C–H acidic pro-nucleophiles (Scheme 4.48). The Michael reaction between nitromethane and N-acylpyrroles has been studied by Soós, reporting excellent results using quinine-based thiourea **71a**.[84] The reaction was found to be fairly general with regard to the use of different β-substituted Michael acceptors and the synthetic utility of the obtained adduct was also demonstrated by several subsequent transformations like transamidation or methanolysis. Alternatively, the reaction of nitroalkanes to 1,1-bis(phenylsulfonyl)ethene has also been reported using in this case quinidine-derived thiourea **72c**.[85] Good yields and moderate to high enantioselectivities were obtained for several nitroalkanes tested.

In the same context, an intramolecular Michael reaction involving a nitroalkane as Michael donor and a simple α,β-unsaturated ester as Michael acceptor has been reported using hydroquinidine-derived thiourea **72a** as

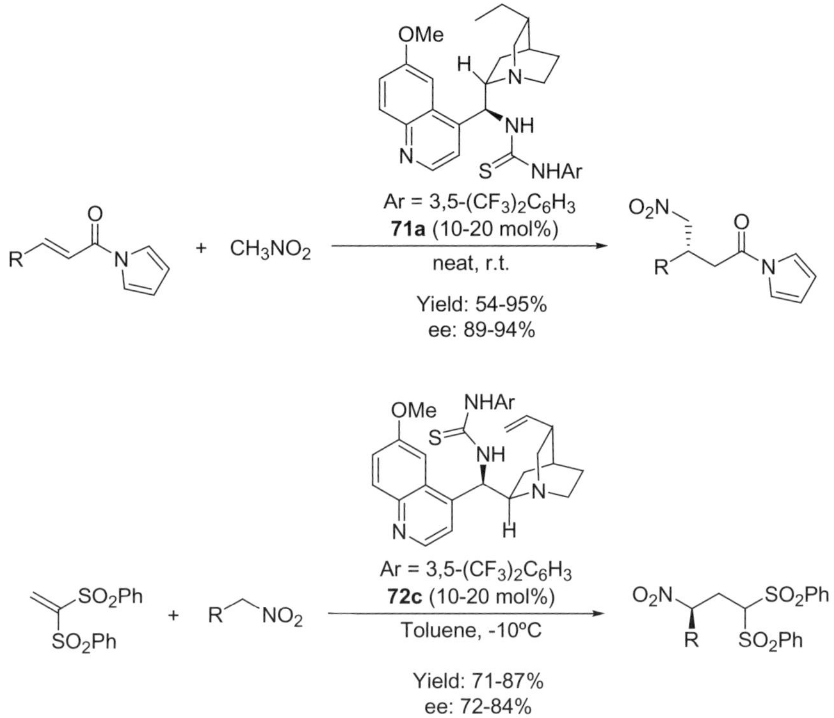

Scheme 4.48 Enantioselective Michael addition of nitroalkanes to *N*-enoylpyrroles and to 1,1-bis(phenylsulfonyl)ethene.

catalyst (Scheme 4.49).[86] The reaction proceeded with moderate yields and diastereoselectivities but with excellent enantioselectivities for a variety of different substrates tested, which also include the use of the sterically hindered and rather challenging α-substituted α,β-unsaturated ester-containing starting materials.

The Michael reaction of β-ketoesters to doubly activated vinyl bis-phosphonates has been carried out using several cinchona alkaloids as catalysts. In this context, many derivatives of quinine were found to perform well as catalysts in the reaction, the commercially available dihydroquinine **84e** being the most efficient one in terms of enantiocontrol (Scheme 4.50).[87] It has to be mentioned that the background uncatalyzed reaction was found to occur very quickly at room temperature because of the highly electrophilic nature of the Michael acceptor, which required the use of low temperatures in order to favor the **84e**-catalyzed pathway. Under the optimized conditions, a variety of different β-ketoesters were employed as Michael donors, yielding the final compounds in good yields and enantioselectivities in all cases in which a cyclic β-ketoester was employed, observing that the use of an acyclic acetoacetate derivative led to the final product in a significantly lower enantiomeric excess.

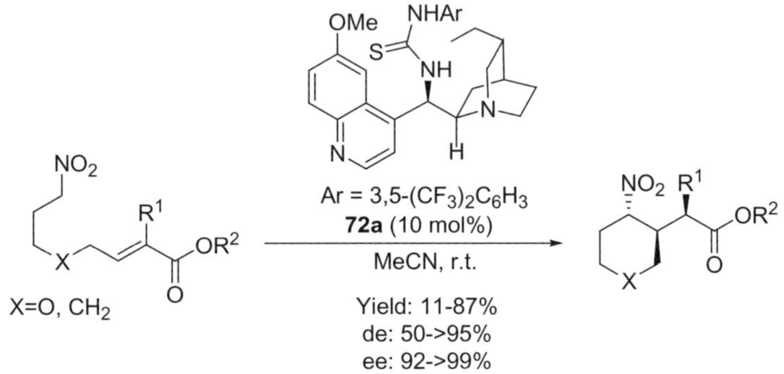

Scheme 4.49 Enantioselective intramolecular Michael addition of 8-nitro-1-octenoates.

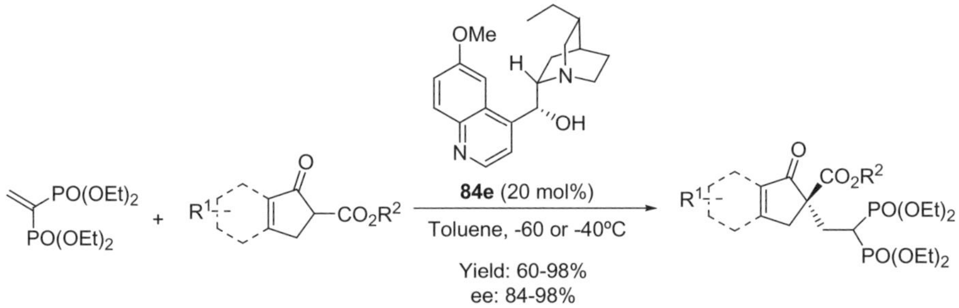

Scheme 4.50 The **84e**-catalyzed enantioselective Michael addition of β-ketoesters to vinyl bis-phosphonates.

Two very interesting examples of Michael reactions with carbon nucleophiles are related to the use of neutral formyl anion equivalents in order to access enantiomerically pure 1,4-dicarbonyl compounds (Scheme 4.51). Lassaletta and Fernandez have employed N,N-dialkylhydrazones as Michael donors based on the extensive experience of the group in the conjugate addition of these compounds to several types of conjugate acceptors under chiral Lewis acid catalysis.[88] An organocatalytic version has also been developed by the same group involving the use of thioureas as catalysts.[89] The key for the success of the reaction relied on the use of an β,γ-unsaturated-α-ketoester as Michael acceptor suitable to engage in H-bonding interactions with the chiral thiourea catalyst through the two adjacent carbonyl oxygen atoms. After screening among a variety of thioureas, bifunctional catalyst **91** was identified as the most efficient one for the reaction, leading to the corresponding addition products with good yields, although enantioselectivities remained around 70–80% ee. A model was proposed explaining the role played by all the H-donor groups present at the catalyst structure, which involved activation of the electrophile

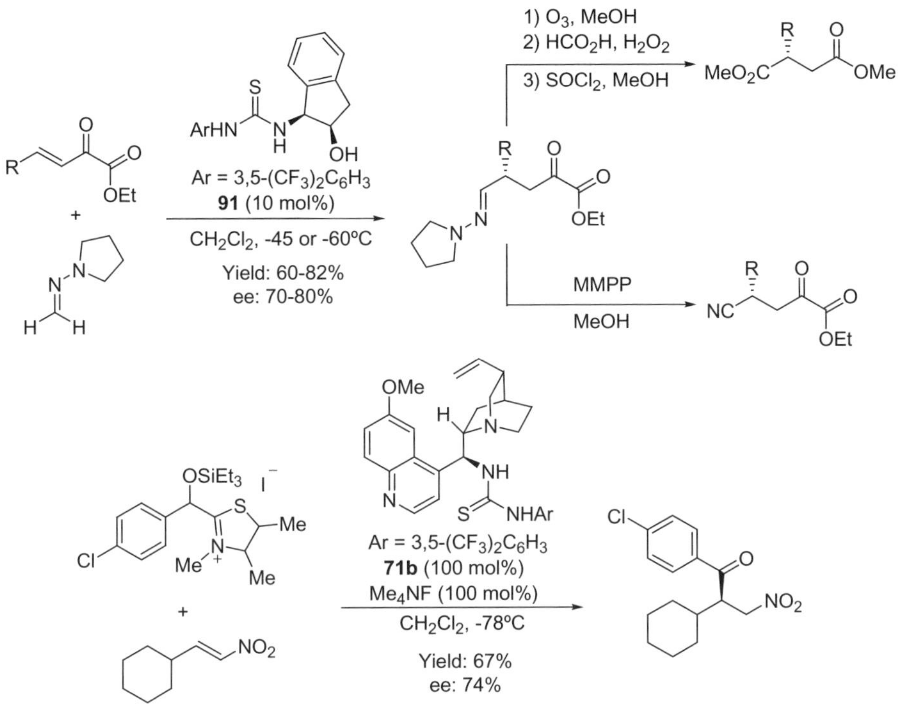

Scheme 4.51 Enantioselective Michael addition of formyl or acyl anion equivalents catalyzed by chiral thioureas.

by the thiourea moiety and a secondary H-bonding interaction between the free OH group at the catalyst and the hydrazone reagent. Importantly, the obtained Michael adducts could be easily converted into other valuable chiral building blocks by manipulation of the hydrazone moiety using simple and high-yielding procedures. On the other hand, Scheidt has employed a silylated thiazolium carbinol as a source of a nucleophilic acyl anion, which is generated in the presence of a fluoride anion and which has been tested in the reaction with a nitroalkene using cinchona-thiourea **71b** as promoter in stoichiometric amounts. Under unoptimized conditions, the reaction furnished the corresponding Michael adduct in 67% yield and with a promising 74% ee.[90]

4.3   Conjugate Friedel–Crafts Reactions

Several organocatalytic variants for the conjugate addition of electron-rich aromatic substrates to electron-deficient olefins have been reported in the last few years which apply the concept of activation of the electrophile by hydrogen-bonding interactions. In this context, chiral bifunctional thiourea ***ent*-91** has been employed as an excellent promoter for the enantioselective conjugate

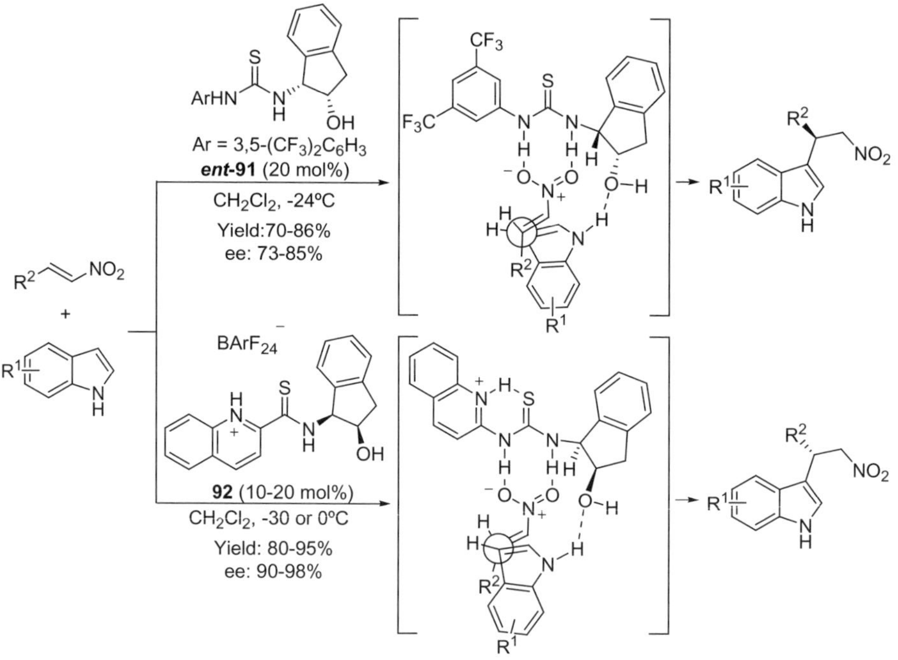

Scheme 4.52   Enantioselective conjugate Friedel–Crafts alkylation of indoles with nitroalkenes.

addition of *N*-unsubstituted indoles to nitroalkenes (Scheme 4.52).[91] The reaction was highly sensitive toward substitution at the aromatic substrate, showing that the installation of electron-withdrawing substituents at the arene was not well tolerated. On the other hand, both aromatic and aliphatic nitroalkenes were found to be suitable Michael acceptors under the optimized reaction conditions if the β-alkyl substituent was not a very bulky one. The fact that *N*-methylindole afforded very low levels of enantioselection pointed toward a dual activation mode exerted by the catalyst, first by interaction of the thiourea moiety with the nitro group of the electrophile and a second inter-action, also by hydrogen bonding, between the indole NH group and the free OH of the indanol moiety. Interestingly, modification of the thiourea moiety by a quinolinium thioamide led to a much more active and efficient catalyst for the same reaction (Scheme 4.52).[92] The better performance of this second catalyst **92** was explained in terms of a more acidic thioamide H-donor group by means of intramolecular H-bonding interaction between the quinolinium salt and the sulfur atom, which would also turn into a more rigid and better orientation of the acidic active thiourea site of the catalyst toward the formation of the H-bonding network involved in the activation of the Michael donor.

The use of *N*-alkyl indoles can therefore be considered as a more challenging situation in this reaction because these compounds are unable to interact with a secondary Brønsted basic site present at the structure of the bifunctional

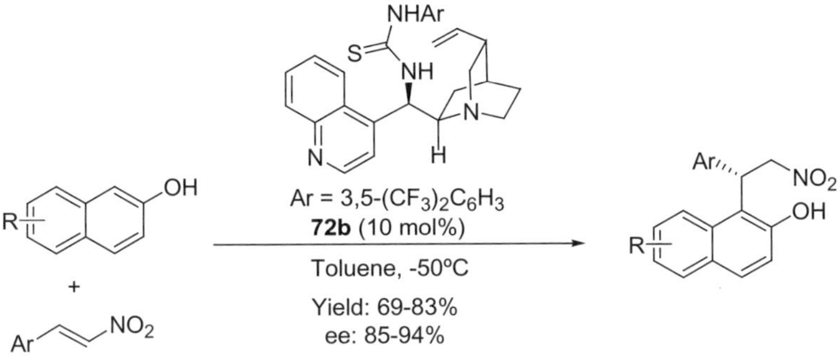

Scheme 4.53 Enantioselective conjugate Friedel–Crafts alkylation of 2-naftols.

catalyst. In fact, the reports existing to date involving the use of thioureas for the activation of the nitroalkane have failed to furnish the final adducts with high enantioselectivities when reacting with indoles which do not incorporate a free NH group.[93]

On the other hand, 2-naphthols have been used with different success as Michael donors in conjugate Friedel–Crafts reactions with nitroalkenes and related substrates (Scheme 4.53). For example, cinchonine-derived thiourea **72b** was identified as an excellent promoter for the reaction of a wide variety of 2-naphthols and nitroolefins, providing excellent yields and enantioselectivities.[94] Remarkably, the more challenging β-alkyl substituted nitroalkenes were also found to undergo the reaction in a highly stereoselective way and with comparable yields to those obtained when nitrostyrene derivatives were employed.

Chiral phosphoric acids have also been identified as effective catalysts for conjugate Friedel–Crafts reactions. For example, the reaction between indoles and nitroalkenes can be carried out in a very efficient way using modified TRYP analogue **60b** as catalyst (Scheme 4.54).[95] As was proposed in other cases, the phosphoric acid plays the role of a bifunctional catalyst in which a double interaction is established between the acidic OH of the catalyst and the nitroalkane on one hand and the acidic NH group of the indole and the P=O moiety on the other. The bulky substituents placed at the 2′-positions of the binaphthyl substructure exert the required stereodifferentiation by steric bias. The privileged structure of this kind of catalyst in terms of higher Brønsted acidity (which turns into a more active catalyst) and excellent stereodirecting ability is especially exemplified in this case, showing that the reaction proceeds with excellent results in all cases tested, which also includes the use of both β-aryl and β-alkyl substituted nitroalkenes as Michael acceptors. The inclusion of 3 Å molecular sieves in the reaction was absolutely necessary for catalytic activity, which was interpreted in terms of competitive binding of water to the catalyst and/or substrates which avoided a clean interaction among the reactive species and therefore leading to a much slower and more sluggish reaction. A very similar catalyst system was employed for the same reaction using 4,7-dihydroindoles as Michael donors (Scheme 4.54),[96] observing that, in this case,

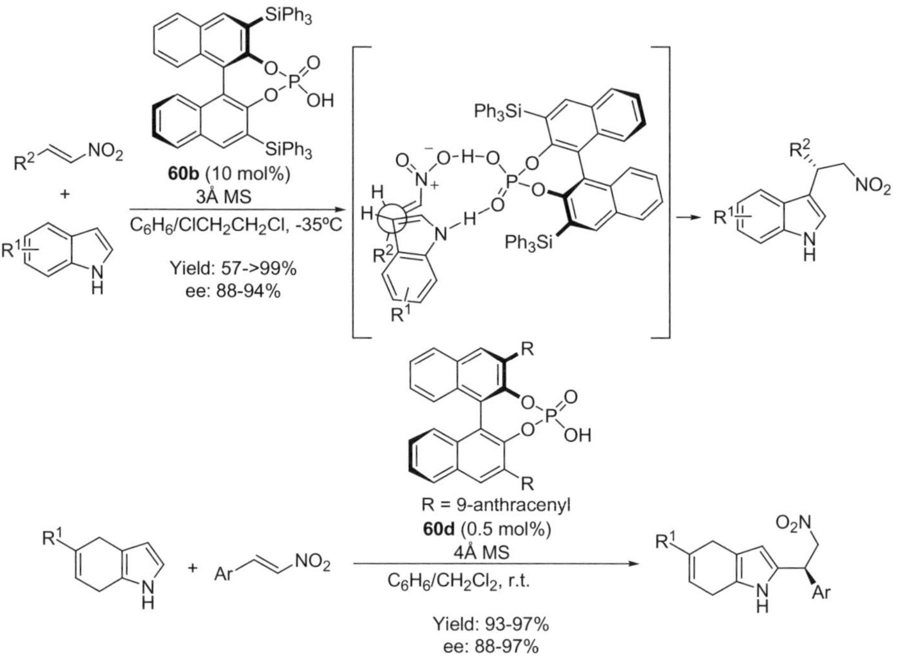

Scheme 4.54 Enantioselective conjugate Friedel–Crafts alkylation of indoles and dihydroindoles catalyzed by phosphoric acids.

the regioselectivity of the reaction changed, obtaining 2-alkylated dihydroindoles good yields and enantioselectivities for a variety of nitrostyrene derivatives tested.

For the reaction of pyrroles with nitroalkenes, phosphoric acid **60d** was found to be the catalyst furnishing the best results. Under the optimized conditions, pyrrole and several 2-substituted pyrroles were reacted with a variety of nitrostyrene derivatives, furnishing the corresponding conjugate addition products in good yields and enantioselectivities (Scheme 4.55).[97] The reaction failed when β-alkyl substituted nitroalkenes were employed. The rate and order of addition of the reagents was an important parameter to be controlled, requiring the slow addition of the nitroalkene to the solution containing the pyrrole and the catalyst by means of a syringe pump. Once again, the role played by the free NH group at the heteroaromatic reagent was also crucial for stereocontrol, suggesting the participation of the phosphoric acid as a bifunctional catalyst, with the OH group activating the nitroalkene and the P=O moiety acting as a Lewis base which activates the pyrrole by a secondary H-bonding interaction.

Nitroalkenes are not the only substrates employed as electrophiles in these conjugate Friedel–Crafts alkylations using chiral phosphoric acids as catalysts. In fact, the first reports in this field were focused on the reaction of indoles with β,γ-unsaturated α-ketoesters as Michael acceptors, which underwent clean

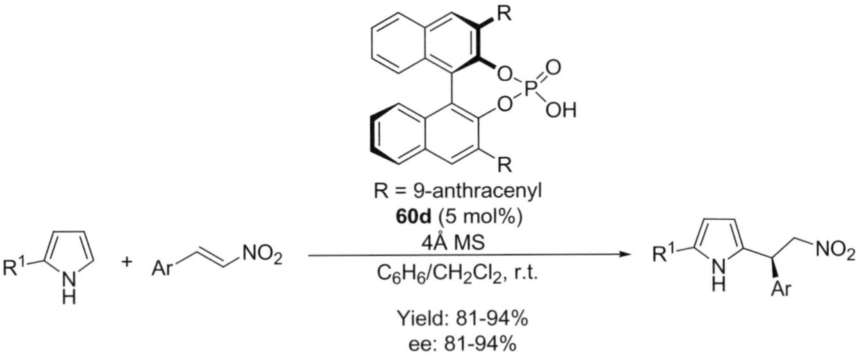

Scheme 4.55 Enantioselective conjugate Friedel–Crafts alkylation of pyrroles with nitrostyrenes catalyzed by **60d**.

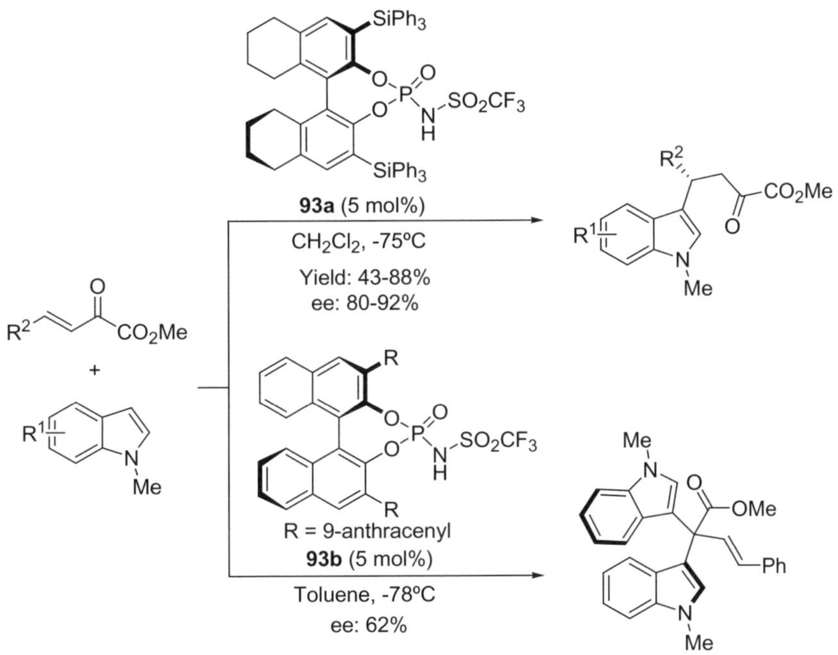

Scheme 4.56 Enantioselective conjugate Friedel–Crafts reaction of indoles with β,γ-unsaturated-α-ketoesters.

reaction with a variety of indoles using partially hydrogenated TRYP-derived phosphoramide **93a** as catalyst (Scheme 4.56).[98] Several important features have to be highlighted from this work. The first one is related to the requirement of a catalyst with a *N*-triflylphosphoramide structure of enhanced Brønsted acidity compared with the parent phosphoric acids, which were found

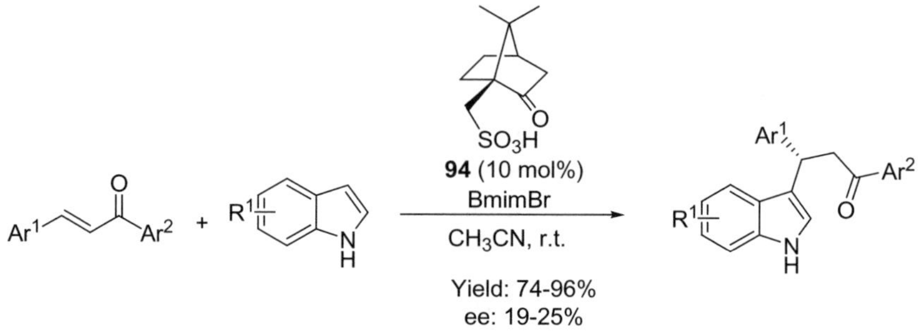

Scheme 4.57   Enantioselective conjugate Friedel–Crafts alkylation of indoles with chalcones.

to be completely inactive in this reaction. The second is related to the fact that N-methyl indoles were found to be very appropriate substrates in the reaction, which indicates that the catalyst is exclusively participating in the activation of the electrophile and therefore no secondary interaction with the Michael donor is thought to exist in the transition state. In addition, the backbone structure of the phosphoramide catalyst had to be carefully optimized in order to control the 1,2- vs. 1,4- regioselectivity of the reaction, requiring the use of catalyst **93a** in which the partially hydrogenated binaphthyl structure together with the spherically arranged phenyl groups at the 2′-silyl substituents increase the steric demand at the catalytic site, resulting in a better shielding of the carbonyl group of the ketoester and giving rise to the preferred 1,4-addition product. On the contrary, binaphthyl-containing catalysts **93b** afforded mainly a product with a bis-indole structure arising from the sequential 1,2-addition to the ketone group followed by substitution of the formed tertiary alcohol by an elimination/addition sequence. This bis-indole product presented axial chirality and was also found to be configurationally stable, being obtained in 62% ee under the best reaction conditions.

To end this section, it has to be mentioned that there is a single example of a conjugate Friedel–Crafts alkylation involving enones as Michael acceptors. In particular, a camphor-based sulfonic acid (**94**) has been used as catalyst in the reaction of indoles with chalcones (Scheme 4.57).[99] It has also to be noted that the best conditions involved the use of catalyst **94** together with an ionic liquid (1-butyl-3-methyl-1H-imidazolium bromide: BmimBr). However, although excellent yields were obtained for a set of different substrates tested, the enantioselectivities remained in rather low values.

4.4   Conjugate Hydrogen-transfer Reactions

Chiral phosphoric acids like TRYP (**60a**) or analogues, in association with primary amines, have already been employed as catalysts in the conjugate reduction of α,β-unsaturated aldehydes using Hantzsch esters as hydride-transfer reagents (see Scheme 3.27 in Chapter 3).[100] However, as pointed out

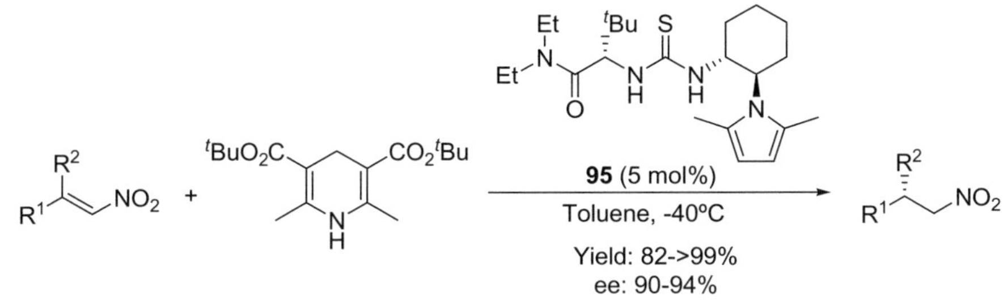

Scheme 4.58 Enantioselective conjugate reduction of β,β-disubstituted nitroalkenes.

there, the main mechanism operating in the activation of the electrophile involved the formation of an iminium ion after condensation of the enal with the primary amine catalyst, the role played by the Brønsted acid being limited to assist the formation of this iminium ion and therefore remaining as a chiral counteranion tightly associated with the activated iminium cation.

Alternatively, the use of thioureas as catalysts for the activation of nitroalkenes has been successfully applied in this context. In particular, List has developed a very effective procedure for the enantioselective conjugate reduction of β,β-disubstituted nitroalkenes using Jacobsen thiourea **95** as catalyst (Scheme 4.58).[101] As is usually found in all organocatalytic versions of this reaction, Hantzsch esters had to be employed as neutral hydride source, compatible with the conditions employed in the reaction and the reagents and intermediates participating in the catalytic cycle. Under the optimized reaction conditions, a variety of different β,β-disubstituted nitroalkenes were successfully reduced furnishing the corresponding adducts incorporating a β-stereogenic center in good yields and enantioselectivities. The reaction behaved very well when one of the β-substituents was either an aromatic or a heteroaromatic moiety or, alternatively, a very bulky alkyl group such as a *tert*-butyl substituent. On the other hand, nitroolefins with two very similar substituents at this position ($R^1 = Me$ and $R^2 = Et$) afforded a somewhat lower level of enantioselection. It has to be pointed out that chiral phosphoric acids such as TRYP (**60a**) were also tested in the reaction and indeed proved to be catalytically active but furnished the final product in rather low enantioselectivity. The methodology was also applied to the conjugate hydrogen transfer of β-nitroacrylates ($R^1 = CO_2Et$), which gave access to α-substituted β-amino acids after convenient reduction of the nitro group. The same thiourea catalyst **95** was also identified as the best promoter for the reaction, which was also found to be remarkably wide in scope, allowing the introduction of different types of substitutents at the nitroacrylate substrate. Importantly, the stereochemical outcome of the reaction was found to be strongly dependent on the double-bond geometry observing that *E* or *Z* nitroolefins furnished opposite enantiomers, which therefore required the use of highly diastereopure starting materials in order to achieve high enantioselectivities. This contrasts with the

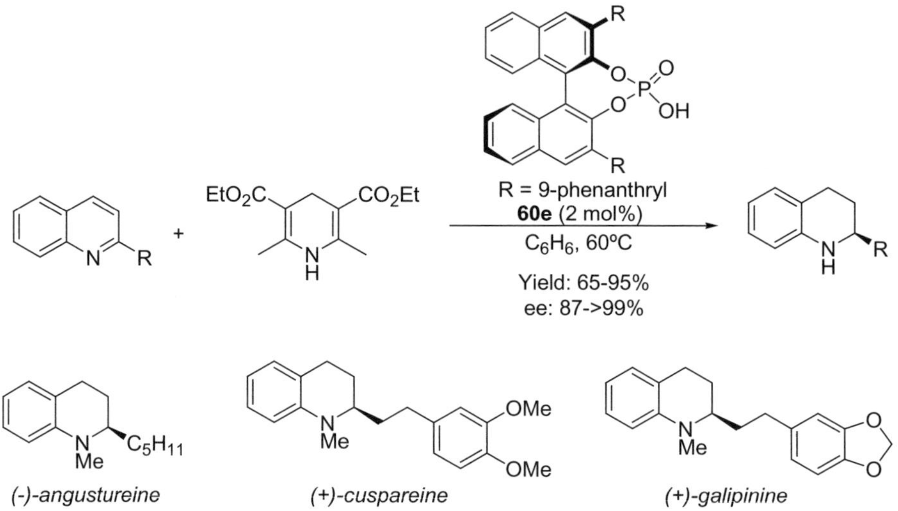

Scheme 4.59 Enantioselective conjugate reduction of 2-substituted quinolines.

stereoconvergency found on the related conjugate reduction of α,β-unsaturated aldehydes proceeding under iminium activation in the presence of chiral imidazolidinone catalysts shown in Scheme 3.26 in Chapter 3.

On the other hand, chiral phosphoric acids have found application as catalysts in the conjugate reduction of several nitrogen-containing heterocyclic compounds such as pyridines and quinolines. In a first approach, the reduction of 2-substituted quinolines was studied (Scheme 4.59),[102] proceeding in a cascade sequence by first 1,4-hydrogen transfer followed by subsequent reduction of the formed enamine intermediate *via* the corresponding activated iminium ion, this final step being the one in which the stereogenic center was generated. Catalyst **60e** was found to be the best one for this reaction among a series of different binaphthol-derived phosphoric acids, furnishing the final 2-substituted tetrahydroquinolines in excellent yields and enantioselectivities and also illustrating the applicability of the methodology with the preparation of several natural products such as (+)-*cuspareine*, (+)-*galipinine* and (−)-*angustureine*.

The transfer hydrogenation of 3-substituted quinolines was also studied, with the particularity that the stereogenic center was generated in this case during the conjugate addition step (Scheme 4.60).[103] The reaction needed for a more sterically demanding catalyst such as partially hydrogenated derivative **60f** and, in addition, the nature of the Hantzsch ester required further optimization in order to achieve the best enantioselectivities, which still remained in values around 80% ee. In this case, the scope of the reaction was limited to the use of quinolines with aromatic or heteroaromatic substituents at the 3-position.

Alternatively, the reduction of highly substituted pyridines was achieved with success under similar reaction conditions, allowing the preparation of a wide range

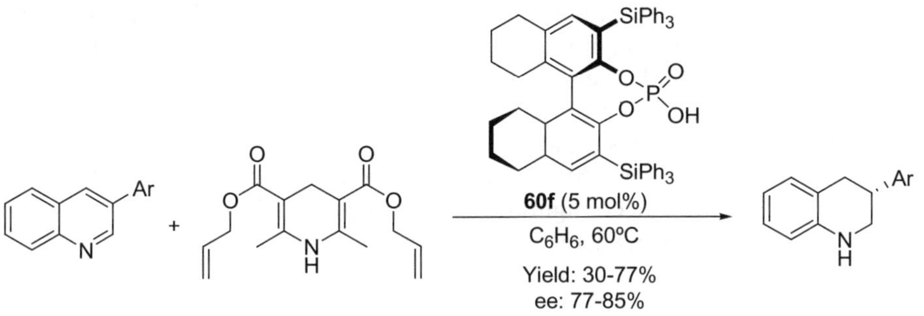

Scheme 4.60   Enantioselective conjugate reduction of 3-substituted quinolines.

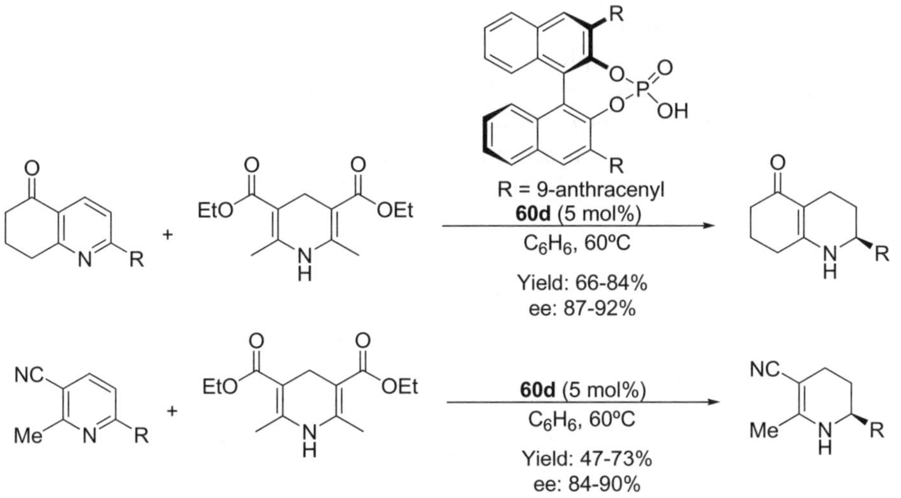

Scheme 4.61   Enantioselective conjugate reduction of 2-substituted quinolines.

of different piperidine compounds containing a stereogenic center at the 2-position (Scheme 4.61).[104] For this reaction, exclusively 2-alkyl substituted pyridines were tested as substrates, with no data reported regarding the possibility to employ starting materials incorporating aryl or heteroaryl substituents at this position.

4.5   Conjugate Addition of Heteronucleophiles

Heteronucleophiles can also undergo conjugate addition to electron-deficient olefins which are activated by a chiral catalyst *via* H-bonding interactions. However, as happened in the case of organocatalysts operating by other different mechanisms of activation (see for example Section 5 in Chapter 3), the intrinsic reversibility of the conjugate addition process leads to configurationally unstable final adducts and therefore to an additional problem to take care of when designing an efficient procedure. This is the main reason for the lack of examples related to this topic, especially compared with the number of methodologies reported for conjugate additions involving C–C bond formation, which proceeds usually *via* an irreversible reaction. The usual way to overcome this problem has been focused on the development of modified pro-nucleophiles leading to a configurationally stable final adduct, which very often results in an intramolecular reaction or a cascade process.

4.5.1   Oxa-Michael Reactions[105]

As pointed out in the previous chapter, the low nucleophilicity of alcohols makes this reaction a very challenging process and, in fact, only a few examples of oxa-Michael reactions proceeding *via* H-bonding activation have been reported to date in the literature, most of them being intramolecular reactions and involving the use of phenol as more acidic *O*-nucleophiles. In an early attempt, cinchonine **83g** was employed to promote the intramolecular oxa-Michael reaction of an *ortho*-substituted phenol containing a functionalized side chain incorporating an α,β-unsaturated ester moiety, conducing to the formation of the chromane skeleton (Scheme 4.62).[106] The reaction proceeded

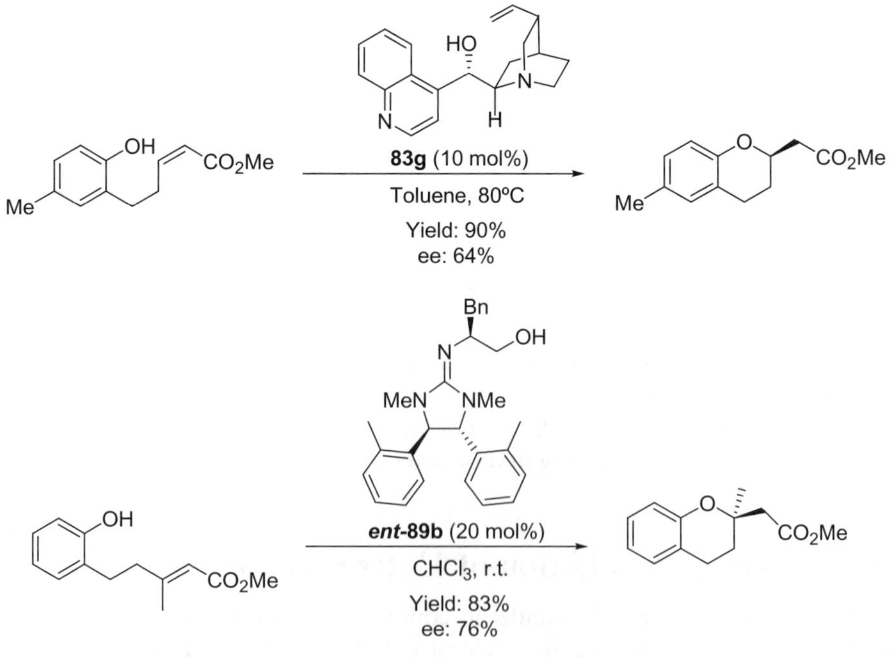

Scheme 4.62   Enantioselective intramolecular oxa-Michael reactions for the synthesis of chromanes.

with good yield and moderate enantioselectivity, the relatively high temperature needed for the reaction to proceed to completion probably being the reason for not achieving a complete degree of stereocontrol. Other modified cinchonine derivatives were tested but with no significant improvements. Alternatively, bicyclic guanidine **ent-89b** was also employed in a related version of this reaction, this time involving a disubstituted α,β-unsaturated ester as Michael acceptor, which led to the formation of the final cyclic compound incorporating a quaternary stereocenter in 83% yield and 76% ee.[107]

On the other hand, a novel cinchona-thiourea catalyst **96** was developed to promote the same reaction,[108] also introducing two electron-withdrawing substituents at the starting material in order to activate the Michael acceptor moiety (and therefore allowing the reaction to be complete using lower temperatures) and also providing an additional basic site able to interact with the bifunctional catalyst, which leads to the formation of a better organized transition state (Scheme 4.63). A subsequent acid-mediated decarboxylation step, which was carried out in a one-pot manner, furnished the corresponding flavanones in a single step with excellent yields and enantioselectivities.

There is also an interesting example of an enantioselective thiourea-catalyzed oxa-Michael reaction using enones as Michael acceptors in which phenylboronic acid was employed as hydroxyl anion equivalent (Scheme 4.64).[109] The authors demonstrated that amine bases were able to activate these kinds of reagents by complexation, thus becoming effective reagents for the transfer of the OH group to the Michael acceptor. The reaction had to proceed in an intramolecular way and, for this reason, γ-hydroxy-α,β-unsaturated ketones had to be employed as substrates. In the enantioselective version, **71b** was identified as a very efficient catalyst, providing a series of β,γ-dihydroxy ketones in excellent yields and enantioselectivities, after oxidative work-up. The process consists of the initial reaction of the boronic acid first with the γ-hydroxy

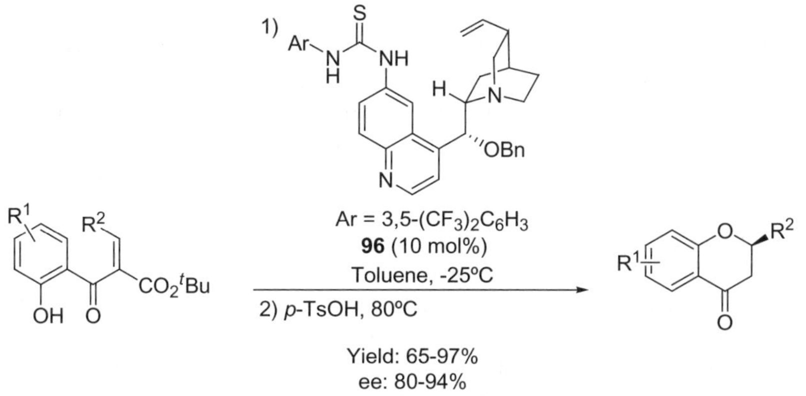

Yield: 65-97%
ee: 80-94%

Scheme 4.63   Enantioselective intramolecular oxa-Michael reaction for the synthesis of chromanones and flavanones.

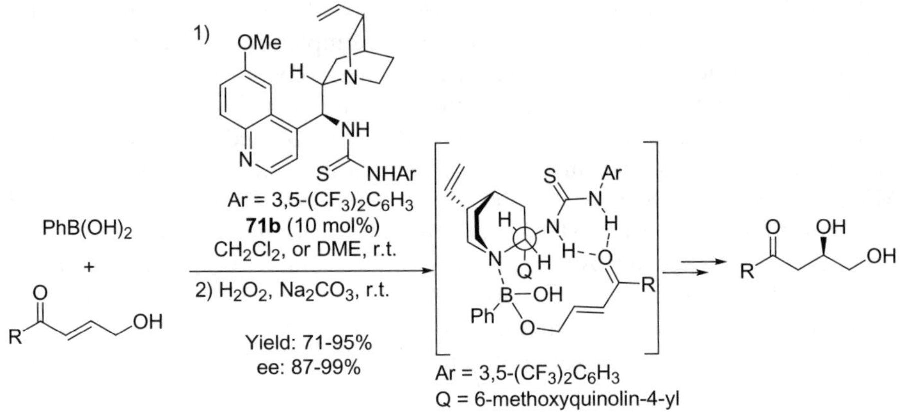

Scheme 4.64 Enantioselective oxa-Michael reaction of phenylboronic acid with γ-hydroxyenones.

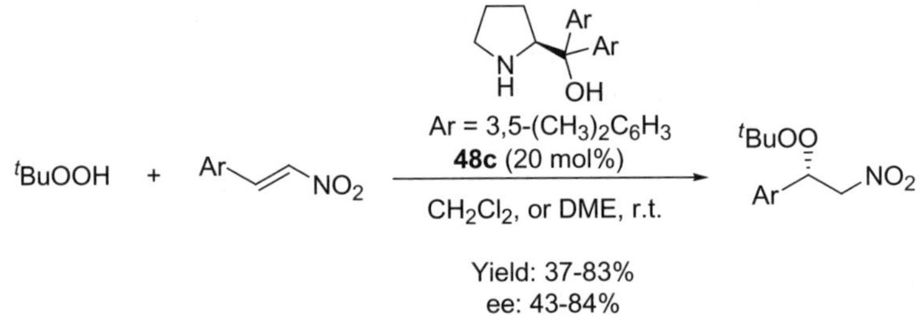

Scheme 4.65 Enantioselective β-peroxidation of nitroalkenes.

moiety of the substrate, forming a boronate intermediate, which, subsequently, proceeded to react intramolecularly with the enone, *via* a transition state in which the bifunctional tertiary amine-thiourea catalyst activated the boronate through the quinuclidine nucleus and the thiourea moiety engaged in double H-bonding interactions with the enone. δ-Hydroxy enones were also found to be suitable substrates, requiring to carry out the reaction at higher temperatures but still furnishing excellent results, although the use of other bulkier and more electron-rich arylboronic acids was required in some cases.

Finally, diarylprolinol **48c** has also been employed as catalyst in the enantioselective β-peroxydation of nitroalkenes with *tert*-butyl hydroperoxide (Scheme 4.65).[110] The catalyst is proposed to engage in a bifunctional activation mode, which on one hand assists the deprotonation of the hydroperoxide and on the other activates the nitroalkene by the formation of one H-bond with the nitro group. Under the optimized conditions, moderate yields and enantioselectivities around 85% ee were achieved for a variety of nitrostyrene derivatives.

4.5.2 Sulfa-Michael Reactions[111]

As pointed out in the introductory section of this chapter, one of the first examples of Michael-type reactions proceeding *via* H-bonding catalysis was the conjugate addition of thiophenols to cyclic enones catalyzed by natural cinchona alkaloids such as cinchonidine (**84d**) reported by Wynberg and Hiemstra in 1981 (Scheme 4.66).[13d] The need for a free OH group at the catalyst for achieving good enantioselection was interpreted in terms of activation of the enone by hydrogen bonding with this hydroxyl group of the catalyst, the tertiary amine moiety at the quinuclidine substructure being responsible for the activation of the thiophenol by deprotonation. A much more recent work has shown that cinchonine **83g** can promote the conjugate addition of aromatic thiols to chalcones with moderate enantioselectivities, obtaining improved values of enantiomeric excess after crystallization.[112]

The combined use of nitroalkenes as Michael acceptors and bifunctional tertiary amine-thioureas as catalyst has been extensively exploited in this reaction. In this context, *N*-sulfinylurea **97** was identified by Ellman as an excellent catalyst for the conjugate addition of thioacetic acid to nitroalkenes, furnishing the corresponding adducts in good yields and enantioselectivities under the optimized conditions, which also involved the use of a rather unusual solvent such as cyclopentyl methyl ether (Scheme 4.67).[113] The reaction showed to have also a very broad substrate scope, allowing the use of both aryl and alkyl substituted nitroalkenes as Michael acceptors, although the latter furnished a slightly lower level of enantioselection. The utility of the methodology

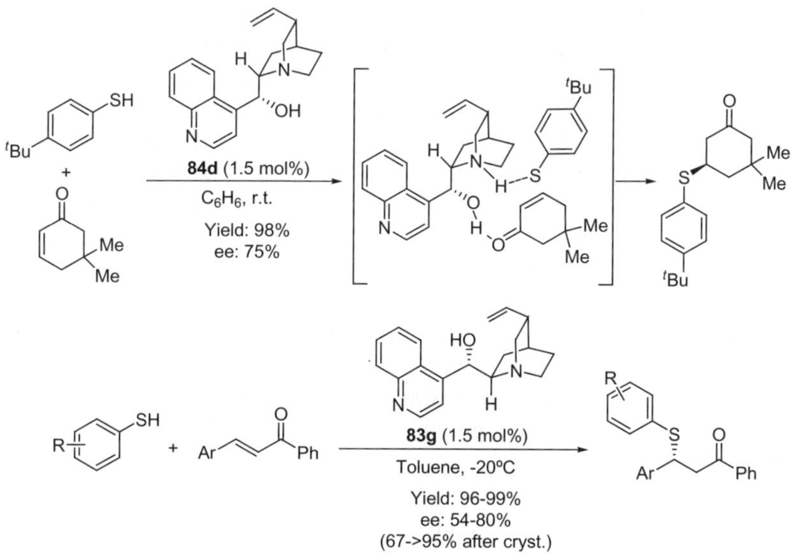

Scheme 4.66 Enantioselective sulfa-Michael reaction of arylthiols with enones catalyzed by natural cinchona alkaloids.

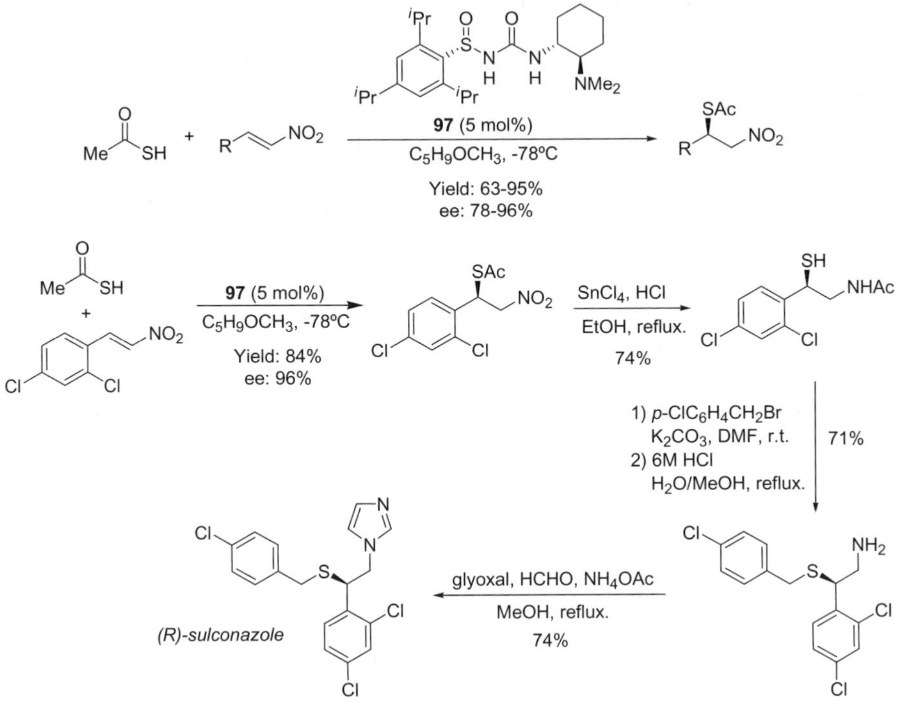

Scheme 4.67 Enantioselective sulfa-Michael reaction of thioacetic acid with nitroalkenes and application to the synthesis of (*R*)-*sulconazole*.

was illustrated with a very straightforward synthesis of (*R*)-*sulconazole*, an antifungal drug. Interestingly, the closely related Takemoto's catalyst was also evaluated in this work and in others,[114] furnishing the conjugate addition product in excellent yield, but with rather low enantioselectivity.

Alternatively, the sulfa-Michael reaction between arylthiols and α-substituted β-nitroacrylates has been described using cinchona-thiourea **71b** as catalyst (Scheme 4.68).[115] This reaction provides access to products with a sulfur-containing quaternary stereocenter, which can also be further manipulated at the nitro group in order to obtain, for example, β,β-disubstituted β-amino acids. The transformation was found to occur in a fully regioselective way, with the nitro group showing up as the dominant element for the activation of the olefin as a Michael acceptor. The reaction was studied on a wide range of different substrates incorporating an aryl substituent at the α-position, furnishing excellent yields and enantioselectivities in almost all the cases. The possibility to introduce a wide variety of arylthiols was also surveyed with success.

The structural and electronic analogies between the nitro and the imide groups have also been surveyed in this transformation (Scheme 4.69). Takemoto's catalyst has been employed in the reaction of thiophenol to imides derived from benzoic acid, furnishing excellent yields of the sulfa-Michael

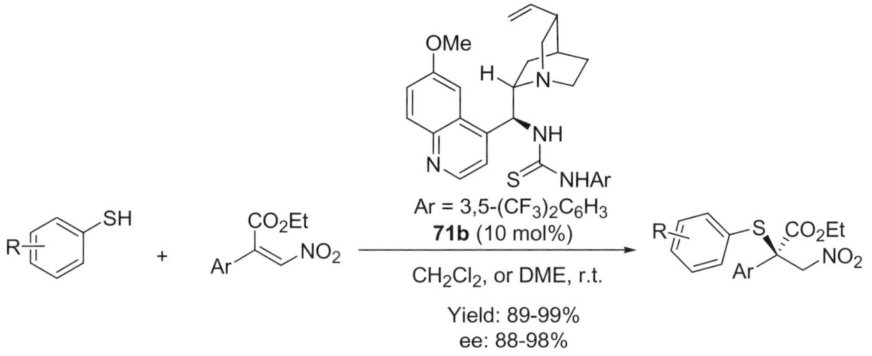

Scheme 4.68 Enantioselective sulfa-Michael reaction of arylthiols with α-aryl-β-nitroacrylates.

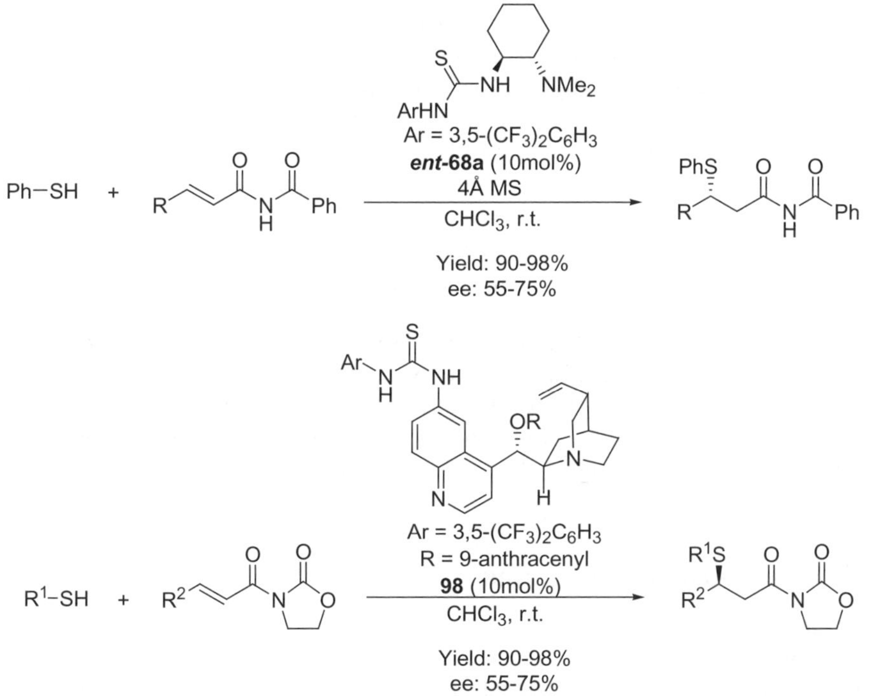

Scheme 4.69 Enantioselective sulfa-Michael reaction of aryl and alkylthiols with α,β-unsaturated imides.

products but with moderate enantioselectivities.[116] On the other hand, quinidine-thiourea catalyst **98** has demonstrated an excellent performance in the conjugate addition of alkyl thiols to *N*-enoyl oxazolidinones, obtaining outstanding results for a variety of both Michael donors and acceptors tested

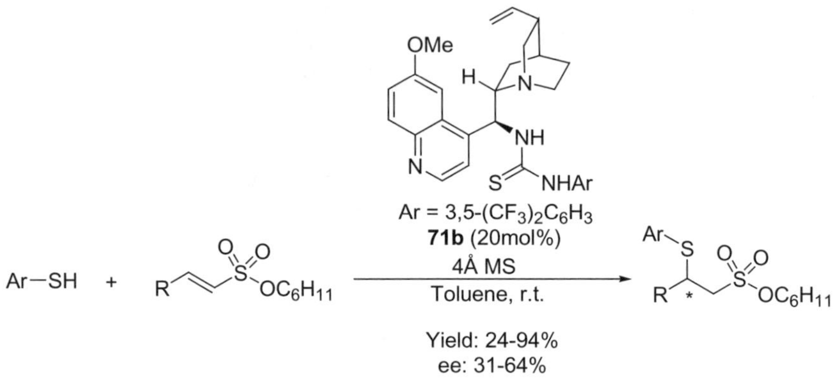

Scheme 4.70   Enantioselective sulfa-Michael reaction of arylthiols with α,β-unsaturated sulfonates.

(Scheme 4.68).[117] The possibility of using benzylic or allylic thiols as pronucleophiles is an important advantage of this methodology due to the easier removal of the benzyl or allyl moiety from the S atom, and therefore allowing the easy liberation of the latent mercapto functionality, if required. The only drawback associated to this methodology is the fact that a three-fold excess of the thiol reagent had to be employed in order to reach to full conversion.

Finally, the use of α,β-unsaturated sulfonates as Michael acceptors has also been surveyed by Enders under the hypothesis that the sulfonate group would behave similarly to the nitro group with regard to the possible interaction with a thiourea catalyst for activation (Scheme 4.70).[118] After intensive efforts directed toward the identification of the best catalyst and reaction conditions for the reaction, a protocol for carrying out the conjugate addition or arylthiols to several β-alkyl substituted α,β-unsaturated sulfonates was established using quinine-based thiourea **71b** as catalyst, affording good yields of the corresponding adducts, although with moderate enantioselectivities.

4.5.3   Aza-Michael Reactions[119]

A general and efficient procedure for the conjugate addition of nitrogen nucleophiles employing organocatalysts which participate under H-bonding catalysis furnishing high enantioselectivities is still to be developed. There is an early report on the conjugate addition of pyrrolidine to γ-crotonolactone for which a set of different chiral thioureas was surveyed as catalysts, but obtaining very low enantioselectivities.[120] Other attempts have been carried out directed toward a more successful aza-Michael reaction under H-bonding catalysis, which involved the use of chalcones as acceptors and a more reactive nitrogen-centered nucleophile such as *O*-benzylhydroxylamine[121] or, alternatively, trying to employ a more NH-acidic pro-nucleophile such as aniline,[122] which would hypothetically enhance the rate of the conjugate addition step by the *in situ* generation of a more reactive amide anion. However, both approaches have failed in providing good enantioselectivities in most cases (Scheme 4.71).

Using benzotriazole as Michael donor and nitroalkenes as acceptors, Wang has found that quinidine-derived catalyst **83g** was able to promote the reaction in a rather efficient way, reaching up to >90% ee in a few examples, although in most cases enantioselectivities remained in moderate values (Scheme 4.72).[123] Remarkably, both β-alkyl and aryl substituted nitroalkenes behaved similarly, therefore showing the broad scope of this reaction. It is interesting to note that thiourea catalysts, which were supposed to be engaged in the formation of a more conformationally rigid H-bonding network with the nitroalkene acceptor, provided the final adducts with lower enantioselectivities than those provided by **83g** under the same reaction conditions. Other *N*-heterocycles like 1*H*-1,2,

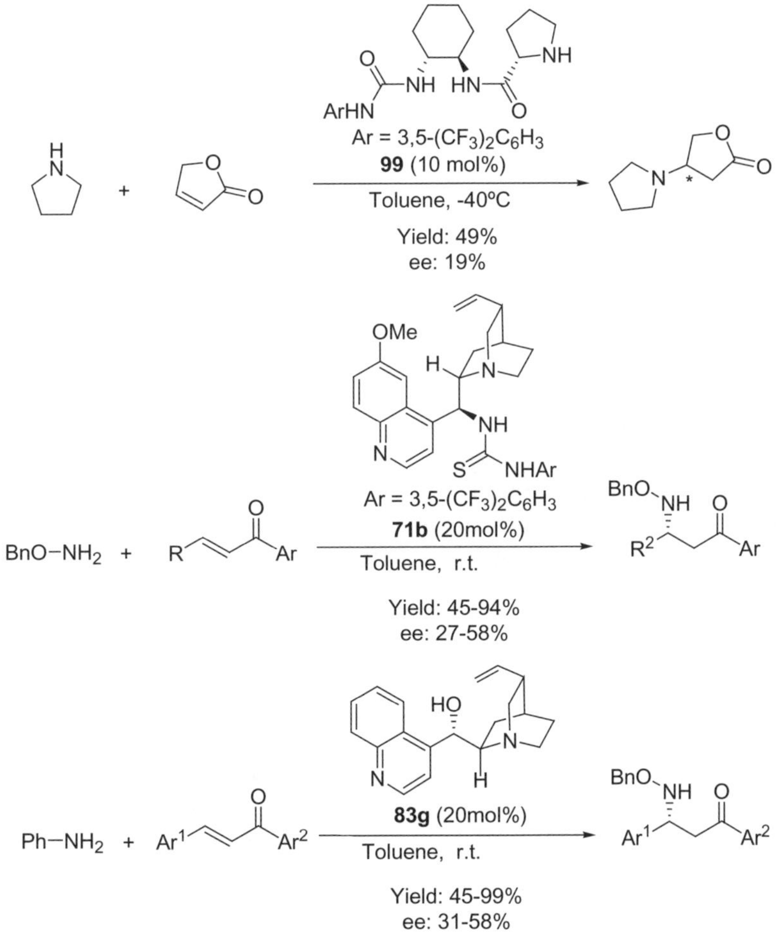

Scheme 4.71   Some attempts to carry out enantioselective aza-Michael reactions under H-bonding catalysis.

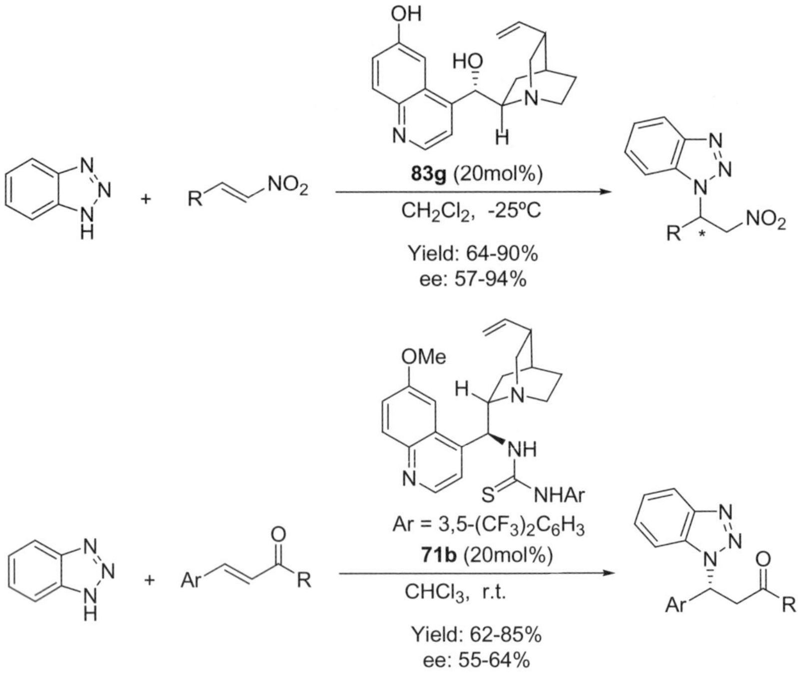

Scheme 4.72 The enantioselective conjugate addition of benzotriazole to nitroalkenes and enones.

3-triazole or 5-phenyl-1*H*-triazole were also surveyed as nucleophiles furnishing also similar results to those obtained with benzotriazole. On the other hand, an attempt to carry out the conjugate addition of benzotriazoles to enones led to different results, observing a slightly higher enantioselectivity when 9-*epi*-amino quinine-derived thiourea **71b** was used instead of the corresponding quinidine-derived catalyst **83g**.[124] Nevertheless, even at the optimized reaction conditions, enantioselectivities remained in the range of 56–64% ee.

Finally, there is one example of an aza-Michael reaction of hydroxylamine ethers with enoylpyrazoles proceeding with excellent enantioselectivities which makes use of bifunctional thiourea ***ent*-91** as promoter (Scheme 4.73).[125] In this case, the pyrazole moiety was chosen as a substrate able to engage in secondary H-bonding interactions with the catalyst, which were also supposed to increase the rigidity of the reaction intermediate. In fact, computational studies demonstrated that the thiourea moiety was involved in a double H-bond formation with the N-2 of the pyrazole unit and that the remaining OH group would be responsible for the activation of the carbonyl group and also for the transference of the *N*-nucleophile *via* an additional H-bonding interaction. Moreover, this study demonstrated that the conformation of the pyrazole amide group was crucial for achieving high stereocontrol, showing that the reaction should proceed *via* an intermediate in which the O=C–N–N

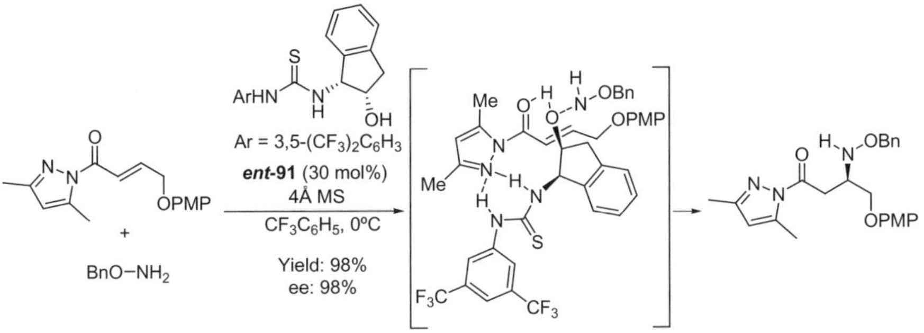

Scheme 4.73   The catalytic enantioselective conjugate addition of hydroxylamine benzyl ether to an *N*-enoylpyrazole.

substructure would remain in a preferred *s-trans* conformation, which was much more stable than the corresponding *s-cis* geometry, the latter furnishing the minor enantiomer. The energy differences between both conformations were much smaller in the transition state than in the substrates due to the easier formation of the H-bonded complex in the *s-cis* conformer, but still high enough to account for the high enantioselectivities obtained. The calculated ΔG values between both transition states correlated well with the experimental data, thus providing a reliable model to explain the behavior of the system. As mentioned before, excellent enantioselectivities were obtained for a set of different β-substituted *N*-enoyl pyrazoles using TBS- and benzyl ether of hydroxylamine as nucleophiles, although it has to be said that the thiourea *ent*-**88** was employed in almost all cases in stoichiometric amounts. Nevertheless, there is one example related to the use of a 30 mol% of *ent*-**91** to promote the reaction, providing an excellent yield and enantioselectivity of the aza-Michael adduct.

4.5.4   Phospha-Michael Reactions[126]

There is only a very limited number of examples of enantioselective conjugate additions of phosphorous-based nucleophiles under H-bonding catalysis, all of them focused on the use of nitroalkenes as Michael acceptors (Scheme 4.74). In one of these reports, the addition of diphenylphosphine to a set of different β-aryl-substituted nitroalkenes was carried out using cinchona alkaloid-based thiourea **71b** as catalyst.[127] The reaction conditions had to be carefully optimized, which included the use of a Et$_2$O/*i*-PrOH mixture as solvent and the one-pot protection of the trisubstituted phosphine moiety at the final adduct by complexation with borane. Although enantioselectivities remained in values around 65% ee, these could be improved to 99% ee in most cases after a single recrystallization process. Other phospha-Michael reactions developed in a similar context involved the use of diphenylphosphite as a more stable and easy to manipulate phosphorous nucleophile and natural quinine **84b** as catalyst, furnishing the final conjugate addition products in slightly higher

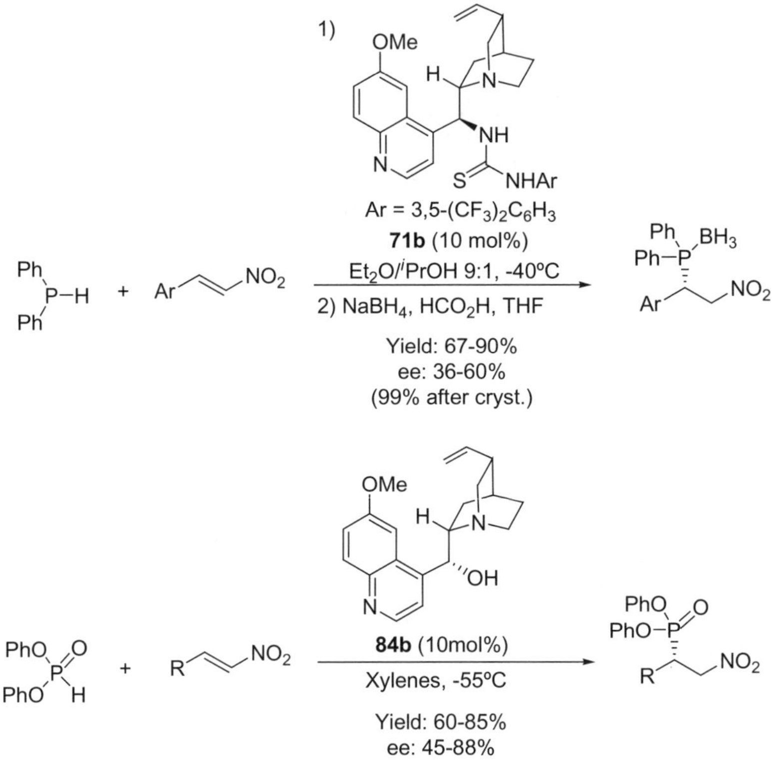

Scheme 4.74 The catalytic enantioselective phospha-Michael reactions using cinchona alkaloid-based catalysts.

enantioselectivities.[128] In this case, the reaction was studied on a broader range of substrates, showing that both β-alkyl and β-aryl nitroalkenes could be successfully used, furnishing similar results.

Alternatively, chiral guanidines have also been successfully employed to catalyze several examples of phospha-Michael reactions (Scheme 4.75). The conjugate addition of diarylphosphine oxides to nitroalkenes has been carried out using **86** as catalyst, furnishing good yields and enantioselectivities of the final products for a variety of nitrostyrene derivatives tested.[129] Importantly, the use of α,β-disubstituted nitroalkenes was also surveyed, observing that the reaction proceeded not only with excellent enantioselectivity but also furnishing a single diastereoisomer, with a high yield. It has to be pointed out that the nature of the aryl substituents at the phosphine had a crucial influence on the stereocontrol requiring the introduction of bulky 1-naphthyl groups for reaching the highest possible ee. In a different report, guanidine **81** has been identified as an excellent catalyst for the conjugate addition of diphenylphosphite to nitroalkenes.[130] In this case, the reaction showed a very wide substrate scope, performing well with both β-aryl and β-alkyl substituted nitroalkenes, observing in all cases excellent yields and enantioselectivities.

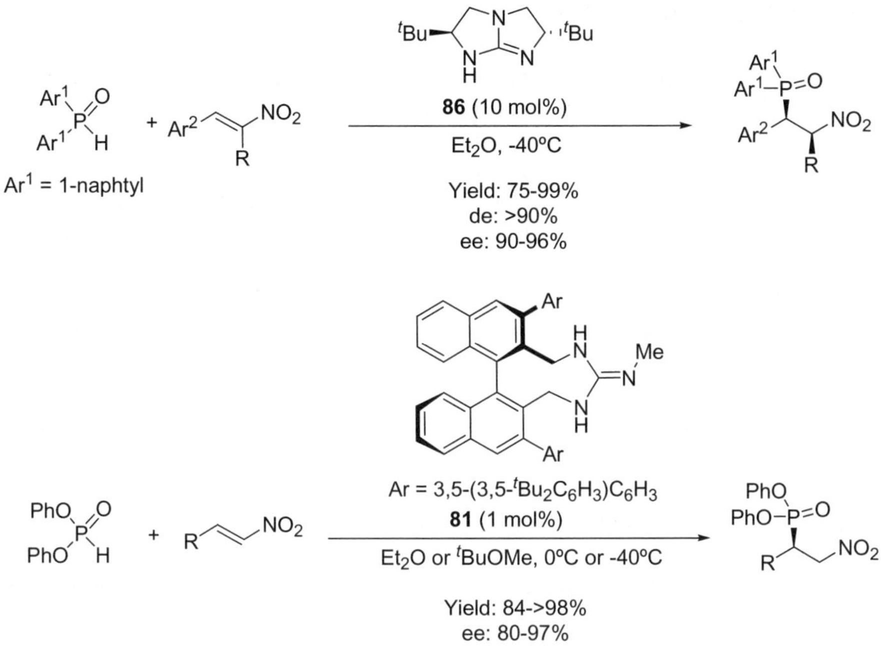

Scheme 4.75 The catalytic enantioselective phospha-Michael reactions using guanidines as catalysts.

4.6 Concluding Remarks

The use of molecules able to activate the Michael acceptor by the formation of a catalyst-substrate complex by means of H-bonding interactions has shown up as a very powerful methodology for the development of enantioselective conjugate additions. In general, a rather broad scope of Michael acceptors is tolerated, although the reaction seems to proceed in a more efficient way when substrates able to engage in multiple H-bonding interactions are employed, due to the fact that these lead to the formation of more conformationally rigid reaction intermediates. On the other hand, while this methodology seems to work exceptionally well with carbon nucleophiles, the use of heteroatom-based Michael donors is still rather undeveloped. Related to the catalyst structure, most of the methodologies reported have focused on the use of chiral thioureas or chiral alcohols derived from cinchona alkaloids as very appropriate promoters for the reaction. In almost all cases, these catalysts contain a basic site in their structure (bifunctional catalyst) which becomes involved in the activation of the pro-nucleophile, by exerting its deprotonation. The use of chiral phosphoric acids or phosphoramides as catalysts of enhanced Brønsted acidity has also opened the way to the activation of substrates that were resistant to activation by other types of catalysts.

References and Notes

1. (a) J. L. Vicario, D. Badía and L. Carrillo, *Synthesis*, 2007, 2065; (b) D. Almaçi, D. A. Alonso and C. Nájera, *Tetrahedron: Asymmetry*, 2007, **18**, 299; (c) S. B. Tsogoeva, *Eur. J. Org. Chem.*, 2007, 1701.
2. For specific reviews covering enantioselective transformations proceeding under H-bonding catalysis see (a) X. H. Yu and W. Wang, *Chem. Asian J.*, 2008, **3**, 516; (b) S. J. Connon, *Chem. Commun.*, 2008, 2499; (c) H. Miyabe and Y. Takemoto, *Bull. Chem. Soc. Jpn.*, 2008, **81**, 785; (d) A. G. Doyle and E. N. Jacobsen, *Chem. Rev.*, 2007, **107**, 5173; (e) M. S. Taylor and E. N. Jacobsen, *Angew. Chem. Int. Ed.*, 2006, **45**, 1520; (f) S. J. Connon, *Chem. Eur. J.*, 2006, **12**, 5418; (g) P. R. Schreiner, *Chem. Soc. Rev.*, 2003, **32**, 289.
3. (a) J. Hine, K. Ahn, J. C. Gallucci and S.-M. Linden, *J. Am. Chem. Soc.*, 1984, **106**, 7980; (b) J. Hine, S.-M. Linden and V. M. Kanagasabapathy, *J. Am. Chem. Soc.*, 1985, **107**, 1082.
4. T. R. Kelly, P. Meghani and V. S. Ekkundi, *Tetrahedron Lett.*, 1990, **31**, 3381.
5. (a) J. F. Blake and W. L. Jørgensen, *J. Am. Chem. Soc.*, 1991, **113**, 7430; (b) D. L. Severance and W. L. Jørgensen, *J. Am. Chem. Soc.*, 1992, **114**, 10966.
6. (a) M. C. Etter, Z. Urbanczyk-Lipkowska, M. Zia-Ebrahimi and T. W. Panunto, *J. Am. Chem. Soc.*, 1990, **112**, 8415; (b) M. C. Etter, *Acc. Chem. Res.*, 1990, **23**, 120.
7. D. P. Curran and L. H. Kuo, *J. Org. Chem.*, 1994, **59**, 3259.
8. D. P. Curran and L. H. Kuo, *Tetrahedron Lett.*, 1995, **36**, 6647.
9. (a) P. R. Schreiner and A. Wittkopp, *Chem. Eur. J.*, 2003, **9**, 407; (b) P. R. Schreiner and A. Wittkopp, *Org. Lett.*, 2002, **4**, 217.
10. T. Okino, Y. Hoashi and Y. Takemoto, *Tetrahedron Lett.*, 2003, **44**, 2817.
11. D. J. Maher and S. J. Connon, *Tetrahedron Lett.*, 2004, **45**, 1301.
12. G. Dessole, R. P. Herrera and A. Ricci, *Synlett*, 2004, 2374.
13. (a) H. Wynberg and R. Helder, *Tetrahedron Lett.*, 1975, **46**, 4057; (b) K. Hermann and H. Wynberg, *J. Org. Chem.*, 1979, **44**, 2238; (c) R. Helder, R. Arends, W. Bolt, H. Hiemstra and H. Wynberg, *Tetrahedron Lett.*, 1977, **25**, 2181; (d) H. Hiemstra and H. Wynberg, *J. Am. Chem. Soc.*, 1981, **103**, 417.
14. (a) M. S. Sigman and E. N. Jacobsen, *J. Am. Chem. Soc.*, 1998, **120**, 4901. See also; (b) P. Vachal and E. N. Jacobsen, *J. Am. Chem. Soc.*, 2002, **124**, 10012.
15. J. Oku and S. Inoue, *J. Chem. Soc. Chem. Commun.*, 1981, 229.
16. (a) U.-H. Dolling, P. Davis and E. J. J. Grabowski, *J. Am. Chem. Soc.*, 1984, **106**, 446; (b) R. S. E. Conn, A. V. Lovell, S. Karady and L. M. Weinstock, *J. Org. Chem.*, 1986, **51**, 4710.
17. D. Leow and C.-H. Tan, *Chem. Asian J.*, 2009, **4**, 488.
18. For some reviews see (a) G. Adair, S. Mukherjee and B. List, *Aldrichimica Acta*, 2008, **41**, 31; (b) T. Akiyama, *Chem. Rev.*, 2007, **107**, 5744.

19. (a) T. Okino, Y. Hoashi, T. Furukawa, X. Xu and Y. Takemoto, *J. Am. Chem. Soc.*, 2005, **127**, 119; (b) T. Okino, Y. Hoashi and Y. Takemoto, *J. Am. Chem. Soc.*, 2003, **125**, 12672.

20. W.-M. Zhou, H. Liu and D.-M. Du, *Org. Lett.*, 2008, **10**, 2817.

21. L. Jiang, H.-T. Zheng, T.-Y. Liu, L. Yue and Y.-C. Chen, *Tetrahedron*, 2007, **63**, 5123.

22. H. Miyabe, S. Tuchida, M. Yamauchi and Y. Takemoto, *Synthesis*, 2006, 3295.

23. (a) A. Hamza, G. Schubert, T. Soós and I. Pápai, *J. Am. Chem. Soc.*, 2006, **128**, 13151. For a similar computational study see; (b) R. Zhu, D. Zhang, J. Wu and C. Liu, *Tetrahedron: Asymmetry*, 2006, **17**, 1611.

24. (a) S. H. McCooey and S. J. Connon, *Angew. Chem. Int. Ed.*, 2005, **44**, 6367. For an account see; (b) S. J. Connon, *Synlett*, 2009, 354. For a related study focused on the Michael reaction of nitromethane to chalcones see ref. 58a.

25. J. Ye, D. J. Dixon and P. S. Hynes, *Chem. Commun.*, 2005, 4481.

26. See for example. (a) O. Bassas, J. Huuskonen, K. Rissanen and A. M. P. Koskinen, *Eur. J. Org. Chem.*, 2009, 1340; (b) P. Elsner, H. Jiang, J. B. Nielsen, F. Pasi and K. A. Jørgensen, *Chem. Commun.*, 2008, 5827; (c) P. S. Hynes, P. A. Stupple and D. J. Dixon, *Org. Lett.*, 2008, **10**, 1389; (d) P. Jakubec, D. M. Cockfield and D. J. Dixon, *J. Am. Chem. Soc.*, 2009, **131**, 16632.

27. X.-J. Li, K. Liu, H. Ma, J. Nie and J.-A. Ma, *Synlett*, 2008, 3242.

28. J. M. Andrés, R. Manzano and R. Pedrosa, *Chem. Eur. J.*, 2008, **14**, 5116.

29. J. Wang, H. Li, H. Duan, L. Zu and W. Wang, *Org. Lett.*, 2005, **7**, 4713.

30. (a) F.-Z. Peng, Z.-H. Shao, B.-M. Fan, H. Song, G.-P. Li and H.-B. Zhang, *J. Org. Chem.*, 2008, **73**, 5202. For the use of this catalyst in the same reaction using 2-fluoro-3-phenyl-3-oxopropanoates see; (b) Y. Oh, S. M. Kim and D. Y. Kim, *Tetrahedron Lett.*, 2009, **50**, 4674. For a computational study regarding a mechanistic proposal see; (c) D. Chen, N. Lu, G. Zhang and S. Mi, *Tetrahedron: Asymmetry*, 2009, **20**, 1365.

31. X. Jiang, Y. Zhang, X. Liu, G. Zhang, L. Lai, L. Wu, J. Zhang and R. Wang, *J. Org. Chem.*, 2009, **74**, 5562.

32. X. Pu, P. Li, F. Peng, X. Li, H. Zhang and Z. Shao, *Eur. J. Org. Chem.*, 2009, 4622.

33. C.-W. Wang, Z.-H. Zhang, X.-Q. Dong and X.-J. Wu, *Chem. Commun.*, 2008, 1431.

34. Z.-H. Zhang, X.-Q. Dong, D. Chen and C.-J. Wang, *Chem. Eur. J.*, 2008, **14**, 8780.

35. J. Lubkoll and H. Wennemers, *Angew. Chem. Int. Ed.*, 2007, **46**, 6841.

36. K. Rabalakos and W. D. Wulff, *J. Am. Chem. Soc.*, 2008, **130**, 13524.

37. X.-Q. Dong, H.-L. Teng and C.-J. Wang, *Org. Lett.*, 2009, **11**, 1265.

38. J. Alemán, A. Milelli, S. Cabrera, E. Reyes and K. A. Jørgensen, *Chem. Eur. J.*, 2008, **14**, 10958.

39. P. S. Hynes, D. Stranges, P. A. Stupple, A. Guarna and D. J. Dixon, *Org. Lett.*, 2007, **9**, 2107.

40. T. Bui, S. Syed and C. F. Barbas III, *J. Am. Chem. Soc.*, 2009, **131**, 8758.
41. (a) Y.-H. Liao, H. Zhang, Z.-J. Wu, L.-F. Cun, X.-M. Zhang and W.-C. Yuan, *Tetrahedron: Asymmetry*, 2009, **20**, 2397. See also; (b) M. Shi, Z.-Y. Lei, M.-X. Zhao and J.-W. Shi, *Tetrahedron Lett.*, 2007, **48**, 5743.
42. D. J. Dixon and R. D. Richardson, *Synlett*, 2006, 81.
43. For a review: D. Leow and C.-H. Tan, *Chem. Asian J.*, 2009, **4**, 488.
44. Z. Yu, X. Liu, L. Zhou, L. Lin and X. Feng, *Angew. Chem. Int. Ed.*, 2009, **48**, 5195.
45. M. Terada, H. Ube and Y. Yaguchi, *J. Am. Chem. Soc.*, 2006, **128**, 1454.
46. Z. Jiang, Y. Pan, Y. Zhao, T. Ma, R. Lee, Y. Yang, K.-W. Huang, M. W. Wong and C.-H. Tan, *Angew. Chem. Int. Ed.*, 2009, **48**, 3627.
47. (a) A. P. Davis and K. J. Dempsey, *Tetrahedron: Asymmetry*, 1995, **6**, 2829; (b) M. T. Allingham, A. Howard-Jones, P. J. Murphy, D. A. Thomas and P. W. R. Caulkett, *Tetrahedron Lett.*, 2003, **44**, 8677; (c) W. Ye, D. Leow, S. L. M. Goh, C.-T. Tan, C.-H. Chian and C.-H. Tan, *Tetrahedron Lett.*, 2006, **47**, 1007; (d) T. Kumamoto, K. Ebine, M. Endo, Y. Araki, Y. Fushimi, I. Miyamoto, T. Ishikawa, T. Isobe and K. Fukuda, *Heterocycles*, 2005, **66**, 347.
48. (a) D. Almasi, D. A. Alonso, E. Gómez-Bengoa and C. Nájera, *J. Org. Chem.*, 2009, **74**, 6163. For a related cinchona alkaloid based benzimidazole catalyst employed in the same reaction see; (b) L. Zhang, M.-M. Lee, S.-M. Lee, J. Lee, M. Cheng, B.-S. Jeong, H. Park and S. Jew, *Adv. Synth. Catal.*, 2009, **351**, 3063.
49. For a review on the use of cupreines and cupreidines as organocatalysts see T. Marcelli, J. H. van Maarseveen and H. Hiemstra, *Angew. Chem. Int. Ed.*, 2006, **45**, 7496.
50. H. Li, Y. Wang, L. Tang and L. Deng, *J. Am. Chem. Soc.*, 2004, **126**, 9906.
51. H. Li, Y. Wang, L. Tang, F. Wu, X. Liu, C. Guo, B. M. Foxman and L. Deng, *Angew. Chem. Int. Ed.*, 2005, **44**, 105.
52. (a) H. Li, L. Zu, H. Xie and W. Wang, *Synthesis*, 2009, 1525. For the same reaction using 2-fluoro-3-oxobutanoates see; (b) H. Li, S. Zhang, C. Yu, X. Song and W. Wang, *Chem. Commun.*, 2009, 2136.
53. J. Wang, H. Li, L. Zu, W. Jiang and W. Wang, *Adv. Synth. Catal.*, 2006, **348**, 2047.
54. B. Tan, X. Zhang, P. J. Chua and G. Zhong, *Chem. Commun.*, 2009, 779.
55. J. Luo, L.-W. Xu, R. A. S. Hay and Y. Lu, *Org. Lett.*, 2009, **11**, 437.
56. A. Lattanzi, *Tetrahedron: Asymmetry*, 2006, **17**, 837.
57. See A. Lattanzi, *Chem. Commun.*, 2009, 1452.
58. (a) B. Vakulya, S. Varga, A. Csámpai and T. Soós, *Org. Lett.*, 2005, **7**, 1967. For a similar version of this reaction using 4-oxoenoates as Michael acceptors see; (b) H.-H. Lu, X.-F. Wang, C.-J. Yao, J.-M. Zhang, H. Wu and W.-J. Xiao, *Chem. Commun.*, 2009, 4251.
59. J. Wang, H. Li, L. Zu, W. Jiang, H. Xie, W. Duan and W. Wang, *J. Am. Chem. Soc.*, 2006, **128**, 12652.

60. C. Gu, L. Liu, Y. Sui, J.-L. Zhao, D. Wang and Y.-J. Chen, *Tetrahedron: Asymmetry*, 2007, **18**, 455.
61. T.-Y. Liu, R. Li, Q. Chai, J. Long, B.-J. Li, Y. Wu, L.-S. Ding and Y.-C. Chen, *Chem. Eur. J.*, 2007, **13**, 319.
62. F. Wu, H. Li, R. Hong and L. Deng, *Angew. Chem. Int. Ed.*, 2006, **45**, 947.
63. A. Russo, A. Perfetto and A. Lattanzi, *Adv. Synth. Catal.*, 2009, **351**, 3067.
64. M. B. Cid, J. López-Cantarero, S. Duce and J. L. García Ruano, *J. Org. Chem.*, 2009, **74**, 413.
65. J. Alemán, B. Richter and K. A. Jørgensen, *Angew. Chem. Int. Ed.*, 2007, **46**, 5515.
66. Z. Jiang, W. Ye, Y. Yang and C.-H. Tan, *Adv. Synth. Catal.*, 2008, **350**, 2345.
67. Z. Jiang, Y. Yang, Y. Pan, Y. Zhao, H. Liu and C.-H. Tan, *Chem. Eur. J.*, 2009, **15**, 4925.
68. T. Akiyama, T. Katoh and K. Mori, *Angew. Chem. Int. Ed.*, 2009, **48**, 4226.
69. T. Inokuma, K. Takasu, T. Sakaeda and Y. Takemoto, *Org. Lett.*, 2009, **11**, 2425.
70. T. R. Wu and J. M. Chong, *J. Am. Chem. Soc.*, 2007, **129**, 4908.
71. F. Wu, R. Hong, J. Khan, X. Liu and L. Deng, *Angew. Chem. Int. Ed.*, 2006, **45**, 4301.
72. M. Bell, K. Frisch and K. A. Jørgensen, *J. Org. Chem.*, 2006, **71**, 5407.
73. C. L. Rigby and D. J. Dixon, *Chem. Commun.*, 2008, 3798.
74. Y. Wang, X. Liu and L. Deng, *J. Am. Chem. Soc.*, 2006, **128**, 3928.
75. B. Wang, F. Wu, Y. Wang, X. Liu and L. Deng, *J. Am. Chem. Soc.*, 2007, **129**, 768.
76. T. Ishikawa, Y. Araki, T. Kumamoto, H. Seki, K. Fukuda and T. Isobe, *Chem. Commun.*, 2001, 245.
77. (a) Y. Hoashi, T. Okino and Y. Takemoto, *Angew. Chem. Int. Ed.*, 2005, **44**, 4032; (b) T. Inokuma, Y. Hoashi and Y. Takemoto, *J. Am. Chem. Soc.*, 2006, **128**, 9413.
78. G. Bartoli, M. Bosco, A. Carlone, A. Cavalli, M. Locatelli, A. Mazzanti, P. Ricci, P. Sambri and P. Melchiorre, *Angew. Chem. Int. Ed.*, 2006, **45**, 4966.
79. C. S. Cucinotta, M. Kosa, P. Melchiorre, A. Cavalli and F. L. Gervasio, *Chem. Eur. J.*, 2009, **15**, 7913.
80. J. Lu, W.-J. Zhou, F. Liu and T.-P. Loh, *Adv. Synth. Catal.*, 2008, **350**, 1796.
81. W. Ye, Z. Jiang, Y. Zhao, S. L. M. Goh, D. Leow, Y.-T. Soh and C.-H. Tan, *Adv. Synth. Catal.*, 2007, **349**, 2454.
82. (a) H. Li, J. Song, X. Liu and L. Deng, *J. Am. Chem. Soc.*, 2005, **127**, 8948. See also; (b) H. Li, J. Song and L. Deng, *Tetrahedron*, 2009, **65**, 3139.
83. T.-Y. Liu, J. Long, B.-J. Li, L. Jiang, R. Li, Y. Wu, L.-S. Ding and Y.-C. Chen, *Org. Biomol. Chem.*, 2006, **4**, 2097.

84. B. Vakulya, S. Varga and T. Soós, *J. Org. Chem.*, 2008, **73**, 3475.

85. Q. Zhu and Y. Lu, *Org. Lett.*, 2009, **11**, 1721.

86. W. J. Nodes, D. R. Nutt, A. M. Chippindale and A. J. A. Cobb, *J. Am. Chem. Soc.*, 2009, **131**, 16016.

87. M. Capuzzi, D. Perdicchia and K. A. Jørgensen, *Chem. Eur. J.*, 2008, **14**, 128.

88. For some reviews see (a) R. Brehme, D. Enders, R. Fernández and J. M. Lassaletta, *Eur. J. Org. Chem.*, 2007, 5629; (b) R. Fernández and J. M. Lassaletta, *Synlett*, 2000, 1228.

89. R. P. Herrera, D. Monge, E. Martín-Zamora, R. Fernández and J. M. Lassaletta, *Org. Lett.*, 2007, **9**, 3303.

90. A. E. Mattson, A. M. Zuhl, T. E. Reynolds and K. A. Scheidt, *J. Am. Chem. Soc.*, 2006, **128**, 4932.

91. R. P. Herrera, V. Sgarzani, L. Bernardi and A. Ricci, *Angew. Chem. Int. Ed.*, 2005, **44**, 6576.

92. M. Ganesh and D. Seidel, *J. Am. Chem. Soc.*, 2008, **130**, 16464.

93. (a) W. Zhuang, R. G. Hazell and K. A. Jørgensen, *Org. Biomol. Chem.*, 2005, **3**, 2566; (b) E. M. Fleming, T. McCabe and S. J. Connon, *Tetrahedron Lett.*, 2006, **47**, 7037.

94. T.-Y. Liu, H.-L. Cui, Q. Chai, J. Long, B.-J. Li, Y. Wu, L.-S. Ding and Y.-C. Chen, *Chem. Commun.*, 2007, 2228.

95. J. Itoh, K. Fuchibe and T. Akiyama, *Angew. Chem. Int. Ed.*, 2008, **47**, 4016.

96. Y.-F. Sheng, G.-Q. Li, Q. Kang, A.-J. Zhang and S.-L. You, *Chem. Eur. J.*, 2009, **15**, 3351.

97. Y.-F. Sheng, Q. Gu, A.-J. Zhang and S.-L. You, *J. Org. Chem.*, 2009, **74**, 6899.

98. (a) M. Rueping, B. J. Nachsteim, S. A. Moreth and M. Bolte, *Angew. Chem. Int. Ed.*, 2008, **47**, 593. For a related example using 4,7-dihydroindoles as Michael donors see; (b) M. Zeng, Q. Kang, Q.-L. He and S.-L. You, *Adv. Synth. Catal.*, 2008, **350**, 2169.

99. W. Zhou, L.-W. Xu, L. Li, L. Yang and C.-G. Xia, *Eur. J. Org. Chem.*, 2006, 5225.

100. (a) S. Mayer and B. List, *Angew. Chem. Int. Ed.*, 2006, **45**, 4193; (b) N. J. A. Martin and B. List, *J. Am. Chem. Soc.*, 2006, **128**, 13368.

101. (a) N. J. A. Martin, L. Ozores and B. List, *J. Am. Chem. Soc.*, 2007, **129**, 8976; (b) N. J. A. Martin, X. Cheng and B. List, *J. Am. Chem. Soc.*, 2008, **130**, 13862.

102. M. Rueping, A. P. Antonchick and T. Theissmann, *Angew. Chem. Int. Ed.*, 2006, **45**, 3683.

103. M. Rueping, T. Theissmann, S. Raja and J. W. Bats, *Adv. Synth. Catal.*, 2008, **350**, 1001.

104. M. Rueping and A. P. Antonchick, *Angew. Chem. Int. Ed.*, 2007, **46**, 4562.

105. For a review on oxa-Michael reactions see C. F. Nising and S. Braese, *Chem. Soc. Rev.*, 2008, **37**, 1218.

106. A. Merschaert, P. Delbeke, D. Daloze and G. Dive, *Tetrahedron Lett.*, 2004, **45**, 4697.
107. N. Saito, A. Ryoda, W. Nakanishi, T. Kumamoto and T. Ishikawa, *Eur. J. Org. Chem.*, 2008, 2759.
108. M. M. Biddle, M. Lin and K. A. Scheidt, *J. Am. Chem. Soc.*, 2007, **129**, 3830.
109. D. R. Li, A. Murugan and J. R. Falck, *J. Am. Chem. Soc.*, 2008, **130**, 46.
110. A. Russo and A. Lattanzi, *Adv. Synth. Catal.*, 2008, **350**, 1991.
111. For a general review on sulfa-Michael reactions see D. Enders, K. Lüttgen and A. A. Narine, *Synthesis*, 2007, 959.
112. (a) J. Skarzewski, M. Zielinska-Blajet and I. Turowska-Tyrk, *Tetrahedron: Asymmetry*, 2001, **12**, 1923.
113. K. L. Kimmel, M. T. Robak and J. A. Ellman, *J. Am. Chem. Soc.*, 2009, **131**, 8754.
114. (a) H. Li, J. Wang, L. Zu and W. Wang, *Tetrahedron Lett.*, 2006, **47**, 2585. For an attempt related to the sulfa-Michael adddition of thioacetic acid to enones using Takemoto's catalyst furnishing low to moderate enantioselectivities see; (b) H. Li, L. Zu, J. Wang and W. Wang, *Tetrahedron Lett.*, 2006, **47**, 3145.
115. H.-H. Lu, F.-G. Zhang, S.-W. Meng, S.-W. Duan and W.-J. Xiao, *Org. Lett.*, 2009, **11**, 3946.
116. B.-J. Li, L. Jiang, M. Liu, Y.-C. Chen, L.-S. Ding and Y. Wu, *Synlett*, 2005, 603.
117. Y. Liu, B. Sun, B. Wang, M. Wakem and L. Deng, *J. Am. Chem. Soc.*, 2009, **131**, 418.
118. D. Enders and K. Hoffman, *Eur. J. Org. Chem.*, 2009, 1665.
119. For general reviews on the aza-Michael reaction see (a) J. L. Vicario, D. Badía, L. Carrillo, J. Etxebarria, E. Reyes and N. Ruiz, *Org. Prep. Proc. Int.*, 2005, **37**, 513; (b) L.-W. Xu and C.-G. Xia, *Eur. J. Org. Chem.*, 2005, 633. For organocatalytic enantioselective aza-Michael reactions see; (c) D. Enders, C. Wang and J. X. Liebich, *Chem. Eur. J.*, 2009, **15**, 11058.
120. Y. Sohtome, A. Tanatani, Y. Hashimoto and K. Nagasawa, *Chem. Pharm. Bull.*, 2004, **52**, 477.
121. D. Pettersen, F. Piana, L. Bernardi, F. Fini, M. Fochi, V. Sgarzani and A. Ricci, *Tetrahedron Lett.*, 2007, **48**, 7805.
122. A. Scettri, A. Massa, L. Palombi, R. Villano and M. R. Acocella, *Tetrahedron: Asymmetry*, 2008, **19**, 2149.
123. J. Wang, H. Li, L. Zu and W. Wang, *Org. Lett.*, 2006, **8**, 1391.
124. J. Wang, L. Zu, H. Li, H. Xie and W. Wang, *Synthesis*, 2007, 2576.
125. (a) M. P. Sibi and K. Itoh, *J. Am. Chem. Soc.*, 2007, **129**, 8064. For a computational study see; (b) L. Simón and J. M. Goodman, *Org. Biomol. Chem.*, 2009, **7**, 483.
126. For a leading review on the asymmetric phospha-Michael reaction see D. Enders, A. Saint-Dizier, M.-I. Lannou and A. Lenzen, *Eur. J. Org. Chem.*, 2005, 29.

127. G. Bartoli, M. Bosco, A. Carlone, M. Locatelli, A. Mazzanti, L. Sambri and P. Melchiorre, *Chem. Commun.*, 2007, 722.
128. J. Wang, L. D. Heikkinen, H. Li, L. Zu, W. Jiang, H. Xie and W. Wang, *Adv. Synth. Catal.*, 2007, **349**, 1052.
129. X. Fu, Z. Jiang and C.-H. Tan, *Chem. Commun.*, 2007, 5058.
130. M. Terada, T. Ikehara and H. Ube, *J. Am. Chem. Soc.*, 2007, **129**, 14112.

CHAPTER 5

Enantioselective Conjugate Addition Reactions via Phase-transfer Catalysis

5.1 Introduction

Quaternary ammonium or phosphonium salts are known to accelerate the reaction between a nucleophile and an electrophile located in different immiscible phases by the formation of an ion-paired intermediate with the nucleophile (which is placed at the aqueous phase), increasing the solubility of this species in the organic solvent and therefore facilitating the reaction with the electrophile (which is located at the organic phase). The foundations of this concept were established by Starks together with Makosza and Brändström in the 1960s,[1] the term "phase-transfer catalysis" (PTC) being introduced by Starks in 1971 in order to explain the more than a thousand-fold rate acceleration observed in the S_N reaction of haloalkanes with aqueous sodium cyanide by the addition of catalytic amounts of hexadecyltributylphosphonium bromide (Scheme 5.1).[2] The rate-enhancing effect was attributed to the formation of a tight ion-paired species between the phosphonium cation with the cyanide anion, which facilitated the dissolution of the latter into the organic phase because of the high lipophilicity of the former. In addition, the large ionic radius of the phosphonium cation was also thought to facilitate the reaction by increasing the nucleophilicity of the cyanide anion.

Once the concept was established, the development of novel methodologies and applications of PTC in many synthetic processes occurred rather fast in the following years. Key to this rapid growth are the true practical advantages associated to the reactions carried out under PTC conditions, which usually

RSC Catalysis Series No. 5
Organocatalytic Enantioselective Conjugate Addition Reactions: A Powerful Tool for the Stereocontrolled Synthesis of Complex Molecules
By Jose L. Vicario, Dolores Badía, Luisa Carrillo and Efraim Reyes
© J. L. Vicario, D. Badía, L. Carrillo and E. Reyes 2010
Published by the Royal Society of Chemistry, www.rsc.org

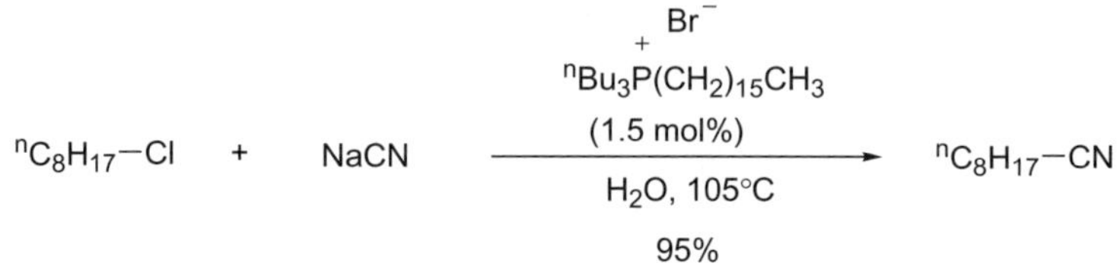

Scheme 5.1 A pioneering example of the use of phosphonium salts as phase-transfer catalysts in S_N reactions.

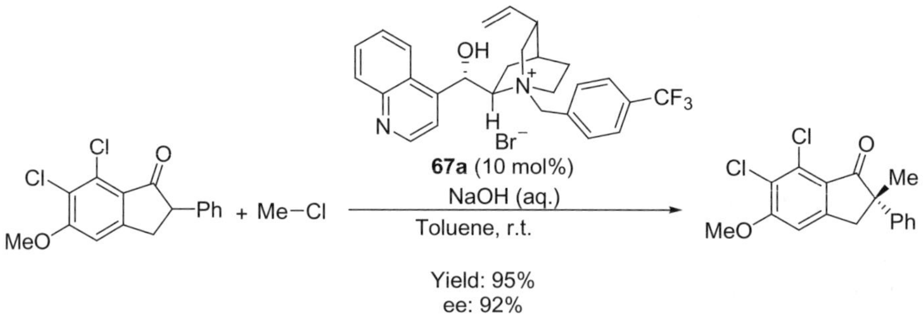

Yield: 95%
ee: 92%

Scheme 5.2 The first reported enantioselective phase-transfer catalyzed reaction.

require very simple experimental protocols and mild reaction conditions. The inclusion of water in the reaction scheme either as solvent or as co-solvent means that the reaction does not require the use of an inert atmosphere or anhydrous reagents/solvents and the possibility to carry out the reaction in water also makes the reactions under PTC catalysis more environmentally benign and advantageous from the economical point of view. Moreover, this kind of reaction can be very easily scaled up, which also increases the interest of this methodology because of its potential for large-scale preparations.

The fact that the phase-transfer catalyst is tightly associated with the nucleophile during the reaction can be employed as a convenient strategy for asymmetric catalysis, allowing achievement of stereocontrol in the reaction by using a chiral ammonium or phosphonium salt as catalyst. Following this hypothesis, the first example of an enantioselective reaction under PTC catalysis was developed by a research group at Merck in 1984 in the context of the asymmetric alkylation of indanones directed toward the preparation of (+)-*indacrinone*, a uricosuric drug (Scheme 5.2).[3] In this pioneering example, a chiral quaternary ammonium salt derived from cinchonine was used as catalyst, which was easily prepared from a commercially available and cheap starting material. It is probably this reaction, the enantioselective alkylation of enolates, which has attracted the highest level of interest among chemists working in phase-transfer catalysis.[4]

The extension of the concept to the conjugate addition reaction needed some more time to appear. Surprisingly, the first reports related to enantioselective Michael reactions under PTC conditions do not involve the use of chiral ammonium or phosphonium salts as catalysts. In the first attempts, chiral crown ethers were employed as phase-transfer catalysts, the Michael reaction of a β-ketoester with methyl vinyl ketone catalyzed by **100a** being one of the first examples reported (Scheme 5.3).[5] Nevertheless, as the progress of the field was advancing, the use of chiral quaternary ammonium salts as catalysts in this context imposed to these chiral crown ether catalysts because of their better performance in this reaction and in many others, as will be shown throughout this chapter. The first chiral ammonium salt-catalyzed Michael reaction reported in the literature described the extension of the alkylation protocol shown in Scheme 5.2 to the use of methyl vinyl ketone as electrophile and was also carried out by another research group at Merck (Scheme 5.3).[6]

Following these initial steps, PTC has become a very powerful tool for achieving stereocontrol in conjugate addition reactions, with several methodologies and catalysts showing excellent performance in several transformations.[4] However, despite all the evident advantages associated to asymmetric PTC reactions already commented on, the number of reports detailing conjugate addition reactions using this methodology is significantly more reduced than those in which enamine, iminium or H-bonding catalysis have been

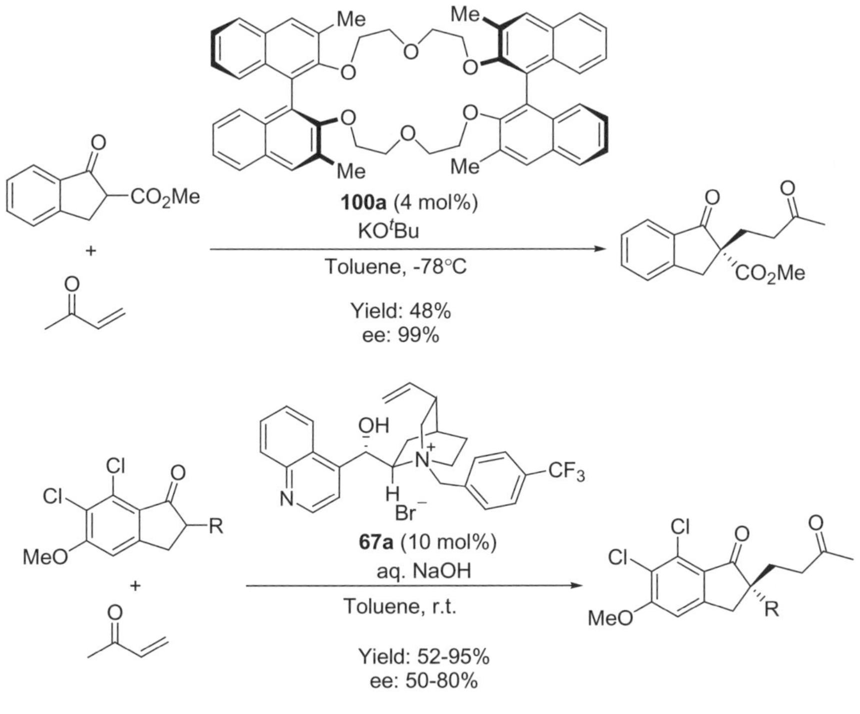

Scheme 5.3   The first enantioselective Michael reactions under PTC conditions.

employed. This is even more surprising considering that the phase-transfer catalysis concept was developed earlier than all these organocatalytic approaches. In fact, conjugate additions under PTC conditions are almost exclusively limited to the Michael reaction using enolates or related carbon nucleophiles. The possibility to carry out conjugate Friedel–Crafts or conjugate reductions is overruled in this case due to the fact that the reactions proceeding under phase-transfer catalysis involve activation of the pro-nucleophile by deprotonation by an external base, and the pro-nucleophiles participating in these particular Michael-type reactions can not be activated in this way. On the other hand, hetero-Michael reactions are extremely difficult to carry out using PTC because of the reversibility of the conjugate addition reaction, which makes the obtained adducts configurationally unstable under the basic conditions employed. This is not a problem in the Michael reactions proceeding with the formation of a C–C bond, which is known to be an irreversible process. Nevertheless, several cascade processes initiated by a hetero-Michael reaction under PTC conditions have been developed in which an irreversible reaction follows the hetero-Michael addition, pushing the equilibria toward the formation of the products and avoids the epimerization.

5.1.1   Mechanistic Considerations Regarding Michael Additions Proceeding *via* Phase-transfer Catalysis

The mechanism operating in enantioselective Michael reactions under PTC conditions involves the formation of a chiral ion-pair between the nucleophile and the catalyst as the key phenomenon operating in the stereocontrolled formation of the new stereogenic center. The reaction typically incorporates an acidic pro-nucleophile (in almost all cases an enolizable carbonyl compound), the Michael acceptor, the catalyst and a Brønsted base, which is typically an inorganic salt such as a hydroxide or a carbonate (Figure 5.1). It starts with the

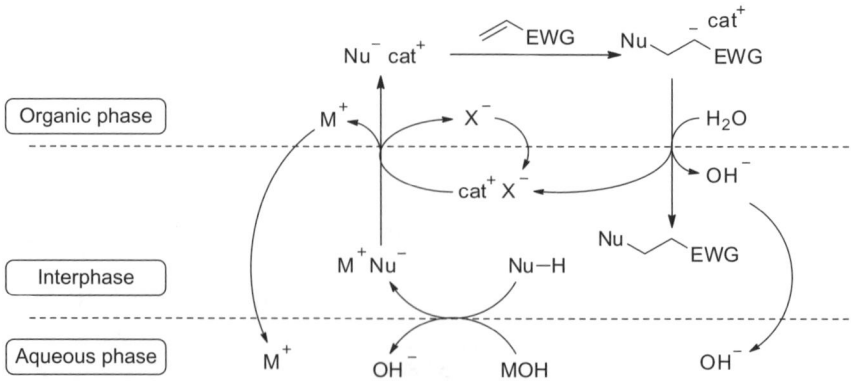

Figure 5.1   General mechanism of a conjugate addition operating under PTC.

activation of the pro-nucleophile (Nu-H) by deprotonation with the Brønsted base (MOH) at the interface between the organic and the aqueous layers, to give the corresponding anion (M^+Nu^-), which stays at the interface between the two layers. Next, ion exchange with the catalyst (cat^+) gives a lipophilic chiral nucleophile, stabilized by the catalyst (cat^+Nu^-), which is able to penetrate into the organic phase, where the conjugate addition reaction takes place with the Michael acceptor. The conjugate addition product still retains its anionic character, therefore remaining linked to the catalyst at the organic phase, but this intermediate subsequently undergoes protonation by water, liberating the catalytic ammonium salt (cat^+OH^-) which returns to the interface, ready to participate in a subsequent catalytic cycle. For the reaction to succeed, the ion-exchange process must be fast enough to allow the formation of the nucleophile-catalyst ion-paired intermediate (cat^+Nu^-) and this also has to participate in the Michael reaction faster than the deprotonated nucleophile (M^+Nu^-) itself. The catalyst is also responsible for the effective shielding of one of the stereotopic faces of the nucleophile, also requiring a tight interaction between the two ionic species, the catalyst cation and the nucleophile anion, during the formation of the stereocenter, which is reinforced by the lipophilic environment afforded by the organic solvent.

Nevertheless, not all the reactions proceeding under PTC conditions are carried out in an aqueous/organic biphasic mixture. In many cases, the inorganic base is incorporated as a solid reagent to a solution of the Michael donor, the Michael acceptor and the catalyst in an organic solvent and therefore the dynamic processes involved in the translation of reagents between the two phases have to occur at the interface of the solid reagent. These conditions are known as solid-liquid phase-transfer catalysis conditions.

As could be seen before, the most widely employed chiral PTC catalysts are quaternary ammonium salts derived from cinchona alkaloids. The quaternary quinuclidine nucleus provides the basis for the interaction with the anionic nucleophile and the organic architecture around the nitrogen cation ensures the required lipophilicity to be soluble in organic solvents. The backbone chirality of the molecule also provides the required steric/electronic bias which allows stereochemical control of the reaction. This general catalyst structure also allows fine-tuning for the preparation of a selective and highly efficient catalyst for the desired transformation (Figure 5.2). For example, it is usually found that the lateral substituent at the quaternary nitrogen atom incorporates an aromatic substituent such as a phenyl or anthracenyl group, which not only increases the solubility of the catalyst in the organic phase, but also can play a crucial role regarding stereocontrol because it can engage in π-stacking interactions with one of the reagents participating in the transformation leading to a more conformationally rigid transition state. The possibility of establishing these π-stacking interactions between the catalyst and the reagents can also happen through the quinoline substituent of the cinchona alkaloid structure. The 6′-position of this quinoline moiety can also be used for modification, in some cases being exploited for anchoring the catalyst to a solid support in the development of recoverable catalysts. Finally, the nature of the substituent at

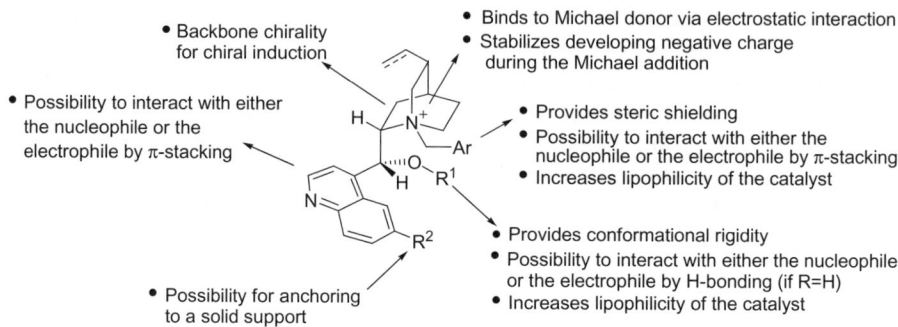

Figure 5.2   General principles applied on the design of chiral cinchona alkaloid-based chiral ammonium salt PTC catalysts.

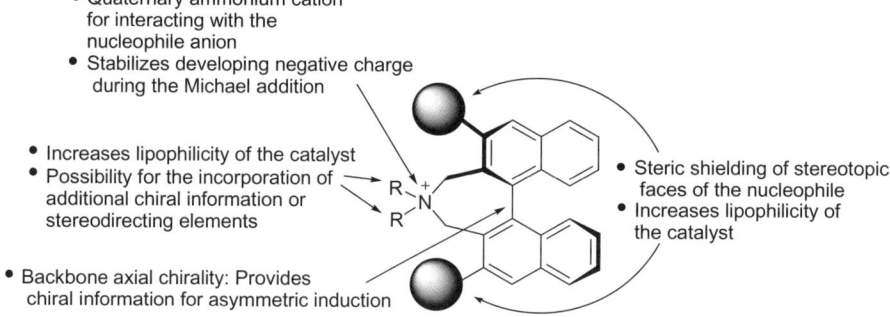

Figure 5.3   Relevant aspects of the design of axially chiral ammonium salts as PTC catalysts.

the C-9 oxygen atom (R^1) also plays an important role for stereocontrol, in the sense that, when this position is maintained as a free OH group, secondary H-bonding interactions can be established with the reagents participating in the Michael reaction. Alternatively, an alkyl side chain can be introduced, which increases the lipophilicity of the catalyst and also leads to a more conformationally rigid compound, with more restricted rotation around the NC-CHOR^1 bond.

Chiral ammonium salts based on different structural architectures have also been developed as active catalysts in Michael-type PTC reactions. Among them, the case of quaternary ammonium salts incorporating axial elements of chirality deserves special attention. The catalyst design resembles that of chiral phosphoric acids presented in the preceding chapter, that is, an axially chiral binaphthylammonium salt basic structure, incorporating bulky substituents at the *ortho* positions of the binaphthyl system, which are responsible for exerting the required steric shielding of one of the stereotopic faces of the nucleophile. The two additional substituents at the quaternary ammonium group (R groups in Figure 5.3) can be employed for introducing additional chiral substituents or stereodirecting elements in order to allow better control of the stereochemical outcome of the reaction. These two substituents and the lateral chains introduced at the binaphthyl system also contribute to the higher lipophilicity of the catalyst, increasing its solubility in organic apolar solvents.

5.2 Conjugate Addition of Stabilized Carbon Nucleophiles

As was pointed out earlier, the use of stabilized carbon nucleophiles in the conjugate addition reaction has been the most widely and, in this case, almost exclusively studied reaction under PTC conditions. In fact, PTC conditions are very convenient for this transformation because there is no need to employ compounds incorporating a highly acidic C–H hydrogen as Michael donor because the reaction scheme itself allows the introduction of an external base for the deprotonation of the donor and, therefore, almost any enolizable carbonyl compound can be employed as nucleophile in this context provided that a suitable base is employed to exert its deprotonation. Nevertheless, it should be pointed out that the nature of the base employed for this purpose has to be carefully elected because it has to be strong enough to exert the desired deprotonation but it must also be compatible with the presence of water. As can be seen in the following, most methodologies reported involve the use of *N*-protected α-aminoesters (usually glycinate-benzophenone imines) as Michael donors, which incorporate an acidic enough α-H atom to undergo deprotonation using weak bases such as aqueous metal hydroxides or carbonates and which deliver a final adduct with an α-amino acid basic architecture, with potential to be used as a chiral building block in synthesis. Alternatively, the use of 1,3-dicarbonyl compounds or nitroalkanes as pro-nucleophiles has also been intensively studied in the same context.

5.2.1 Enones as Michael Acceptors

As has already been mentioned, the first reported asymmetric Michael reactions under PTC conditions involved the use of chiral crown ethers as catalysts. These initial attempts consisted of the **100a**-catalyzed conjugate addition of a cyclic β-ketoester to methyl vinyl ketone and proceeded with excellent enantioselectivity, although in a moderate 48% yield (see Scheme 5.3). A more complex crown ether **101** derived from *chiro*-inositol has been tested afterwards with success in the reaction of ethyl glycinate benzophenone imine with methyl and ethyl vinyl ketone (Scheme 5.4).[7] Under the optimized reaction conditions, the expected Michael addition products were isolated as highly enantioenriched compounds in rather high yields. A different series of aza-crown ethers derived from glucose have also been tested in the conjugate addition of 2-nitropropane to chalcones. The structure of the catalyst was optimized with regard to the influence of the substituents at the glucose unit and at the nitrogen atom,

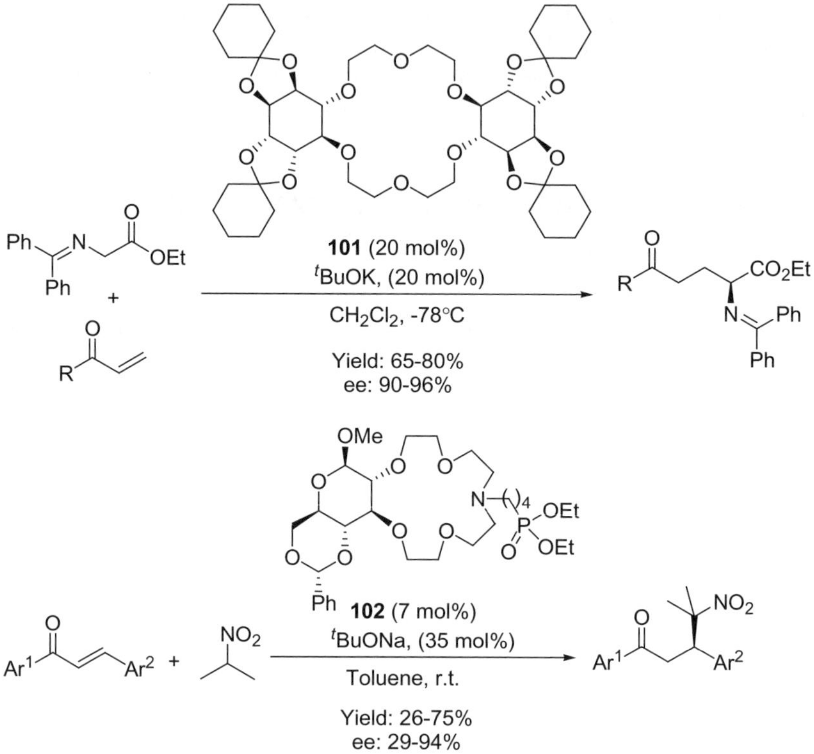

Scheme 5.4   Enantioselective Michael reaction of ethyl glycinate benzophenone imine with several enones using chiral crown ethers as catalysts.

identifying **102** as the most efficient promoter of the reaction, which provided moderate to good enantioselectivities of the final Michael adducts.[8] Nevertheless, it has to be pointed out that the use of these kinds of chiral crown ethers as PTC catalysts has been intensive using acrylates as Michael acceptors and therefore these cases will be discussed in Section 5.2.2.

Another relevant pioneering report in this context was carried out by Corey, using cinchona-based chiral ammonium salt **103a** as catalyst (Scheme 5.5).[9] In this case, the conjugate addition of *tert*-butyl glycinate benzophenone imine to cyclohexanone and ethyl vinyl ketone was found to proceed with excellent yields and enantioselectivities and, in the cyclohexanone case, also with almost complete diastereoselectivity. Cesium hydroxide was selected for this particular reaction as the most appropriate base to generate the reactive enolate species.

Interestingly, the authors also found that when enolizable aryl alkenyl ketones were stirred in the presence of catalyst **103b**, a self-condensation reaction took place furnishing a series of dimerization products with high yields and enantioselectivities (Scheme 5.6).[10] This reaction had to take place by, first, deprotonation of the γ-CH group on one molecule of the starting material,

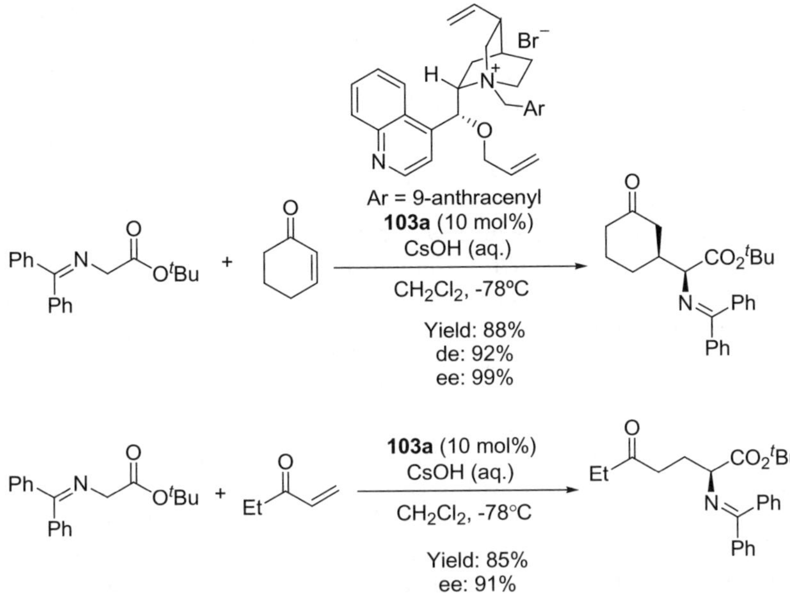

Scheme 5.5   Enantioselective **103a**-catalyzed Michael reaction of *tert*-butyl glycinate benzophenone imine with several enones under PTC conditions.

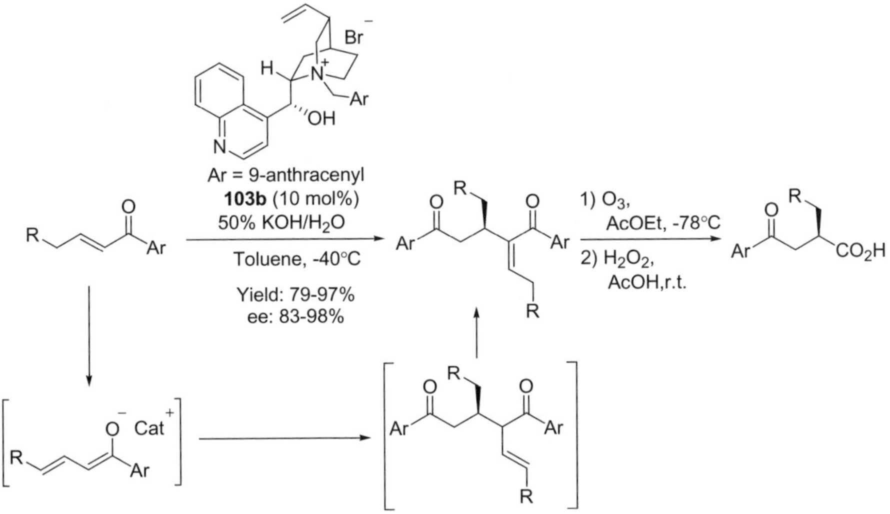

Scheme 5.6   Enantioselective dimerization of enolizable enones catalyzed by **103b**.

followed by conjugate addition through the α-carbon to a second molecule of the same substrate and a final C=C bond transposition to afford a more stable conjugated product. The obtained dimeric products were subsequently transformed into useful chiral building blocks such as enantioenriched α-substituted γ-oxoacids by ozonolytic cleavage of the C=C double bond followed by oxidative cleavage of the formed α-diketone intermediate.

The same group has also reported the enantioselective conjugate addition of nitromethane to chalcones, which was carried out using a cinchonine-based chiral phase transfer catalyst **67b**. Although a somewhat lower enantioselectivity was observed, highly enantioenriched products could be obtained by a single recrystallization. The high acidity of nitromethane allowed the use of a very weak base such as CsF for the activation of the nucleophile. The methodology was applied for the enantioselective synthesis of (*R*)-*baclofen* (Scheme 5.7) by carrying out a subsequent Baeyer–Villiger oxidation, reduction and hydrolysis under standard conditions.[11] As can be seen, the use of catalyst **67b** with a basic pseudoenantiomeric structure with regard to the quinidine-based catalyst **103a** previously employed in the reaction with *tert*-butyl glycinate benzophenone imine shown in Scheme 5.5 led to a different sense of stereoinduction in the reaction, as is usually found in almost all reactions using cinchona alkaloid-based catalysts. This feature was also exploited by Corey for the synthesis of (*S*)-*baclofen* using the quinidine-based analogue catalyst **103c**.[11a]

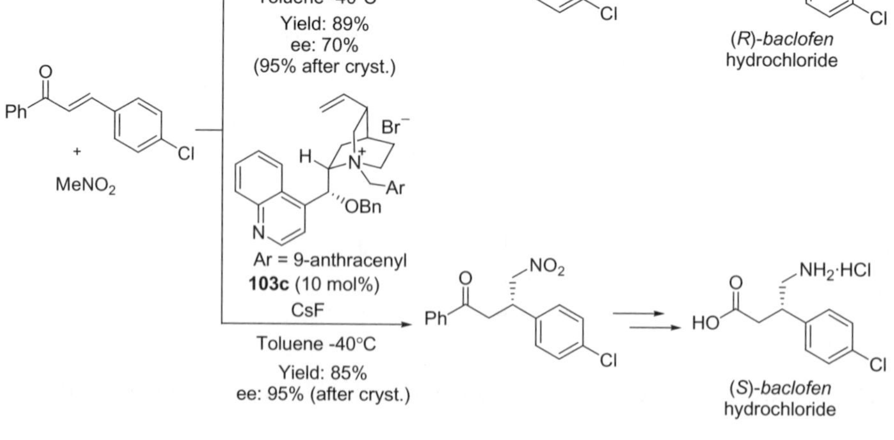

Scheme 5.7   Enantioselective Michael reaction of nitromethane to chalcones applied to the enantioselective synthesis of *baclofen*.

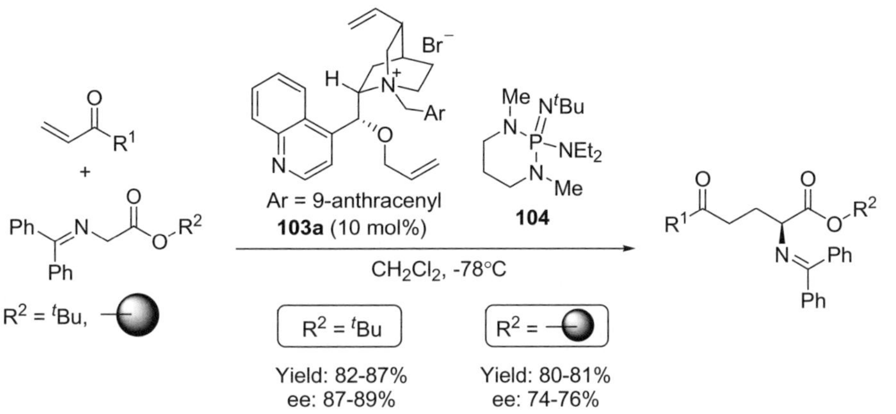

Scheme 5.8   Enantioselective Michael reaction of glycinate benzophenone imines with several enones in solid phase and in solution.

The conjugate addition of glycinate benzophenone imines to methyl vinyl ketone and ethyl vinyl ketone can be carried out in solution and in solid-phase using cinchonidine-based ammonium salt **103a** as catalyst (Scheme 5.8).[12] An interesting contribution results from the introduction of a non-ionic organic base **104** in the reaction scheme for the generation of the enolate. In the solid-phase reaction, a resin-bound glycinate imine was employed as nucleophile, which allowed an easy purification and isolation of the final product. The results obtained were similar to those observed by Corey in the same reaction,[11a] also using catalyst **103a** but in the presence of an inorganic base. Remarkably, the resin-bound substrates afforded similar results to those obtained in the solution phase reaction, with only a small decrease on the enantioselectivity for the solid-phase reactions.

Malonates can also be successfully employed as Michael donors in the reaction with enones under PTC catalysis. For example, the reaction of dimethyl malonate with cyclopentenone using **103d** as catalyst has been shown to proceed with excellent yield and enantioselectivity, which was used by the authors for achieving the preparation of both enantiomers of *methyl dihydrojasmonate*, unnatural compounds which are constituents of famous fragrances because of their important olfactory properties (jasmine-like odor).[13] The reaction consists of the Michael reaction of dimethyl malonate to a cyclic enone incorporating an alkyl substituent in the α position, therefore generating two stereogenic centers in a fully diastereo- and enantioselective way, which are in the correct configuration for the synthesis of the fragrance compound (Scheme 5.9). Subsequent monohydrolysis/decarboxylation afforded the target product in good overall yield and as a highly enantiopure material.

Other *N*-benzyl quaternary ammonium salts derived from cinchona alkaloids have been used for the direct addition of malonates to chalcones (Scheme 5.10). In the screening of the reaction, the structure of the catalyst needed some optimization and, after some experiments, the authors identified

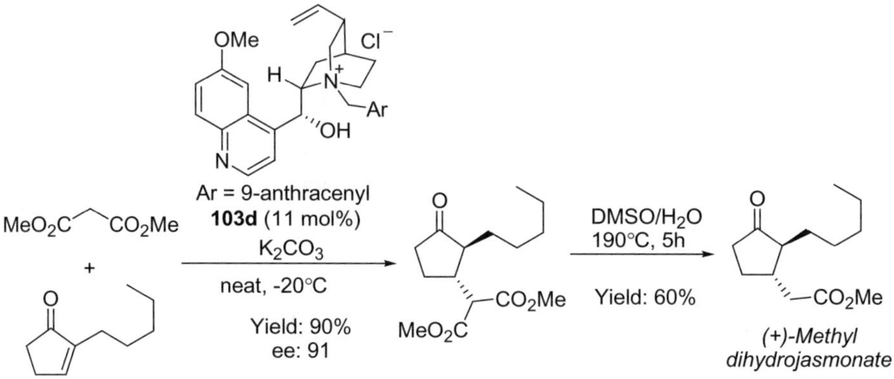

Scheme 5.9   Enantioselective synthesis (+)-*methyl dihydrojasmonate* using enantio-selective phase-transfer catalysis.

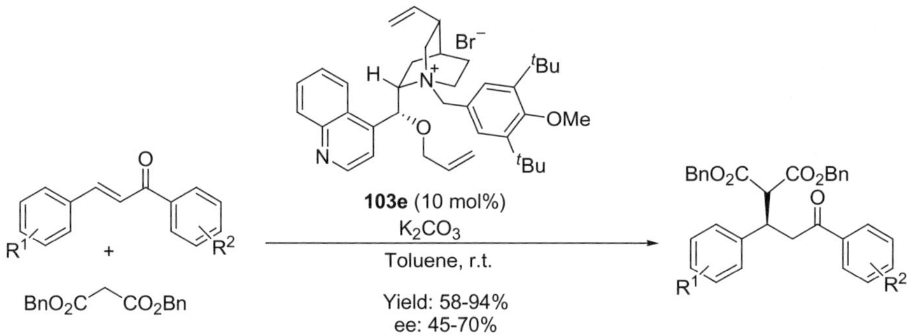

Scheme 5.10   Enantioselective Michael reaction of malonates to enones catalyzed by **103e**.

cinchonidinium salt **103e** as the best promoter of the reaction,[14] in which the OH group of cinchonidine was modified as the corresponding allyl ether and a benzyl group incorporating bulky substituents at the aromatic ring was introduced as the lateral substituent at the nitrogen atom of the quinuclidine nucleus when preparing the quaternary ammonium salt. Nevertheless, despite intensive efforts, only moderate enantioselectivities were obtained for a series of Michael acceptors tested.

   Much more recently, the conjugate addition of 1-fluoro-bis(phenylsulfo-nyl)methane to chalcones has been carried out using another different cinchona alkaloid-based quaternary ammonium salt as catalyst (Scheme 5.11).[15] In this case, the catalyst was obtained from quinidine and incorporated a different bulky arylmethyl group attached to the quinuclidine nitrogen, and maintained the OH group without being modified. Good yields and enantioselectivities were obtained in the majority of the cases. A model was proposed to account for the stereochemical outcome of the reaction based on the known crystal

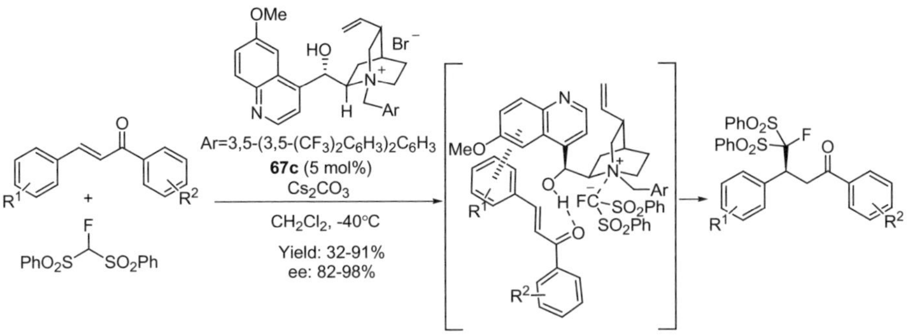

Scheme 5.11 Enantioselective Michael reaction of 1-fluoro bis(phenylsulfonyl)methane to enones.

structure of a complex between a similar catalyst and malonic acid together with some orientative conformational analysis carried out by PM3 calculations. This model proposes that the free OH-group of the catalyst would participate in the formation of a hydrogen bond with the enone carbonyl, leading to a more rigid transition state and also that π-stacking interactions between the β-aryl substituent at the Michael acceptor and the quinolyl substituent of the catalyst on one hand and between the aromatic ketone moiety of the acceptor and the *N*-substituent of the catalyst on the other would also contribute to the stereocontrol. These proposals were also supported by the poor results furnished by a similar *O*-methylated catalyst tested and also by the fact that better results were obtained with enones containing β-aryl groups incorporating electron-releasing substituents, while β-alkyl substituted enones afforded somewhat lower levels of enantioselection. The need for a very bulky substituent at the quaternary nitrogen atom was explained in terms of a better shielding of one of the enantiotopic faces of the enone.

In this context, there is a relevant example of a newly designed cinchona-alkaloid derived bis-ammonium salt **105** employed as catalyst in the Michael reaction of cyclic β-ketoesters with methyl vinyl ketone (Scheme 5.12).[16] Excellent yields and moderate enantioselectivities of the corresponding Michael adducts were obtained under the best reaction conditions, which also allowed the use of an organic base (Hünig base) for the deprotonation of the β-ketoester. However, perhaps the most relevant feature associated to the use of this catalyst is the fact that it can be easily separated from the reaction medium by precipitation in ether, which allowed its recycling for further uses without loss of activity.

The behavior of polyunsaturated systems such as α,β,γ,δ-unsaturated ketones as electrophiles in the Michael reaction under PTC conditions has been the subject of a thoughtful study (Scheme 5.13).[17] The conjugate addition of β-ketoesters to this kind of particular conjugated systems has been found to occur with complete 1,6-selectivity, affording a final β,γ-unsaturated adduct, which in some exceptional cases can isomerize to the more stable conjugated

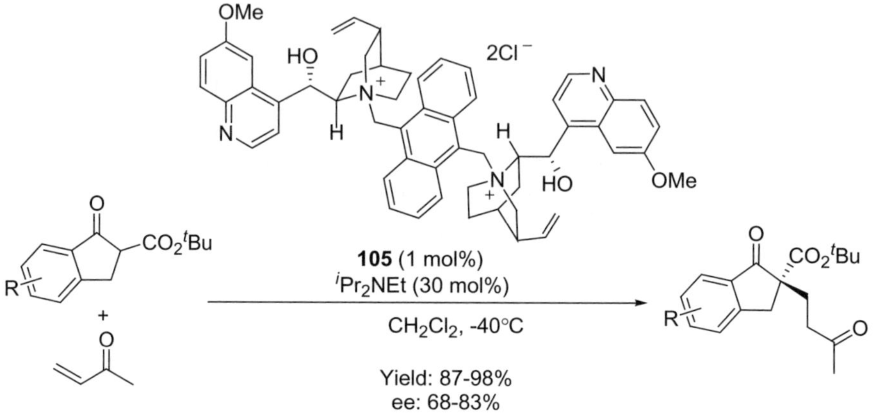

Scheme 5.12 Enantioselective Michael reaction of cyclic β-ketoesters with methyl
vinyl ketone using bis-quinidinium salt **105** as catalyst.

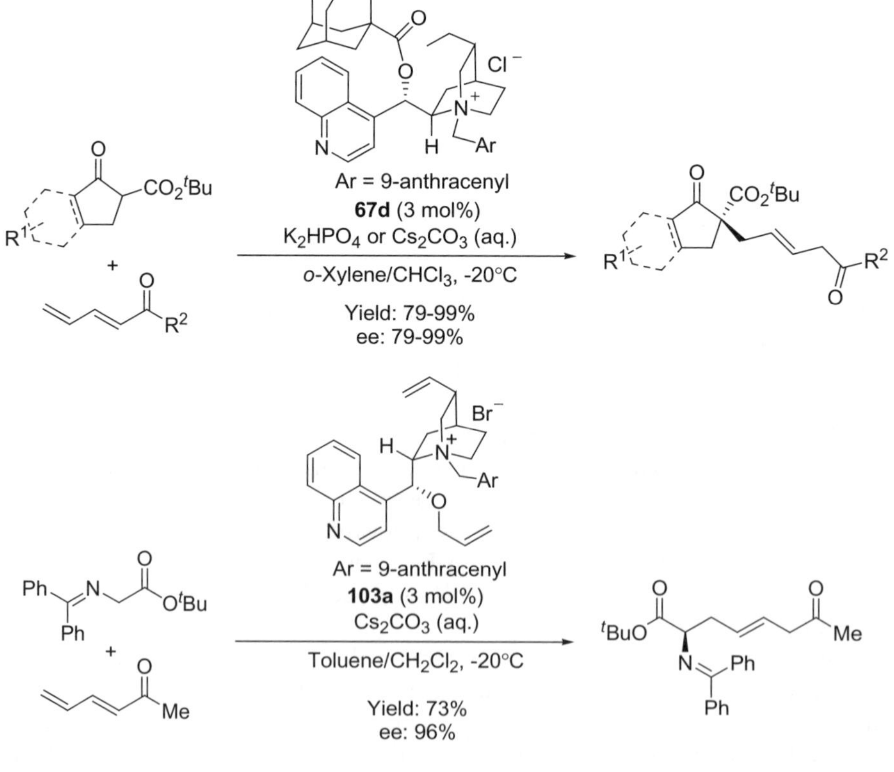

Scheme 5.13 Enantioselective PTC Michael reaction using α,β,γ,δ-unsaturated
ketones.

enone. *O*-Adamantoyl modified dihydrocinchonium derivative **67d** was identified as a very efficient catalyst for this transformation, affording excellent yields and enantioselectivities for a variety of polyunsaturated ketones tested. It should be noted that the nature of the base employed in the reaction also had an important influence in the final result, the mild K_2HPO_4 being the most effective one in the majority of cases, although in some of them the stronger Cs_2CO_3 was found to be necessary. The use of *tert*-butyl glycinate benzophenone imine as Michael donor was also surveyed, requiring a different catalyst (cinchonidine-derived ammonium salt **103a**) and also an even stronger base like KOH for the deprotonation of the glycinate. Under the optimized reaction conditions good yields and outstanding enantioselectivities were obtained. It also has to be noted that other different Michael acceptors such as α,β,γ,δ-unsaturated esters and sulfones were also employed with success in this reaction.

The same group has also surveyed the use of allenic ketones as Michael donors in the reaction with cyclic β-ketoesters (Scheme 5.14).[18] In this case, the reaction took place at the β-position of the allenone electrophile, also furnishing a β,γ-unsaturated ketone product and generating a quaternary stereogenic center. As happened in the previous reaction with α,β,γ,δ-unsaturated ketones, the same *O*-adamantyl containing ammonium salt **67d** was identified as the best catalyst for the reaction, which allowed the preparation of a variety of differently substituted Michael adducts with good yields and enantioselectivities. Moreover, when γ-substituted Michael acceptors were employed, excellent diastereoselectivity control was also achieved regarding the geometry of the C=C double bond remaining at the final product, favoring the formation of the less sterically hindered *E* diastereoisomer. As happened in the previous case, allenic esters were also employed with success as Michael acceptors in the same context.

The Mukaiyama–Michael strategy is a very useful tool for performing Michael addition using compounds of lower acidity which can not be deprotonated using

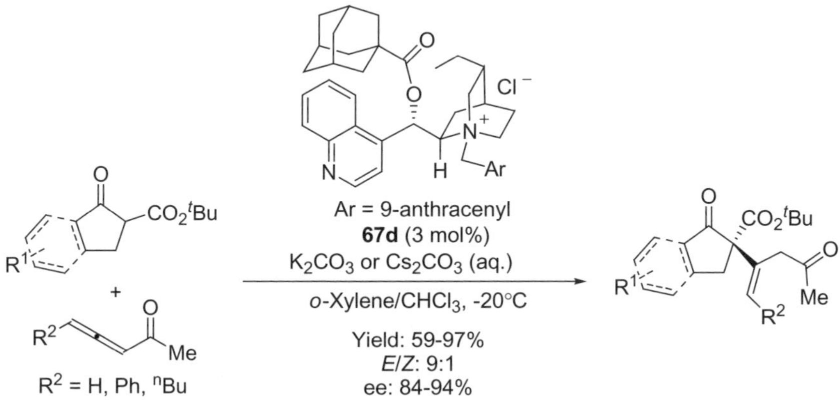

Scheme 5.14 Enantioselective PTC Michael reaction of β-ketoesters with allenic ketones.

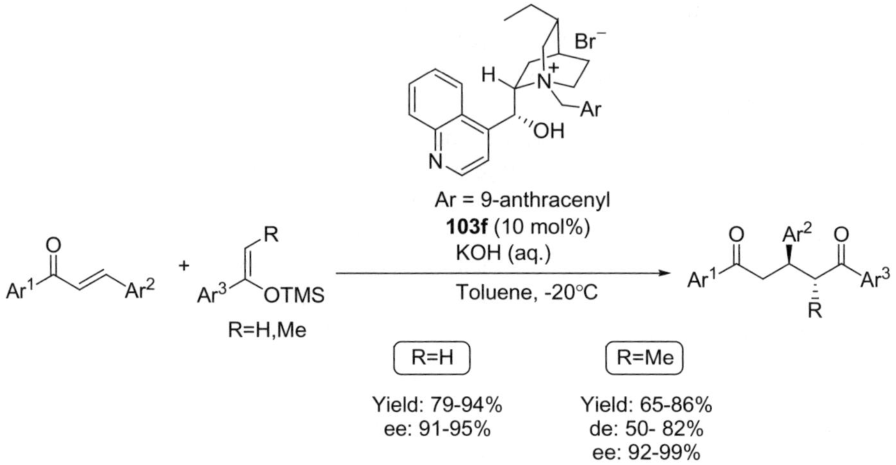

Ar = 9-anthracenyl
103f (10 mol%)
KOH (aq.)

Toluene, -20°C

R=H,Me

R=H	R=Me
Yield: 79-94% ee: 91-95%	Yield: 65-86% de: 50- 82% ee: 92-99%

Scheme 5.15 Enantioselective Mukaiyama–Michael reaction under PTC conditions.

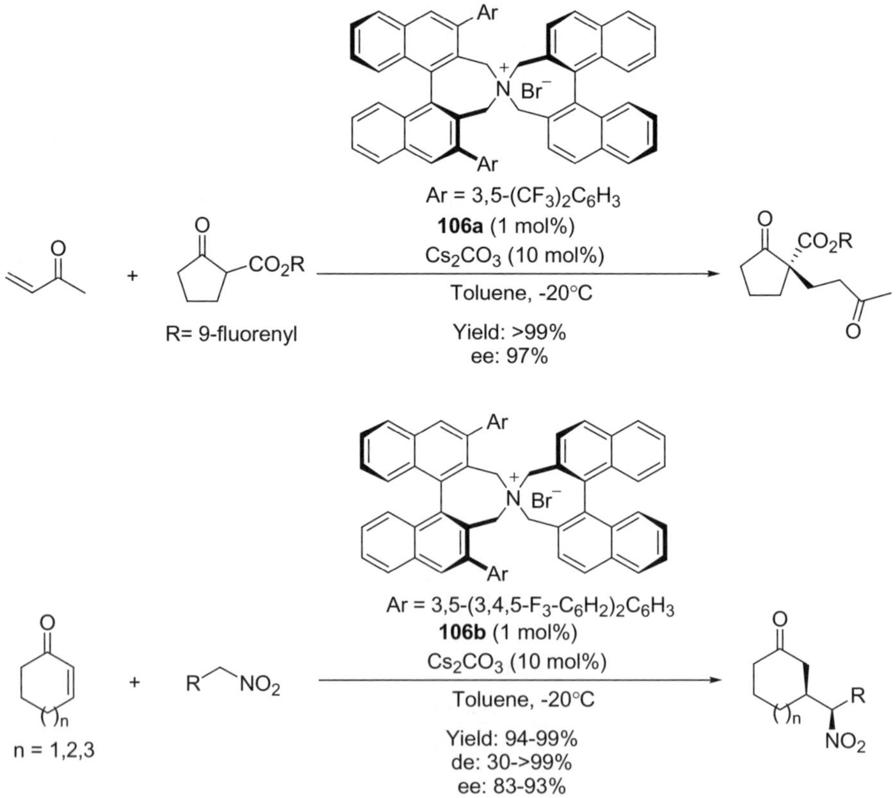

Ar = 3,5-(CF$_3$)$_2$C$_6$H$_3$
106a (1 mol%)
Cs$_2$CO$_3$ (10 mol%)

Toluene, -20°C

R= 9-fluorenyl

Yield: >99%
ee: 97%

Ar = 3,5-(3,4,5-F$_3$-C$_6$H$_2$)$_2$C$_6$H$_3$
106b (1 mol%)
Cs$_2$CO$_3$ (10 mol%)

Toluene, -20°C

n = 1,2,3

Yield: 94-99%
de: 30->99%
ee: 83-93%

Scheme 5.16 Enantioselective Michael reactions catalyzed by *N*-spiro ammonium salts **106a** and **106b**.

one of the weak bases usually employed in PTC reactions. In this sense, Corey and coworkers have developed a procedure for carrying out the conjugate addition of silyl enol ethers to chalcones using **103f** as catalyst, which has shown up as a very useful protocol to perform the formal conjugate Michael reaction of simple enolizable ketones to enones.[19] As can be envisioned in Scheme 5.15, differently substituted chalcones can be used in this transformation as Michael acceptors, obtaining the final adducts with good to excellent yields and very high enantioselectivities. Moreover, α-substituted silyl enol ethers could also be employed as Michael donors, conducing to the formation of conjugate addition products containing two contiguous stereocenters and observing that, in this case, the reaction proceeded with very high diastereoselectivity, favoring the formation of the *anti* diastereoisomer and maintaining the excellent efficiency observed in the same reactions with α-unsubstituted silyl enol ethers.

Chiral ammonium salts other than those derived from cinchona alkaloids have also been developed for carrying out this reaction with great success (Scheme 5.16). For example, *N*-spiro ammonium bromide **106a** incorporating an axially chiral binaphthyl backbone has been identified as an outstanding catalyst for the conjugate addition of a cyclic β-ketoester to methyl vinyl ketone, requiring the use of a fluorenyl ester as pro-nucleophile in order to obtain a high enantioselectivity.[20] A closely related catalyst **106b** has also been successfully employed in the Michael reaction of nitroalkanes with cyclic enones.[21] Almost quantitative yields of the final adducts were obtained for a variety of nitroalkanes and cyclic enones of different ring size. The diastereoselectivity of the reaction was highly dependent upon the size of the substituent at the nitroalkane, observing the exclusive formation of the *syn* diastereoisomer in the cases in which rather bulky substituents such as *i*-Pr or cyclohexyl were incorporated. Regarding enantioselection, the major diastereoisomer was obtained in higher than 80% ee in all cases tested.

The same group has also studied this family of spiro-ammonium catalysts in the reaction of α-substituted α-cyanoacetates with acetylenic ketones (Scheme 5.17).[22] In this case, intensive efforts were directed toward the identification of the best catalyst for the reaction, leading to a new ammonium salt **106c** as the best promoter for the reaction, which allowed the final adducts to be obtained in almost quantitative yield in most cases. However, enantioselectivity remained generally in values around 85% ee and rather poor diastereoselectivities were observed regarding stereocontrol on the geometry of the C=C double bond arising in the final products formed after the conjugate addition process.

The conjugate addition of malonates to chalcones has been carried out using modified *N*-spiro quaternary ammonium salt **107**,[23] in which only one binaphthyl group was attached to the quaternary nitrogen atom, the remaining two substituents at this nitrogen atom being attached to a biphenyl-type moiety. The introduction of two OH groups at the benzylic positions of the two 2′ substituents of this biphenyl backbone was crucial to achieve good enantioselection, although no rationale was given to explain this effect (Scheme 5.18). The nature of the alkoxy substituent of the malonate also played an important

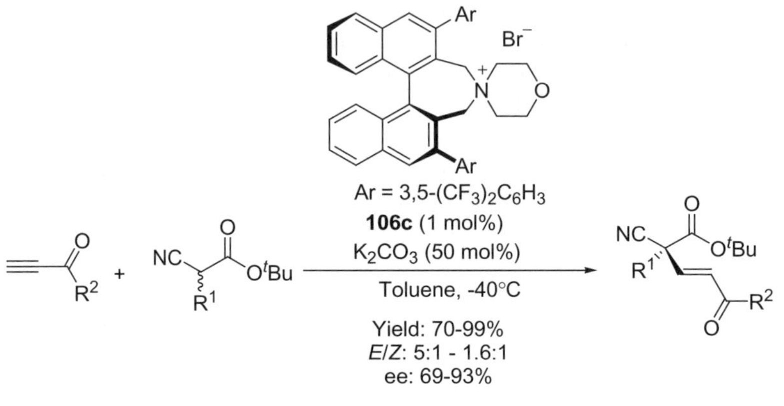

Ar = 3,5-(CF$_3$)$_2$C$_6$H$_3$
106c (1 mol%)
K$_2$CO$_3$ (50 mol%)

Toluene, -40°C

Yield: 70-99%
E/Z: 5:1 - 1.6:1
ee: 69-93%

Scheme 5.17   Enantioselective Michael reaction of cyanoacetates with acetylenic ketones.

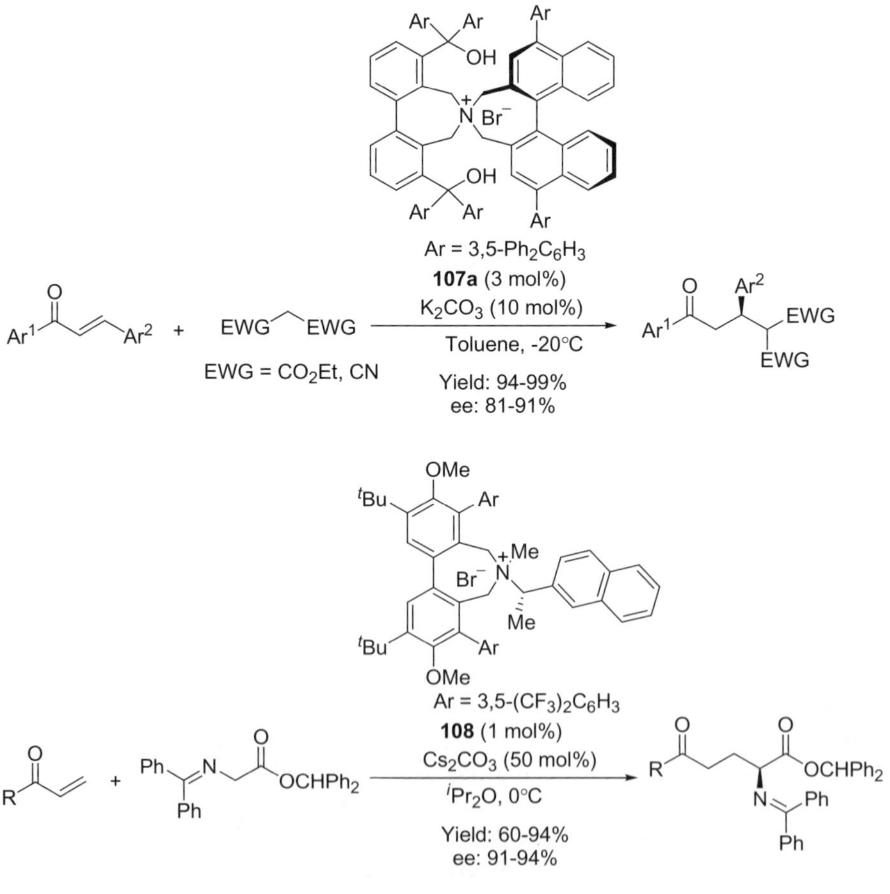

Ar = 3,5-Ph$_2$C$_6$H$_3$
107a (3 mol%)
K$_2$CO$_3$ (10 mol%)

Toluene, -20°C

Yield: 94-99%
ee: 81-91%

EWG = CO$_2$Et, CN

Ar = 3,5-(CF$_3$)$_2$C$_6$H$_3$
108 (1 mol%)
Cs$_2$CO$_3$ (50 mol%)

^{i}Pr$_2$O, 0°C

Yield: 60-94%
ee: 91-94%

Scheme 5.18   Enantioselective Michael reactions catalyzed by **107** and **108**.

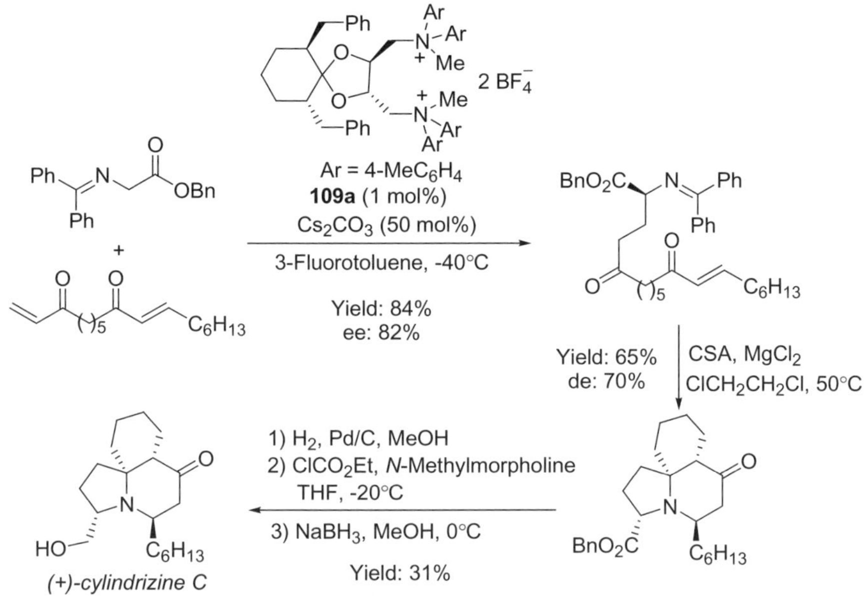

Scheme 5.19 The total synthesis of *cylindrizine C* using a **109a**-catalyzed Michael reaction as key step.

role in the reaction obtaining good enantioselectivities for methyl and ethyl malonates, but very poor results were observed with other malonates derived from bulkier alcohols. Under the optimized conditions, the obtained yields and enantioselectivities were remarkably high for the reaction of a variety of chalcones with diethyl malonate and also the performance of the catalyst was tested in the conjugate addition of malononitrile, furnishing an excellent result. A different chiral ammonium salt **108** incorporating a biphenyl-type basic structure has also been developed for the reaction of glycinate imines to alkyl vinyl ketones furnishing excellent results.[24]

On the other hand, bis-ammonium salt **109a** derived from TADDOL has been expressly designed by Shibasaki for the conjugate addition of benzyl glycinate benzophenone imine to heptadec-1,10-dien-3,9-dione as the key step toward the enantioselective total synthesis of (+)-*cilindrizine C* (Scheme 5.19),[25] a natural product isolated from marine sources which exhibits bioactivity against a DNA-repair-deficient yeast strain and also displaying important anti-tumor activities. After extensive work, the authors found that an excellent yield and a synthetically useful 84% ee of the Michael adduct could be obtained using 10 mol% of **109a**, with Cs_2CO_3 as the base required for the activation of the nucleophile and using the rather exotic 3-fluorotoluene as the reaction solvent. Next, the obtained Michael adduct was subjected to a highly diastereoselective cascade reaction consisting of the hydrolysis of the benzophenone imine followed by iminium formation and subsequent intramolecular Mannich reaction and a final aza-Michael process. The target

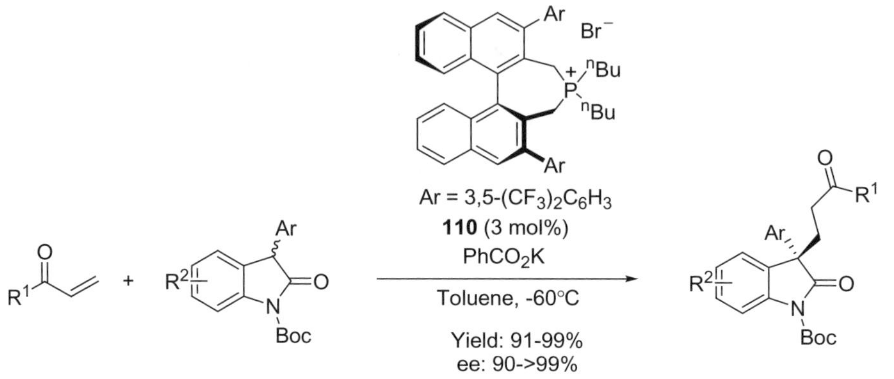

Scheme 5.20 Enantioselective Michael reaction of 3-aryl oxindoles with enones catalyzed by phosphonium salt **110**.

natural product was obtained after chemoselective reduction of the benzyloxycarbonyl moiety.

Finally, it should also be pointed out that enantioselective Michael reactions to enones under PTC conditions do not exclusively rely on the use of chiral ammonium salts as catalysts and, for example, chiral phosphonium salts can also be successfully employed in this context. This is the case of binaphthyl-containing phosphonium salt **110**, which was demonstrated to be an outstanding catalyst in the conjugate addition of oxindoles to enones under PTC conditions (Scheme 5.20).[26] The Michael adducts were obtained in excellent yields and enantioselectivities for a wide variety of differently substituted 3-aryl oxindoles tested as Michael donors. Remarkably, the high acidity of the 3-aryloxindoles employed as Michael donors allowed the use of a very mild base such as potassium benzoate for the activation of the nucleophile.

5.2.2 Acrylic Acid Derivatives as Michael Acceptors

Almost all the catalysts presented up to this point have also been tested in the Michael reaction with acrylic acid derivatives, typically acrylate esters and acrylonitrile. The usual situation found is that these kinds of acceptors furnish similar results to those observed in the same reaction with vinyl ketones, although in some cases a different behavior is observed. On the other hand, there are also some catalytic systems which have been exclusively tested on acrylate-type electrophiles, as will be shown in the following.

In this context, crown ether **101**, which performed rather well in the reaction of glycine imines with vinyl ketones (see Scheme 5.4), has been employed in the addition of the same kinds of nucleophiles to different alkyl acrylates, furnishing good yields and enantioselectivities (Scheme 5.21).[7] The same catalyst performed poorly in the same reaction with acrylonitrile, furnishing a rather low degree of enantioselection. Similarly, derivative **100b** was employed to

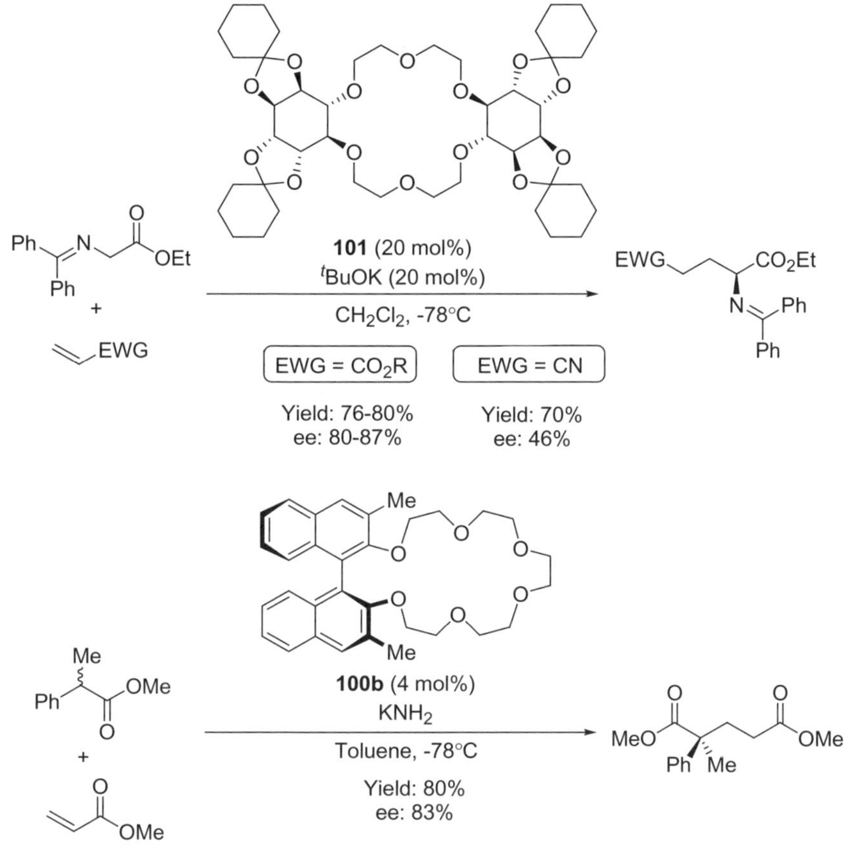

Scheme 5.21 Some Michael additions to acrylic acid derivatives catalyzed by chiral crown ethers.

catalyze the reaction of methyl 2-phenylpropanoate with methyl acrylate, delivering the corresponding adduct in 80% yield and 83% ee.[4] The best reaction conditions required the use of a rather strong base such as potassium amide for the deprotonation of the substrate.

Many other different efforts have been directed toward the synthesis and design of an efficient crown ether catalyst able to catalyze this reaction with better results. In general, the Michael addition of methyl phenylacetate to methyl acrylate has been employed as the test reaction in order to evaluate the performance of the new catalysts prepared. As can be seen in Scheme 5.22, variable results have been obtained for many different crown ethers employed in this reaction but, as a general rule, it can be said that all these newly developed systems have been demonstrated to participate as rather efficient promoters of the reaction with regard to catalytic activity, but enantioselectivities have remained in moderate or low values. In all cases, the size of the crown ether macrocycle has been chosen according to the nature of the cation

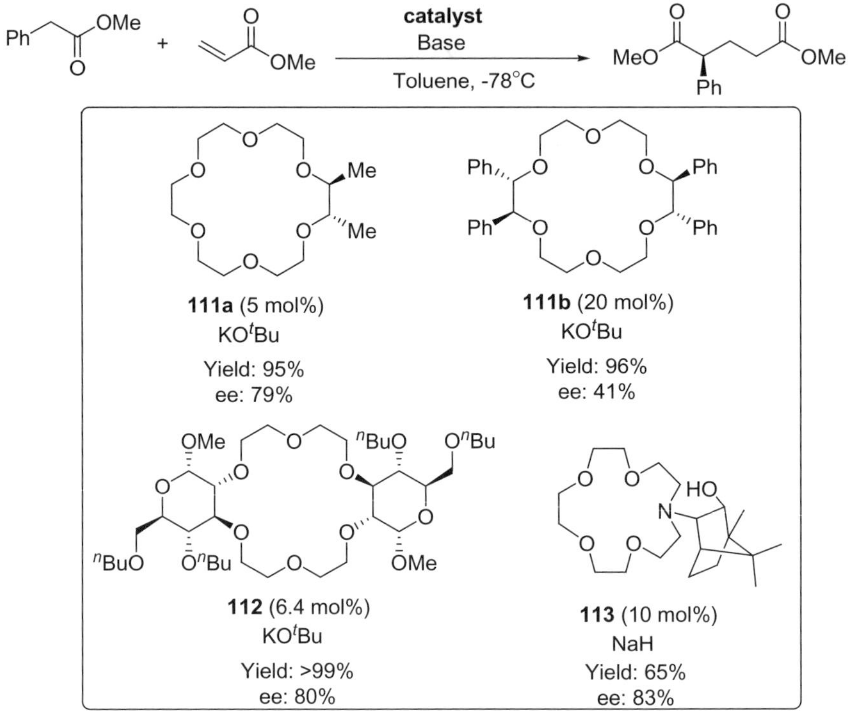

Scheme 5.22   Some chiral crown ethers employed as catalysts in the Michael reaction of methyl phenylacetate with methyl acrylate.

of the inorganic base employed to activate the nucleophile. Relevant examples of these chiral crown ether catalysts include the use of simple C_2-symmetric chiral 18-crown-6 derivatives **111a**[27] and **111b**,[28] the more complex C_2-symmetric sugar-based 18-crown-6 compound **112**[29] and camphor-derived aza-crown ether **113**.[30] A model explaining the stereochemical outcome of the reaction has also been provided for catalyst **113** using computational methods which correlated well with the experimental results.

With respect to the use of chiral quaternary ammonium salts derived from cinchona alkaloids as catalysts, the addition of *tert*-butyl glycinate benzophenone imine to methyl acrylate had already been studied by Corey in 1998 in his first report on cinchonidinium salt **103a**-catalyzed enantioselective Michael reactions using enones as Michael acceptors shown in Scheme 5.5.[9] Under the same reaction conditions, both methyl acrylate and acrylonitrile reacted efficiently, furnishing the final Michael adducts in excellent yields and enantioselectivities (Scheme 5.23).[31] The same reaction using a solid-supported version of the glycine imine Michael donor and methyl acrylate or acrylonitrile as acceptors has also been studied under the same conditions as those shown in Scheme 5.8 for the reaction using enones as Michael acceptors, which also

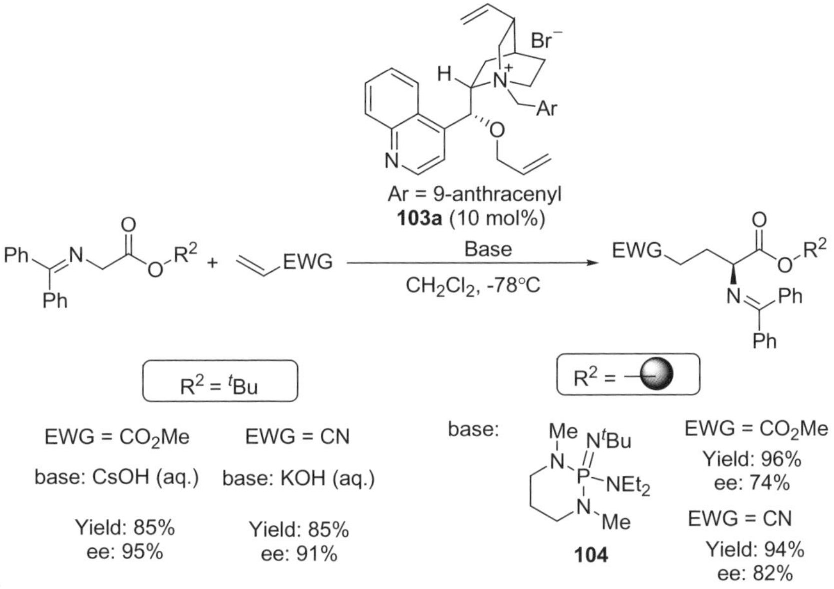
Br⁻
H
N
Ar
O
N
Ar = 9-anthracenyl
103a (10 mol%)
Ph
N
O
O
R²
+
EWG
Base
CH₂Cl₂, -78°C
EWG
O
O
R²
N
Ph
Ph
Ph
R² = ᵗBu
R² =
EWG = CO₂Me
base: CsOH (aq.)
Yield: 85%
ee: 95%
EWG = CN
base: KOH (aq.)
Yield: 85%
ee: 91%
base:
Me
NᵗBu
N
P
NEt₂
N
Me
104
EWG = CO₂Me
Yield: 96%
ee: 74%
EWG = CN
Yield: 94%
ee: 82%

Scheme 5.23 Enantioselective Michael reactions glycinate imines with acrylic acid derivatives catalyzed by **103a**.

involved the incorporation of **104** as a more convenient base for the activation of the pro-nucleophile.[12]

In a much more recent report, conveniently *O*-protected α-hydroxymethyl aryl ketones were also successfully employed as Michael donors in the conjugate addition to methyl acrylate and acrylonitrile using a cinchonine-based ammonium salt as catalyst.[32] Bis-quinidine-based ammonium salt **105** has also shown to be useful in the reaction of cyclic β-ketoesters to methyl and ethyl acrylate.[16]

Other chiral ammonium salts not derived from cinchona alkaloids have also been used in this context. This is the case, for example, of the tartaric acid-derived quaternary ammonium salts having a spiroammonium functionality introduced by Arai and coworkers for the reaction of *tert*-butyl glycinate benzophenone imines with *tert*-butyl acrylate (Scheme 5.24).[33] In this particular case, several spiroammonium catalysts bearing different substituents have been synthesized and studied in the reaction, obtaining in general high yields of the conjugate addition product, although the enantioselectivities remained moderate in most cases, compound **114** being the catalyst which showed the best performance. The solvent was also found to play an important role in the reaction with regard to stereocontrol, leading to the best 77% ee with the less common *tert*-butyl methyl ether.

On the other hand, axially chiral *N*-spiro ammonium salt **106d** has been found to catalyze the Michael reaction of a serine nucleophile equivalent with ethyl

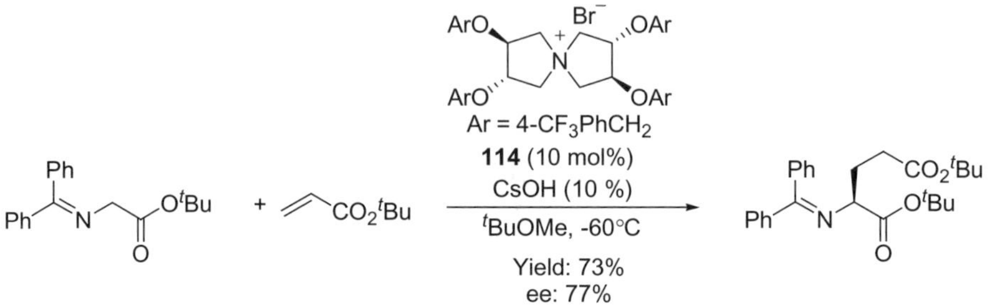

Scheme 5.24   Enantioselective **114**-catalyzed Michael reaction of a glycine imine with *tert*-butyl acrylate.

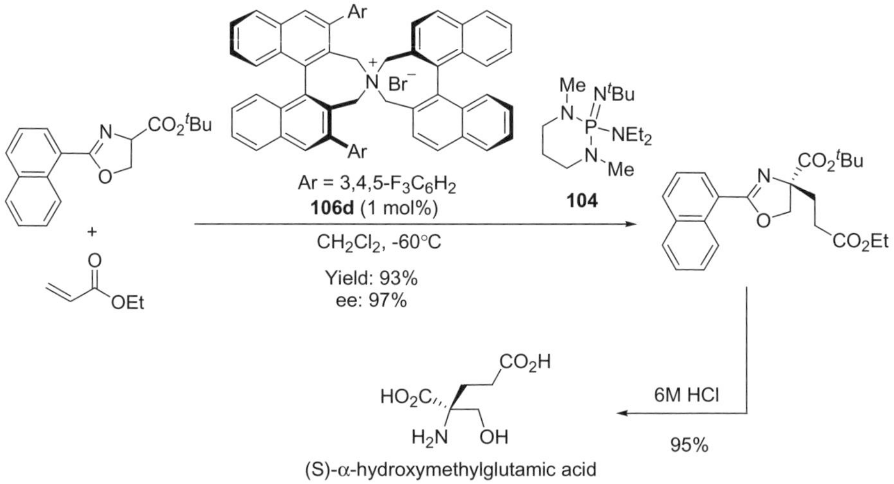

Scheme 5.25   Enantioselective Michael reaction for the synthesis of (*S*)-α-hydroxymethylglutamic acid.

acrylate (Scheme 5.25).[34] The authors found that enantioselectivity was strongly dependent on the nature of the base employed to deprotonate the substrate and after an extensive research it was found that the final Michael adduct could be isolated in excellent yield and enantioselectivity using phosphazene base **104**. In addition, the aromatic group linked to the oxazoline moiety was also found to play an important role in the reaction, not only with respect to the yield but also regarding the enantioselection. The best results were obtained using a 1-naphthyl substituent. The authors also carried out the synthesis of (*S*)-α-hydroxymethylglutamic acid, a potent glutamate neurotransmitter receptor ligand, by simple hydrolytic cleavage of the oxazoline group.

The Michael reaction of α-substituted cyanoacetates with acetylenic esters can also be carried out in a very efficient way using a different

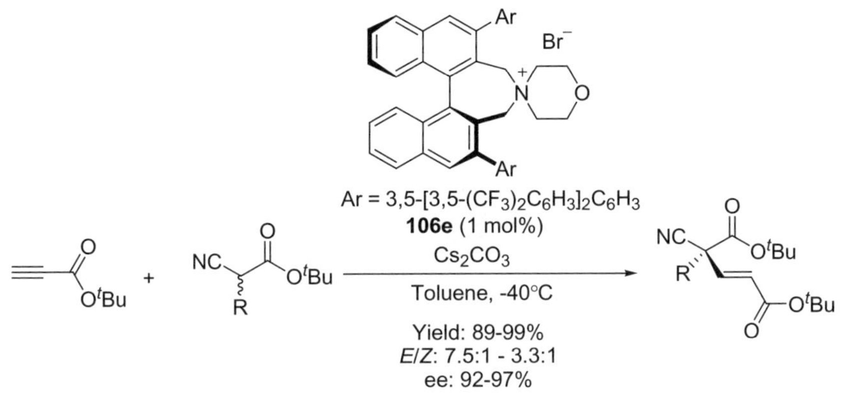

Ar = 3,5-[3,5-(CF$_3$)$_2$C$_6$H$_3$]$_2$C$_6$H$_3$
106e (1 mol%)
Cs$_2$CO$_3$

Toluene, -40°C

Yield: 89-99%
E/*Z*: 7.5:1 - 3.3:1
ee: 92-97%

Scheme 5.26 Enantioselective Michael reaction of cyanoacetates with acetylenic ketones.

N-spiroammonium catalyst of type **106** but incorporating a binaphthyl chiral backbone and a morpholine unit. Under the optimized reaction conditions, a variety of *tert*-butyl cyanoacetates were reacted with *tert*-butyl propiolate providing the final adducts in almost quantitative yields and with excellent enantioselectivities using **106e** as the most efficient catalyst (Scheme 5.26).[35] However, regarding the geometry control at the newly generated C=C double bond, the diastereoselectivities remained in moderate values, ranging from 7:1 to 2:1 *E/Z* ratios, favoring the more stable *E* diastereoisomer. Interestingly, the enantiomeric excesses observed for the minor *Z* diastereoisomers, when isolated, were found to be around 10% lower than those obtained in the corresponding *E* diastereoisomer. It should be remembered that a very closely related catalyst **106c** has also been presented to be an efficient catalyst in the same reaction with acetylenic ketones (see Scheme 5.17), the difference between this and **106e** being the bulkier aryl substituents introduced at the 2' positions of the binaphthyl backbone, which were found to be necessary to achieve high enantioselectivities in this particular case.

Finally, the tartrate-derived quaternary ammonium salts developed by Shibasaki have also been demonstrated to perform very well in the Michael reaction of glycine imines with acrylates (Scheme 5.27).[36] In this case, compound **109b**, which is a modified analogue of **109a** already employed by the same authors in the Michael reaction of glycine imines to enones directed to the total synthesis of *cyclindrizine C* (see Scheme 5.19), was identified as the best catalyst for the reaction among a set of different families of TADDOL-based bis-ammonium salts tested. It is important to note that the presence of two quaternary ammonium salt moieties on the catalyst structure was revealed to be crucial for achieving good stereoselection, observing that a closely related catalyst such as **109c**, with a single quaternary ammonium center, provided the Michael adduct in very low ee. The nature of the anionic counterion was also somewhat important, obtaining the best results with the bis-trifluoroborate

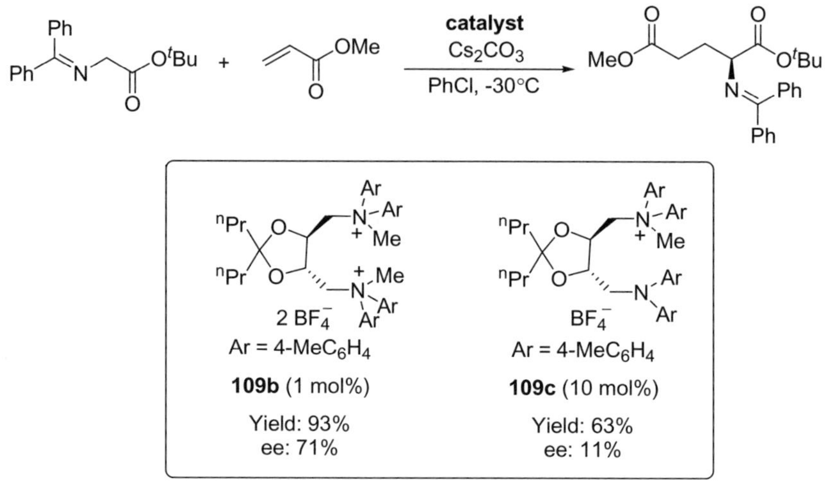

Scheme 5.27   Enantioselective Michael reaction of *tert*-butyl glycinate benzophenone imine with methyl acrylate catalyzed by TADDOL-based ammonium salts.

salt, compared with a set of experiments using other counteranions such as I^-, ClO_4^-, TfO^-, PF_6^- or SbF_5^-.

5.2.3   α,β-Unsaturated Aldehydes as Michael Acceptors

The use of α,β-unsaturated aldehydes as Michael acceptors always represents a challenging situation because of the tendency of enals to undergo 1,2- rather than the desired 1,4- addition reaction. Moreover, working under phase-transfer catalysis conditions incorporates an additional element of difficulty, because of the propensity of enolizable enals to undergo self-condensation side reactions. For this reason, there are only a few examples reporting enantioselective Michael reactions with α,β-unsaturated aldehydes as Michael acceptors under PTC conditions, both coming from the Maruoka research team and also both making use of chiral *N*-spiro quaternary ammonium salts as catalysts.

A first approach described the use of catalyst **106a** in the Michael reaction of a cyclic β-ketoester with acrolein (Scheme 5.28).[20] The reaction proceeded satisfactorily, furnishing quantitatively the desired conjugate addition product in excellent enantioselectivity, requiring the *in situ* protection of the formyl moiety as the corresponding cyclic acetal derivative. However, the authors reported the need of a 9-fluorenyl ester Michael donor and the reaction was not studied further, with no data reported about the scope and limitations of the methodology.

The same year, this group reported an alternative protocol for this reaction, which involved the use of silyl nitronates as activated Michael donors and a novel fluoride ammonium salt **106f** as catalyst (Scheme 5.29).[37] This catalyst

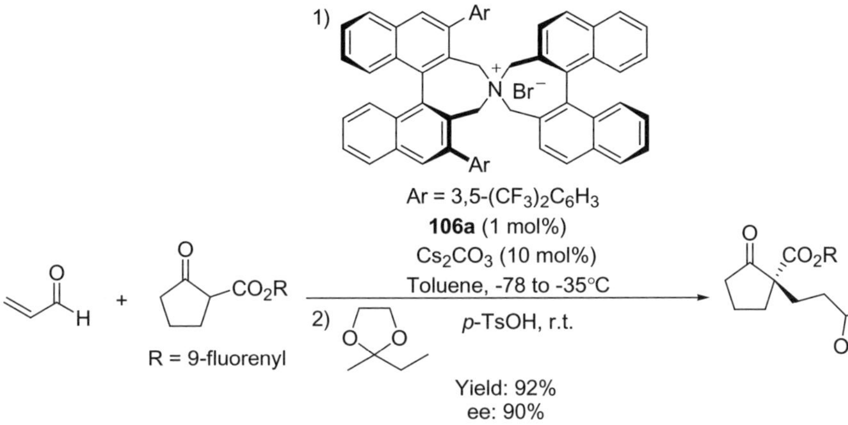

Scheme 5.28   Enantioselective Michael reaction of a cyclic β-ketoester with acrolein catalyzed by **106a**.

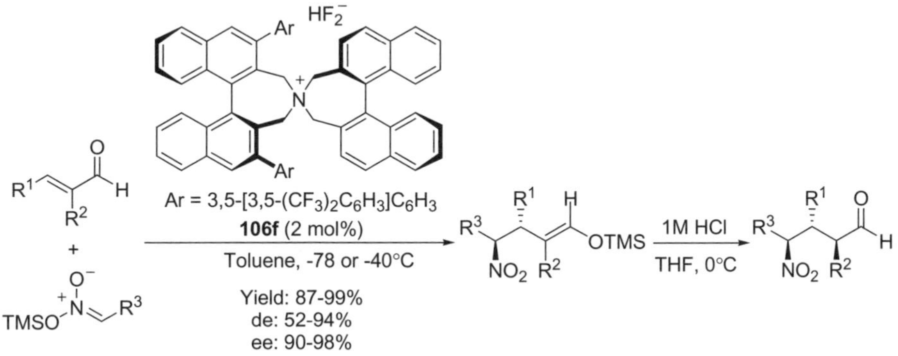

Scheme 5.29   Enantioselective **106f**-catalyzed Michael reaction of silyl nitronates with α,β-unsaturated aldehydes.

incorporated bulkier aryl substituents at the 2′-positions of the binaphthyl substructure than those of **106a** and also a bifluoride HF_2^- anion as counterion was included, the latter required for the *in situ* activation of the silyl nitronate reagent. Working under these typical Mukaiyama–Michael conditions, a variety of differently substituted nitronates and enals reacted with each other in a highly efficient manner, providing the corresponding conjugate addition products in excellent yields and enantioselectivities. More importantly, the use of α-alkyl substituted nitronates led to the formation of products containing two contiguous stereogenic centers with excellent diastereoselectivity, overcoming the limitations typically associated to the enantioselective nitro-Michael reaction under Lewis acid catalysis. Moreover, α,β-disubstituted enals

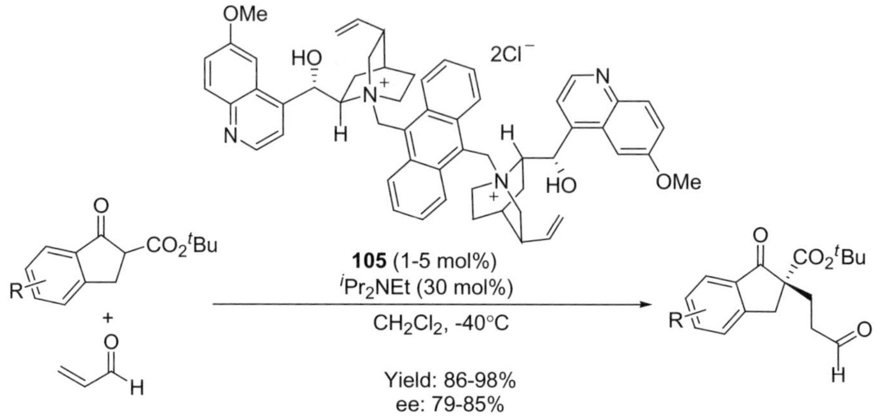

Scheme 5.30 Enantioselective **105**-catalyzed Michael reaction of cyclic β-ketoesters with acrolein.

were also tested with success, leading to the formation of final products containing three contiguous stereocenters after acid hydrolysis with excellent results regarding both yields and stereoselectivities.

Finally, it should also be pointed out that the dimeric quinidine-based bis-ammonium salt **105** developed by Nájera has also been shown to perform quite well in the reaction of cyclic β-ketoesters with acrolein, furnishing the corresponding adducts in good yields and enantioselectivities (Scheme 5.30).[16] As was pointed out earlier, one of the main advantages associated with the use of this catalyst is the possibility of its recycling for further uses in other reactions without loss of activity after precipitation with diethyl ether.

5.2.4 Nitroalkenes as Michael Acceptors

Nitroalkenes are another kind of challenging Michael acceptor to be used under PTC conditions. Contrary to what was presented in the preceding chapters, in which weak nucleophiles were employed in the Michael reaction, therefore resulting in nitroalkenes very often being found to be exceptionally useful Michael acceptor counterparts, the high electrophilicity of these reagents makes the uncatalyzed background reaction extremely competitive in this case. This is due to the fact that, under PTC conditions, the formation of an active enolate-type species occurs rather easily in the usual basic media employed for the reaction and these enolates are highly reactive toward conjugate addition with such a highly activated olefin.

For this reason, the only literature example regarding the use of nitroalkenes as Michael acceptors in enantioselective Michael reactions under PTC conditions is related to the use of oxindoles as pro-nucleophiles (Scheme 5.31).[38] In this context, using deuterium labeling experiments, it was found that oxindoles underwent fast deprotonation in neutral aqueous media only in the

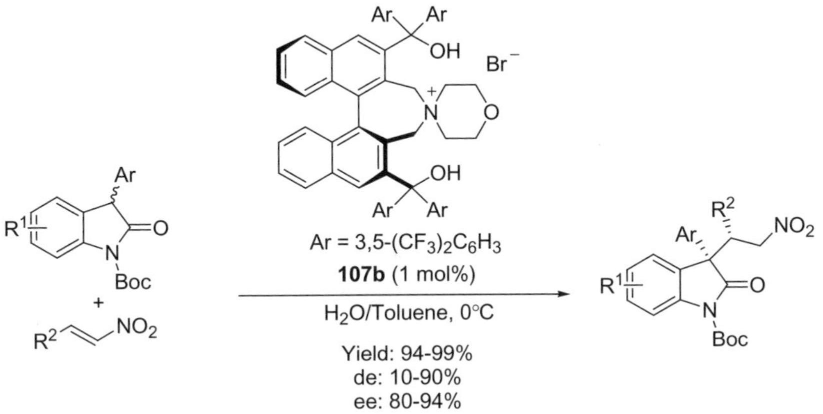

Scheme 5.31 Enantioselective **107b**-catalyzed Michael reaction of α-aryl oxindoles with nitroalkenes.

presence of ammonium salt catalyst **107b**, while they remained essentially unchanged when this catalyst was not present or when other quaternary ammonium salts like tetrabutyl ammonium chloride were added to the mixture. In the same sense, incorporation of water as co-solvent was also found to be necessary for this deprotonation process to occur. With these premises in mind, the possible uncatalyzed background reaction involving achiral nucleophilic species was absolutely ruled out and, in fact, the reaction of a variety of α-substituted oxindoles with a series of nitroolefins was found to occur smoothly in the presence of **107b**, furnishing the expected conjugate addition products containing two stereogenic centers, one of them a quaternary one, in excellent yields and enantioselectivities. The diastereoselectivity of the reaction was, however, found to be highly dependent on the nature of the α-substituent of the oxindole donor and on the β-substituent of the nitroalkene acceptor. The reaction tolerated well the substitution at the nitroalkene reagent, allowing the use of both aromatic and aliphatic nitroalkenes with similar results. On the other hand, only α-aryl substituted oxindoles could be used as Michael donors, observing very poor results with α-alkyl substituted derivatives, probably due to the lower acidity of these compounds which underwent a very slow deprotonation under the reaction conditions employed.

5.2.5 Other Michael Acceptors

Some other activated olefins have been employed as acceptors in Michael reactions with enolates under PTC conditions, using different chiral ammonium salts as catalysts. For example, a very recent and elegant report has shown that conveniently substituted heterocyclic moieties can also play the role of suitable electron releasing groups for the activation of an olefin toward conjugate addition. In particular 4-nitro-5-styrylisoxazoles were found to undergo clean

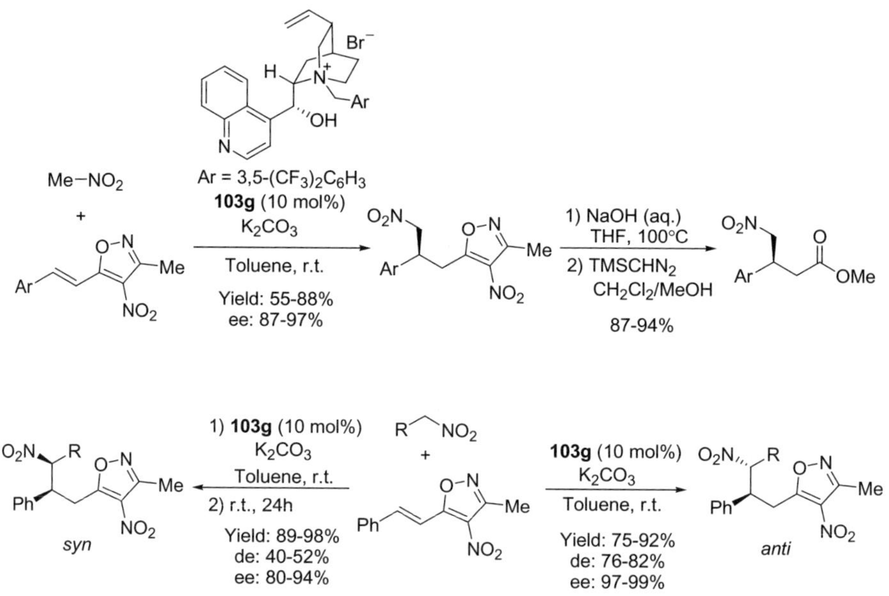

Scheme 5.32 Enantioselective Michael reaction using 4-nitro-5-styrylisoxazoles as Michael acceptors.

and smooth addition with nitromethane in the presence of cinchonidine-based ammonium salt **103g** as catalyst (Scheme 5.32).[39] The reaction proceeded in moderate to good yields for a variety of differently β-aryl substituted substrates, which in almost all cases furnished the final adducts in more than 90% ee. In addition, the nitroisoxazole heterocycle could be very easily converted into a carboxylic group by simple hydrolytic treatment. Moreover, the use of other nitroalkenes leading to the formation of a second stereogenic center was also surveyed with success, maintaining similar high yields and enantioselectivities to those observed when nitromethane was employed; the reaction proceeded with a very high degree of diastereoselection with the preferential formation of the *anti* diastereoisomer. Interestingly, the authors were also able to obtain the corresponding *syn* diastereoisomer by a simple modification of the experimental procedure, which consisted of stirring the reaction for 24 hours at room temperature after consumption of the starting materials. By this modification, the Michael adducts underwent fast *in situ* epimerization because of the basic reaction medium, enabling thermodynamic equilibrium, which favored the formation of the more stable *syn* diastereoisomers.

Continuing with the use of cinchona alkaloid-based quaternary ammonium salts as catalysts, phenyl vinyl sulfones have also been employed as Michael acceptors in the reaction with glycine imines using cinchonidinium salt **103a** as catalyst both in solution or in a solid-supported version (Scheme 5.33), furnishing similar results to those provided by the corresponding vinyl ketones and acrylates shown in Schemes 5.8 and 5.23.[12]

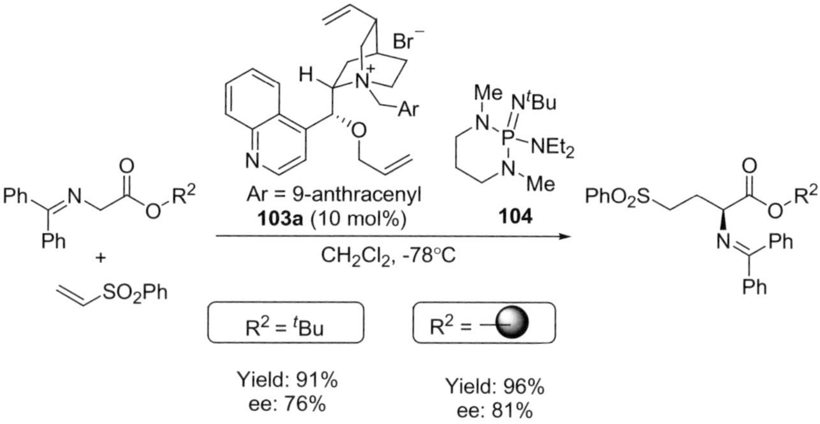

Scheme 5.33 Enantioselective Michael reactions of glycinate imines with phenyl vinyl sulfone catalyzed by **103a**.

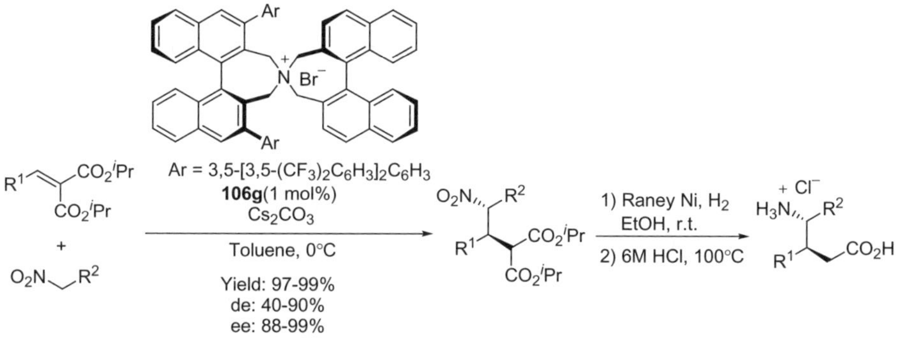

Scheme 5.34 Enantioselective **106g**-catalyzed Michael reaction of nitroalkanes with alkylidenemalonates.

On the other hand, *N*-spiro ammonium bromide **106g** has also been successfully employed to catalyze the Michael reaction of nitroalkanes to alkylidenemalonates (Scheme 5.34).[40] In this case, the reaction has been tested with different nitroalkanes, all of them conducing to the formation of final products containing two contiguous stereogenic centers. This catalyst afforded similar results to those reported in the previous reaction, that is, the final products were obtained in good yields and as mixtures of diastereoisomers in ratios ranging from 7:3 to 95:5, the *anti* isomer being the major one. In all cases, the enantioselectivity of the reaction was found to be excellent, with enantiomeric excesses typically higher than 95% ee. Interestingly, β-alkyl substituted Michael acceptors could also be employed in this case, furnishing comparable results to those observed in the corresponding β-aryl substituted alkylidenemalonates. The authors also demonstrated that these adducts could be converted into highly enantioenriched β,γ-disubstituted γ-amino acids by chemoselective reduction of the nitro group followed by a hydrolysis/decarboxylation procedure.

5.3 Conjugate Addition of Heteroatom-centered Nucleophiles

As was mentioned earlier, carrying out conjugate addition reactions of heteronucleophiles in an enantioselective way using phase-transfer catalysis is a challenging task because of the inherent reversibility of the 1,4-addition process, which usually happens spontaneously in basic media as it is typically used in PTC reactions. This means that, under these reaction conditions, the obtained adducts are not configurationally stable, undergoing fast racemization *via* the retro-Michael/Michael process. In this context, the typical way to overcome this problem has been to incorporate a suitable electrophile able to react with the intermediate enolate-type nucleophile which is generated after the conjugate addition process, in a typical cascade reaction which avoids the retro-Michael reaction. Very often, this hetero-Michael addition-initiated cascade process involves an intramolecular reaction after the conjugate addition, leading to the stereocontrolled formation of complex polyheterocyclic structures. As cascade reactions will be specifically covered in Chapter 7, all the examples pertaining to hetero-Michael reactions followed by a subsequent transformation under PTC conditions will be presented here.

In fact, the only example of a simple hetero-Michael reaction under PTC conditions which does not involve a cascade process consists of an intramolecular aza-Michael reaction of conveniently 2-substituted indole compounds incorporating an α,β-unsaturated ester moiety at the side chain (Scheme 5.35).[41] This methodology shows up as a very straightforward and efficient method for building up the 1,2,3,4-tetrahydropyrazino[1,2-*a*]indole

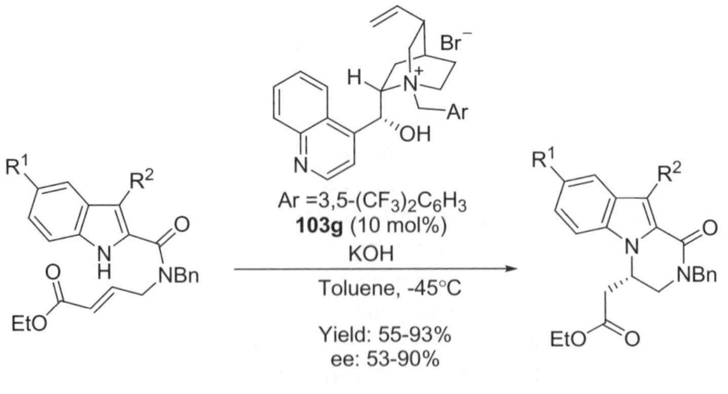

Scheme 5.35 Enantioselective intramolecular aza-Michael reaction under PTC conditions.

architecture, which is the key basic substructure of several interesting biologically active products. After screening for the best catalyst among a wide number of different cinchona alkaloid-based quaternary ammonium salts, the authors found that cinchonidinium bromide **103g** was an excellent promoter for the reaction. The temperature control was also crucial for enantiocontrol, observing a dramatic drop in the enantioselectivity when working at room temperature. The reason for the configurational stability of these particular Michael adducts was not provided but, possibly, the non-basic nature of the indolyl nitrogen atom in the final product might be the key factor for the success of this reaction.

5.4   Concluding Remarks

The possibility to exert asymmetric induction by using chiral cations tightly associated to the corresponding nucleophile anion participating in a Michael-type reaction under phase-transfer conditions can be successfully applied as an alternative methodology to other organocatalytic enantioselective Michael reactions. In this case, the catalyst participates by interacting with the Michael donor and it is in this context in which the complementarity between this approach and other organocatalytic methodologies reaches the maximum level. PTC conditions allow the use of enolizable carbonyl compounds of low acidity as Michael donors because of the possibility to employ a base strong enough to exert their deprotonation, a possibility which is ruled out in the enamine/iminium catalysis or in the reactions under H-bonding activation. As a result of this, many simple carbonyl compounds (especially glycinate imines) have been successfully employed in the reaction with a broad scope of Michael acceptors. On the other hand, while this methodology seems to work exceptionally well with carbon nucleophiles able to form enolates, the use of heterocycles or hydride equivalents as nucleophiles is not possible under this activation mechanism and the reactions with heteroatom-based Michael donors are still rather undeveloped because of the difficulties associated with the hetero-Michael reaction. Finally, related to the catalyst structure, most of the methodologies reported have focused on the use of chiral ammonium salts, either derived from cinchona alkaloids or, alternatively, consisting of *N*-spiro compounds incorporating a binaphthyl axial backbone. All these catalysts have been shown to perform well in many cases but it should be noted that the optimum catalyst structure is usually found after many experiments, and there is not a clear rationalization of the structural requirements that a good PTC catalyst has to fulfill in order to achieve good enantioselection in a given reaction.

References and Notes

1. For pioneering reports see (a) M. Makosza, *Tetrahedron Lett.*, 1966, **7**, 4621; (b) M. Makosza, *Tetrahedron Lett.*, 1966, **7**, 673; (c) M. Makosza,

Tetrahedron Lett., 1969, **10**, 677. For several accounts see; (d) A. Brädström, *Adv. Phys. Org. Chem.*, 1977, **15**, 267; (e) M. Makosza, *Pure Appl. Chem.*, 1975, **43**, 439; (f) M. Makosza, *Pure Appl. Chem.*, 2000, **72**, 1399.

2. C. M. Starks, *J. Am. Chem. Soc.*, 1971, **93**, 195.
3. U.-H. Dolling, P. Davis and E. J. J. Grabowski, *J. Am. Chem. Soc.*, 1984, **106**, 446.
4. For some reviews on asymmetric PTC reactions see (a) S. Jew and H. Park, *Chem. Commun.*, 2009, 7090; (b) T. Hashimoto and K. Maruoka, *Chem. Rev.*, 2007, **107**, 5656; (c) T. Ooi and K. Maruoka, *Aldrichimica Acta*, 2007, **40**, 77; (d) T. Ooi and K. Maruoka, *Angew. Chem. Int. Ed.*, 2007, **46**, 4222; (e) M. J. O'Donnell, *Acc. Chem. Res.*, 2004, **37**, 506; (f) B. Lygo and B. Andrews, *Acc. Chem. Res.*, 2004, **37**, 518; (g) K. Maruoka and T. Ooi, *Chem. Rev.*, 2003, **103**, 3013; (h) D. Martyres, *Synlett*, 1999, 1508; (i) A. Nelson, *Angew. Chem. Int. Ed.*, 1999, **38**, 1583.
5. D. J. Cram and G. D. Y. Sogah, *J. Chem. Soc. Chem. Commun.*, 1981, 625.
6. R. S. E. Conn, A. V. Lovell, S. Karady and L. M. Weinstock, *J. Org. Chem.*, 1986, **51**, 4710.
7. T. Akiyama, M. Hara, K. Fuchibe, S. Sakamoto and K. Yamaguchi, *Chem. Commun.*, 2003, 1734.
8. (a) T. Bakó, P. Bakó, A. Szöllõsy, M. Czuger, G. Keglevich and L. Tõke, *Tetrahedron: Asymmetry*, 202, **13**, 203; (b) T. Novák, J. Tatai, P. Bakó, M. Czuger, G. Keglevich and L. Tõke, *Synlett*, 2001, 424; (c) P. Bakó, T. Novák, K. Ludányi, L. Tõke and G. Keglevich, *Tetrahedron: Asymmetry*, 1999, **10**, 2373; (d) P. Bakó, K. Vizvardi, Z. Bajor and L. Tõke, *Chem. Commun.*, 1998, 1193.
9. E. J. Corey, M. C. Noe and F. Xu, *Tetrahedron Lett.*, 1998, **39**, 5347.
10. F.-Y. Zhang and E. J. Corey, *Org. Lett.*, 2004, **6**, 3397.
11. (a) E. J. Corey and F.-Y. Zhang, *Org. Lett.*, 2000, **2**, 4257. For a closely related later example see; (b) D. Y. Kim and S. C. Huh, *Tetrahedron*, 2001, **57**, 8933.
12. M. J. O'Donnell, F. Delgado, E. Domínguez, J. de Blas and W. L. Scott, *Tetrahedron: Asymmetry*, 2001, **12**, 821.
13. T. Perrard, J.-C. Plaquevent, J.-R. Desmurs and D. Hébrault, *Org. Lett.*, 2000, **2**, 2959.
14. (a) D. Y. Kim, S. C. Huh and S. M. Kim, *Tetrahedron Lett.*, 2001, **42**, 6299.
15. T. Furukawa, N. Shibata, S. Mizuta, S. Nakamura, T. Toru and M. Shiro, *Angew. Chem. Int. Ed.*, 2008, **47**, 8051.
16. S. Tarí, R. Chinchilla and C. Nájera, *Tetrahedron: Asymmetry*, 2009, **20**, 2651.
17. L. Bernardi, J. López-Cantanero, B. Niess and K. A. Jorgensen, *J. Am. Chem. Soc.*, 2007, **129**, 5772.
18. P. Elsner, L. Bernardi, G. D. Salla, J. Overgaard and K. A. Jorgensen, *J. Am. Chem. Soc.*, 2008, **130**, 4897.
19. F.-Y. Zhang and E. J. Corey, *Org. Lett.*, 2001, **3**, 639.
20. T. Ooi, T. Miki, M. Taniguchi, M. Shiraishi, M. Takeuchi and K. Maruoka, *Angew. Chem. Int. Ed.*, 2003, **42**, 3796.

21. T. Ooi, S. Takada, S. Fujioka and K. Maruoka, *Org. Lett.*, 2005, **7**, 5143.
22. Q. Lan, X. Wang and K. Maruoka, *Tetrahedron Lett.*, 2007, **48**, 4675.
23. T. Ooi, D. Ohara, K. Fukumoto and K. Maruoka, *Org. Lett.*, 2005, **7**, 3195.
24. B. Lygo, B. Allbutt and E. H. M. Kirton, *Tetrahedron Lett.*, 2005, **46**, 4461.
25. T. Shibuguchi, H. Mihara, A. Kuramochi, S. Sakuraba, T. Ohshima and M. Shibasaki, *Angew. Chem. Int. Ed.*, 2006, **45**, 4635.
26. R. He, C. Ding and K. Maruoka, *Angew. Chem. Int. Ed.*, 2009, **48**, 4559.
27. S. Aoki, S. Sasaki and K. Koga, *Tetrahedron Lett.*, 1989, **30**, 7229.
28. J. Crosby, J. F. Stoddard, X. Sun and M. R. W. Venner, *Synthesis*, 1993, 141.
29. L. Tõke, P. Bakó, G. M. Keserti, M. Albert and L. Fenichel, *Tetrahedron*, 1998, **54**, 213.
30. E. Brunet, A. Poveda, D. Rabasco, E. Oreja, L. M. Font, M. S. Batra and J. C. Rodríguez-Ubis, *Tetrahedron: Asymmetry*, 1994, **5**, 935.
31. F.-Y. Zhang and E. J. Corey, *Org. Lett.*, 2000, **2**, 1097.
32. M. B. Andrus and Z. Ye, *Tetrahedron Lett.*, 2008, **49**, 534.
33. S. Arai, R. Tsuji and A. Nishida, *Tetrahedron Lett.*, 2002, **43**, 9535.
34. Y.-J. Lee, J. Lee, M.-J. Kim, B.-S. Jeong, J.-H. Lee, T.-S. Kim, J. Lee, J.-M. Ku, S. Jew and H. Park, *Org. Lett.*, 2005, **7**, 3207.
35. X. Wang, M. Kitamura and K. Maruoka, *J. Am. Chem. Soc.*, 2007, **129**, 1038.
36. (a) T. Ohshima, T. Shibuguchi, Y. Fukuta and M. Shibasaki, *Tetrahedron*, 2004, **60**, 7743; (b) T. Shibuguchi, Y. Fukuta, Y. Akachi, T. Ohshima and M. Shibasaki, *Tetrahedron Lett.*, 2002, **43**, 9539.
37. T. Ooi, K. Doda and K. Maruoka, *J. Am. Chem. Soc.*, 2003, **125**, 9022.
38. H. He, S. Shirakawa and K. Maruoka, *J. Am. Chem. Soc.*, 2009, **131**, 16620.
39. A. Baschieri, L. Bernardi, A. Ricci, S. Suresh and M. F. A. Adamo, *Angew. Chem. Int. Ed.*, 2009, **48**, 9342.
40. T. Ooi, S. Fujioka and K. Maruoka, *J. Am. Chem. Soc.*, 2004, **126**, 11790.
41. M. Bandini, A. Eichholzer, M. Tragni and A. Umani-Ronchi, *Angew. Chem. Int. Ed.*, 2008, **47**, 3238.

CHAPTER 6

Enantioselective Conjugate Addition Reactions Proceeding via *Other Types of Activation*

6.1 Introduction

The rich reactivity profile of activated olefins and nucleophiles participating in a conjugate addition reaction opens the way for multiple approaches in which an organic molecule can activate either the Michael donor or the Michael acceptor. As a consequence of this, many different methodologies have been reported to carry out organocatalytic enantioselective conjugate addition reactions which were not covered in the preceding chapters. Relevant examples include the catalysis by *N*-heterocyclic carbenes, which has reached a meritorious level of sophistication in the last few years and which allows carrying out transformations which are not able to be performed by means of enamine, iminium, H-bonding or PTC activation. Another important class of reactions are those promoted by chiral tertiary amines participating in the activation of the Michael donor by deprotonation, and in which the participation of other interactions like H-bonding interactions is not observed. A final section will also cover other miscellaneous reports, which include some very important and relevant examples like the catalysis by small peptides or by an *N*-protected version of a simple α-amino acid such as cysteine.

RSC Catalysis Series No. 5
Organocatalytic Enantioselective Conjugate Addition Reactions: A Powerful Tool for the Stereocontrolled Synthesis of Complex Molecules
By Jose L. Vicario, Dolores Badía, Luisa Carrillo and Efraim Reyes
Published by the Royal Society of Chemistry, www.rsc.org

6.2 Enantioselective Conjugate Addition Reactions Catalyzed by *N*-heterocyclic Carbenes

Carbenes have been known since the very early foundations of organic chemistry and have been extensively studied for many years. The first evidence of the existence and structure of carbenes was provided by Buchner, Curtius, Staudinger and Kupfer,[1] between 1885 and 1912 and, for a long time, these were considered simple reaction intermediates and were believed to be too unstable to be isolated, the first isolation of a transition-metal carbene complex being reported in 1964 by Fischer and Maasbol.[2] However, since the pioneering reports by Bertrand[3] and Arduengo,[4] in which the first stable *N*-heterocyclic carbenes were isolated, the application of these compounds in organic synthesis has witnessed impressive progress, although most of the chemistry developed in this context has focused on the use of *N*-heterocyclic carbenes as ligands for metal complexes.[5] In fact, *N*-heterocyclic carbenes are used by several natural enzymes to catalyze some relevant processes like, for example, acylation reactions, which are catalyzed by transketolases. These natural catalysts incorporate thiamine (vitamin B1) as a cofactor, which contains a thiazole group on its structure participating as the precursor of a nucleophilic carbene, which is the real catalytically active species involved in the activation of the acylating reagent.

The mechanism operating in *N*-heterocyclic carbene-mediated reactions was proposed initially by Breslow in 1958 for the thiazolium salt-catalyzed benzoin condensation (Scheme 6.1).[6] This proposal involves the formation of a carbene as the catalytically active species by deprotonation of the thiazolium cation, which subsequently adds to one molecule of the aldehyde, generating a nucleophilic intermediate known as the Breslow intermediate. Next, this

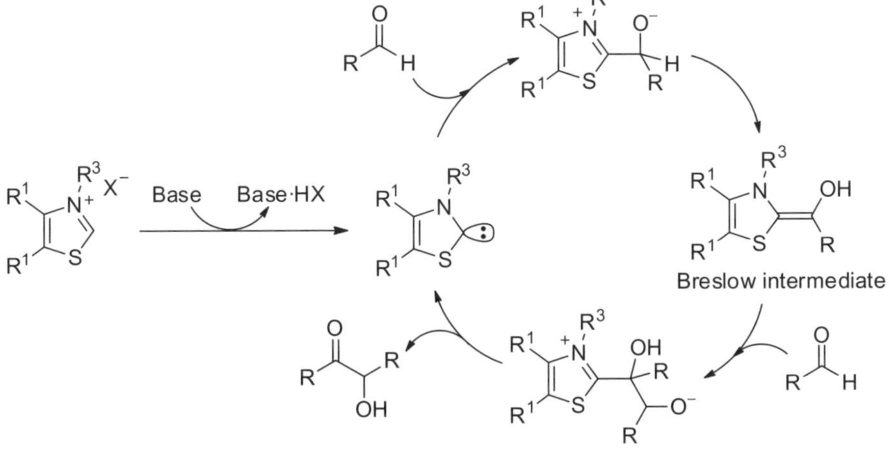

Scheme 6.1 The mechanism proposed for the thiazolium salt-catalyzed benzoin condensation proposed by Breslow.

intermediate undergoes addition with an electrophile (a second molecule of the aldehyde in the case of the benzoin condensation) and a final elimination step releases the product and allows the turnover of the catalyst. Possibly the most important and relevant mechanistic implication behind this activation profile is the fact that the carbene catalyst is promoting an umpolung on the typical reactivity profile of the aldehyde, which, according to the terminology coined by Seebach for the classification of synthons,[7] changes after activation from the standard a[1] behavior to behave as a d[1] synthon, in the form of the Breslow intermediate.

In this context, the first attempts to employ chiral *N*-heterocyclic carbenes as organocatalysts in an enantioselective reaction were carried out in the context of the benzoin condensation, which combines two molecules of an aldehyde (usually an aromatic one). The first pioneering examples were developed by Sheehan and coworkers,[8] which were followed by several authors,[9] highlighting the report by Enders in 2002, in which, for the first time, a highly enantiose-lective *N*-heterocyclic carbene-catalyzed benzoin condensation was reported (Scheme 6.2). As can be seen in this scheme, the general principle applied to the design of these catalysts relied on a thiazolium central unit as the precursor of

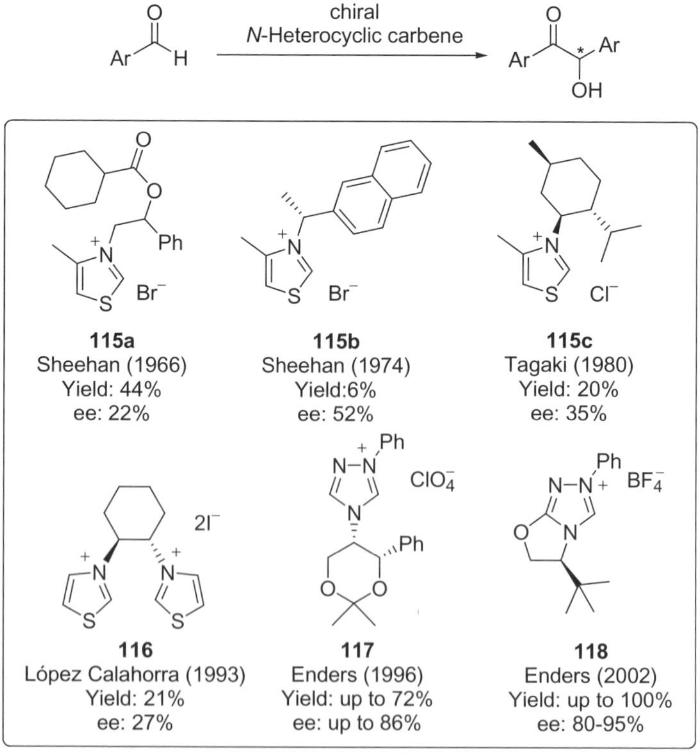

Scheme 6.2   The first enantioselective benzoin condensations catalyzed by chiral *N*-heterocyclic carbenes.

the reactive carbene and the chiral information was placed at the lateral chain linked to the nitrogen atom of the heterocycle. The qualitative advances in the catalyst design by Enders pointed toward the triazolium salt moiety as the most efficient carbene precursor and also to the need for a more conformationally rigid structure, which placed the chiral information closer to the reaction center, using bicyclic triazolium salt **118** derived from *tert*-leucine.

The application of this ability of *N*-heterocyclic carbenes to activate aldehydes (or ketones) for their participation as nucleophiles in polar reactions to the conjugate addition reaction was discovered by Stetter in 1975,[10] when he found that aldehydes underwent 1,4-addition to electron-deficient olefins in the presence of a carbene generated *in situ* from a thiazolium salt in the presence of a base (also known as the Stetter reaction). Many attempts have been made since then to develop enantioselective variants of this reaction, although it has to be pointed out that the most important advances have been reported quite recently, as will be shown in this section.

6.2.1 The Intramolecular Enantioselective Stetter Reaction

The first intramolecular non-stereoselective version of the Stetter reaction was not reported until 1995.[11] Nevertheless, most of the initial studies and reports toward an enantioselective version of the Stetter reaction were focused on intramolecular variants of this reaction because of the favorable entropic factors generally associated to intramolecular reactions. In fact, the first steps were carried out by Enders in the context of an intermolecular version obtaining very low yields and poor enantioselectivities,[12] which was attributed to the formation of stable adducts between the carbene and the Michael acceptor, leading to catalyst deactivation. Better results were reported afterwards by the same group when they changed to an intramolecular Stetter reaction using **117** as precatalyst, which allowed the preparation of enantioenriched chromanones in moderate to good yields and enantioselectivities in the range of 41–74% ee (Scheme 6.3).[13]

After this pioneering report, probably the most successful enantioselective intramolecular version of this reaction was developed by Rovis[14] in 2002

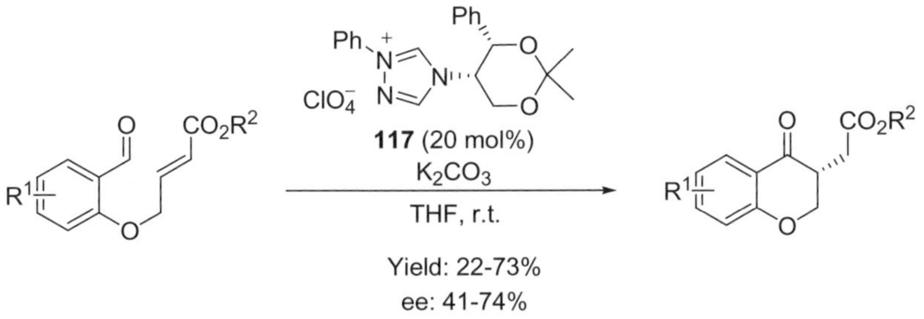

Yield: 22–73%
ee: 41–74%

Scheme 6.3 The first enantioselective intramolecular Stetter reaction.

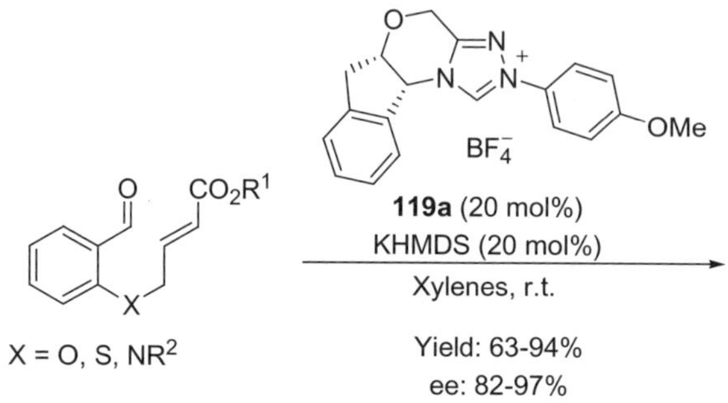

Scheme 6.4 The **119a**-catalyzed intramolecular Stetter reaction developed by Rovis.

(Scheme 6.4). After examining a variety of *N*-heterocyclic carbene catalysts derived from different β-amino alcohols, aminoindanol-derived pre-catalyst **119a** was identified as the most appropriate one with regard to yield and stereochemical control. After the best reaction conditions were established, the authors demonstrated the remarkably wide scope of the reaction, allowing the easy preparation of quinolinones, chromanones and thiochromanones in very good yields and enantioselectivities by simply changing the nature of the tether chain linking the formyl group and the enoate Michael acceptor. On the other hand, the length of the tether was found to have a great influence on the reaction; it was observed that similar substrates which would generate a benzo-fused oxepine or furane ring were either unreactive or furnished racemic compounds because of their pronounced tendency to racemization due to the high acidity of the H-atom in the stereogenic center.

Further studies were carried out in order to expand the scope of the methodology to other different functionalities as Michael acceptors. In this sense, using modified catalyst **119b** it was demonstrated that good results could be obtained using different α,β-unsaturated esters (methyl allyl or *tert*-butyl esters), ketones, thioesters, Weinreb amides or nitriles (Scheme 6.5).[15] The use of dry toluene as solvent was also found to be more appropriate than xylene, which was the solvent chosen for the initial report. On the other hand, substrates incorporating an α,β-unsaturated aldehyde or a nitroalkene as Michael acceptor moiety were found to perform poorly in the reaction. On the contrary, alkenylphosphine oxides and alkenylphosphonates were found to be outstanding substrates for this reaction, furnishing excellent results for a variety of different examples tested, including those conducing to the formation of a benzofurane-type final compound.[16]

However, even though this reaction proceeds in a very efficient way when an aromatic aldehyde is employed as the Michael donor, the use of aliphatic aldehydes is much more problematic due to the intrinsic instability of enolizable aldehydes in the basic media required in the reactions catalyzed by *N*-heterocyclic carbenes. In fact, pre-catalyst **119a** performed poorly in this case, but this limitation was overcome with the use of bicyclic triazolium salt **120a** derived from phenylalanine as catalyst precursor.[14] This new *N*-heterocyclic

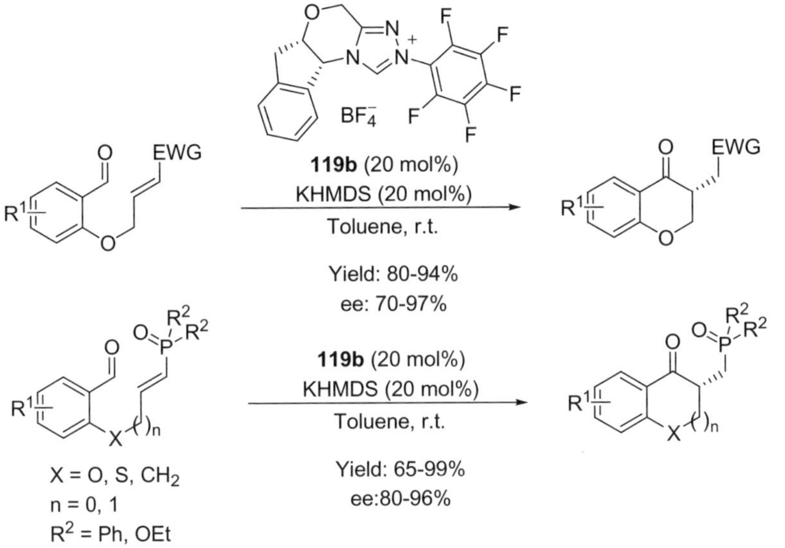

Scheme 6.5 Scope of the **119b**-catalyzed intramolecular Stetter reaction.

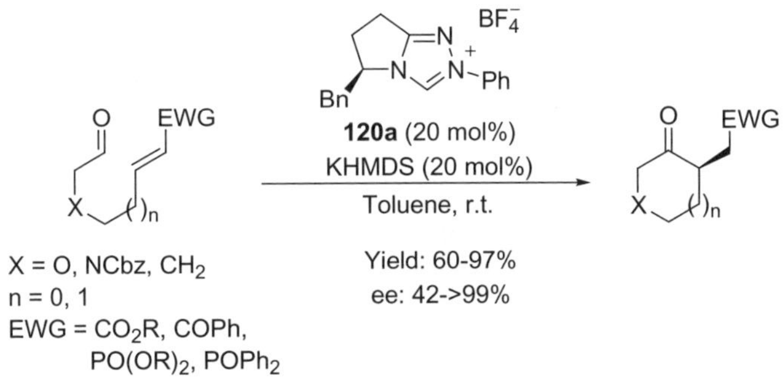

Scheme 6.6 The intramolecular Stetter reaction using aliphatic aldehydes as Michael donors.

carbene catalyst was found to be, in general, much more active than **119a** or **119b** for the same reactions but furnishing slightly lower enantioselectivities. On the other hand, this catalyst allowed reactions, which were not possible to perform using any of these **119**-type pre-catalysts, to be carried out in a satisfactory way. In the particular case related to the use of aliphatic aldehydes as Michael donors, the reaction proceeded satisfactorily for a variety of substrates, furnishing good yields and enantioselectivities when **120a** was used as the catalyst precursor (Scheme 6.6).[14,15]

Later, the same author used the same type of carbenes for introducing an all-carbon quaternary stereocenter in the final product by using trisubstituted electron-deficient alkenes as Michael acceptors (Scheme 6.7).[17] Different

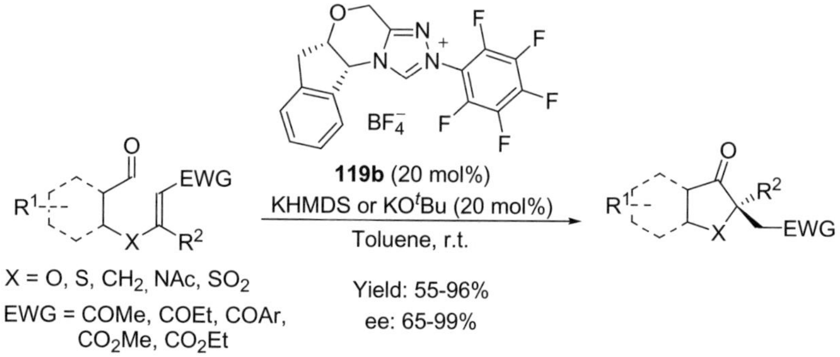

Scheme 6.7 The intramolecular Stetter reaction generating quaternary stereocenters.

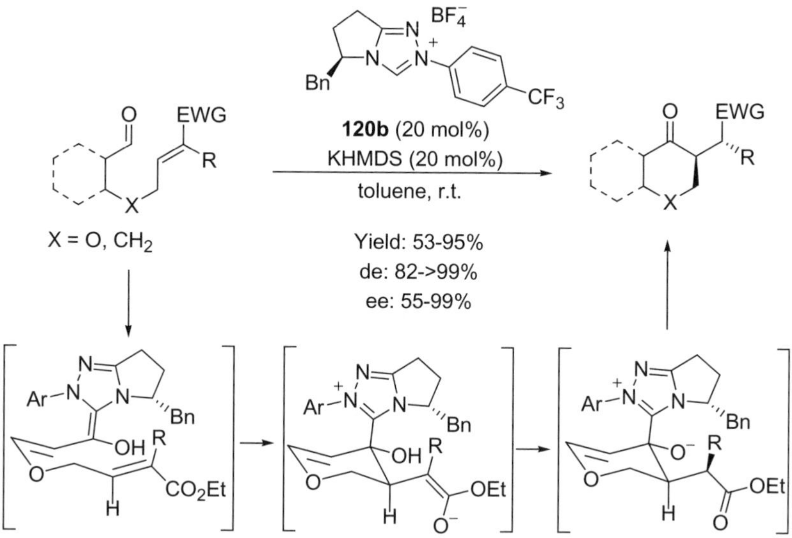

Scheme 6.8 The diastereo- and enantioselective intramolecular Stetter reaction.

substrates incorporating aromatic or aliphatic aldehydes as Michael donors reacted in a very efficient way using pre-catalyst **119b** and KO^{t}Bu or KHMDS respectively as base. The geometry of the alkene had an important influence on the reaction, observing that, generally, *E* alkenes reacted faster and provided better enantioselectivities than those observed in the same reaction with the corresponding *Z* isomer. However, under suitable reaction conditions also substrates containing *Z*-configured Michael acceptors reacted with satisfactory yields and enantioselectivities.

Alternatively, substrates incorporating an α-substituted Michael acceptor have also been studied in this reaction, leading to the generation of two contiguous stereocenters after the intramolecular Stetter reaction (Scheme 6.8).[18]

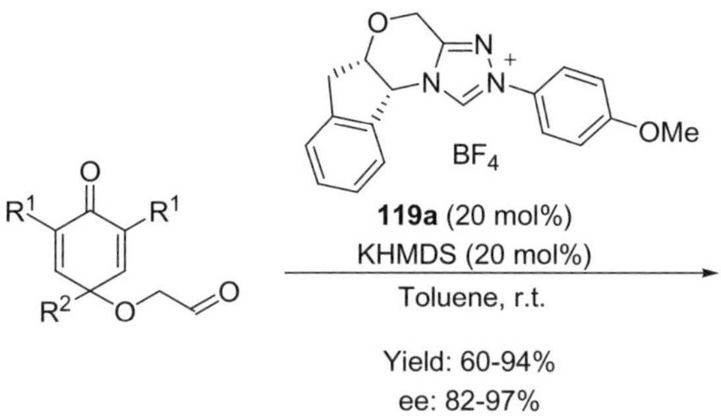

Scheme 6.9 The desymmetrization of cyclohexadienones by the **119a**-catalyzed intramolecular Stetter reaction.

In this case, bicyclic triazolium salt **120b** was identified as the most efficient pre-catalyst, furnishing the final compounds in excellent yields, diastereo- and enantioselectivities. A model has also been provided to explain the high ste-reoselectivities obtained, which implied the formation of the Breslow inter-mediate in a preferred *E* geometry, which underwent addition to the Michael acceptor through its less hindered face, forming the first stereocenter in the observed configuration. The diastereoselectivity of the reaction was explained in terms of a diastereoselective fast intramolecular proton transfer occurring before release of the catalyst, which has to occur faster than rotation around the $C_{enolate}$–C bond in order to achieve good stereocontrol.

Desymmetrization of conveniently substituted cyclohexadienones was also carried out employing the *N*-heterocyclic carbene-catalyzed Stetter methodol-ogy. As is shown in Scheme 6.9, using enones as acceptors in the reaction and in the presence of the same type of triazolium salt **119a** originally employed in their first report, Rovis described the synthesis of benzofuranone derivatives containing three contiguous stereocenters.[19] The final products were obtained in all cases with completely diastereo- and enantioselectivity in good to excel-lent yields and tolerating a wide variety of different substituents at the starting materials.

Other different chiral carbene catalysts have been developed by other authors and tested specifically in intramolecular Stetter reactions (Scheme 6.10). This is the case of menthol-derived thiazolium salt **121**[20] and tripeptide **122** incorporating a thiazolylalanine amino acid[21] as constituents, which were used as pre-catalysts in the same reaction shown in Schemes 6.3 and 6.4 leading to chromanones. However, these two new catalytic systems, although active and able to promote rather efficiently the reaction, only furnished moderate levels of enantioselection. In a different approach, a C_2-symmetric imidazoli-dinium salt **123** has been employed to generate the corresponding catalytically active carbene species and employed in the cyclization of 7-oxo-2-pentenoates leading to chiral cyclopentenones.[22] In this case, this intramolecular Stetter reaction proceeded with moderate to good yields and enantioselectivities up to 80% ee.

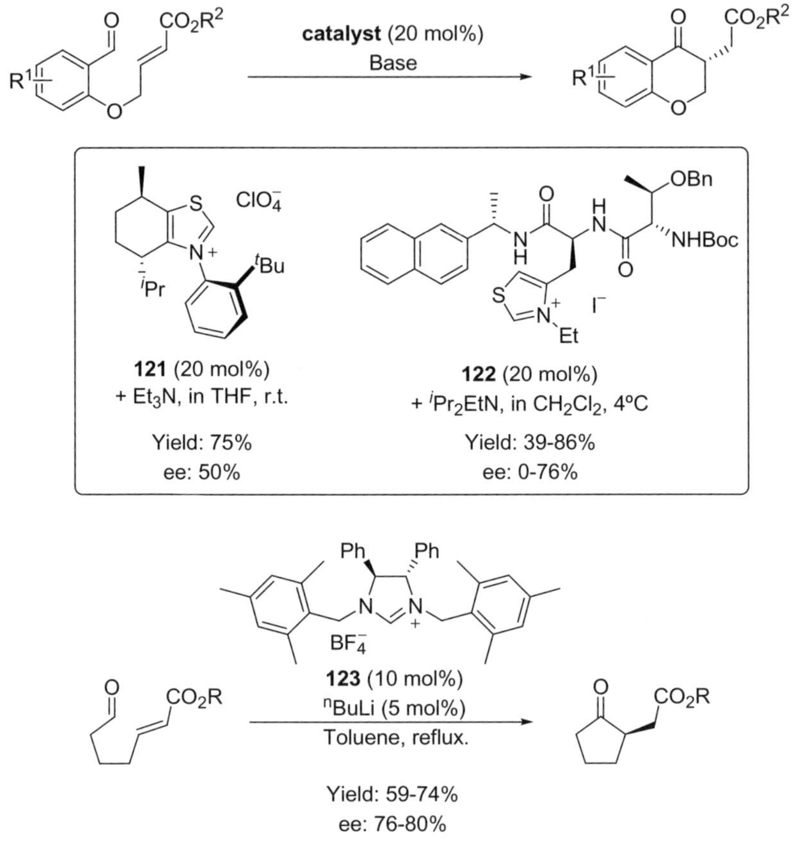

O   CO₂R²

R¹

catalyst (20 mol%)
Base

O   CO₂R²

R¹
O

ClO₄⁻

121 (20 mol%)
+ Et₃N, in THF, r.t.

Yield: 75%
ee: 50%

OBn

O   H
N   N
H   NHBoc
O

I⁻

Et

122 (20 mol%)
+ ᶦPr₂EtN, in CH₂Cl₂, 4°C

Yield: 39-86%
ee: 0-76%

Ph   Ph

N   N
+

BF₄⁻

123 (10 mol%)
ⁿBuLi (5 mol%)
Toluene, reflux.

O   CO₂R

O   CO₂R

Yield: 59-74%
ee: 76-80%

Scheme 6.10   Other chiral carbene catalysts developed for the intramolecular Stetter reaction.

6.2.2   The Intermolecular Enantioselective Stetter Reaction

The intermolecular version of the Stetter reaction needed some more time to be developed. The first reports correspond independently to the groups of Enders and Rovis, both using bicyclic triazolium-based chiral carbenes as catalysts (Schemes 6.11 and 6.12). The report by Enders consisted of the conjugate addition of heteroaromatic aldehydes to arylidinemalonates using triazolium salt ***ent*-120c** as the catalyst precursor, which is easily prepared from proline.[23] The optimized reaction conditions required the use of cesium carbonate as the base for the generation of the carbene and THF as solvent. Under these conditions, furaldehyde and pyridinecarbaldehyde reacted with different β-aryl substituted alkylidenemalonates, furnishing the corresponding conjugate addition products in good yields and moderate enantioselectivities, although these could be raised to >90% ee after crystallization in most cases. Remarkably, the reaction could also be scaled-up easily, observing the same results with regard to yields and enantioselectivities. Applying these conditions

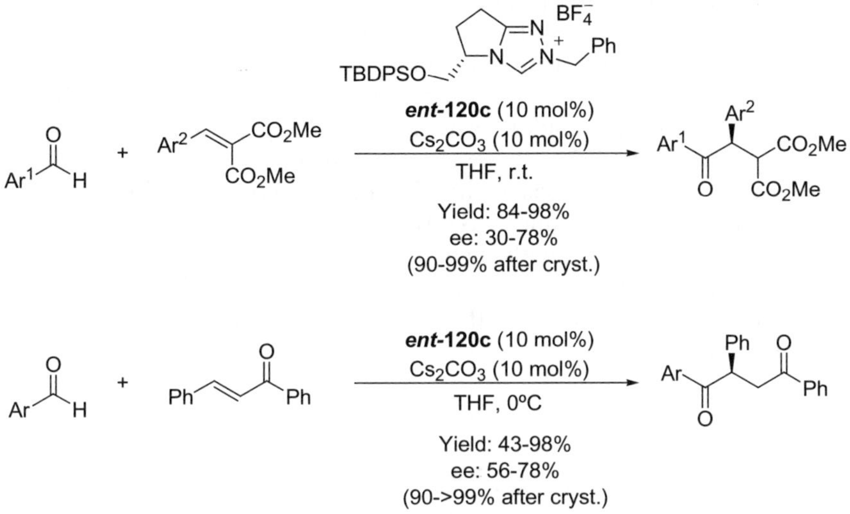

Scheme 6.11   The ent-**120c**-catalyzed intermolecular Stetter reaction developed by Enders.

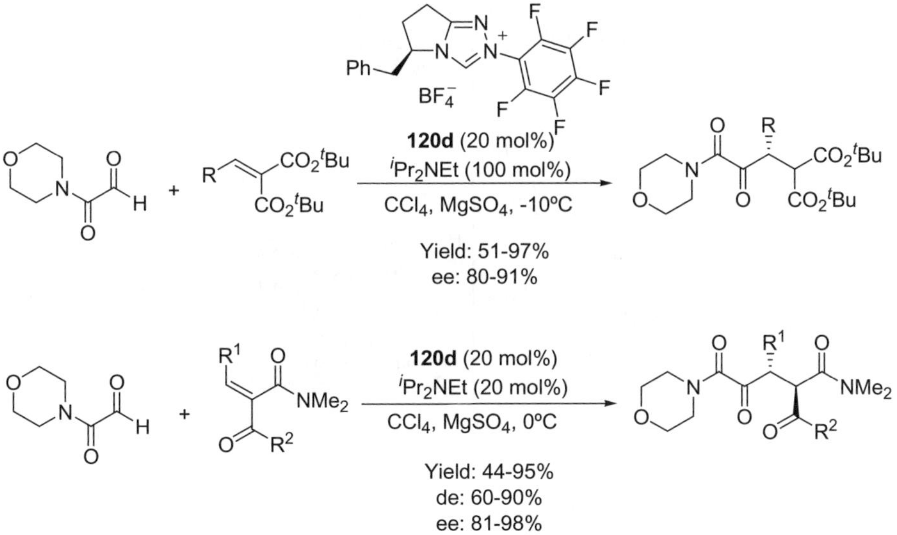

Scheme 6.12   The **120d**-catalyzed intermolecular Stetter reaction.

to the reaction using chalcone as Michael acceptor proceeded with better results.[24]

In the report by Rovis, triazolium salt **120d** derived from phenylalanine was identified as the best catalyst for the conjugate addition of glyoxylamides with a variety of β-substituted alkylidenemalonates (Scheme 6.12).[25] Moderate to excellent yields of the final products were obtained and an excellent level of enantioselectivity was observed in almost all cases, with values typically around 90% ee. Importantly, the introduction of anhydrous MgSO$_4$ as water scavenger was found to have a very positive effect on catalytic activity and therefore this was incorporated as an additive in the optimized protocol. Also in this case, the authors demonstrated that the reaction could be scaled up without any decrease in yield or enantioselectivity. The methodology was also extended to the use of alkylidene ketoamides as Michael acceptors,[26] generating an additional stereogenic center with high diastereoselectivity control. In these two reports, the scope of the reaction was restricted to the use of β-alkyl-substituted Michael acceptors, with no data reported about the possibility to use arylidenemalonates.

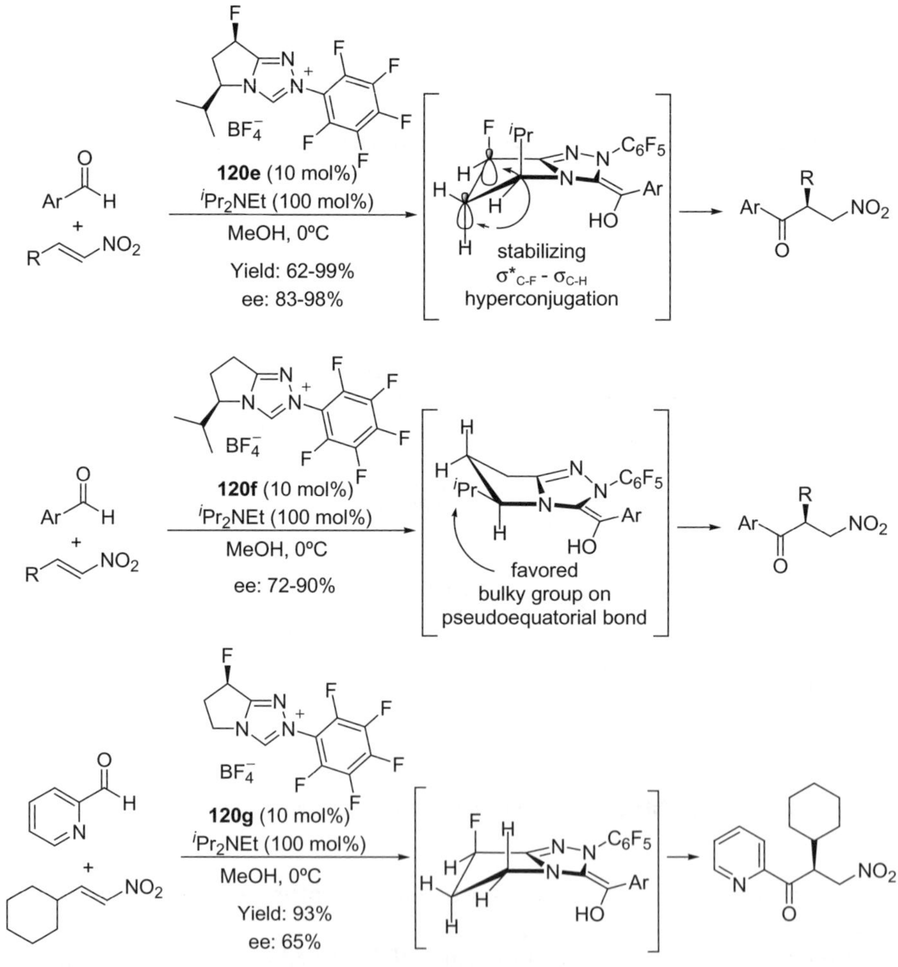

Scheme 6.13 The intermolecular Stetter reaction using nitroalkenes as Michael acceptors.

Finally, there is an interesting new concept for catalyst design reported by Rovis with a new triazolium salt **120e** developed expressly for the inter-molecular Stetter reaction of aromatic aldehydes with nitroalkenes (Scheme 6.13).[27] This consists of a modification of catalyst **120c** by the substitution of the benzyl group at the stereogenic center by an *iso*-propyl group and the introduction of a fluorine atom on the catalyst backbone in the appropriate 1,3-*cis* relative configuration between the ^{i}Pr and the F substituents. The incor-poration of this fluorine substituent led to a much more conformationally rigid arrangement in the reactive Breslow intermediate and therefore to a much more enantioselective reaction compared to the results afforded by the corresponding analogue **120f** without this fluorine substituent. This effect was explained in terms of a very effective stabilization of a conformation in which both sub-stituents, the fluorine and the *iso*-propyl, remained in a 1,3-diaxial arrangement because of the *gauche* effect caused by the maximization of σ-σ* hyperconju-gative interactions when the σ*$_{C-F}$ orbital and one of the adjacent σ$_{C-H}$ align essentially parallel. This stabilizing hyperconjugative effect was also demon-strated from the fact that the use of triazolium salt **120g** incorporating a single F-based stereogenic center led to the final adducts in slightly lower enantio-selectivities, which were comparable to those afforded by catalyst **120f**. It should also be pointed out that the reaction required the use of heteroaromatic aldehydes as Michael donors, observing that benzaldehyde was completely unreactive under the reaction conditions employed. This points toward an active role of this heteroatom in the catalytic cycle.

6.3 Enantioselective Conjugate Addition Reactions Catalyzed by Chiral Brønsted Bases

Chiral bases can be employed to promote conjugate additions by activating the pro-nucleophile exerting its deprotonation. This leads to the formation of an anionic species (typically an enolate) which should remain tightly bounded to the corresponding cation (the protonated chiral base), therefore allowing the latter to control the stereochemical outcome of the reaction. The main principle underlying this activation mode is therefore the same as that operating in chiral PTC, although in this case the reaction proceeds in an homogeneous envir-onment and it is the same base that deprotonates the pro-nucleophile which remains linked to the nucleophile during the conjugate addition step, with no phase-transfer or ion exchange phenomenon involved. In this context, although several chiral Brønsted bases have been developed and employed as catalysts in several enantioselective transformations, for the particular case of conjugate additions, all cases reported refer to the use of dimeric cinchona alkaloid-based compounds as base catalysts (two representative examples are shown in Figure 6.1). These compounds are known chiral ligands employed in several metal-catalyzed reactions like the Sharpless aminohydroxylation and most of them are commercially available. The structure of the catalyst consists of two identical cinchona alkaloids coupled together by an aryl-type linker through

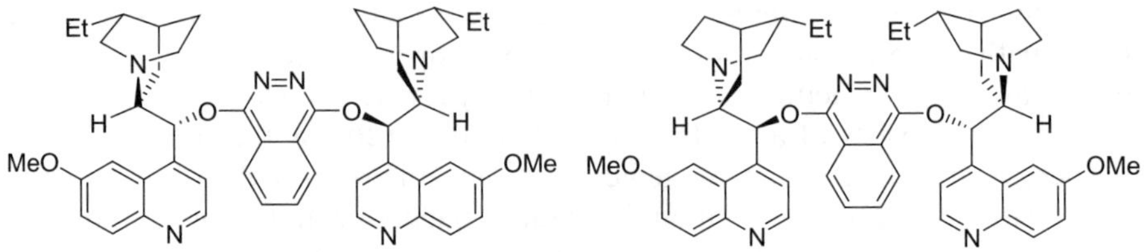

Figure 6.1   Two examples of dimeric cinchona alkaloid-based Brønsted base catalysts.

the 9-OH group and, as is usually found when these types of alkaloids are employed as catalysts, the use of pseudoenantiomeric quinine- or quinidine-based substructures leads to catalysts with opposite stereoinducing effects. As can be seen in the two examples in Figure 6.1, one of them derives from dehydroquinide (DHQ₂PHAL) **124a** and the other (DHQD₂PHAL) **125a** from the pseudoenantiomeric dehydroquinidine, both incorporating a phthalazine group as the linker employed to join the two cinchona alkaloids together. These types of catalysts incorporate multiple Brønsted basic sites in their structures, capable of interacting with the substrates participating in the reaction, generating a chiral environment for the reaction to proceed stereoselectively, although it is still not completely clear which types of interactions are especially relevant to stereoselectivity or to the way that the reagents and the catalyst arrange in the transition state.

It should be pointed out that participation of chiral bases in enantioselective Michael reactions is very often found in the literature as a part of a bifunctional catalyst, in particular cooperating with catalysts operating by H-bonding interactions. As we have considered in this case and in the other cases of bifunctional catalysts operating by enamine/iminium activation together with Brønsted base-type secondary interactions that it is indeed this H-bonding/enamine/iminium activation mechanism that is the main contributor for the catalytic effect, all of these examples have been included in each of the corresponding preceding chapters. For this reason, only catalysts exclusively operating by deprotonation of the nucleophile will be covered in this section, the examples being organized according to the nature of the Michael acceptor, as has been done in the other chapters.

6.3.1   Conjugate Addition of Carbon Nucleophiles

The Michael reaction of 2-alkyl substituted 1,3-diketones with alkynones has been carried out by Jørgensen using chiral base **124a** as the most efficient catalyst (Scheme 6.14).[28] Importantly, this reaction not only generates a quaternary stereocenter but also delivers a highly functionalized final product containing an α,β-unsaturated ketone side chain, suitable for further

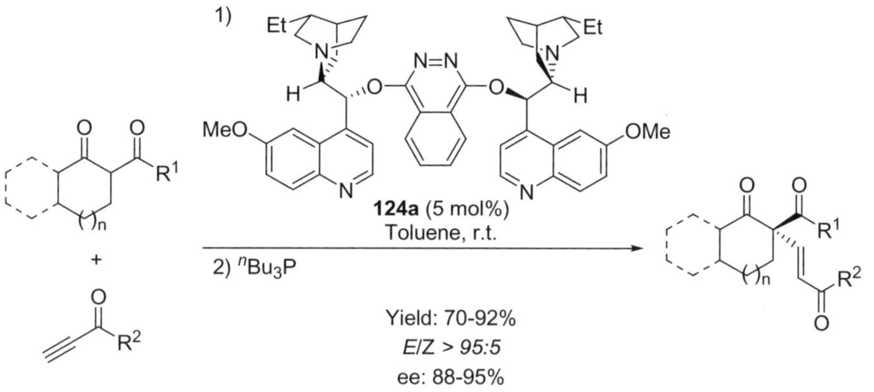

Scheme 6.14 Enantioselective **124a**-catalyzed Michael reaction of 1,3-diketones with alkynones.

transformations. After identifying the best catalyst and reaction conditions for the reaction to proceed with the highest possible enantioselectivity, it was still found that the Michael adducts were obtained as a mixture of *E* and *Z* isomers. However, the addition of a catalytic amount of nBu_3P to the crude reaction mixture promoted an isomerization process, furnishing exclusively the *E* diastereoisomers. Therefore, running the reaction under these optimized conditions delivered the final products with very high yields and enantioselectivities and as single diastereoisomers.

In a different context, a vinylogous conjugate addition reaction using alkenylmalononitriles as Michael donors and nitroalkenes as acceptors has been reported, independently by two different research groups, both reaching almost the same conclusions (Scheme 6.15).[29] In this particular case, the reaction occurred by the initial formation of an allylic anion after deprotonation of the acidic γ-proton by the chiral base, which subsequently underwent the Michael reaction. Importantly, in both cases, the reaction proceeded cleanly with a variety of differently substituted alkylidenemalononitriles and nitroalkenes, the main difference between both methodologies being the solvent employed in the reaction (CH_2Cl_2 or acetone). For this reaction, chiral base catalyst **125b** different from that previously used (see Scheme 6.14) was employed in which the aryl linker was changed from phthalazine to a 2,5-diphenylpyrimidine-type scaffold and dihydroquinidine was the basic cinchona alkaloid architectural element. The final products were obtained with excellent yields, diastereo- and enantioselectivities and the formation of regioisomers arising from the competitive possible reaction through the α-carbon of the nucleophile was not observed in either of the two reports. Later, the same strategy was applied to the use of quinones as Michael acceptors,[30] in this case identifying the phthalazine-based bis-dihydroquinine catalyst **124a** as the most efficient one. The results were slightly inferior to those observed in the reaction with nitroalkenes regarding the yield, diastereo- and enantioselectivity, probably due to the

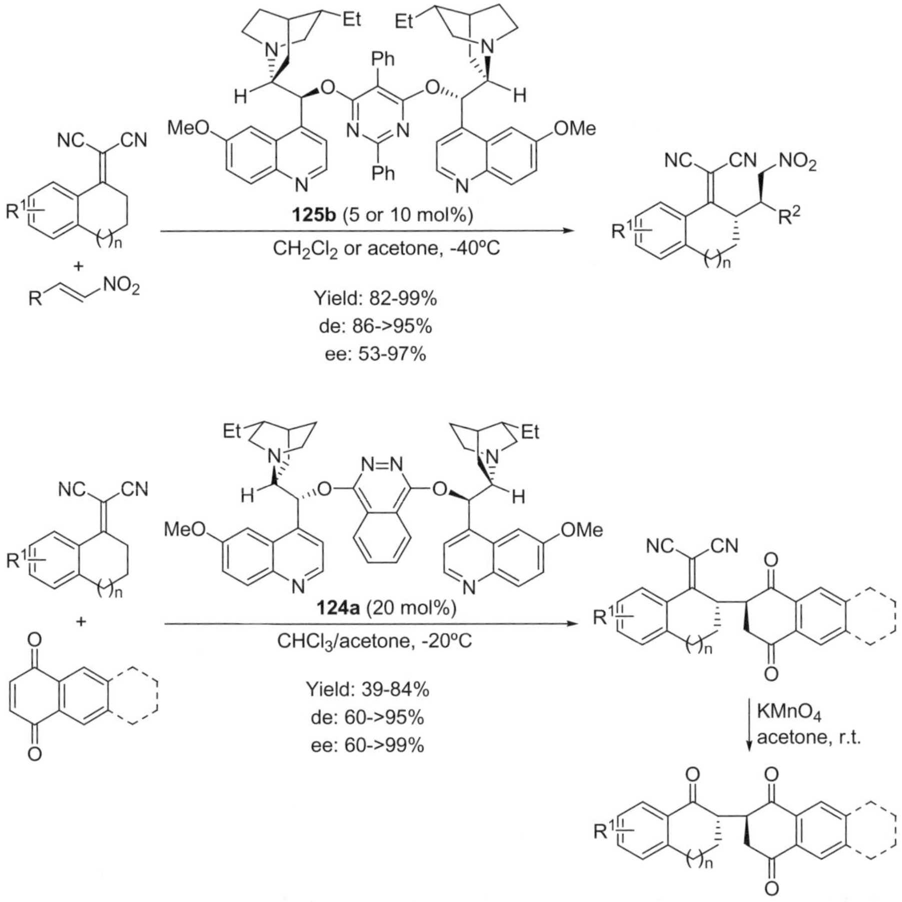

Scheme 6.15 Enantioselective vinylogous Michael reactions catalyzed by chiral Brønsted bases.

poorer abilities of quinones as electrophiles compared to nitroalkenes. The KMnO$_4$-mediate oxidative of the alkylidenemalononitrile moiety was carried out, demonstrating the utility of this group as carbonyl surrogate.

6.3.2   Conjugate Addition of Heteroatom-based Nucleophiles

Two examples of hetero-Michael reactions have been reported using these kinds of bis-cinchona alkaloid-based chiral Brønsted bases as catalysts. One of them refers to a sulfa-Michael reaction and the other is a case of an aza-Michael reaction.

In this context, Deng has shown that the conjugate addition of arylthiols to cyclic enones proceeded in a very highly enantioselective fashion using **125b** as catalyst (Scheme 6.16).[31] This protocol furnished the conjugate addition

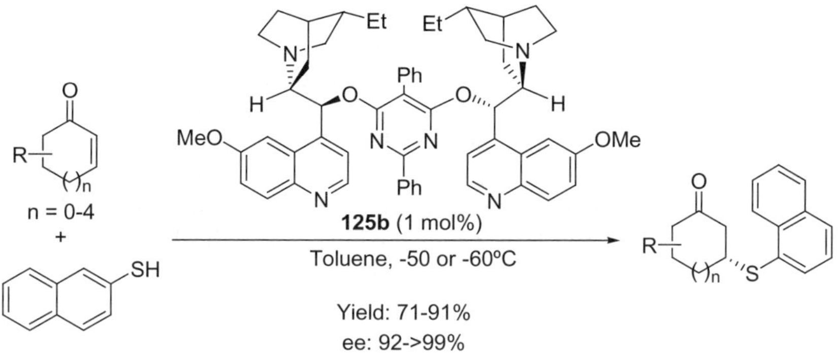

Scheme 6.16 Enantioselective sulfa-Michael reaction catalyzed by **125b**.

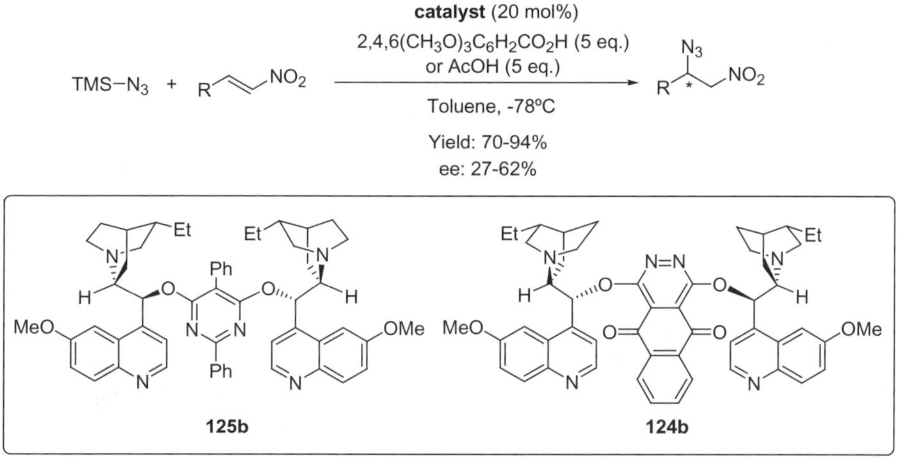

Scheme 6.17 Enantioselective aza-Michael reaction catalyzed by **125b** or **124b**.

products in excellent yields and with very high levels of enantioinduction in almost all cases studied and using a remarkably low catalyst loading. The scope of the reaction was remarkably wide with respect to the cyclic enone substrate, obtaining good results regardless of its substitution pattern or ring size. On the other hand, the nature of the arylthiol employed was crucial for achieving high stereocontrol, observing that 2-thionaphthol was the most appropriate Michael donor.

The other report regarding a base-catalyzed hetero-Michael reaction was a contribution from the group of Jørgensen regarding the conjugate addition of azide to nitroalkenes (Scheme 6.17).[32] In this case, the authors carried out an intensive effort directed toward the identification of the best catalyst and reaction conditions which would provide high yields and enantioselectivities. However, while the reaction was found to proceed with a good efficiency level with respect to yield or conversion, the enantioselectivities remained in moderate values. The reaction design made use of trimethylsilylazide as the source of hydrazoic acid, which was generated by the addition of a carboxylic acid. At the end, the best reaction conditions comprised the use of pyrimidine- or anthraquinone-based catalysts **125b** or **124b** and also a different carboxylic acid for the generation of the azide anion (acetic acid or 2,4,6-trimethoxybenzoic acid). Surprisingly, the reaction was studied exclusively on β-alkyl-substituted nitroalkenes and no example is provided illustrating the behavior of nitrostyrene derivatives as Michael donors in this reaction.

6.4   Other Miscellaneous Enantioselective Conjugate Addition Reactions

Small peptides can also promote several organic reactions in a stereoselective way.[33] With respect to conjugate addition reactions, in most of the cases, the peptide participates in the catalytic cycle by some of the previously covered activation modes, like enamine or iminium catalysis or additionally by activating the Michael acceptor H-bonding catalysis (see for example Schemes 2.8, 2.14 and 2.17 in Chapter 2, Scheme 3.20 in Chapter 3, Scheme 4.58 in Chapter 4, Scheme 5.6 in Chapter 5 and Scheme 6.10 in this Chapter). However, there are some other examples in which a small peptide has been proposed to catalyze a conjugate addition reaction in which the activation mechanism is not completely clear. Although the possibility of the participation of H-bonding interactions with the substrate can not be ruled out, the authors usually propose that the formation of an H-bonding network is mostly involved in the stabilization of a particular conformation of the peptide, which therefore could generate a chiral pocket in which the reagents participating in the reaction are incorporated. Although this is only a proposal and H-bonding interactions with the substrates could probably also operate, we have decided to cover these particular cases in this section.

In fact, the only known organocatalytic procedure for carrying out a conjugate addition reaction using azide as nucleophile which provides good levels of enantioselection is one of the most representative examples to illustrate this case. In a first report, it was found that tripeptide **126a** was a very effective catalyst in the conjugate addition of azide to α,β-unsaturated imides (Scheme 6.18).[34] The nucleophile was also in this case generated *in situ* by using trimethylsilyl azide as HN_3 precursor, being hydrolyzed by the presence of one equivalent of a carboxylic acid. The structure and conformation of the catalyst had to be optimized in order to reach a conformationally locked secondary structure at the peptide, which translated into exceptionally high enantioselectivities. This was achieved by using the combination of two amino acids, proline and *tert*-leucine, as β-turn inducers, in which the formation of an intramolecular H-bond was found to contribute importantly to the stabilization of the catalytically active conformation. Additionally a modified N1-alkylated histidine terminal residue was also found to be required for catalytic activity and the modification of the acid terminus of the *tert*-leucine residue as

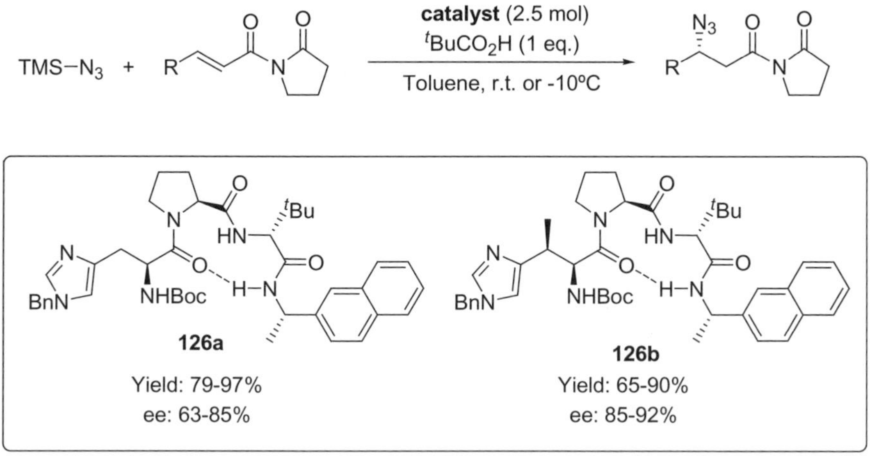

Scheme 6.18 Enantioselective aza-Michael reaction catalyzed by **126a** and **126b**.

the corresponding (*S*)-α-methylnaphthylamide was also necessary for obtaining high enantioselectivity. After identifying this *hit* catalytic peptide, the reaction of a variety of β-substituted α,β-unsaturated imides with TMSN₃ and tri-methylacetic acid proceeded smoothly providing the corresponding β-azido imides in good yields and enantioselectivities around 70–80% ee. Further studies led to a better catalyst architecture,[35] mainly focusing on searching for a more conformationally rigid peptide with respect to rotation around the side chain of the *N*-benzyl histidine residue. This was accomplished by introducing an additional methyl substituent at this side chain, observing that this new peptide **126b** was able to promote the reaction much more efficiently, providing the final compounds with higher optical purity.

There is also a later report by the same group regarding the use of a related catalyst **127** in the Michael addition of α-nitro ketones to enones (Scheme 6.19).[36] In this case, a variety of peptides were tested as catalysts but the best results were obtained with pentapeptide **127**, in which, among many other different changes, the α-amino *iso*-butyric acid was employed instead of *tert*-leucine for the β-turn inducer sequence; an additional modified arginine and a phenylalanine methyl ester were incorporated at the NH₂ and the CO₂H ends respectively. This catalyst furnished the final Michael adducts with moderate enantioselectivities, albeit in good yields in many cases.

In a completely different context, an *N*-protected version of the amino acid cysteine has been found to be an excellent promoter for the intramolecular Rauhut-Currier reaction (Scheme 6.20),[37] in which an enolizable enone played the role of the Michael acceptor, adding to another α,β-unsaturated ketone moiety in already present at the substrate structure. The mechanism of the reaction involved the activation of the enone which has to play the role of the donor by the catalyst *via* sulfa-Michael addition through the mercapto functionality of the *N*-protected cysteine derivative. The formed enolate

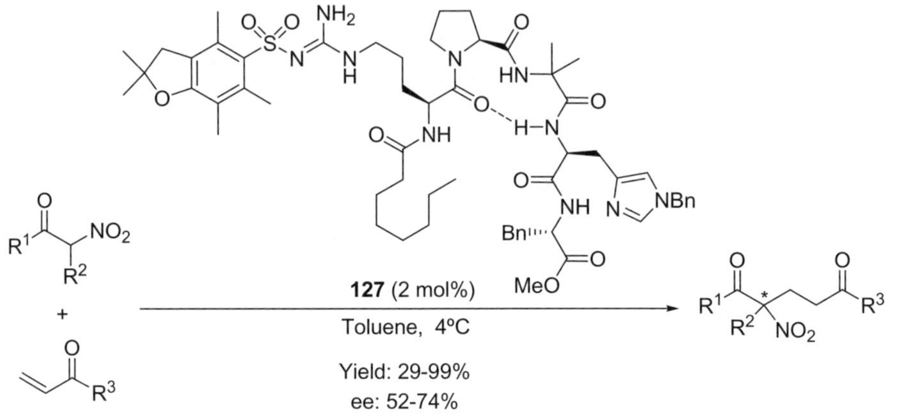

Scheme 6.19 Enantioselective Michael reaction of α-nitroketones with enones catalyzed by **127**.

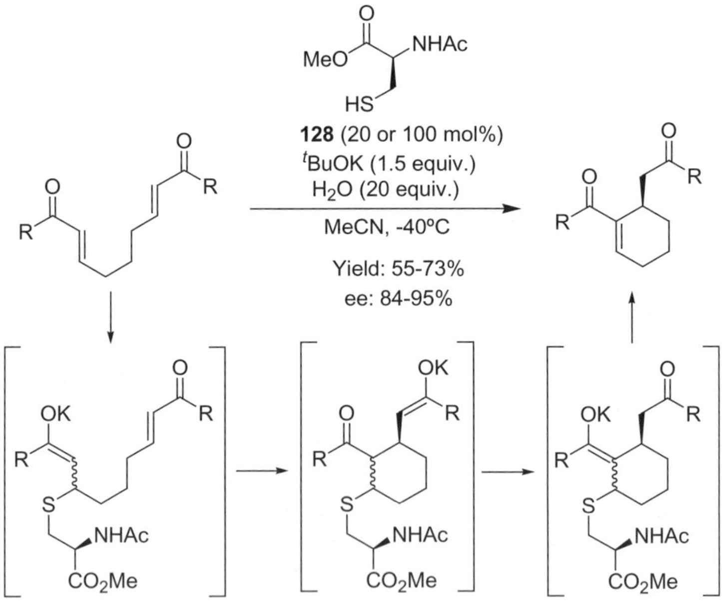

Scheme 6.20 Enantioselective intramolecular Rauhut–Currier reaction catalyzed by
N-acetyl cysteine.

undergoes intramolecular Michael reaction with the internal enone electrophile, and a final retro-sulfa-Michael reaction released the catalyst generating
an endocyclic C=C double bond. It should be pointed out that the catalyst
required activation by deprotonation of the SH group by one equivalent of an
external base additive (¹BuOK) and also that the reaction was carried out in

most cases using stoichiometric amounts of the promoter **128**, although in some cases the reaction was found to proceed satisfactorily using a 20 mol% of the catalyst. The reaction was tested on a variety of symmetrical bis-enone substrates, furnishing the final cyclohexenone cyclization products with excellent yields and enantioselectivities. A substrate incorporating an α,β-unsaturated ester moiety as internal electrophile was also tested but furnishing poorer results. It should be remembered at this point that another version of the intramolecular Rauhut–Currier reaction, furnishing compounds with a cyclopentene structure, has also been described using dienamine catalysis (see Scheme 2.25 in Chapter 2).

6.5 Concluding Remarks

In conclusion, there are other alternative mechanistic ways different from enamine/iminium, H-bonding or PTC, for which an organic molecule can activate one of the reagents participating in an enantioselective Michael reaction. Some of them, although covering a rather specific transformation (the Stetter reaction, the Rauhut–Currier reaction, *etc.*) have shown up as extremely powerful alternative methodologies, which also allow the inversion of the typical reactivity displayed by several functional groups in a typical unpolung process, as is the case of the *N*-heterocyclic carbenes. The use of chiral Brønsted bases or peptides is still limited to a few transformations, but is extremely efficient, for which there are no other organocatalytic asymmetric methodology available furnishing better performances. This is the case of the sulfa-Michael reaction to cyclic enones catalyzed by dimeric cinchona alkaloids or the conjugate addition of azide to α,β-unsaturated imides catalyzed by a synthetic tripeptide. In view of the state-of-the-art in this field, there is a lot of room for improvement or for the development of novel methodologies and reactions and, in this context, impressive advances are expected to occur in the following years.

References and Notes

1. (a) E. Buchner and C. Curtius, *Ber. Dtsch. Chem. Ges.*, 1885, **8**, 2377; (b) H. Staudinger and O. Kupfer, *Ber. Dtsch. Chem. Ges.*, 1912, **45**, 501.
2. E. O. Fischer and A. Maasbol, *Angew. Chem. Int. Ed. Engl.*, 1964, **3**, 580.
3. A. Iguau, H. Grutzmacher, A. Baceiredo and G. Bertrand, *J. Am. Chem. Soc.*, 1988, **110**, 6463.
4. A. J. Arduengo III, R. L. Harlow and M. Kline, *J. Am. Chem. Soc.*, 1991, **113**, 361.
5. Reviews: (a) D. Enders, O. Niemeyer and A. Henseler, *Chem. Rev.*, 2007, **107**, 5606; (b) K. Zeitler, *Angew. Chem. Int. Ed.*, 2005, **44**, 7506. For accounts see; (c) D. Enders and T. Balensiefer, *Acc. Chem. Res.*, 2004, **37**, 534; (d) J. Read de Alaniz and T. Rovis, *Synlett*, 2009, 1189.
6. R. Breslow, *J. Am. Chem. Soc.*, 1958, **80**, 3719.

7. D. Seebach, *Angew. Chem. Int. Ed. Engl.*, 1979, **18**, 239.

8. (a) J. Sheehan and D. H. Hunnemann, *J. Am. Chem. Soc.*, 1966, **88**, 3666; (b) J. S. Sheehan and T. Hara, *J. Org. Chem.*, 1974, **39**, 1196.

9. (a) W. Tagaki, Y. Tamura and Y. Yano, *Bull. Chem. Soc. Jpn.*, 1980, **53**, 478; (b) J. Martí, J. Castells and F. López-Calahorra, *Tetrahedron Lett.*, 1993, **34**, 521; (c) D. Enders, K. Brener and J. H. Teles, *Helv. Chim. Acta*, 1996, **79**, 1217; (d) D. Enders and U. Kallfass, *Angew. Chem. Int. Ed.*, 2002, **41**, 1743.

10. (a) H. Stetter and H. Kuhlmann, *Synthesis*, 1975, 379; (b) H. Stetter, *Angew. Chem. Int. Ed. Engl.*, 1976, **15**, 639.

11. E. Ciganek, *Synthesis*, 1995, 1311.

12. D. Enders, *Stereoselective Synthesis*, Springer-Verlag, Heidelberg, 1993, p. 63.

13. D. Enders, K. Breuer, J. Runsink and J. H. Teles, *Helv. Chim. Acta*, 1996, **79**, 1899.

14. M. S. Kerr, J. Read de Alaniz and T. Rovis, *J. Am. Chem. Soc.*, 2002, **124**, 10298.

15. (a) J. Read de Alaniz, M. S. Kerr, J. L. Moore and T. Rovis, *J. Org. Chem.*, 2008, **73**, 2033; (b) M. S. Kerr and T. Rovis, *Synlett*, 2003, 1934.

16. S. C. Cullen and T. Rovis, *Org. Lett.*, 2008, **10**, 3141.

17. (a) M. S. Kerr and T. Rovis, *J. Am. Chem. Soc.*, 2004, **126**, 8876; (b) J. L. Moore, M. R. Kerr and T. Rovis, *Tetrahedron*, 2006, **62**, 11477.

18. J. Read de Alaniz and T. Rovis, *J. Am. Chem. Soc.*, 2005, **127**, 6284.

19. Q. Liu and T. Rovis, *J. Am. Chem. Soc.*, 2006, **128**, 2552.

20. J. Pesch, K. Harms and T. Bach, *Eur. J. Org. Chem.*, 2004, 2025.

21. S. M. Mennen, J. T. Blank, M. B. Tran-Dubé, J. E. Inbriglio and S. J. Miller, *Chem. Commun.*, 2005, 195.

22. Y. Matsumoto and K. Tomioka, *Tetrahedron*, 2006, **47**, 5843.

23. D. Enders and J. Han, *Synthesis*, 2008, 3864.

24. D. Enders, J. Han and A. Henseler, *Chem. Commun.*, 2008, 3989.

25. Q. Liu, S. Perreault and T. Rovis, *J. Am. Chem. Soc.*, 2008, **130**, 14066.

26. Q. Liu and T. Rovis, *Org. Lett.*, 2009, **11**, 2856.

27. D. DiRocco, K. M. Oberg, D. M. Dalton and T. Rovis, *J. Am. Chem. Soc.*, 2009, **131**, 10872.

28. M. Bella and K. A. Jørgensen, *J. Am. Chem. Soc.*, 2004, **126**, 5672.

29. (a) D. Xue, Y.-C. Chen, Q.-W. Wang, L.-F. Cun, J. Zhu and J.-G. Deng, *Org. Lett.*, 2005, **7**, 5293; (b) T. B. Poulsen, M. Bell and K. A. Jørgensen, *Org. Biomol. Chem.*, 2006, **4**, 63.

30. J. Aleman, C. B. Jacobsen, K. Frisch, J. Overgaard and K. A. Jørgensen, *Chem. Commun.*, 2008, **31**, 632.

31. P. McDaid, Y. Chen and L. Deng, *Angew. Chem. Int. Ed.*, 2002, **41**, 338.

32. M. Nielsen, W. Zhuang and K. A. Jørgensen, *Tetrahedron*, 2007, **63**, 5849.

33. Reviews: (a) E. A. Colby, S. M. Davie, Y. X. Mennen and S. J. Miller, *Chem. Rev.*, 2007, **107**, 5759; (b) E. R. Jarvo and S. J. Miller, *Tetrahedron*, 2002, **58**, 2481.

34. T. E. Horstmann, D. J. Guerin and S. J. Miller, *Angew. Chem. Int. Ed.*, 2000, **39**, 3635.
35. D. J. Guerin and S. J. Miller, *J. Am. Chem. Soc.*, 2002, **124**, 2134.
36. B. R. Linton, M. H. Reutershan, C. M. Aderman, E. A. Richardson, K. R. Brownell, C. W. Ashley, C. A. Evans and S. J. Miller, *Tetrahedron Lett.*, 2007, **48**, 1993.
37. C. E. Aroyan and S. J. Miller, *J. Am. Chem. Soc.*, 2007, **129**, 256.

CHAPTER 7

Enantioselective Cascade Reactions Initiated by Conjugate Addition

7.1 Introduction

One of the most attractive features of the Michael-type reactions is the fact that an intermediate nucleophilic species is generated after the conjugate addition step, which is able to react with an electrophilic reagent present in the reaction medium either in an inter- or an intramolecular fashion, in a typical domino or cascade process. Domino or cascade reactions represent a highly efficient approach for the straightforward construction of complex molecules in the shortest and most efficient way because of their ability to build up complex molecules in a very efficient way and minimizing the number of laboratory operations and the subsequent generation of waste chemicals.[1] Moreover, if stereochemical control is desired, domino processes also show up as a very effective approach for building up the target molecule with good stereo-selectivity because once a first stereocenter is generated, the subsequent dia-stereoselective processes of the cascade normally occur together with an exponential growth on the enantiopurity of the compound formed at the end.

In this context, the use of small organic molecules as catalysts in domino processes initiated by conjugate addition reactions shows up as a very useful and competitive tool for the generation of molecular complexity from readily available and cheap starting materials and also displaying an exceptional per-formance with regard to stereochemical control.[2] Almost all the types of organocatalysts covered in the previous chapter have been employed in this context, with a variety of many different and useful methodologies with

RSC Catalysis Series No. 5
Organocatalytic Enantioselective Conjugate Addition Reactions: A Powerful Tool for the Stereocontrolled Synthesis of Complex Molecules
By Jose L. Vicario, Dolores Badía, Luisa Carrillo and Efraim Reyes
© J. L. Vicario, D. Badía, L. Carrillo and E. Reyes 2010
Published by the Royal Society of Chemistry, www.rsc.org

enormous potential to be applied in the synthesis of valuable compounds. In this way, many cascade processes in which catalysts operating by enamine, iminium, H-bonding, PTC or other types of mechanisms can be found in the literature for the stereoselective preparation of complex molecules. As has been done in the previous chapters, the reactions will be presented according to the mechanism involved in the activation of the reagents participating in the cascade-initiating Michael-type reaction. Within each section, the reported methodologies will be classified according to the type of reactions which follow the conjugate addition process.

7.2 Cascade Processes Initiated by Conjugate Addition *via* Enamine Activation

The enamine activation strategy has been employed in several very efficient processes for the synthesis of complex molecules, especially focused in the stereoselective preparation of highly functionalized carbocyclic compounds. For this reason, the most commonly found situation is that the amine catalyst is involved in the activation of an enolizable aldehyde *via* enamine formation, which undergoes the first Michael-type reaction delivering an iminium intermediate from which in most cases the catalyst is released by hydrolysis, leading to the formation of a multifunctional acyclic intermediate compound. The cascade process therefore continues by some kind of intramolecular reaction with another functionality appropriately introduced in the starting materials, without the participation of the catalyst and therefore under substrate control. In some other cases, substrates capable of being activated by the amine catalyst by the formation of both an enamine or iminium intermediate undergo a cascade process in which it is indeed the first step which proceeds under enamine activation although some of the subsequent steps might involve some of the other reagents incorporated into the reaction scheme being activated by the catalyst *via* iminium ion formation.

7.2.1 Cascade Michael/Aldol Reactions

A representative example of a cascade Michael-nitroaldol process initiated by enamine activation is shown in Scheme 7.1, in which pentanedial reacted with a variety of nitroalkenes in the presence of **31a** leading to the formation of highly functionalized tetrasubstituted cyclohexanes containing four stereocenters in a single step with excellent yields and as single diastereoisomers of excellent optical purity.[3] The reaction consisted of a first conjugate addition of pentadienal to the nitroalkene, proceeding under enamine activation and in which the chiral catalyst controlled the stereochemistry of two stereogenic centers, the stereochemical outcome of this reaction being according to what was proposed for the general **31a**-catalyzed intermolecular Michael addition of aldehydes to nitroalkenes (see Scheme 2.12 in Chapter 2). Next, the intramolecular nitroaldol (Henry) reaction had to take place affording the final compound and

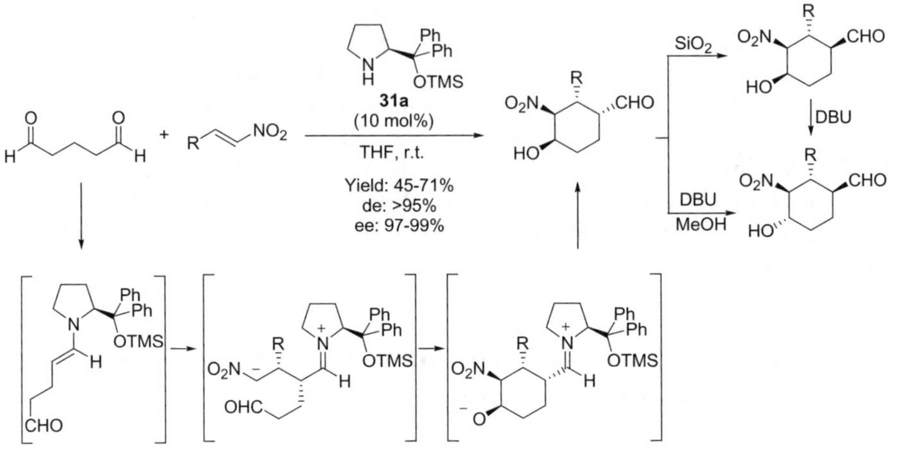

Scheme 7.1   Enantioselective synthesis of tetrasubstituted cyclohexanes by Michael/
nitroaldol cascade reaction.

generating the remaining two stereocenters at this step. Interestingly, the
obtained compounds were observed to epimerize during purification when
treated with silica gel, obtaining a major diastereoisomer in which the stereocenter containing the formyl group had isomerized in order to reach a more
stable conformation in which this group moved from axial to equatorial
position. Moreover, stirring one of the cascade products in the presence of a
base also led to the formation of a different diastereoisomer, in which both the
formyl and the hydroxy group had isomerized *via* a retro-Henry/Henry process,
furnishing the most stable diastereoisomer in which a conformation exists with
all substituents arranged in equatorial positions. This last isomerization
experiment points toward the presence of the catalyst incorporated in the
intermediate during the intramolecular nitroaldol reaction, which would
explain the formation of the obtained product and not the thermodynamically
most stable one.

   A very similar approach was reported some years later, in which a doubly
activated olefin was employed as the Michael acceptor which also makes use of
31a as catalyst (Scheme 7.2).[4] In this context, the reaction of pentanedial with an
arylidenemalonate proceeded with low yield but excellent enantio- and diastereoselectivity but, when α-cyano-α,β-unsaturated esters were employed as substrates, the yield of the resulting reaction notably increased, although together
with an important decrease in the diastereoselectivity of the process regarding
stereocontrol at the quaternary stereocenter being formed. Importantly, this
reaction led to the formation of tetrasubstituted cyclohexanes incorporating a
quaternary stereocenter, which was formed with a notably high degree of stereocontrol, with diastereoselectivities ranging from 6:1 to 10:1 with respect to the
configuration at this quaternary stereogenic center. Any other diastereoisomer
which could be potentially formed in the reaction was not observed.

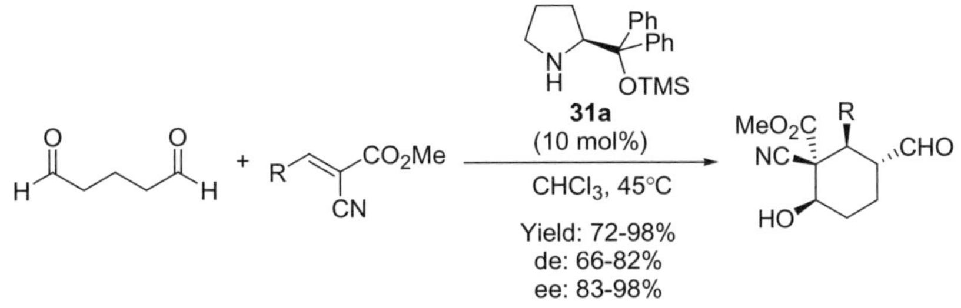

Scheme 7.2 Enantioselective Michael/aldol cascade reaction generating a quaternary stereocenter.

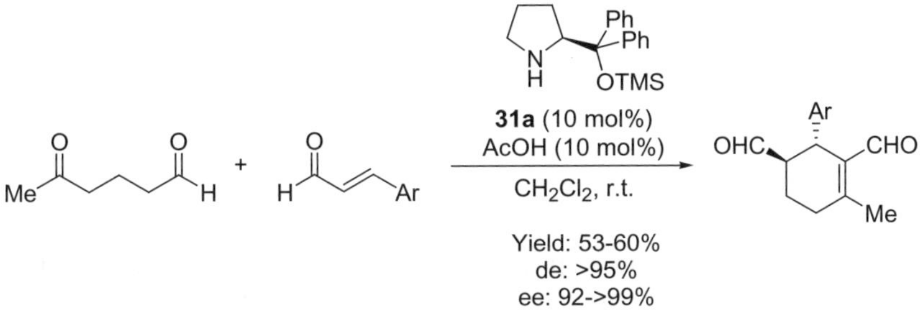

Scheme 7.3 Enantioselective Michael/aldol/dehydration cascade reaction for the synthesis of functionalized cyclohexenes.

There is also another similar case in which 5-oxohexanal was employed as functionalized Michael donor undergoing Michael addition/intramolecular aldol reaction with aromatic enals (Scheme 7.3), which also ended up with a final dehydration step leading to the formation of functionalized cyclohexenes. Under the optimized reaction conditions, the final compounds were obtained in moderate yields but with excellent enantioselectivities and as single diastereoisomers.[5] It should be pointed out that, from the mechanistic point of view, a dual activation of the 5-oxohexanal (*via* enamine formation) and the α,β-unsaturated aldehyde (*via* iminium ion formation) might operate in this case in the catalytic cycle, although no mechanistic proposal was provided by the authors.

This dual enamine/iminium activation profile in cascade Michael/aldol reactions can also be found even in some early reports, mostly focused on the self-dimerization of enals catalyzed by proline or analogues derived thereof, which generally proceeded with low enantioselectivities.[6] There is not a clear and definitive mechanistic pathway confirmed for these reactions, although the most widely accepted proposal for the dimerization of enals (Scheme 7.4)[6c] involved sequential activation of one molecule of the substrate as a dienamine (Michael donor) and another molecule as iminium ion (Michael acceptor).

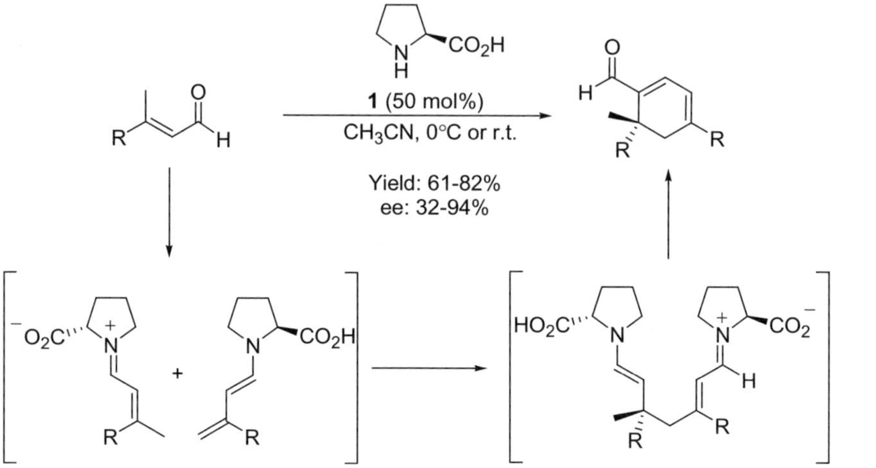

Scheme 7.4 Possible mechanism for the self dimerization of enals catalyzed by proline.

After the Michael reaction step had taken place, an intermediate containing both an iminium and an enamine moiety was generated and their direct reaction between each other followed by elimination was proposed to account for the generation of the final compound.

7.2.2   Cascade Michael/Michael Reactions

In a similar approach to the strategy shown in these previous examples, Michael donors incorporating an activated olefin in their structure in addition to the required formyl group have been employed in cascade reactions as functionalized reagents. This is the case of the example shown in Scheme 7.5, in which a 7-oxo-2-heptenoate reacted with a variety of nitroalkenes using **31a** as catalyst, furnishing once again a series of highly functionalized cyclohexane compounds containing multiple stereocenters.[7] The reaction consisted of a first Michael addition of the aldehyde moiety present at the Michael donor *via* enamine activation, followed by intramolecular Michael reaction. However, it should be pointed out that this is not a real cascade process, requiring the addition of an external base in order to deprotonate the intermediate nitro-derivative formed after the initial Michael reaction, in order to generate an activated nitronate nucleophile which assisted the second intramolecular conjugate addition reaction, in a typical one-pot procedure. As a consequence, it is clear that the catalyst did not participate in any case during this final cyclization. It should be pointed out that the geometry of the enoate moiety at the functionalized donor was crucial with respect to the diastereoselectivity of the reaction, observing the formation of mixtures of diastereoisomers when the

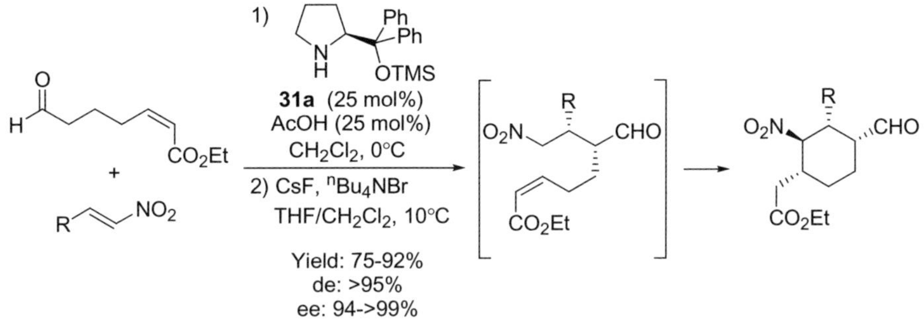

Scheme 7.5 Enantioselective synthesis of tetrasubstituted cyclohexanes by Michael/ Michael tandem reaction.

E-α,β-unsaturated ester was employed, while a single diastereoisomer was obtained using the corresponding *Z* substrate. This reaction design has also been extended to the use of an α,β-unsaturated aldehyde as the initial Michael acceptor, in this case it being necessary to consider the possible dual activation of the Michael donor and acceptor by the catalyst through the formation of the corresponding enamine and iminium intermediates.[8]

The possibility of using ketones as Michael donors in a similar reaction design has also been surveyed for the preparation of complex and highly substituted bicyclic skeletons (Scheme 7.6).[9] In this case, several cyclohexanone derivatives reacted with 2-nitroallyl acetates in the presence of pyrrolidine-thiourea catalyst **6b** in a process consisting of a first Michael reaction of the cyclohexanone to the nitroalkene, followed by elimination, which generated *in situ* a second nitroalkene moiety ready to participate as an internal electrophile with an enamine formed after iminium/enamine interconversion. Yields and enantioselectivities were consistently high for a variety of different aryl-substituted allyl nitroacetates, although the reaction failed when β-alkyl substituted substrates were employed. The reaction scope with regard to the ketone employed was also found to be rather narrow, with six- and five-membered cyclic ketones performing excellently in the reaction, but observing that cycloheptenone or an acyclic ketone such as acetone were unreactive under the reaction conditions employed. In all cases, the final compounds were obtained as single diastereoisomers, showing the excellent stereochemical control exerted by the catalyst during the whole process, although the strained nature of the final products might also contribute to the high diastereoselectivity observed.

However, it is in this context where the use of primary amines as catalysts demonstrates the high performance of this concept. As has already been mentioned in Chapter 2, primary amines are much more efficient catalysts for the activation of ketones than the corresponding secondary amines due to the formation of a less sterically congested enamine or iminium ion and also because of the more efficient geometry control in their formation. For example, primary amine **28a** was found to be able to activate α-enolizable enones toward

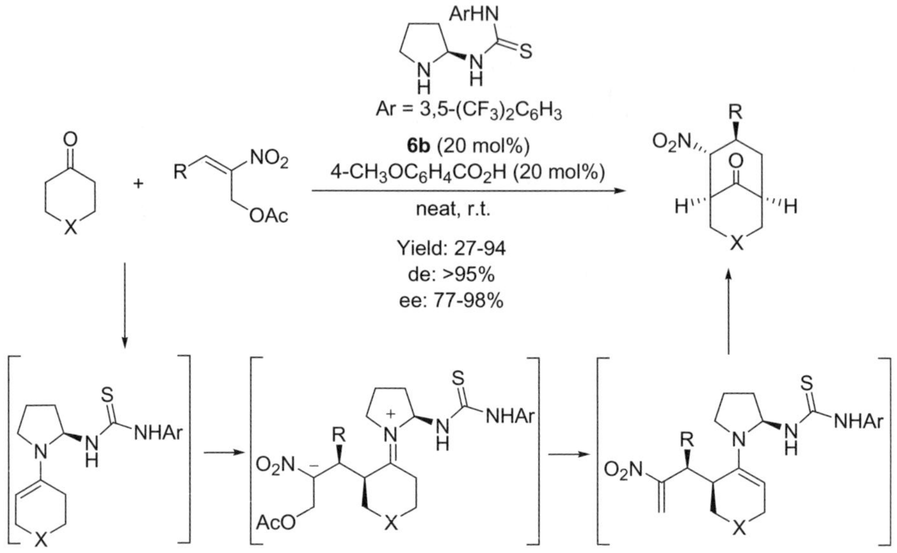

Scheme 7.6 Enantioselective synthesis of bicyclic skeletons by Michael/Michael cascade reaction.

reaction with a nitroalkene, also furnishing cyclohexanones containing four stereocenters very cleanly and in a very diastereo- and enantioselective fashion (Scheme 7.7).[10] In this reaction, the α,β-unsaturated ketone substrate was activated by the primary amine catalyst in the form of a dienamine intermediate, which next underwent Michael addition to the nitroalkene delivering a nitronate/iminium ion ready to participate in the subsequent intramolecular Michael reaction. The possibility of a mechanistic pathway involving Diels–Alder reaction between the dienamine and the nitroalkene was discarded after isolating one of the intermediates of the reaction in a control experiment. Perhaps the only limitation of the process is related to the need for a β-aryl substituted enone as Michael donor but, on the other hand, the reaction was extremely general with regard to the Michael acceptor employed, allowing not only the use of nitroalkenes but also other highly activated olefins like α-cyanocinnamates or maleimides.

7.2.3 Michael/Michael/Aldol Triple Cascade Reactions

The triple Michael/Michael/aldol sequence developed by Enders shown in Scheme 7.8 can probably be considered as one of the most impressive demonstrations of the ability and power of enantioselective organocatalytic cascade reactions for the generation of molecular complexity from very simple and cheap starting materials.[11] In this reaction, a nitroalkene, an enolizable aldehyde and an α,β-unsaturated aldehyde reacted with each other in the

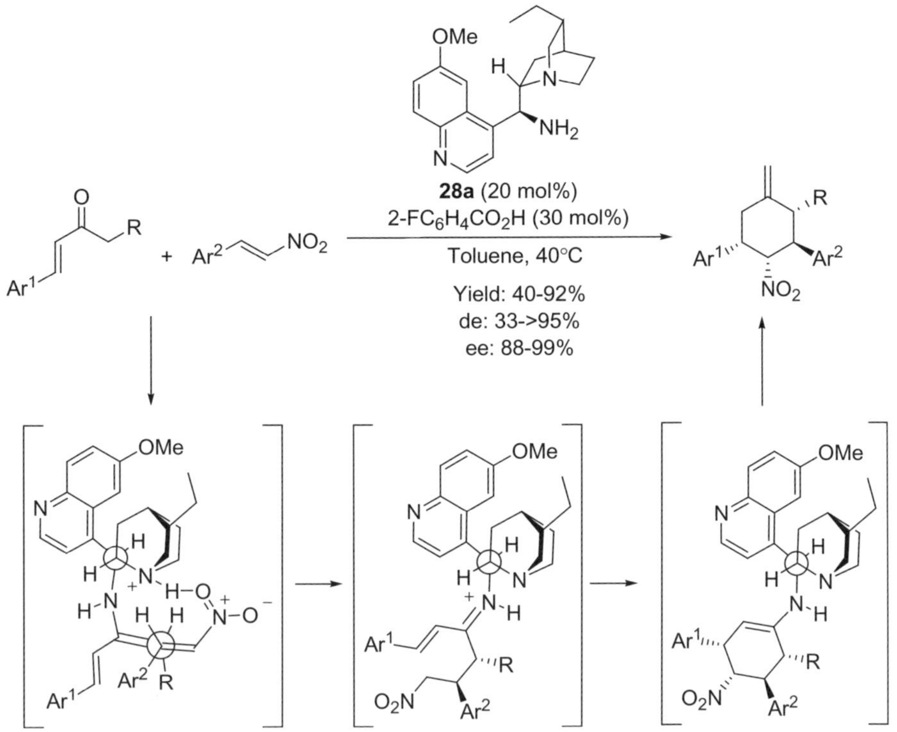

Scheme 7.7 Enantioselective synthesis of cyclohexanones by Michael/Michael cascade reaction between enones and nitroalkenes.

presence of chiral amine **31a**, furnishing a cyclohexenecarbaldehyde final product containing four contiguous stereogenic centers. One major diastereoisomer was obtained out of the eight possible ones, which was also obtained in almost enantiomerically pure form, together with minor amounts of the C-5 epimer. This cascade process involved a first step in which the enolizable aldehyde underwent Michael addition to the nitroalkene reagent requiring the activation of the nucleophile by the chiral amine catalyst *via* enamine formation. In this step, the chiral catalyst was able to exert a very efficient degree of stereocontrol in the two contiguous stereogenic centers generated in this step and, moreover, this reaction was found to occur with complete chemoselectivity favoring the reaction of the enamine Michael donor to the nitroalkene in the presence of the α,β-unsaturated aldehyde electrophile because of the higher electrophilicity of the former. Next, the catalyst had to be released from the adduct, delivering a functionalized and a highly acidic nitroalkane, which next operated as a C-nucleophile in a subsequent Michael addition step to the enal and which also required activation by the catalyst *via* iminium ion formation. Also in this case, two additional stereogenic centers were formed and it should be at this point that the formation of the minor diastereoisomer occurred.

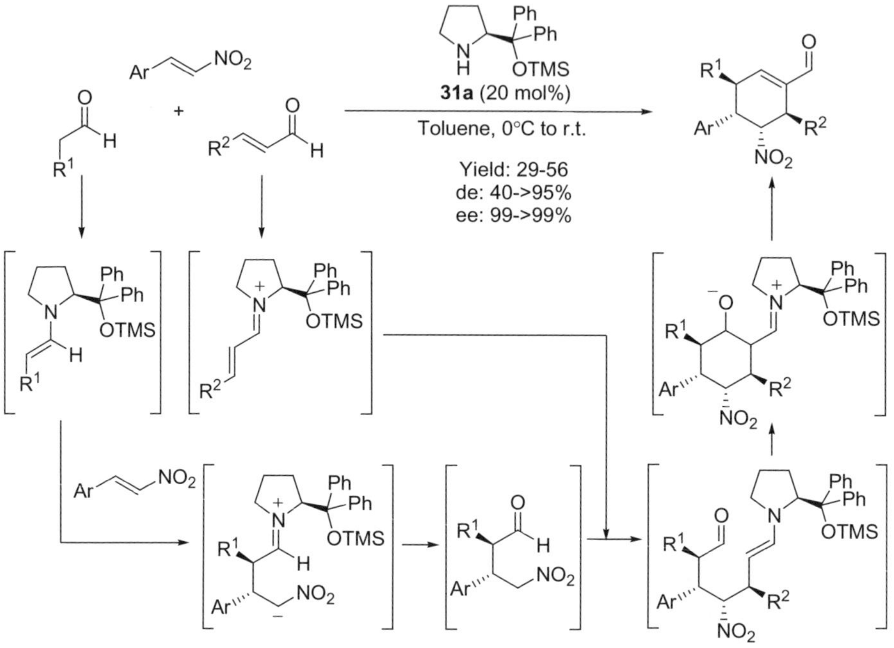

Scheme 7.8 Enantioselective triple Michael/Michael/aldol cascade reaction developed
by Enders.

Finally, the enamine intermediate generated after this second Michael reaction
underwent intramolecular aldol reaction followed by dehydration, delivering
the final highly functionalized cyclohexenecarbaldehyde product.

This transformation proved to be rather general with respect to the enolizable aldehyde and the enal, with higher yields obtained when aliphatic α,β-
unsaturated aldehydes were employed. On the other hand, all the nitroalkenes
tested were nitrostyrene derivatives, with no data provided with regard to the
use of the more problematic aliphatic nitroalkenes. In addition, several selective
transformations were carried out on the compounds obtained by this methodology,[12] therefore proving the possibility of chemical manipulation of the
different functionalities present in their structure. A version involving the use of
recyclable polystyrene-supported prolinol-based catalysts has also been
reported furnishing similar yields to those obtained by Enders in solution for a
comparative example.[13]

In an extension of this work, a one-pot procedure has also been set up for the
conversion of the adducts obtained in this cascade reaction into more complex
tricyclic compounds containing multiple stereogenic centers and several functionalities amenable to be modified into a wide variety of potentially useful
chiral building blocks. The approach consisted of the use of an enolizable
aldehyde reagent incorporating a conjugate diene moiety at an appropriate

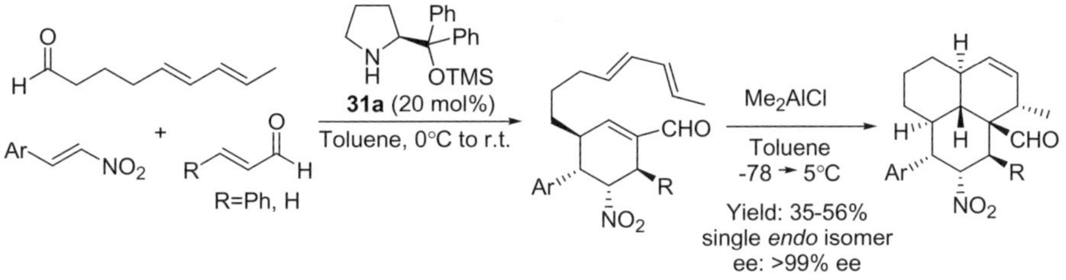

Scheme 7.9 Triple Michael/Michael/aldol cascade followed by one-pot intramolecular Diels–Alder reaction.

position as substrate for the organocatalytic Michael/Michael/aldol process and, therefore, after the **31a**-catalyzed condensation with the corresponding aromatic nitroalkene and α,β-unsaturated aldehyde, an excess of a Lewis acid such as Et$_2$AlCl was added to the crude reaction mixture inducing a highly diastereoselective intramolecular Diels–Alder reaction (Scheme 7.9).[14] By this process, a tricyclic framework containing eight stereogenic centers was formed in moderate yield and high stereoselectivity in a single operation. Depending on the length of the tether alkyl chain used between the formyl and the diene moieties in the initial Michael donor reagent, different tricyclic frameworks were obtained but, while the intramolecular Diels–Alder reaction conducing to the formation of a six-membered ring at the final adducts proceeded with excellent diastereoselectivity, when the length of the tether resulted in the formation of a five-membered ring mixtures of diastereoisomers were obtained. On the other hand, this reaction sequence was only tested using cinnamaldehyde and acrolein as the α,β-unsaturated aldehyde reagents.

This triple Michael/Michael/aldol cascade design reaction has been applied to other combinations of reagents and, for example, the substitution of the nitroalkene reagent by other highly electrophilic α,β-unsaturated carbonyl compounds like 2-cyanoacrylates results in a highly efficient protocol for the preparation of cyclohexenecarbaldehydes similar to those obtained by Enders but containing a quaternary stereocenter (Scheme 7.10).[15] The chemoselectivity of the process was in this case guaranteed because of the higher reactivity of the doubly activated 2-cyanoacrilate Michael acceptor with respect to the enal reagent. Also in this case, *O*-TMS diphenylprolinol **31a** was identified as the most efficient catalyst for this transformation, which, in turn, allowed the preparation of the final cyclic compounds in yields and enantioselectivities similar to those obtained by Enders. With respect to the diastereoselectivity of the reaction, also in this case one major diastereoisomer was obtained out of the eight possible ones, together with minor amounts of the C-5 epimer. Very recently, alkylideneoxindoles have also been employed as suitable Michael acceptors in this reaction design, providing a direct and efficient approach to spirocyclic oxindoles,[16] which is a structural feature of several naturally occurring alkaloids.

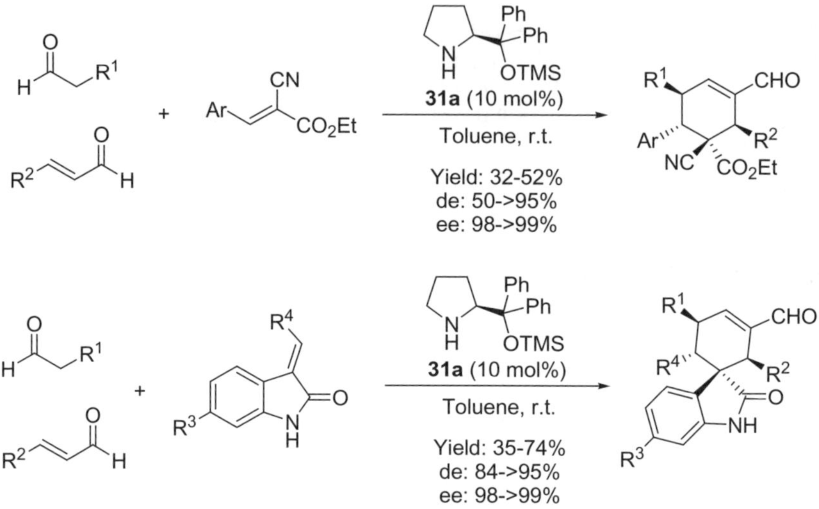

Scheme 7.10   Triple Michael/Michael/aldol cascade using 2-cyanoacrilates or alkylideneoxindoles as highly active electrophiles.

7.2.4 Other One-pot Procedures Initiated by Michael Reactions Proceeding under Enamine Activation

There are some other reported methodologies that rely on sequential reactions
which are started by a conjugate addition reaction *via* enamine activation of the
Michael donor. It should be mentioned that all these examples can not be
strictly considered as cascade processes, because the reaction conditions involve
a multistep sequence starting with the Michael addition reaction which is followed by a subsequent process, induced by the addition of further reagents
after the Michael reaction is observed to be complete. Nevertheless, the fact
that the reaction is carried out in a one-pot fashion rather than happening as a
cascade process does not represent a disadvantage from the point of view of the
chemical efficiency, considering that, in both cases, a complex product is
obtained after a sequence of different reactions proceeding in the right order,
without the need to isolate any of the intermediates generated after each
reaction.

A very good and representative example of this strategy is depicted in
Scheme 7.11, which shows an excellent and highly efficient synthetic approach
to (–)-*oseltamivir*, the precursor of the anti-influenza drug (–)-*oseltamivir
phosphate* or *Tamiflu*[®].[17] The synthesis started with a one-pot process consisting
initially of the **31**-catalyzed Michael reaction of (2-pentyloxy)acetaldehyde to
tert-butyl 2-nitroacrylate followed by a cascade process consisting of a second
intermolecular Michael reaction with a 2-etoxycarbonyl vinylphosphonate
Michael acceptor and a subsequent intramolecular Horner–Wadsworth–
Emmons reaction, forming a cyclohexenedicarboxylate intermediate and a final

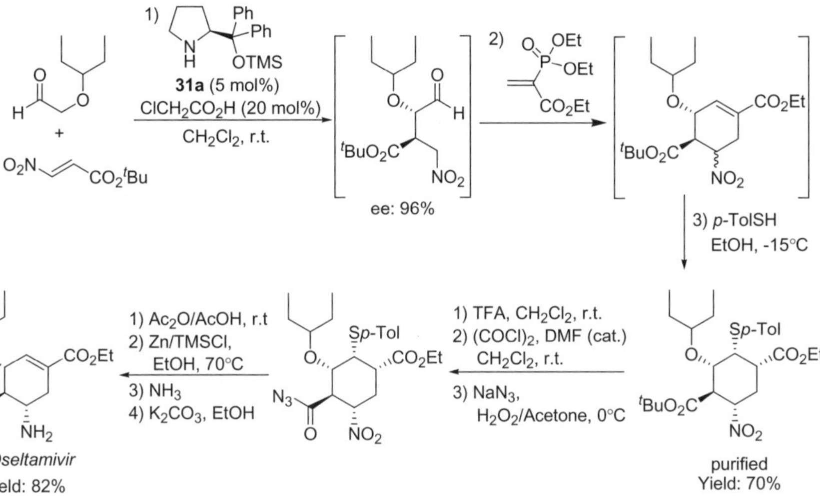

Scheme 7.11 Asymmetric synthesis of (–)-*oseltamivir* by three consecutive one-pot operations.

sulfa-Michael reaction with *p*-toluenethiol, leading to the formation of a highly functionalized hexasubstituted cyclohexane product, which was obtained after column chromatography purification in 70% yield, calculated from the starting materials. Importantly, the cyclohexenedicarboxylate intermediate generated after intramolecular Horner–Wadsworth–Emmons reaction was observed to be formed as a mixture of diastereoisomers at C-5 (the nitro-containing stereogenic center), the major diastereoisomer being the one with the wrong configuration at this stereocenter. This was solved by the final sulfa-Michael step, which was carried out in the presence of a basic additive, which promoted the isomerization of this C-5 stereocenter to the thermodynamically more stable (5*S*) diastereoisomer with the required configuration for the synthesis of the target compound. The second one-pot operation consisted of hydrolysis of the *tert*-butoxycarboxylate group followed by conversion into the corresponding acyl azide and the final process entailed a cascade Curtius rearrangement/*N*-acetylation followed by a second Zn-mediated reduction of the nitro group and a final base-promoted retro-sulfa-Michael reaction, delivering (–)-*oseltamivir* in 82% yield from the hexasubstituted cyclohexane product previously isolated. The exceptional performance of this approach can be clearly appreciated from the very high (57%) overall yield of the synthesis.

There is also an example of a one-pot reaction in which an initial Michael addition of a functionalized aldehyde containing an olefin moiety at the convenient position to a nitroalkene as Michael acceptor was followed by the formation of a 1,3-dipole and a subsequent intramolecular [3 + 2] cycloaddition (Scheme 7.12).[18] In this case, ethyl 7-oxo-2-heptenoate was reacted with a series of nitroalkenes using **31a** as catalyst and next *N*-hydroxylphenylamine was added to the reaction mixture, promoting the formation of the corresponding

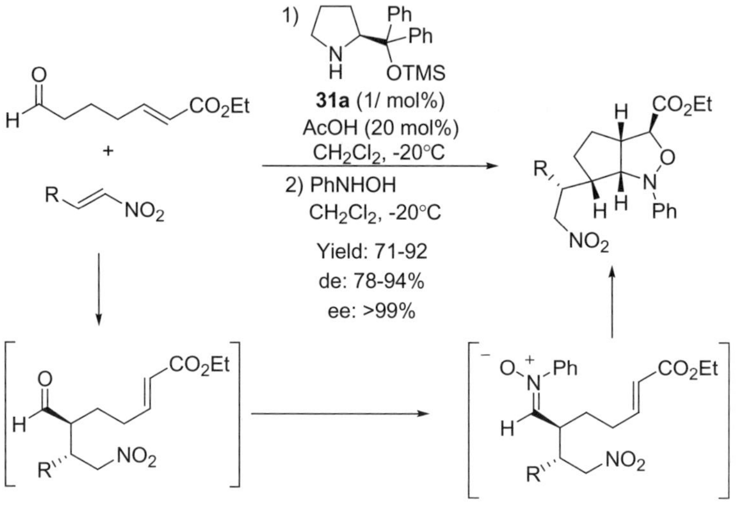

Scheme 7.12   One-pot Michael/intramolecular [3 + 2] cycloadditions for the synthesis of complex bicyclic isoxazolidines.

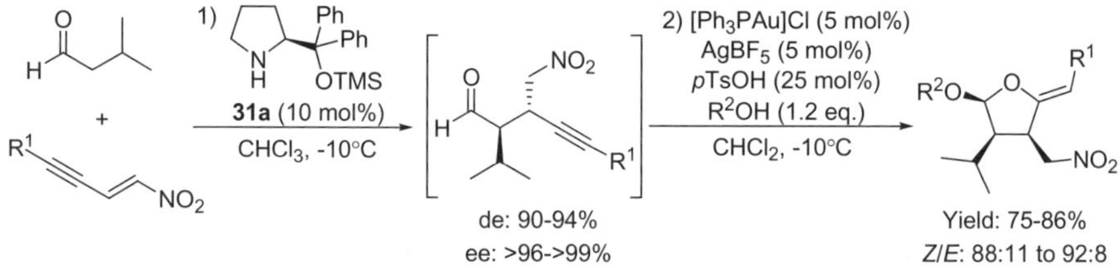

Scheme 7.13   Combining an organocatalytic Michael reaction with gold catalysis.

nitrone, which subsequently underwent intramolecular 1,3-dipolar cycloaddition with the lateral enoate functionality. This results in a very flexible and straightforward approach for the construction of complex bicyclic isoxazolidine frameworks containing five stereocenters, which were obtained in good yields and very high diastereo- and enantioselectivities.

The following example depicted in Scheme 7.13 deserves special attention because it illustrates how organocatalysis and metal catalysis can be combined for the development of very efficient new chemical processes.[19] In this example, Alexakis and Krause have combined an organocatalytic Michael addition of aldehydes to β-alkynyl-substituted nitroalkenes with a gold-catalyzed cascade process consisting of hemiacetal formation followed by intramolecular alkyne hydroalkoxylation. This tandem process was possible because of the compatibility of gold catalysis with the presence of water, alcohol or amine reagents in the reaction medium. Therefore, stereocontrol was achieved by the organocatalytic Michael reaction, which also proceeded with complete 1,4- *vs.*

1,6-selectivity, as had already been shown in previous reports from the same group (see Scheme 2.16 in Chapter 2) and the subsequent gold-catalyzed reaction allowed the activation of both the formyl group in a first instance to promote the hemiacetalization reaction and secondly activated the alkyne toward intramolecular oxymetallation by forming the corresponding π-complex.

7.3 Cascade Processes Initiated by Conjugate Addition *via* Iminium Activation

A particularly interesting situation arises when applying the iminium activation strategy in cascade processes initiated by Michael-type reactions, because the intermediate enamine generated after the conjugate addition step is ready to participate in a subsequent reaction, provided that a conveniently functionalized substrate or an additional electrophile is included in the reaction design and resulting in the active participation of the catalyst in the second step of the cascade process. Nevertheless, in some cases, the cascade process occurs after the catalyst is released and the functionalities present in the Michael adduct react intramolecularly with each other in a process in which the stereochemistry of the newly generated stereogenic centers is controlled by the chirality present at the substrate.

7.3.1 Michael/Aldol Cascade Reactions

The Robinson annulation can be considered as the classical Michael/aldol cascade reaction and, therefore, this transformation has also been the subject of some attempts to exploit the iminium activation approach, with a special interest in the L-proline-catalyzed condensation of 1,3-diketones with methyl vinyl ketone. In this case, the activation of the nucleophile was not necessary because of the highly acidic nature of the Michael donor, which delivered directly the required enolate nucleophile, which indicated that the role played by the amine catalyst was supposed to be exclusively limited to the activation of the electrophile *via* iminium ion formation. The first example of this reaction was reported by Barbas in 2000 (Scheme 7.14),[20] showing that the Robinson annulation reaction between methyl vinyl ketone and a 2-methylcyclohex-1,3-dione could be carried out in an enantioselective manner using proline as catalyst (49% yield, 76% ee). A more elaborate version of this reaction has been also reported consisting of two cascade processes carried out in a one-pot manner, the first being a Knoevenagel reaction/hydrogenation cascade followed by the Robinson annulation. This reaction design allowed the stereo-controlled preparation of Wieland–Miescher and Hajos–Parrish ketone analogues with different substitution patterns.[21]

Following a similar reaction design, a cascade Michael/aldol reaction has been studied using other different 1,3-dicarbonyl compounds for the building up of highly functionalized cyclohexanes containing multiple stereogenic

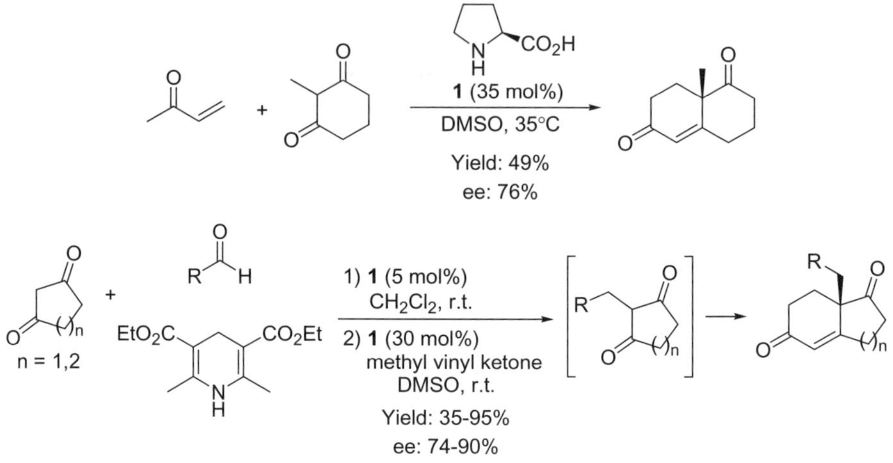

Scheme 7.14   Several examples of the L-proline-catalyzed enantioselective Robinson annulation.

centers. In this context, Jørgensen has applied his imidazoline catalyst **53a** to the reaction between β-ketoesters and enones (Scheme 7.15).[22a] The proposed mechanism involved a first step in which the Michael reaction between both reagents under iminium activation established the first stereogenic center, after which the catalyst was released and a mixture of diastereomeric Michael adducts was obtained. Next, an intramolecular aldol reaction would take place with participation of the imidazoline catalyst as a base assisting the deprotonation of the α-hydrogen of the ketone, although the possible participation of an intermediate enamine species was not completely ruled out. Interestingly, the final compounds were obtained as single diastereoisomers even though a mixture of diastereoisomers was suggested to be formed in the initial Michael reaction. This was interpreted in terms of the formation of the most stable isomer, in which a preferred conformation exists containing all bulky groups in equatorial bonds at the final cyclohexane derivatives. The final compounds were obtained in variable yields, showing dependence on the nature of the ester moiety (better yields were obtained with benzyl esters than with ethyl or methyl esters) and also limiting the range of substrates tested to β-aryl substituted enones as Michael acceptors and to 3-aryl-substituted β-ketoesters as Michael donors. This reaction was further extended to other acidic carbon nucleophiles such as β-ketosulfones and 1,3-diketones,[22b] the latter reaction also being possible using L-proline (**1**) as catalyst, although with lower enantioselectivity.[22c]

In a related work, a useful procedure for the synthesis of cyclohexenones was also developed by means of the reaction of ethyl benzoylacetate with methyl alkenyl ketones, using primary amine **129** as catalyst in this case.[23] This reaction also consisted of a Michael/intramolecular aldol reaction sequence starting

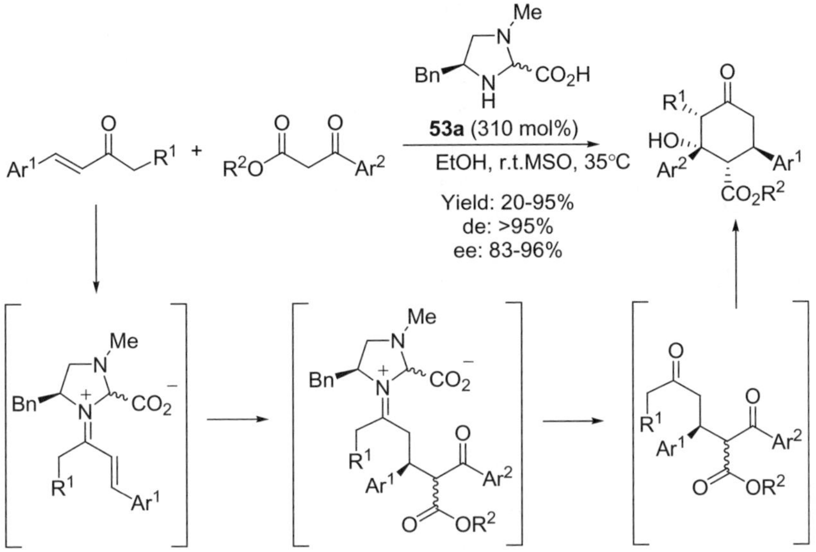

Scheme 7.15 Synthesis of polysubstituted cyclohexanones by cascade Michael/aldol reaction of β-ketoesters with α,β-unsaturated ketones.

with the activation of the enone acceptor *via* iminium ion formation but, contrary to the previous examples, a dehydration final step took place after the aldol reaction, delivering a cyclohexene-type final product (Scheme 7.16). Remarkably, the reaction proceeded with improved efficiency when *L-N*-Boc phenylglycine was employed as Brønsted acid co-catalyst, although the fact that this additive was chiral did not have any influence on the stereochemical outcome of the reaction. Under the optimized reaction conditions the final adducts were obtained in moderate to excellent yields and as mixtures of the two possible diastereoisomers in ratios around 3:1, although each of them was obtained as highly enantiopure material.

When this strategy has been applied to the use of α,β-unsaturated aldehydes as initial Michael acceptors, the reaction proceeded in a similar way. In this context, an interesting variant has also been set up which involved the use of *tert*-butyl β-cetoesters as pro-nucleophiles in the Michael reaction that initiates the cascade process (Scheme 7.17).[24] Under the optimized reaction conditions, the Michael/aldol cascade process took place as expected, using *O*-TMS diarylprolinol **31c** as the most efficient catalyst for this transformation. Nevertheless, after substrate consumption was observed, *p*-TSA had to be added to the crude reaction mixture and heating for several hours was also necessary, which induced a dehydration reaction and also hydrolysis of *tert*-butoxycarbonyl moiety followed by a decarboxylation process, delivering, in a one-pot operation, 5-alkyl or 2,5-dialkyl substituted 2-cyclohexenones containing a single stereogenic center in excellent yields and enantioselectivities. The reaction was also carried out using a β-trialkylsilyl substituted α,β-unsaturated

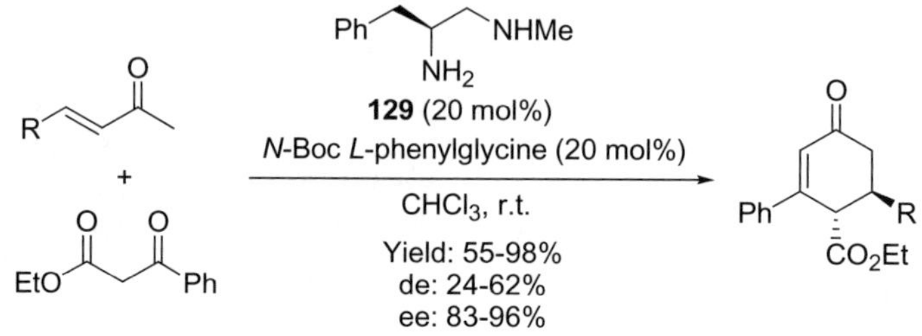

Scheme 7.16 Synthesis of cyclohexenones by cascade Michael/aldol/dehydration sequence.

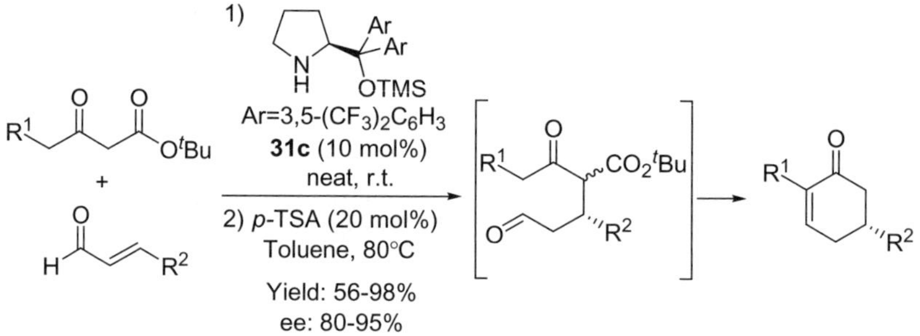

Scheme 7.17 One-pot procedure for the synthesis of cyclohexenones by Michael/aldol/dehydration/decarboxylation sequence.

aldehyde as substrate, allowing the elaboration of the obtained adducts by exploiting the reactivity of the trialkylsilyl substituent at the stereogenic center.[25]

Alternatively, when the reaction between γ-chloro-β-cetoesters and enals using the same catalyst was carried out, a one-pot Michael/Darzens reaction was observed, delivering α,β-epoxycyclohexanones in excellent yields, diastereo- and enantioselectivities (Scheme 7.18).[26] In this case, treating both reagents in the presence of the catalyst at r.t. initiated the Michael/intramolecular aldol cascade reaction and, subsequently, K₂CO₃ and DMF had to be added to the reaction mixture followed by stirring for several hours in order to accomplish the final intramolecular nucleophilic substitution needed for the formation of the epoxide moiety. Interestingly, the product isolated after the initial Michael reaction was found to be formed as a mixture of diastereoisomers and, therefore, the high diastereoselectivity of the reaction was explained in terms of the reversibility of the intramolecular aldol reaction step, which would lead to the participation of a single isomer during the final epoxide formation (if the configuration of the configurationally unstable CO₂ᵗBu-containing stereocenter is not considered). This intermediate would be organized in

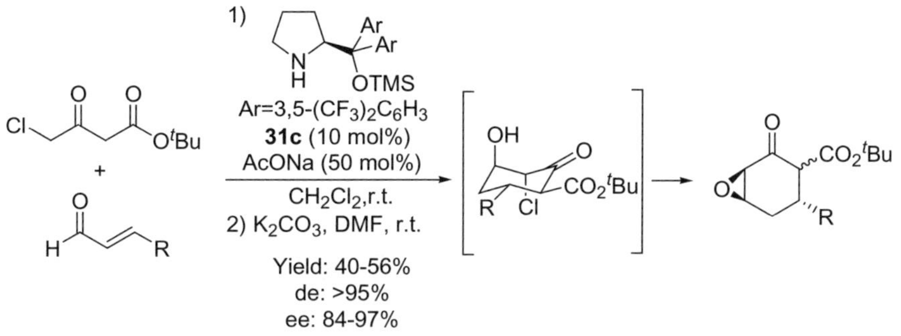

Scheme 7.18 One-pot Michael/Darzens reaction for the synthesis of α,β-epoxycyclohexanones.

a reactive arrangement in which the R group would fix the conformation of the cyclohexane ring by remaining in an equatorial position and the OH group and chlorine atom would remain in *anti*-periplanar disposition both in axial positions, fulfilling the stereochemical constraints required for the intramolecular S_N2 reaction to occur. The stereogenic center between both carbonyl groups was not configurationally unstable and, therefore, the products were isolated as equilibrating mixtures of the two possible epimers.

Changing from 1,3-dicarbonyl compounds to 1,2-diketones in this reaction design led to the formation of a different type of product with a more complex bicyclic structure. In this context, the reaction of 1,2-cyclohexanedione with a set of different α,β-unsaturated aldehydes has been carried out using catalyst **31a**, isolating a final adduct with a bicyclo[3.2.1]octane arrangement with different substituents (Scheme 7.19).[27] The reaction proceeded *via* a first Michael addition of the cyclohexanedione to the enal under iminium activation and the generated enamine intermediate underwent intramolecular aldol reaction selectively with one of the carbonyl groups. The final products were obtained in good yields, and as single diastereoisomers of >90% enantiomeric purity. The very high stereoselectivity level achieved can most probably be attributed to the exceptional facial selectivity control exerted by the catalyst in the first iminium-mediated Michael addition together with the highly strained nature of the final bicycle, which favored a well-defined orientation of the reactive sites participating in the intramolecular aldol reaction step. The authors also demonstrated that these bicyclic compounds could be easily converted into highly substituted enantioenriched cycloheptanones by means of oxidative cleavage of the α-hydroxyenone moiety.

Alternatively, a cascade Michael/aldol process has also been devised for the preparation of cyclopentenes also using this iminium/enamine manifold. In this reaction, a malonate reagent containing a functionalized side chain incorporating a formyl group at the appropriate position has been used as the nucleophile initiating the cascade process (Scheme 7.20).[28] The reaction started with the Michael addition of the malonate to the α,β-unsaturated aldehyde

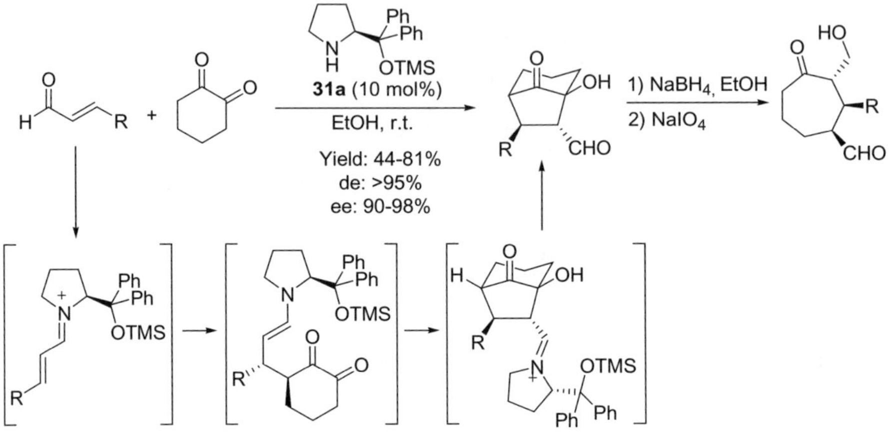

Scheme 7.19 Cascade Michael/aldol reaction for the synthesis of bicyclo[3.2.1]octanes.

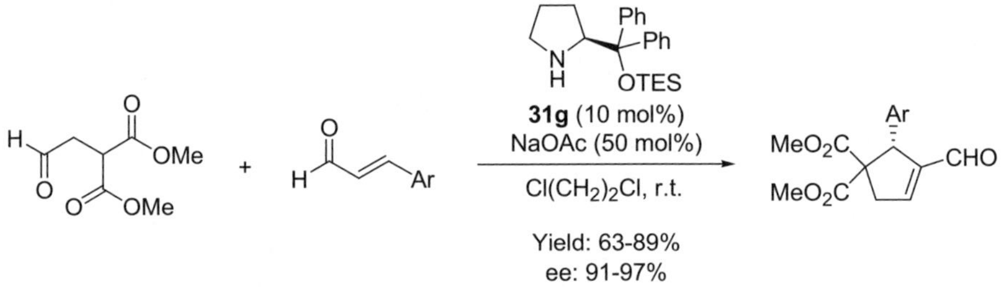

Scheme 7.20 Cascade Michael/aldol/dehydration reaction for the formation of cyclopentenes.

under iminium activation and the enamine intermediate formed immediately afterwards underwent subsequent intramolecular aldol reaction with the formyl moiety present at the functionalized malonate reagent. Next, the catalyst was released and a final dehydration reaction took place delivering a conjugate cyclopentene adduct as reaction product. The optimal conditions found for the reaction involved O-TES diphenylprolinol **31g** as catalyst, and the inclusion of a basic additive such as NaOAc for the activation of the malonate nucleophile by deprotonation. Under these conditions, the reaction took place smoothly for a wide range of α,β-unsaturated aldehydes containing aromatic substituents, providing excellent yields and enantioselectivities of the final compounds. However, the reaction failed completely when aliphatic enals were tested as substrates.

Functionalized nitroalkanes containing an aldehyde or ketone functionality at a convenient position can also be employed as substrates in the reaction with α,β-unsaturated aldehydes. In this context, Enders has developed two related procedures for the synthesis of functionalized cyclohexenes and 3,4-dihydronaphthalenes by using γ-nitroketones and 2-nitromethyl benzaldehydes respectively as functionalized Michael donors (Scheme 7.21).[29] The reaction proceeded under the expected mechanistic pathway, starting with the stereo-controlled Michael reaction of the nitroalkane to the activated α,β-unsaturated aldehyde followed by intramolecular aldol reaction/dehydration. Yields and stereoselectivities were found to be higher in the reaction leading to the dihydronaphthalene skeleton, although in the majority of the cases enantiomeric excesses were higher than 90% or, if not, they could be raised to >99% ee after crystallization.

The high performance of the iminium/enamine portfolio in cascade processes for the formation of highly functionalized chiral compounds in an easy and stereocontrolled way has been clearly demonstrated with a very cleverly designed cascade nitro-Michel/nitroaldol cascade such as that shown in Scheme 7.22.[30] In this reaction, a 2-substituted 1,3-dinitroalkane reagent was employed in combination with an α,β-unsaturated aldehyde in the presence of **31c**, leading to the direct formation of 2,4-dinitrocyclohexanols containing five stereogenic centers, which were generated with a very high degree of stereocontrol. This reaction proceeded by a first nitro-Michael reaction of the dinitro compound to the activated iminium ion generated by condensation between the enal and the catalyst. Based on the configuration of the final adducts, the authors assumed that the cyclization step proceeded without catalyst control and, therefore, it was proposed that hydrolysis of the intermediate enamine had

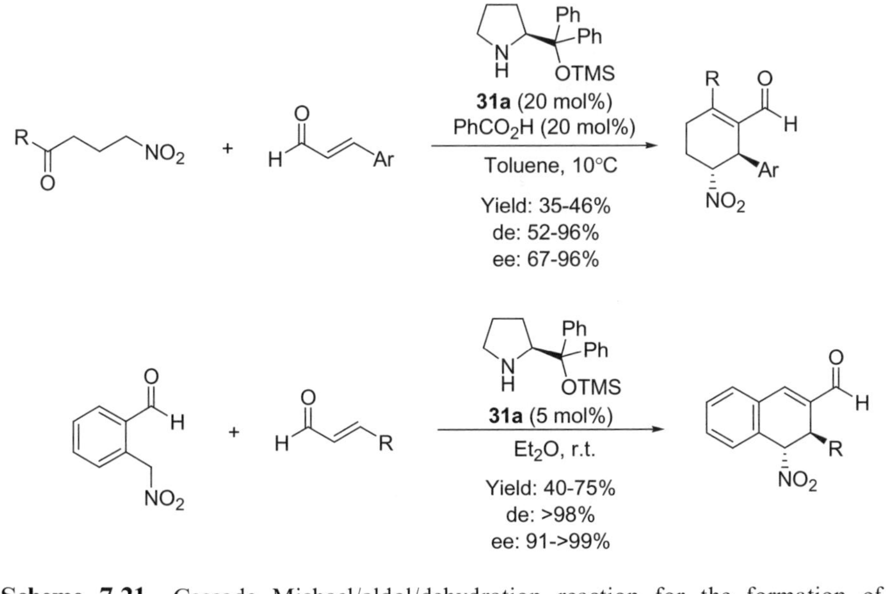

Scheme 7.21 Cascade Michael/aldol/dehydration reaction for the formation of cyclohexenes.

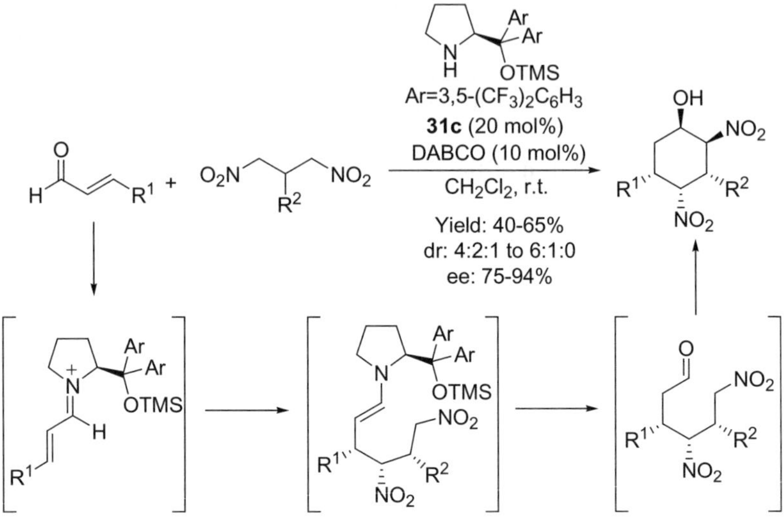

Scheme 7.22 Cascade nitro-Michael/Henry reaction for the formation of cyclohexanes with five stereogenic centers.

to occur after the nitro-Michael reaction, delivering a dinitroaldehyde intermediate, which finally had to engage in an intramolecular Henry reaction. Remarkably, one major diastereoisomer was obtained out of the 32 possible ones, with two other minor diastereoisomers being formed in variable amounts. These minor diastereoisomers were isolated for a single case and were identified, one of them as the C-3 epimer and the other with the opposite configurations at C-1, C-2 and C-3, which was formed in almost negligible amounts. These results indicate that the catalysts exerted an excellent stereodifferentiation in the first Michael addition reaction with respect to the stereocenters at C-4 and C-5, while the stereocontrol C-3, which was generated because of the desymmetrization of the starting dinitroalkane, was controlled by the nature of the substituents incorporated both at the Michael donor and the Michael acceptor (R^1 and R^2). The formation of the third very minor diastereoisomer was due to the different orientation of the substituents in the intramolecular Henry reaction. In general, the reaction showed a remarkably high substrate scope with regard to the enal reagent, allowing the incorporation of alkyl-, aryl- and functionalized alkyl substituents, and also in the 2-aryl substituted dinitroalkane reagent, leading to the final compounds with enantioselectivities in the range of 75 to 94% ee.

1,3,5-Tricarbonyl compounds have also been employed as potential 1,3-dinucleophiles able to undergo a first Michael addition step with α,β-unsaturated aldehydes and which can afterwards react with the remaining formyl group for the formation of a cyclohexane ring (Scheme 7.23). This has led to the development of a couple of cascade processes consisting of a Michael reaction proceeding in an asymmetric fashion by iminium activation of the enal, followed by releasing of the catalyst and intramolecular condensation. Following this reaction design, dimethyl 3-oxo-pentanodioate[31] and ethyl 5-diethoxyphosphoryl-3-oxobutanoate[32] have been employed as dinucleophiles, using *O*-trialkylsilyl diarylprolinols **31f** and **31c** respectively as the best catalyst for each case (Scheme 7.23). The first reaction provided the final compounds with excellent yields and enantioselectivities after diastereoselective reduction of the cascade product, which was presumably obtained as an equilibrating mixture of diastereoisomers due to the inherent configurational instability of the stereogenic center at the β-ketoester moiety. However, the reaction scope was limited to the use of aromatic enals, reporting poorer results with the β-alkyl-substituted substrates. In the reaction using ethyl 5-diethoxyphosphoryl-3-oxobutanoate, the final compounds were directly isolated as single diastereoisomers with excellent yields and enantioselectivities and for a wide range of different alkyl- and aryl-substituted α,β-unsaturated aldehydes. However, modified conditions had to be found for the use of aliphatic enals, which included the incorporation of hydroquinine as co-catalyst (5 mol%) and also the use of a smaller amount of Brønsted acid additive (2.5 mol%).

In a different work, Jørgensen has also reported that carrying out the reaction between two equivalents of dimethyl 3-oxo-pentanodioate and an α,β-unsaturated aldehyde using **31a** as catalyst resulted in the fully diastereo- and enantioselective formation of a compound with a bicyclo[3.3.1]non-2-ene structure containing six stereocenters (Scheme 7.24).[33] This reaction consisted of two cascade reactions proceeding in a one-pot manner, starting with a Michael/intramolecular aldol condensation cascade process and next piperidine was added to the reaction mixture, which promoted a second Michael/intramolecular aldol cascade process, delivering cleanly the final tricyclic compounds.

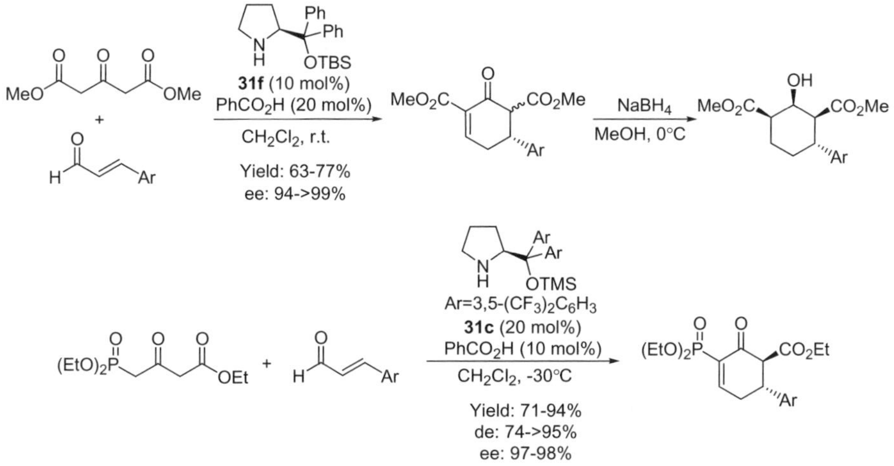

Scheme 7.23 Cascade Michael/intramolecular condensation for the synthesis of cyclohexenes.

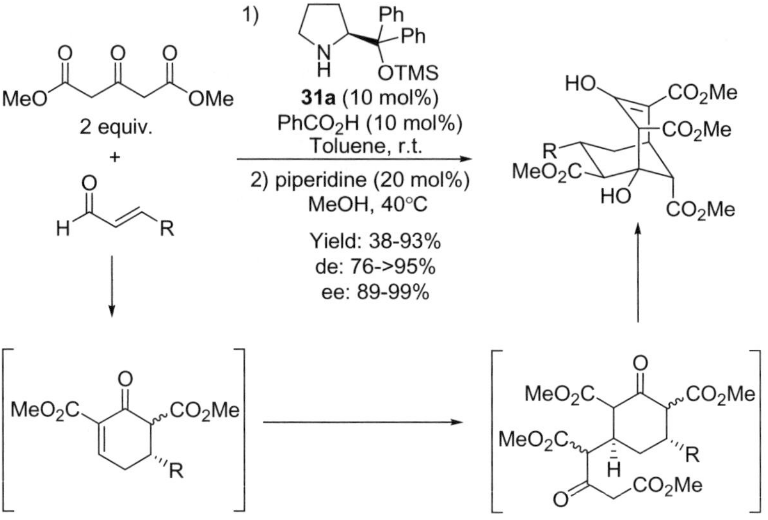

Scheme 7.24 Asymmetric synthesis of optically active bicyclo[3.3.1]non-2-enes containing six stereocenters.

Finally, there is also a very straightforward methodology for building up pentasubstituted cyclohexanes starting from a 1,3-diketone, an α,β-unsaturated aldehyde and nitromethane developed very recently (Scheme 7.25).[34] This reaction consisted of a tandem procedure initiated by the conjugate addition of the 1,3-diketone to the enal under iminium activation and, next, nitromethane had to be added to the reaction mixture, promoting a double intermolecular/intramolecular Henry reaction, which delivered the final product. As a consequence, participation of the catalyst in the process was exclusively limited to the initial Michael reaction, in which the first stereocenter was generated, with the subsequent two nitroaldol reaction occurring in a highly diastereoselective fashion and exclusively under substrate control. A remarkable finding associated to this report is the fact that tetrabutylammonium fluoride was identified as a very efficient promoter of the intermolecular Henry reaction, which was included together with nitromethane in the second step of this tandem process.

7.3.2 Michael/Michael Cascade Reactions

Functionalized reagents containing a C–H acidic moiety able to participate as carbon pro-nucleophile in a conjugate addition with an α,β-unsaturated aldehyde or ketone under iminium activation and which also incorporates an activated double bond on its structure capable of reacting intramolecularly with the enamine intermediate generated after the conjugate addition step has been employed as bifunctional compounds in cascade reactions. A good example can be found in the elegant asymmetric synthesis of highly

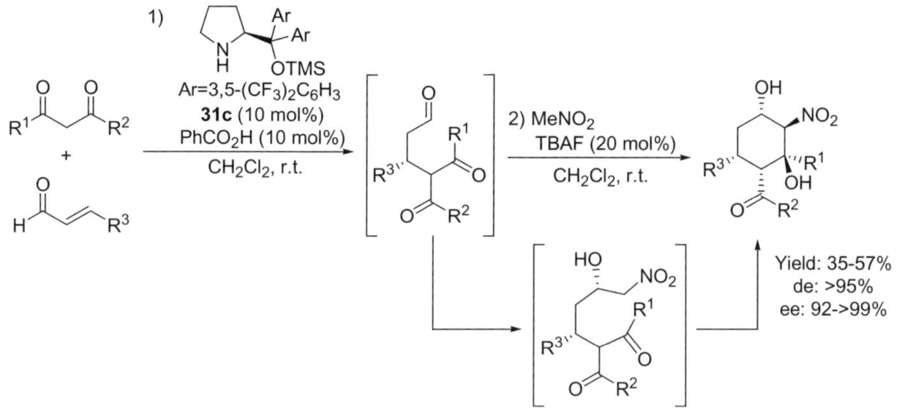

Scheme 7.25   Asymmetric synthesis of pentasubstituted cyclohexanes.

functionalized cyclopentenes shown in Scheme 7.26.[35] In this report, *O*-TMS diphenylprolinol **31a** was identified as an excellent catalyst for the Michael/Michael cascade reaction between α,β-unsaturated aldehydes and a functionalized malonate reagent containing an α,β-unsaturated ester moiety located at a convenient position. The reaction appeared to be fairly tolerant with regard to substitution at the enal reagent and also with respect to the nature of the alkoxide substituents of the malonate reagent. In all cases the reaction proceeded with high diastereoselectivity and enantioselectivity, providing the final cyclopentenes with excellent yields. Furthermore, a modified version of this reaction using a δ-nitro-α,β-unsaturated ester as bifunctional reagent has also been explored.[36] In the latter case, the reaction consisted of a nitro-Michael/Michael cascade reaction, delivering cyclopentanes with four contiguous stereogenic centers, which were isolated in good yields and with high diastereo- and enantioselectivities.

A very interesting vinylogous Michael/Michael cascade sequence has been reported between substituted alkylidenemalononitriles and enones, using in this case a chiral primary amine catalyst for the more effective activation of the enone reagent (Scheme 7.27).[37] The reaction consisted of the first Michael addition of the alkylidenemalononitrile reagent to the iminium ion derived from the enone, which proceeded through the deprotonation of the more acidic γ-position of the α,α-dicyanoalkene reagent. Next, equilibration of the two possible regioisomers of the enamine intermediate formed after the Michael addition step had to occur before the second intramolecular Michael reaction took place delivering the final cyclic derivatives after releasing the catalyst in a final hydrolysis step. However, the possibility of the catalyst releasing before the cyclization step and consequently the possibility that the intramolecular Michael reaction proceeded without participation of the catalyst should also be considered as a plausible mechanism operating in this reaction. Interestingly, this behavior was only observed when cyclohexylidenemalononitriles were employed, while simple β-alkyl-substituted alkylidenemalonitrile only

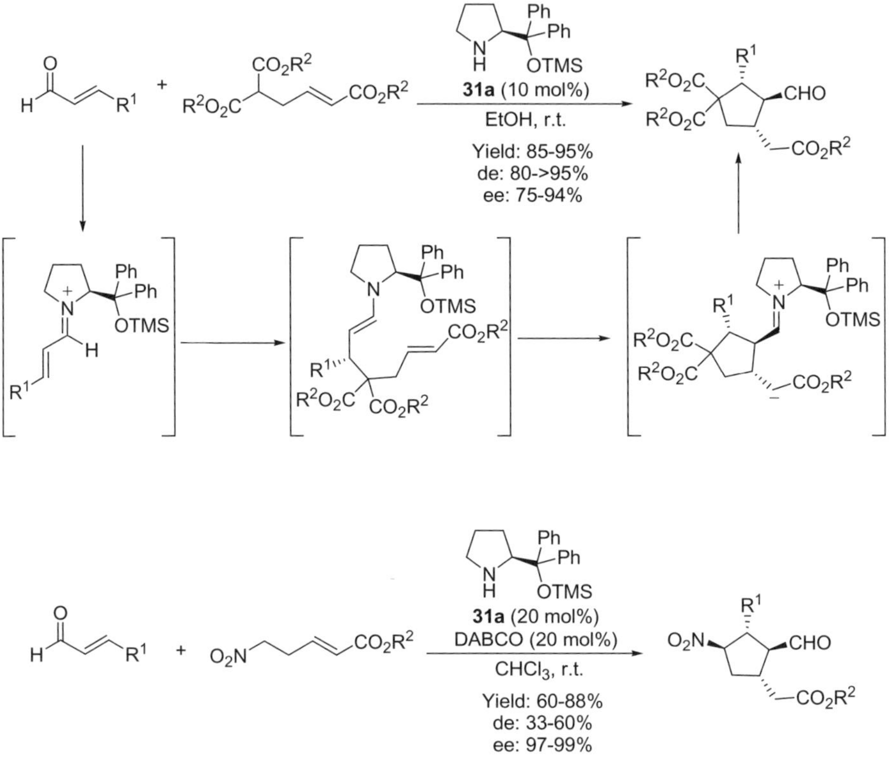

Scheme 7.26 Asymmetric synthesis of cyclopentanes by Michael/Michael cascade reaction.

underwent the first vinylogous Michael addition reaction. In this context, when 4-substituted cyclohexylidenemalononitriles were employed, the process occurred together with the desymmetrization of the starting material, therefore generating a new stereogenic center with complete stereocontrol. Under the optimal conditions, a wide range of different reagents was employed obtaining the final adducts with good yields and excellent diastereo- and enantioselectivities, although the applicability of this methodology was shown to be limited to the use of β-aryl substituted enones (Scheme 7.27).

In a different context, a very interesting and rich chemistry has been developed around cascade Michael/Michael processes using γ,δ-unsaturated β-ketoesters (Nazarov reagents) as functionalized Michael donors in the reaction with α,β-unsaturated aldehydes using *O*-TMS diarylprolinols as catalysts. The first contribution was made from Jørgensen trying to achieve the synthesis of functionalized cyclohexanes in which the Nazarov reagent was expected to undergo Michael addition to the enal after being activated by the catalyst *via* iminium ion formation and, next, the intermediate enamine would have to undergo intramolecular Michael reaction to the remaining enone moiety present at the intermediate. In its place, an unexpected product was isolated, the

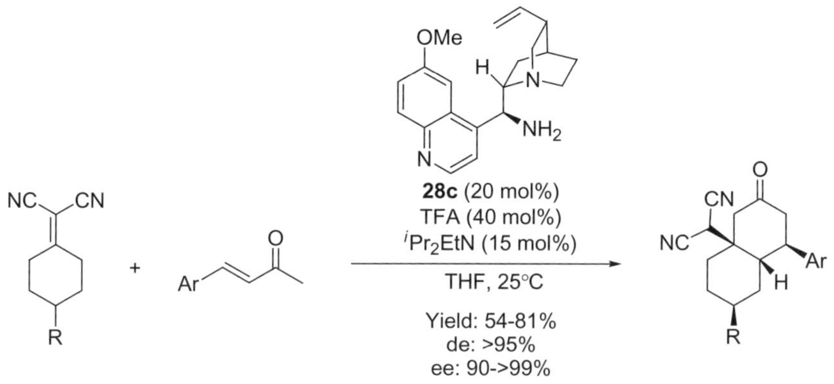

Scheme 7.27 Asymmetric cascade vinylogous Michael/Michael reaction.

formation of which was interpreted by assuming a Michael/Morita–Baylis–Hillman cascade process (Scheme 7.28). Further experimentation led to the finding of an outstanding protocol for the preparation of polysubstituted cyclohexenols containing multiple functionalities and two stereogenic centers with good chemical efficiency and an outstanding stereochemical control.[38] As shown in Scheme 7.28, the reaction started with the Michael reaction between the malonate moiety of the Nazarov reagent and the enal under iminium activation by catalyst **31a**, which had to be followed by hydrolysis affording the corresponding Michael adduct, which, subsequently, should have to engage in a second catalytic cycle in which the amine catalyst also had to participate by promoting an intramolecular Morita–Baylis–Hilman reaction. In this second cycle, the catalyst underwent aza-Michael addition to the enone moiety and the intermediate enolate next reacted intramolecularly with the formyl group delivering the final compound after a final retro-aza-Michael reaction which released the amine catalyst forming the final compound (isolated as the more stable enol tautomer). The high stereochemical control obtained in the reaction is due to the highly enantioselective Michael reaction occurring in the first stage of the cascade process, the formation of the stereocenter containing the OH moiety being a substrate-controlled cyclization process. Control experiments were carried out in order to confirm the proposed mechanism, which led to the isolation of the intermediate Michael adduct and the verification that this underwent a fully diastereoselective intramolecular Morita–Baylis–Hillman reaction in the presence of **31a** and also when other catalysts typically employed for this kind of reaction (PPh$_3$ or DABCO) were employed. Otherwise, no cyclization occurred in the presence of acid additives. Many interesting transformations were also carried out on the obtained adducts in order to prove their potential as chiral building blocks in the synthesis of complex cyclohexane architectures.

However, when an aryl substituent was introduced at the end of the Michael acceptor moiety of the Nazarov reagent employed, the reaction was found to

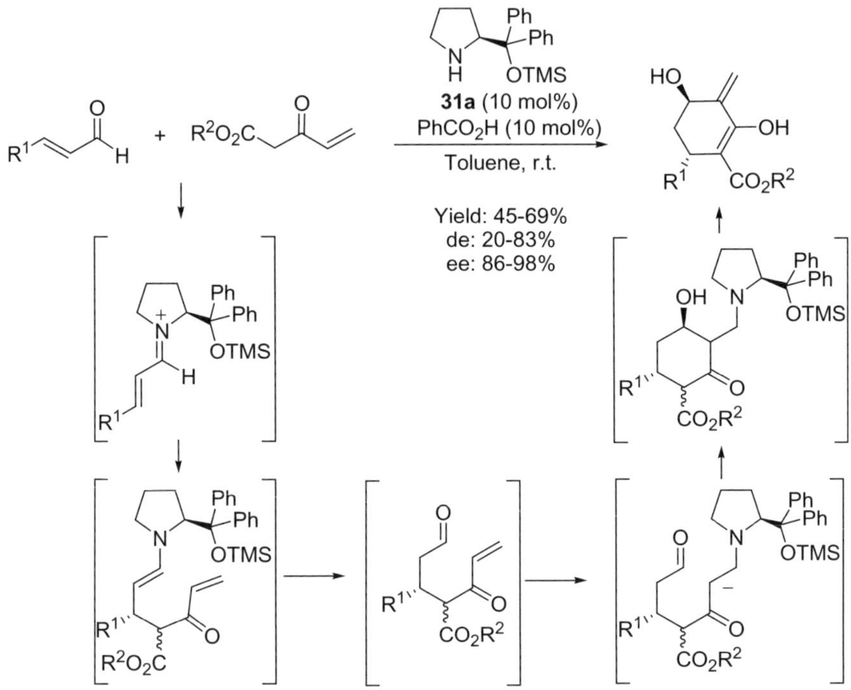

Scheme 7.28 Enantioselective Michael/Morita–Baylis–Hillman cascade reaction.

take place by a different pathway, furnishing a completely different product, which resulted from a formal [3 + 3] cycloaddition (Scheme 7.29).[39] The formation of this product was explained by assuming that, after the first Michael addition to the iminium ion and the subsequent hydrolytic cleavage of the catalyst, the intermediate would undergo a competitive intramolecular hemiacetal formation between the remaining β-ketoester moiety (in the form of the corresponding enol) and the formyl group. This pathway was favored in the case of this γ-aryl substituted Nazarov reagent because of its lower electrophilic character, which hampered the first aza-Michael addition of the catalyst operating in the Morita–Baylis–Hillman pathway. Moreover, the aromatic nature of this substituent also contributed to the stabilization of the enol form of the β-ketoester moiety. After optimizing the reaction conditions, the authors also decided to carry out the *in situ* oxidation of the hemiacetal group for the better isolation and characterization of the obtained compounds.

Alternatively, when Nazarov substrates in which the Michael acceptor moiety contained the activated C=C double bond as a part of a five-membered ring were employed, the reaction took place as it was initially devised by Jørgensen and coworkers in their initial attempt depicted in Scheme 7.28, furnishing the corresponding bicyclic compound resulting from the expected Michael/Michael cascade process (Scheme 7.30).[40] In this case, the final adducts were obtained as the corresponding enol tautomers with regard to the

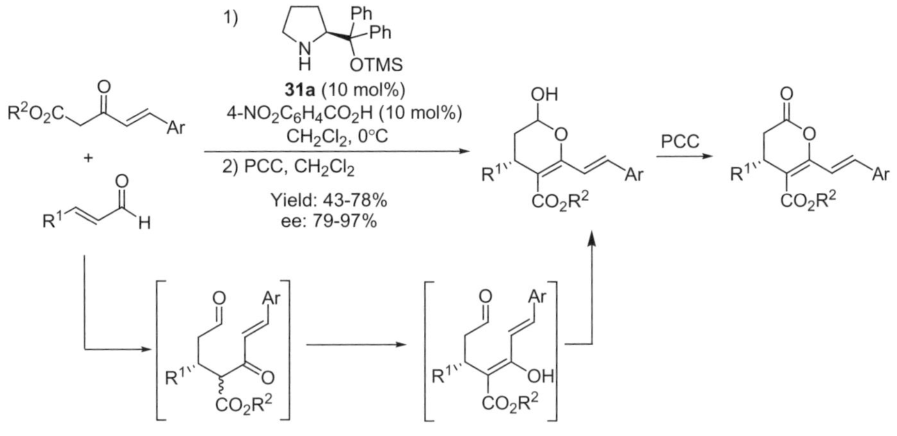

Scheme 7.29 Enantioselective formal [3 + 3] cycloaddition between Nazarov reagents and α,β-unsaturated aldehydes.

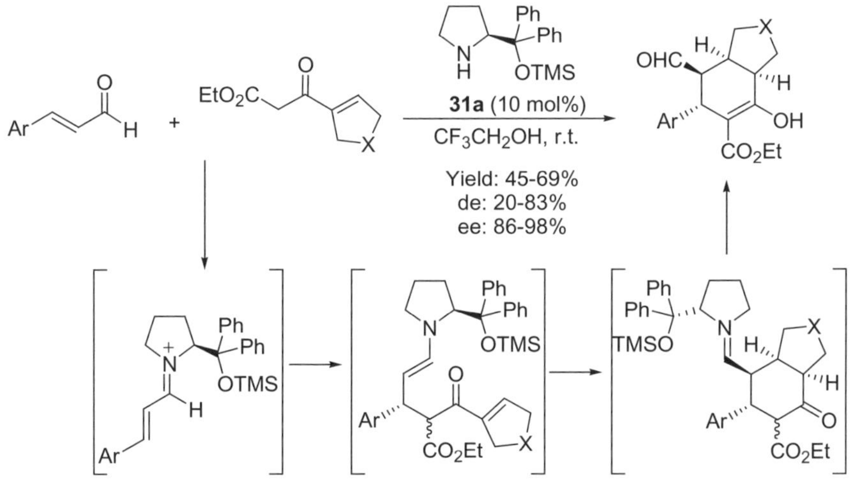

Scheme 7.30 Enantioselective Michael/Michael cascade using Nazarov reagents.

β-ketoester moiety remaining in the structure of the bicycle. The reaction proceeded with excellent yield, furnishing almost a single diastereoisomer and with excellent enantioselectivity, in a process in which up to four stereocenters were generated simultaneously together with the formation of a rather challenging bicyclic *cis*-fused structure.

On the other hand, the group of Jørgensen has also reported an alternative procedure in which two molecules of an α,β-unsaturated aldehyde and an active methylene compound reacted together in a triple Michael/Michael/aldol sequence, yielding substituted 1-formylcyclohexenones containing up to three

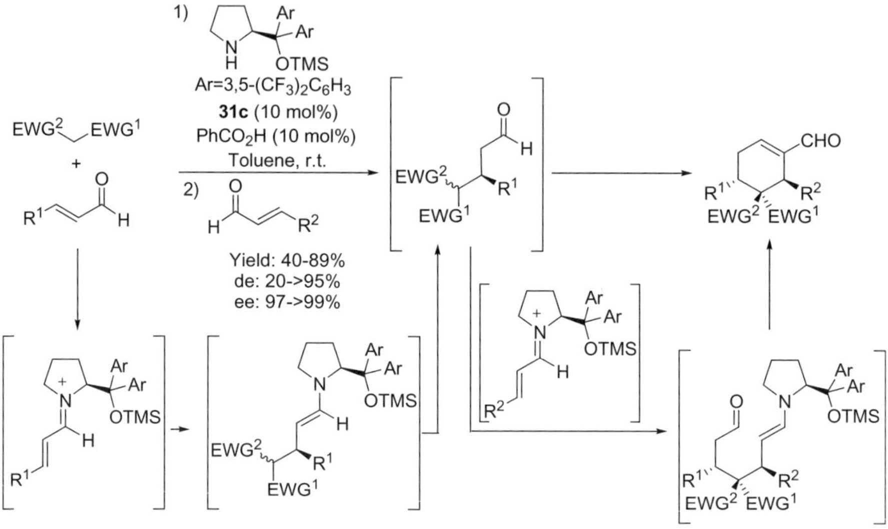

Scheme 7.31   Enantioselective triple Michael/Michael/aldol cascade reaction.

stereogenic centers (Scheme 7.31).[41] In this reaction, the cascade process started
with the Michael addition of the active methylene compound to the enal under
iminium activation and the generated enamine intermediate underwent a second Michael reaction with another molecule of the α,β-unsaturated aldehyde,
which was also activated by the catalyst as the corresponding iminium ion.
Finally, the enamine formed after this second Michael reaction underwent
intramolecular aldol condensation followed by hydrolysis in order to finish the
catalytic cycle. Remarkable features that illustrate the high efficiency of this
process are the fact that two unsaturated aldehydes with different substituents
and hence of different reactivity toward the two Michael reactions taking place
in the cascade process could be used in this reaction. This allowed the
sequential incorporation of two different alkyl chains at the 3 and 5 positions of
the cyclohexene ring, although the experimental procedure had to be modified
and the second enal reagent had to be added after consumption of the first α,β-
unsaturated aldehyde was observed. Remarkably, the yield and the diastereoselectivity of this reaction were found to be significantly higher than those
found in both previously presented examples, which should be attributed to the
higher efficiency in controlling the diastereoselectivity of the second Michael
reaction of the cascade process because, in this case, the catalyst was involved in
the activation of both the donor and the acceptor reagents, which did not occur
at the other cases. An additional interesting feature of this reaction is the fact
that active methylene compounds with two different electron-withdrawing
groups could also be employed, leading to the formation of a quaternary stereocenter, although, in this case, the major diastereoisomer was formed together with significant amounts of the corresponding C-5 epimer.

A similar example of a triple Michael/Michael/Aldol reaction has also been developed afterwards, using in this case nitromethane as the initial Michael donor and two equivalents of an α,β-unsaturated aldehyde reagent, which also proceeds with an excellent degree of chemical efficiency and stereocontrol.[42]

7.3.3　Michael/α-Alkylation Cascade Reactions

Employing a functionalized Michael donor containing a good leaving group on its structure able to promote an intramolecular alkylation reaction after the conjugate addition takes place has also been surveyed by several authors. Nevertheless, it has to be pointed out that, when the iminium activation approach is required in this reaction, a chemoselectivity problem has to be solved, which arises from the possibility of the amine catalyst to react with the electrophilic moiety of the functionalized acceptor, delivering the corresponding *N*-alkylation product which would eventually lead to catalyst deactivation. Nevertheless, several important and efficient methodologies have been developed in this field, which is also another example of the enormous possibilities opened by the iminium activation concept for synthetic organic chemists.

In this context, the most studied reaction has been the enantioselective cyclopropanation of α,β-unsaturated aldehydes. The reaction design consisted of the use of bromomalonates or bromonitromethane as functionalized C–H acidic Michael donors able to undergo a subsequent intramolecular alkylation after the conjugate addition step, therefore delivering a chiral cyclopropane derivative. This design was applied for the first time by Ley to the enantioselective nitrocyclopropanation of cyclohexenone using pyrrolidine tetrazole **2a** as catalyst, which, under the best reaction conditions, furnished the final compound in 90% ee and, remarkably, as a single diastereoisomer (Scheme 7.32).[43] An improved version of this reaction was reported afterwards in which changing to primary amine catalyst **28c** led to a more efficient reaction, furnishing the final nitrocyclopropanes in excellent yields, diastereo- and enantioselectivities in the reaction of cyclic enones of different ring sizes.[44] The use of acyclic enones required modified conditions, which included carrying out the reaction in two steps, but showed to be equally efficient, although the substrate scope was restricted to 1,3-diaryl substituted enones (Scheme 7.32).

Following this reaction design, the nitrocyclopropanation of α,β-unsaturated aldehydes has also been reported, in this case using *O*-TMS or *O*-TES diphenylprolinols **31a** or **31g** as the best catalysts.[45] However, while excellent enantioselectivities were generally obtained, the yields were moderate and mixtures of diastereoisomers in almost 1:1 ratio were obtained in most cases. On the contrary, catalyst **31a** allowed carrying out the cyclopropanation reaction using dialkyl bromomalonates as functionalized Michael donors (Scheme 7.33).[46] This reaction has been developed simultaneously by the groups of Córdova and Wang and proceeded with good yields, high diastereoselectivities and excellent enantioselectivities for a wide range of α,β-unsaturated aldehydes, especially when

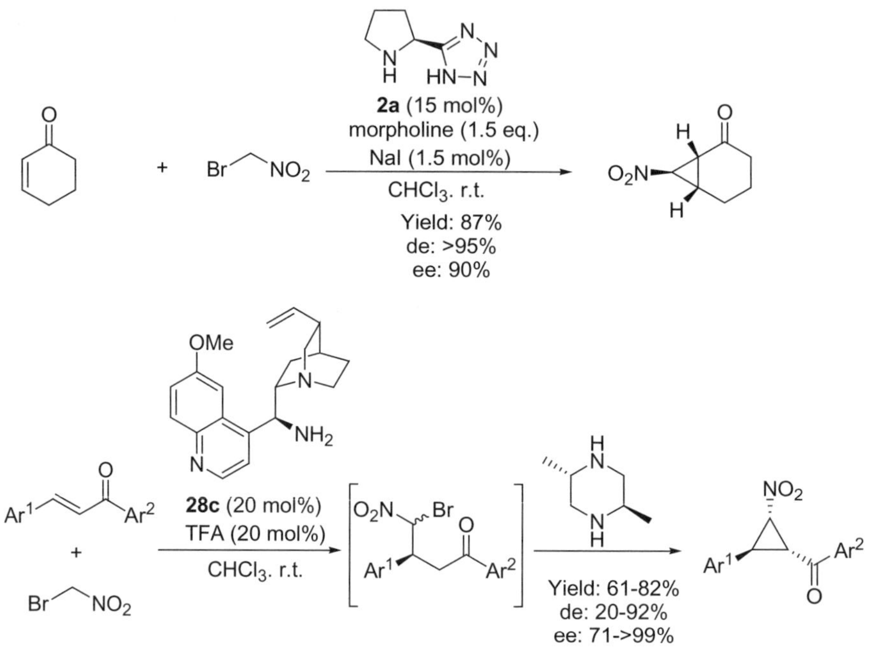

Scheme 7.32 Enantioselective cyclopropanation of enones by Michael/α-alkylation cascade reaction.

aryl-substituted enals were employed as substrates. The corresponding aliphatic α,β-unsaturated aldehydes furnished slightly lower yields and needed the use of a two-fold excess of the enal, although the reaction proceeded with similarly high diastereo- and enantioselectivity. The nature of the alkyl substituents of the malonate reagent also had some influence on the reaction, showing that increasing the size led to the need for longer reaction times, and even resulted in a significant decrease of the yield of the reaction when, for example, diisopropyl bromomalonate was employed. Extension of the methodology to the use of ethyl 2-bromoacetoacetate has also been successfully achieved, leading to the formation of a quaternary stereocenter in the final cyclopropane products with excellent yields, diastereo- and enantioselectivities.[46a,b,47] Alternatively, this concept has also been applied to the synthesis of cyclopentane structures by using γ-bromo-β-cetoesters as functionalized Michael donors with excellent results (Scheme 7.33).[46b,48]

It is important to note that all these cyclopropanation reactions require the incorporation of a stoichiometric amount of a base additive for the reaction to proceed to completion or in order to reach full conversion in a reasonable time, which is thought to operate as scavenger of the HBr generated as the reaction proceeds. However, the presence of this basic additive has been found to promote a competitive retro-Michael process on the cyclopropane products obtained in the reaction of bromomalonates with enals, leading to

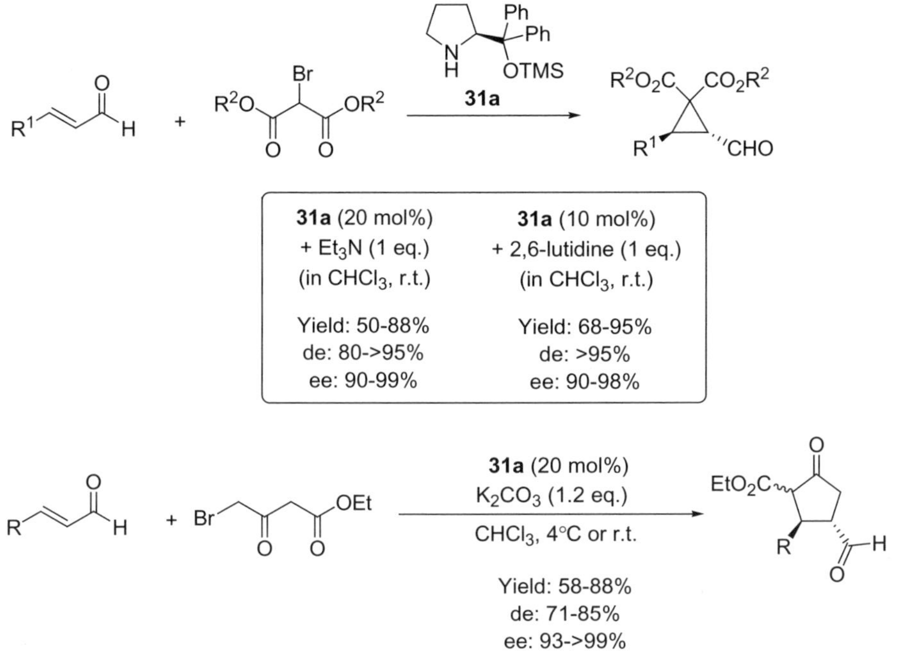

Scheme 7.33 Synthesis of cyclopropanecarbaldehdyes and cyclopentanecarbalde-
hydes *via* Michael/α-alkylation cascade reaction.

α,β-disubstituted α,β-unsaturated aldehydes, which might be the explanation
for the lower yields observed in some of the reactions presented before.[46c]

A very different approach to the cyclopropanation of α,β-unsaturated
aldehydes has also been reported by MacMillan, involving the use of sulfonium
ylides as functionalized Michael donors and using indolinecarboxylic acid **130**
as chiral secondary amine catalyst (Scheme 7.34).[49] The first experiments were
carried out using proline as catalyst but obtaining moderate enantioselectivities. For this reason, a new catalyst design was proposed based on the
introduction of a fused aromatic ring at the catalyst structure, directed to
provide additional steric requirements that should allow better control of the
iminium geometry and hence a much higher enantioselectivity compared to
proline. The hypothesis turned out to be completely right as illustrated with the
use of indolinecarboxylic acid **130** as catalyst, which led to the cyclopropanation products in excellent yields, diastereo- and enantioselectivities. Moreover,
the reaction was shown to be very tolerant toward substitution at the enal
reagent, providing excellent results independently of the nature of the β-substituent. In addition, in this paper an interesting novel activation mechanism
has been suggested in which the formation of a carboxylate iminium ion
together with the use of a zwitterionic reagent such as the sulfonium ylide
opened the possibility for the catalyst to exert its stereodirecting ability

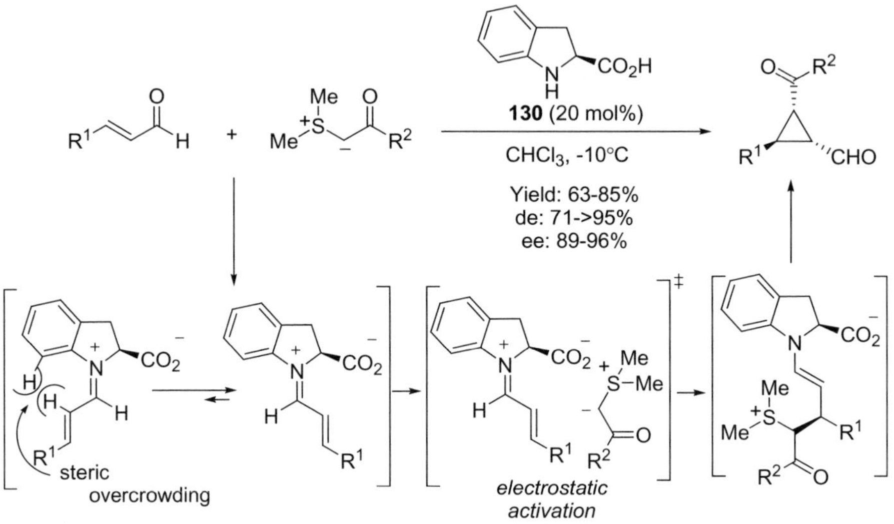

Scheme 7.34   Enantioselective cyclopropanation of α,β-unsaturated aldehydes using sulfonium ylides.

via electrostatic interactions between the carboxylate group and the ylide. This novel activation mechanism has been confirmed by several kinetic experiments and by comparing the behavior of the iminium ion derived from **130** with a set of different iminium ions unable to participate in this electrostatic activation.[50] Arsonium ylides have also been used in this reaction, also achieving good results, in this case by using *O*-TMS diphenylprolinol **31a** as catalyst.[51]

7.3.4   Other Cascades Initiated by Michael Reactions Using Stabilized Carbon Nucleophiles

Several other methodologies have been developed for the synthesis of relatively complex molecules using cascade reactions which start with the Michael reaction of an enolate-type nucleophile with an α,β-unsaturated aldehyde under iminium activation of the latter. This is the case of the example depicted in Scheme 7.35, in which a phosphorous ylide-containing β-ketoester was employed as functionalized pro-nucleophile in the reaction with α,β-unsaturated aldehydes using **31h** as catalyst.[52] This cascade process consisted of an initial Michael addition of the functionalized ylide to the enal followed by hydrolysis of the catalyst and a final intramolecular Wittig reaction. Under the optimized conditions, the reaction was carried out using a series of α,β-unsaturated aldehydes, furnishing a variety of differently substituted cyclohexenes containing two contiguous stereogenic centers in good yields and excellent diastereo- and enantioselectivities.

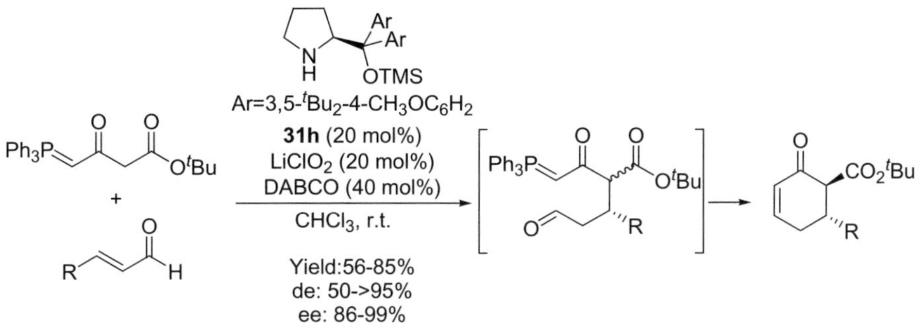

Scheme 7.35 Cascade Michael/Wittig reaction for the synthesis of cyclohexenones.

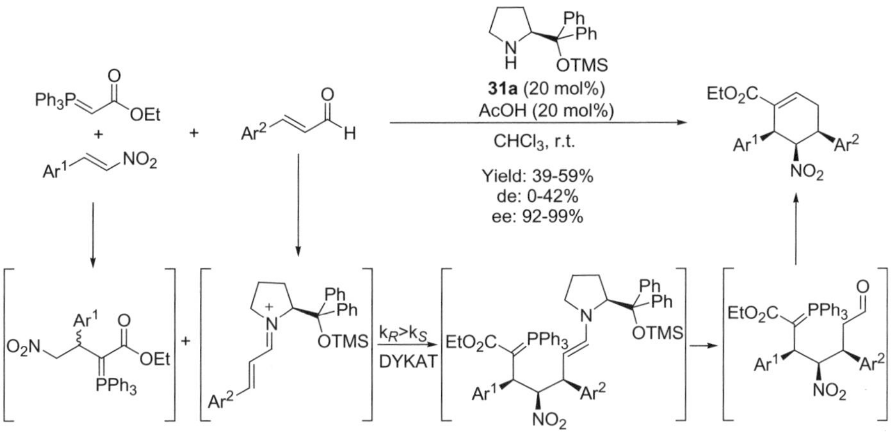

Scheme 7.36 Michael/Michael/Wittig triple cascade.

A more elaborate version of this reaction has also been reported in which ethyl triphenylphosphoranylideneacetate, a nitroalkene and an α,β-unsaturated aldehyde reacted together furnishing a final product with a cyclohexenecarboxylate structure containing three contiguous stereocenters (Scheme 7.36).[53] In this case, the cascade process consisted of an initial Michael addition of the phosphorous ylide to the more reactive nitroalkene electrophile without participation of the catalyst and therefore delivering a chiral intermediate in a racemic form, which subsequently underwent Michael reaction with the remaining α,β-unsaturated aldehyde, the latter being activated as the corresponding iminium ion after condensation with the catalyst. At this point, this Michael reaction had to take place as a dynamic kinetic asymmetric transformation (DYKAT), in which one of the enantiomers reacted faster with the chiral iminium ion intermediate, and the remaining unreactive enantiomer would racemize quickly *via* retro-Michael/Michael reaction. The cascade process ended with a subsequent intramolecular Wittig reaction after releasing the

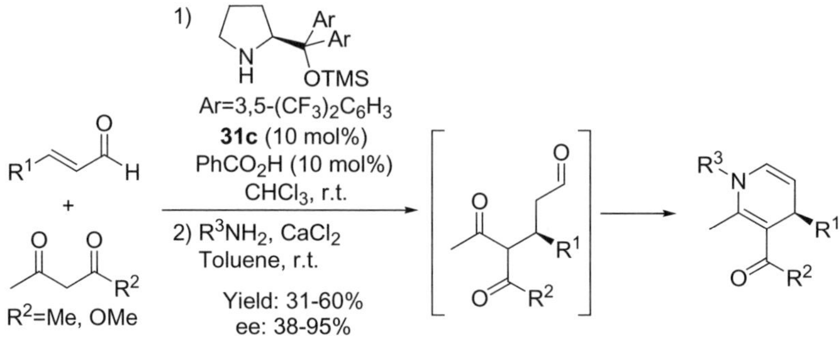

Scheme 7.37   Enantioselective synthesis of 1,4-dihydropyridines by one-pot Michael/
                   enamine formation/intramolecular condensation reaction.

catalyst. Although yields were only moderate and mixtures of diastereoisomers
were obtained in most cases, the enantiomeric excess of the major all-*syn* diastereoisomer was excellent.

In a different context, Jørgensen has developed an easy and straightforward
method for the asymmetric synthesis of 1,4-dihydropyridines using a one-pot
operation (Scheme 7.37).[54] This procedure involved an initial Michael reaction
between methyl acetoacetate or 2,4-pentanedione and an α,β-unsaturated
aldehyde, which was followed by the addition of a primary amine, the latter
condensing with the formyl moiety of the Michael adduct formed in the first
step, thus delivering an enamine intermediate which subsequently underwent
intramolecular addition/dehydration, building up the final chiral 1,4-dihy-
dropyridine compounds. *O*-TMS diarylprolinol **31c** was identified as the best
catalyst for the reaction, which was also found to have a remarkably wide
substrate scope with regard to the α,β-unsaturated aldehyde employed in the
first Michael addition reaction and also tolerating very well the variation of the
substituent introduced with the primary amine reagent employed subsequently,
allowing the use of both aromatic and aliphatic amines with success.

On the other hand, a very interesting approach to densely functionalized
indolo[2,3*a*]quinolizidine systems has been developed involving the reaction of
an α,β-unsaturated aldehyde and a modified tryptamine-derived malonate-type
reagent using **31a** as catalyst (Scheme 7.38).[55] The reaction consisted of an
initial cascade process starting with a Michael addition of the malonate reagent
to the enal under iminium activation and release of the catalyst by hydrolysis,
delivering the addition product in the form of a more stable cyclic hemiaminal,
which was generated after intramolecular addition of the amide nitrogen to the
formyl group. Next, in a one-pot operation, a catalytic amount of strong
Brønsted acid such as HCl was added, which promoted the formation of an *N*-
acyliminium ion by dehydration of this hemiacetal intermediate which finally
underwent intramolecular reaction with the electron-rich indole moiety. The
reaction proceeded with the generation of a tetracyclic system after the formation of two C–C and one C–N bond and also involving the formation of

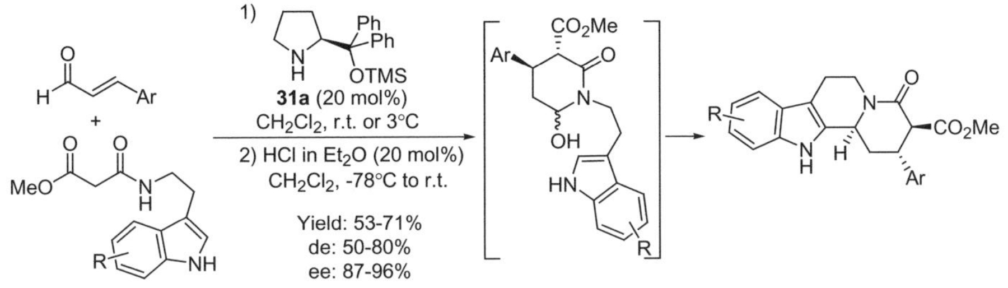

Scheme 7.38 One-pot enantioselective synthesis of indolo[2,3*a*]quinolizidines.

three new stereogenic centers. Moderate yields of the final compounds were obtained for a variety of different aromatic α,β-unsaturated aldehydes and also showing that the introduction of electron-donating substituents such as methoxy groups at the indole ring was also tolerated, although no examples were provided illustrating the use of aliphatic enals in this transformation. The products were also obtained as highly enantioenriched compounds and as mixtures of diastereoisomers ranging from 3:1 to 9:1 resulting from a less effective stereochemical control in the cyclization reaction *via* the *N*-acyliminium ion intermediate.

Finally, iminium catalysis has also been combined with the use of *N*-heterocyclic carbenes in a cascade sequence in which an α,β-unsaturated aldehyde and a 1,3-diketone or a β-ketoester reacted with each other in the presence of catalytic amounts of *O*-TMS diarylprolinol **31c** and *N*-heterocyclic carbene precursor **120h** (Scheme 7.39).[56] The cascade process involved sequential and chemoselective initial activation of the enal by **31c** in the form of the corresponding iminium ion followed by conjugate addition of the 1,3-dicarbonyl compound, delivering the corresponding Michael adduct after releasing of the first catalyst. This Michael adduct was subsequently activated by the carbene, forming the corresponding Breslow intermediate, which reacted intramolecularly with one of the keto groups present at this intermediate, generating the final cyclopentanone-type product after releasing of the second catalyst. Under the optimized reaction conditions, a series of different mono- and bicyclic cyclopentanones was prepared with excellent yields and enantioselectivities and with the preferential formation of one diastereoisomer up to the four possible ones.

7.3.5 Cascades Initiated by Conjugate Friedel–Crafts Reaction

Several different cascades have been developed in which the first conjugate addition step involved the use of an electron-rich arene or heteroarene as Michael donor. Interestingly, a very important number of the cases reported in this context refer to intermolecular variants rather than the more favored and therefore easier intramolecular approach. A representative example can be

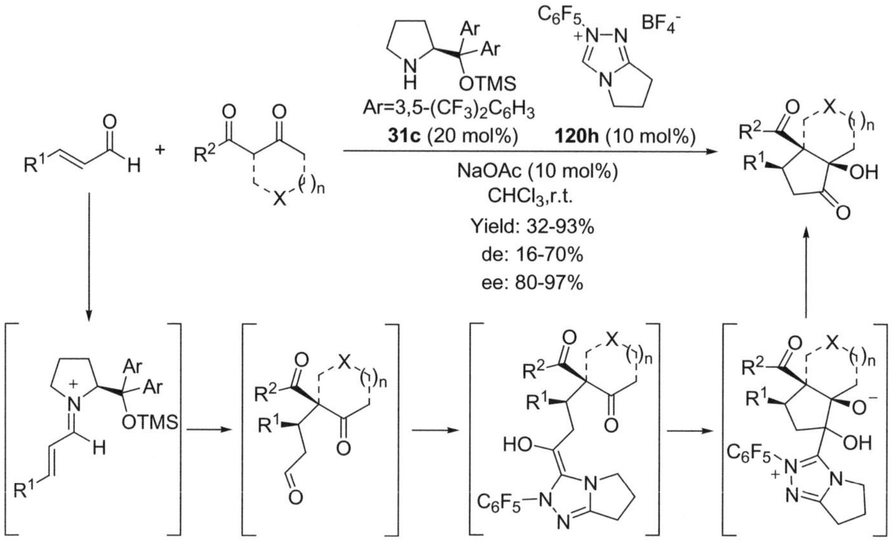

Scheme 7.39 Combining iminium and *N*-heterocyclic cabene catalysis in a cascade reaction.

found in Scheme 7.40, in which a very interesting and highly efficient protocol for carrying out a sequential Friedel–Crafts conjugate addition/chlorination sequence is presented.[57] The reaction design involved two consecutive reactions proceeding under the iminium/enamine activation manifold, starting with the conjugate addition of a heteroaromatic compound to an α,β-unsaturated aldehyde, followed by the subsequent electrophilic chlorination of the generated enamine intermediate using a pentachlorocyclohexanone derivative as the source of electrophilic chlorine atom. In this particular case, imidazolidinone **50b** was identified as the best promoter with regard to catalytic activity and stereocontrol. The reaction furnished a variety of β-aryl-α-chloro substituted aldehydes in excellent yields and stereoselectivities for a variety of electron-rich aromatic compounds and differently substituted enals tested.

In a different approach, a cascade conjugate Friedel–Crafts/intramolecular aminal formation has been developed and applied to the enantioselective total synthesis of (–)-*flustramine B* (Scheme 7.41).[58] The reaction design relied on the use of a 1,3-disubstituted indole reagent derived from tryptamine as electron-rich heterocycle undergoing enantioselective conjugate Friedel–Crafts reaction with acrolein under iminium activation. The intermediate formed after the conjugate addition could not aromatize in this case because of its particular substitution pattern and therefore underwent intramolecular nucleophilic addition with the participation of the amine group incorporated at the functionalized side chain, therefore delivering the tricyclic basic skeleton of the target compound in excellent yield, as a single diastereoisomer and with excellent enantiomeric excess. In this case, the final compound was reduced *in situ*, furnishing directly the corresponding alcohol, which was converted in

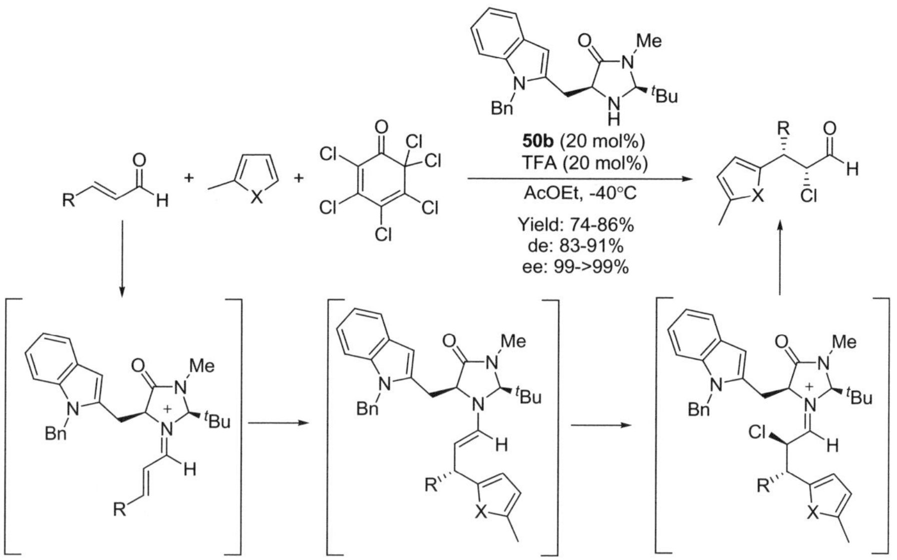

Scheme 7.40 A representative example of the Friedel–Crafts/chlorination cascade reaction.

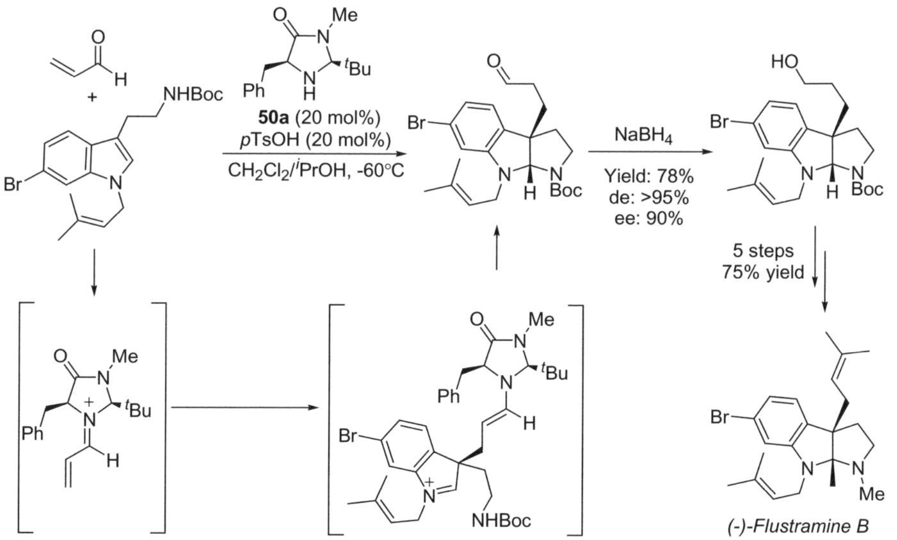

Scheme 7.41 Enantioselective total synthesis of (–)-*flustramine-B*.

only five steps into the target natural product using very high-yielding procedures. This protocol has been extended to other similar heterocyclic derivatives of tryptamine with different substituents at the aromatic core and with a variety of other enals different from acrolein, observing in all cases that this cascade process proceeded with excellent results.

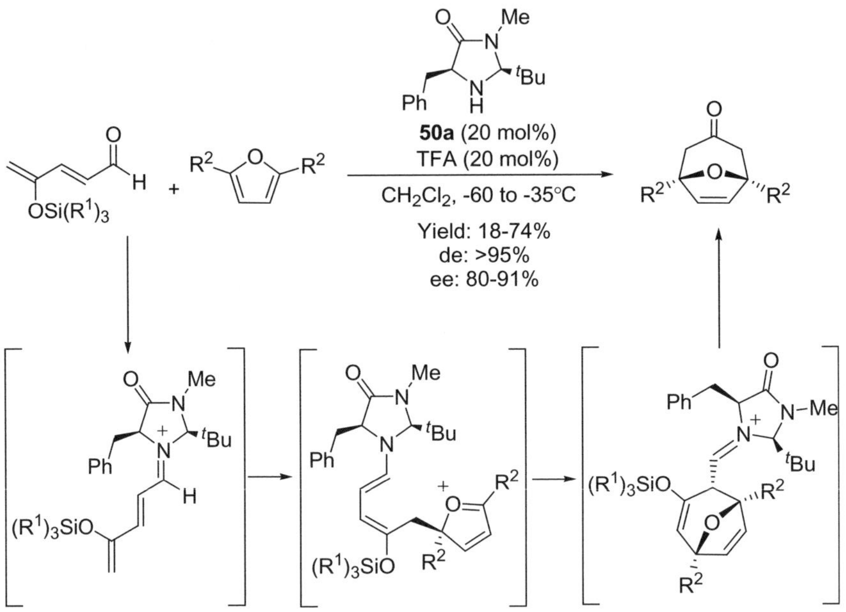

Scheme 7.42 Formal [4 + 3] cycloaddition reaction between silyloxydienals and furans catalyzed by **50a**.

A very interesting cascade process between a conjugated 4-silyloxy-2,4-dienal and 2,5-dialkyl-substituted furans resulting in a formal [4 + 3] cycloaddition has also been developed in an asymmetric version using imidazolidinone **50a** as catalyst (Scheme 7.42).[59] This reaction consisted of a first conjugate Friedel–Crafts alkylation followed by an intramolecular aldol reaction with the oxonium ion remaining at the furan moiety, which could not recover its aromaticity because of the presence of the two alkyl substituents at C-2 and C-5, which prevented elimination and therefore stabilized the oxonium ion during the time required to allow the intramolecular aldol reaction to occur. Different trialkylsilyl groups were tested at the silyloxydienal reagent and also several alkyl substituents were introduced at the furan reagent, showing that the final cycloaddition products could be obtained in low to good yields, with excellent *endo*-selectivity and enantioselectivities between 80 and 90% ee. Aromatic substituents at the furan led to the formation of mixtures of diastereoisomers and to a significant decrease in the enantioselectivity.

Finally, there is also a relevant example of an intermolecular conjugate Friedel–Crafts/electrophilic amination cascade developed very recently by Melchiorre in which 2-methyl-1H-indole reacted with a variety of α,β-unsaturated aldehydes and dialkyl azodicarboxylates in the presence of a primary amine catalyst, leading to the formation of a family of products containing two contiguous stereocenters, one of them a quaternary one (Scheme 7.43).[60] This reaction started with the conjugate addition of 2-methyl-1H-indole to the enal

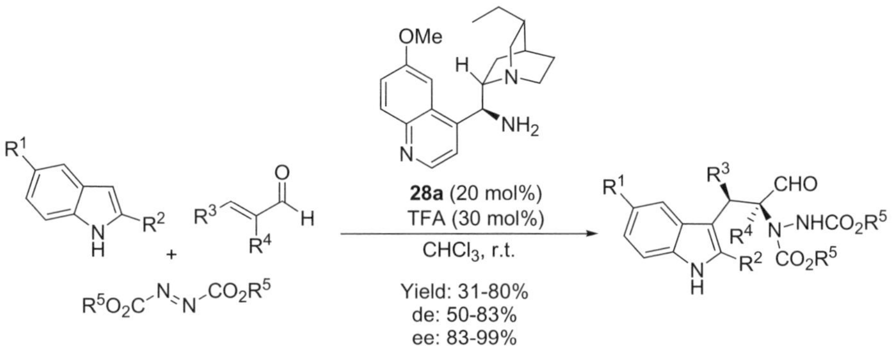

Scheme 7.43 Cascade conjugate Friedel–Crafts/electrophilic amination using α,β-disubstituted aldehydes.

under iminium activation and the intermediate enamine subsequently underwent addition to the dialkylazodicarboxylate reagent. Under the optimized reaction conditions, the cascade products were obtained in moderate yields, as mixtures of *syn/anti* diastereoisomers ranging from 3:1 to 11:1 and with excellent enantioselectivities. Perhaps the most interesting feature of this report is related to the possibility to use α,β-disubstituted α,β-unsaturated aldehydes as substrates for iminium activation, in addition to the interest of the highly functionalized compounds obtained in the reaction as chiral building blocks in synthesis and to the fact that a quaternary stereocenter is generated in the second electrophilic amination step. The key to the success of the process is associated to the use of a primary amine such as **28a** as catalyst, able to activate challenging substrates like the already mentioned α,β-disubstituted α,β-unsaturated aldehydes because of the release of steric congestion associated to the formation of an iminium ion derived from a primary amine.

7.3.6 Cascades Initiated by Conjugate Hydrogen-transfer Reaction

The use of Hantzsch esters as hydride donors for carrying out conjugate hydrogen transfer reactions under iminium activation has also been extended to initiate a cascade process. An interesting application of this concept to the synthesis of substituted cyclohexanes and cyclopentanes has been developed by List in a cascade reaction consisting of a conjugate reduction followed by intramolecular Michael reaction using substrates containing both an enone and an enal communicated by an all-carbon tether (Scheme 7.44).[61] The chemoselectivity of the reaction relied on the higher reactivity of the enal moiety toward the formation of the iminium ion, which resulted in its preferential activation for undergoing the conjugate reduction with the Hantzsch ester. Next, an intramolecular Michael reaction of the intermediate enamine with the

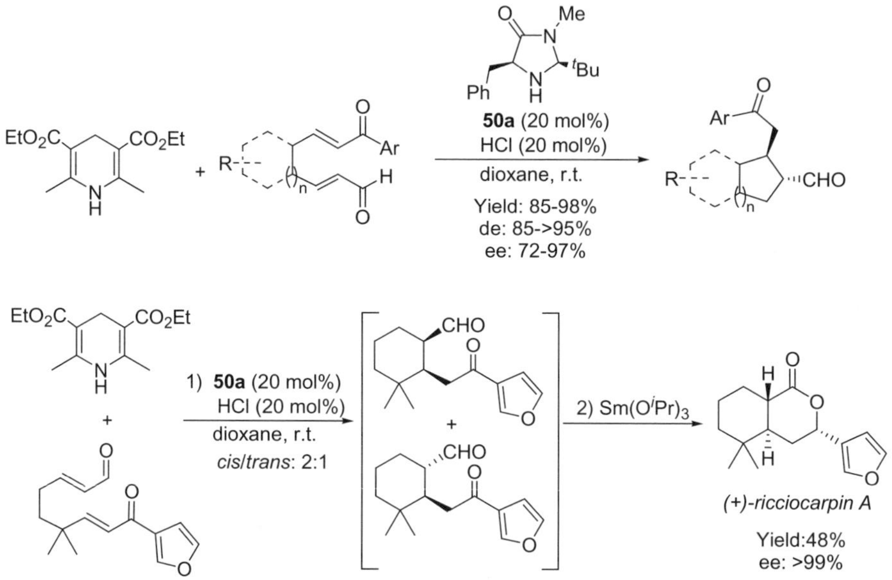

Scheme 7.44 Cascade conjugate hydrogen transfer/intramolecular Michael reaction and application to the total synthesis of (+)-*ricciocarpin A*.

enone moiety had to take place, delivering the final cyclic compounds. Substrates incorporating an arene moiety as the linker between the enone and the enal reacted very efficiently conducing to the formation of a cyclopentane ring with excellent yields, diastereo- and enantioselectivities. When the arene tether was changed by a saturated alkyl chain the reaction proceeded satisfactorily when a six-membered ring was formed but in the reaction conducing to the formation of a cyclopentane resulted to be significantly less enantioselective. This methodology has been successfully applied to the total synthesis of (+)-*ricciocarpin A*, a potent moluscicidal natural product, which was accomplished in a single one-pot operation starting from a conveniently substituted substrate incorporating the required enal and the enone (Scheme 7.44).[62] The synthesis consisted of the **50a**-catalyzed cascade conjugate hydrogen transfer/intramolecular Michael reaction followed by a Sm(O^{i}Pr)$_3$ catalyzed Tischenko reaction/lactonization. The first **50a**-catalyzed cascade proceeded smoothly, furnishing an intermediate adduct as an almost enantiopure material, although as 2:1 mixture of diastereoisomers, predominating the undesired *cis* isomer. However, the subsequent Tischenko reaction was observed to occur faster with the minor (and desired) *trans* diastereoisomer and, in addition, it was also found that Sm(O^{i}Pr)$_3$ was able to promote the epimerization of the formyl-containing stereocenter leading to the formation of the target natural product in 48% yield and >99% ee.

This reaction manifold has also been brilliantly exploited by MacMillan in an intermolecular conjugate reduction/halogenation reaction, which resulted in a formal addition of HCl or HF to α,β-unsaturated aldehydes.[57] In this case, imidazolidinone catalyst **ent-50c** was found to promote efficiently the reaction for a model substrate. However, perhaps the most interesting concept developed when studying these reactions was the possibility of employing two different imidazolidinone catalysts in this transformation, each of them promoting its own catalytic reaction *via* different ways of selective activation. For example, the reaction between 3-phenyl-2-butenal and a Hantzsch ester and also incorporating *N*-fluorobenzenesulfonimide as the source of electrophilic fluorine in the presence of different amounts of imidazolidinone catalysts **46b** and **ent-50c** furnished the expected addition product in much higher yield and diastereoselectivity than that provided for the same reaction in which only catalyst **ent-50c** was employed (Scheme 7.45). This indicated that two different catalytic cycles took place in the reaction, each of them involving its own catalyst, which also means that these imidazolidinones were also able to engage selectively in a specific type of activation. In this case, it is believed that the less sterically demanding imidazolidinone **ent-50c** promoted the conjugate reduction of the substrate *via* iminium activation and, next, imidazolidinone **46b** catalyzed the α-fluorination reaction by activating the saturated aldehyde generated after the conjugate reduction *via* enamine formation. Also in this case, the addition of reagents was carried out separately, incorporating the second imidazolidinone catalyst **46b** and the electrophilic fluorine source after substrate consumption was observed, therefore classifying this transformation as a tandem process rather than a cascade reaction. Confirmation of the specificity of the catalysts toward different ways of activation was obtained when carrying out the tandem sequence using separately the two enantiomers of imidazolidinone catalyst **46b** and **ent-46b**, which led to the isolation of each of the two different diastereoisomers, therefore indicating that the second electrophilic fluorination step took place by exclusive catalyst control exerted by **46b** (Scheme 7.45).

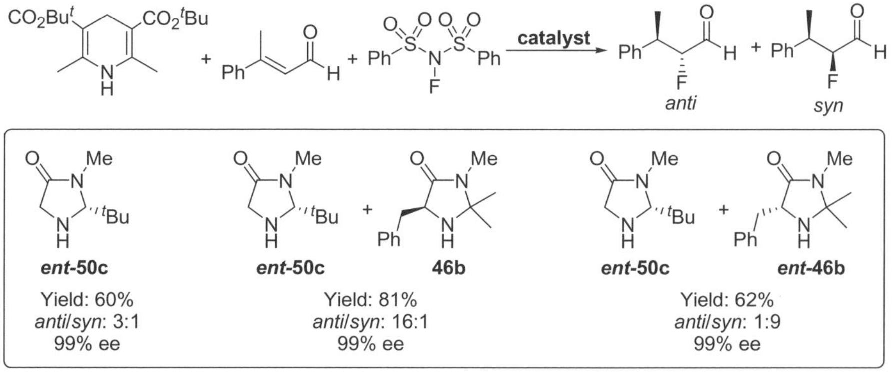

Scheme 7.45 Tandem conjugate hydrogen-transfer/electrophilic fluorination using cycle-specific imidazolidinone catalysts.

This concept of cycle-specific catalysis has been furthermore extended by MacMillan himself in a series of tandem reactions initiated by conjugate hydrogen transfer. In these cases, imidazolidinone ***ent*-50c** was employed as the catalyst participating preferentially in iminium activation of the enal, as in the previous case. However, it was found that incorporating proline **1** as the enamine-specific catalyst led to an outstanding performance in the process, with improved results with respect to imidazolidinone **46b** employed in the previous examples. Making use of this design, MacMillan developed a highly efficient procedure for carrying out hydroamination, hydrooxidation and reductive Mannich reactions using respectively dibenzyl azodicarboxylate, nitrosobenzene and a glyoxylate imine as electrophilic reagents in the second enamine-mediated step of the tandem process (Scheme 7.46).[63] Similarly to what had been made before, changing the configuration of the enamine-specific catalyst form L-proline to D-proline also allowed the two possible *syn* or *anti*

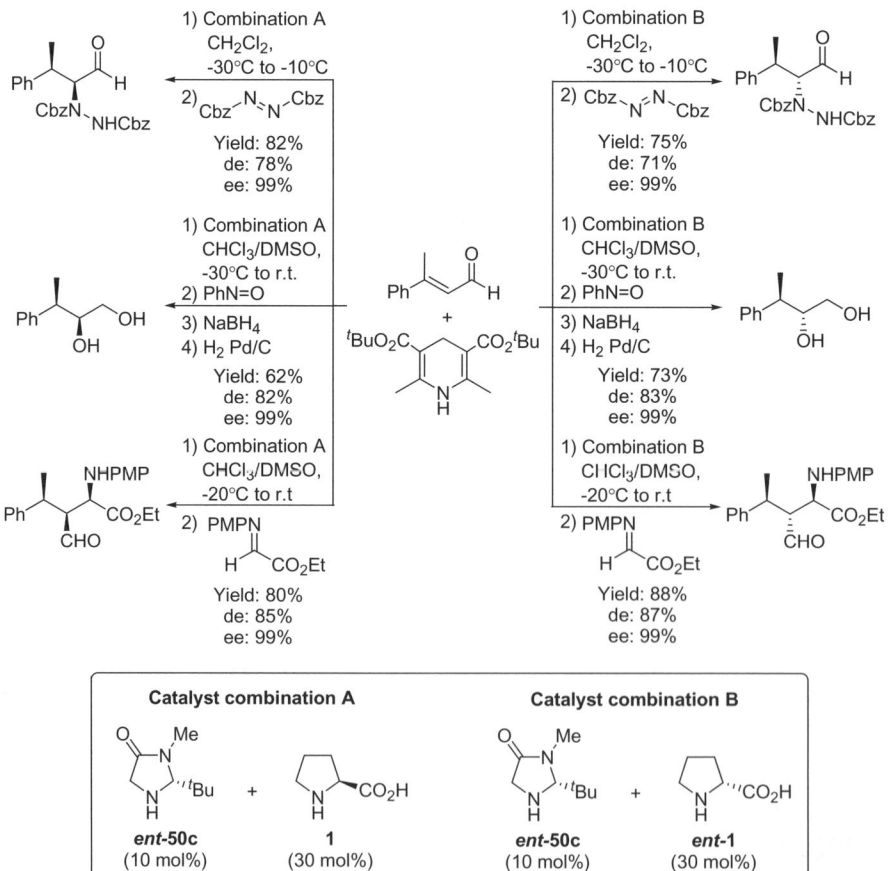

Scheme 7.46 Cycle-specific catalysis in several tandem processes initiated by conjugate hydrogen-transfer reaction.

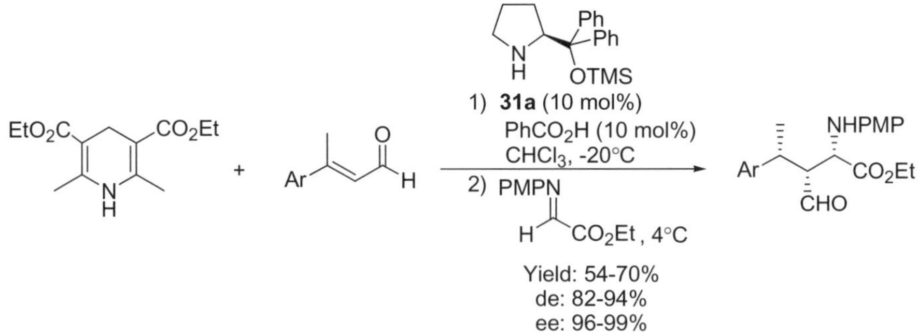

Scheme 7.47   Tandem conjugate reduction/intermolecular Mannich reaction.

isomers to be obtained, in a good example of a very efficient diastereodivergent procedure.

Finally, there is also another example of an intermolecular process involving conjugate hydrogen-transfer reduction of α,β-unsaturated aldehydes followed by Mannich reaction using *O*-TMS diphenylprolinol **31a** as catalyst (Scheme 7.47).[64] This reaction also consisted of a tandem reaction rather than a cascade process, therefore requiring the addition of the imine electrophile after consumption of the starting α,β-unsaturated aldehyde. This reaction was applied to the synthesis of highly functionalized acyclic α-amino acid derivatives containing up to three stereocenters favoring the formation of the all-*syn* diastereoisomer, which were obtained in good to moderate yields, good diastereoselectivities and with enantiomeric excesses higher than 95%.

7.3.7   Cascades Initiated by Hetero-Michael Reaction

As has already been pointed out in most of the preceding chapters, carrying out a cascade process avoids the inherent reversibility of the hetero-Michael reactions and pushes forward all the equilibria participating in the catalytic cycle, which turns into better conversions and solves the problem of the configurational instability of the hetero-Michael adducts. For this reason, intensive research has been carried out in developing such cascade processes, typically employing heteronucleophiles containing an electrophilic functionality able to react with the enamine intermediate generated after the conjugate addition to the iminium ion.

One of the first attempts in this field refers to a multicomponent intermolecular sulfa-Michael/amination cascade process in which an α,β-unsaturated aldehyde and a thiol were reacted in the presence of a dialkyl azodicarboxylate reagent, giving access to β-amino-γ-thioalcohol derivatives with excellent yields and enantioselectivities. Isolation of the final products was accomplished by *in situ* reduction followed by base-promoted cyclization (Scheme 7.48).[65] *O*-Trimethylsilyl diarylprolinol **31c** was identified as the best

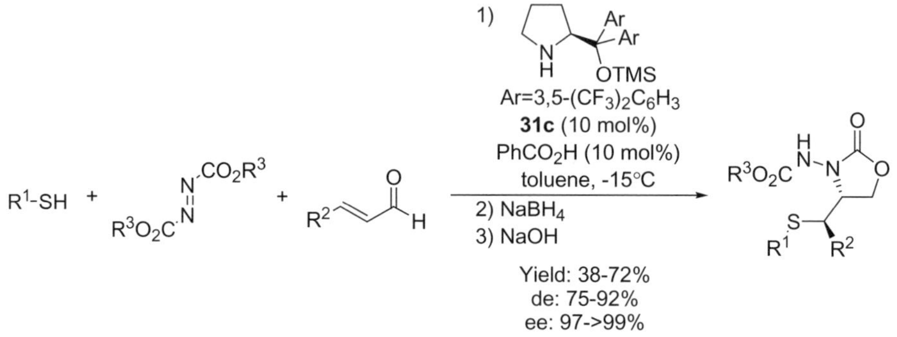

Scheme 7.48   Intermolecular sulfa-Michael/electrophilic amination reaction catalyzed by **31c**.

catalyst for this transformation. In fact, the authors started optimizing the reaction with the initial intermolecular sulfa-Michael addition, but they found that the hydrolysis of the enamine intermediate generated after the iminium-mediated sulfa-Michael reaction proceeded very slowly, leading to low concentrations of free catalyst available in the reaction medium and therefore slowing down the process. It was found afterwards that the addition of the azodicarboxylate electrophile accelerated remarkably the sulfa-Michael process, which was attributed to an easier releasing of the catalyst favored by the formation of a highly sterically encumbered iminium intermediate after the electrophilic amination step. With all these evidences in hand, the cascade sequence was developed leading to a highly efficient process for the generation of chemical complexity from very simple and cheap reagents.

A different concept has been devised very recently in which, in contrast to the previously presented reaction and other related ones, the intermolecular cascade process does not involve the use of two different reagents (a nucleophile for the hetero-Michael reaction and an electrophile for the subsequent reaction with the enamine intermediate) but one single reagent, which, after the Michael process starts, participates as the electrophile reagent but incorporates the nucleophile counterpart as leaving group. After the electrophilic addition step, the nucleophile reagent is regenerated *in situ* in the reaction medium, incorporating into the subsequent catalytic cycle without the need for the addition of one equivalent of this reagent. This approach also reduces the amount of waste chemicals generated in the process. This design has been applied to the enantioselective aminosulfenylation of α,β-unsaturated aldehydes catalyzed by *O*-TMS diphenylprolinol **31a** and using an *N*-alkylthiosuccinimide as electrophilic reagent that regenerates the nucleophile (succinimide) after the electrophilic addition step (Scheme 7.49).[66] The reaction required a catalytic amount of free succinimide in order to start the process but this nucleophile was continuously delivered to the reaction medium after the reaction of the intermediate enamine with the *N*-alkylthiosuccinimide electrophile had taken place. The final highly

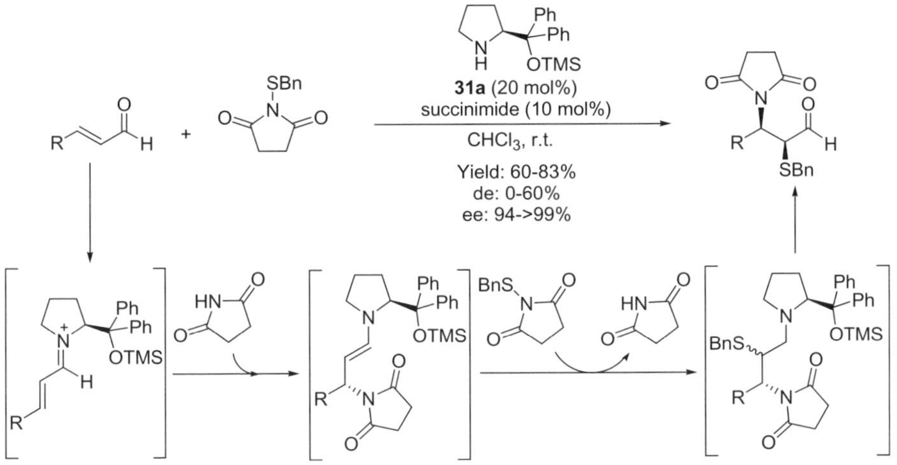

Scheme 7.49 Enantioselective aminosulfenylation of α,β-unsaturated aldehydes.

functionalized acyclic products were obtained with good yields, and very high enantioselectivities, although mixtures of *anti/syn* diastereoisomers were obtained in ranges varying from 1:1 to 4:1.

However, despite these two impressive examples of intermolecular cascade processes initiated by hetero-Michael reactions, the majority of the developed methodologies in this field involve intramolecular versions, leading to the formation of highly functionalized cyclic structures. A very illustrative example can be found in Scheme 7.50, consisting of an enantioselective sulfa-Michael/aldol reaction using 2-mercapto-1-phenylethanone as the functionalized Michael donor leading to the formation of enantioenriched tetrahydrothiophenes in a single step.[67] Interestingly, the authors found that two different regioisomers could be selectively obtained by introducing slight modifications in the reaction conditions. If the reaction was carried out in the presence of an acid co-catalyst 2-benzoyl-5-alkyltetrahydrotiophene-4-ols were formed but, under basic conditions, a completely different isomer was obtained. These results were interpreted in terms of an iminium-enamine manifold operating in the first case, with the acid co-catalyst facilitating the formation of the iminium intermediate and the catalyst participating in the second intramolecular aldol reaction, the latter proceeding *via* enamine activation. On the contrary, under basic conditions, hydrolysis of the enamine intermediate might occur and, therefore, a final intramolecular base-catalyzed aldol reaction with the more reactive aldehyde moiety as electrophile and without the participation of the catalyst would lead to the final product. In this second case, the basic conditions employed could also promote the uncatalyzed sulfa-Michael reaction which was interpreted as the reason for the slightly lower enantioselectivity observed in this case.

Hetero-Michael/aldol cascade reactions have been studied by several research groups using *ortho*-substituted benzaldehydes as functionalized reagents. In this context, different benzo-fused heterocyclic architectures have

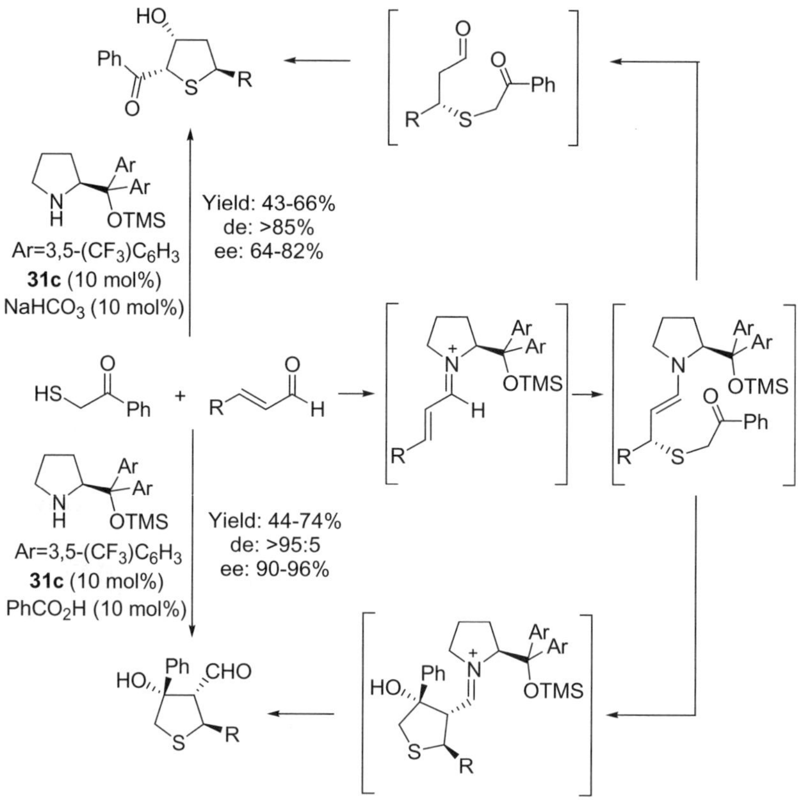

Scheme 7.50 Sulfa-Michael/aldol cascade for the synthesis of tetrahydrothiophenes.

been prepared by the reaction of phenols, thiophenols and anilines incorporating a formyl group at the *ortho* position in combination with α,β-unsaturated aldehydes and using a chiral amine catalyst, usually a diarylprolinol silyl ether such as **31a** or **31c**. Due to the particular structure which is built up, these reactions usually finish with a dehydration reaction after the intramolecular aldol step. The most studied transformation is the one involving an initial oxa-Michael addition step, leading to the formation of final compounds with a benzopyrane structure (Scheme 7.51). An initial report used **31a** as catalyst and furnished moderate yields and enantioselectivities[68a] but soon afterwards, several more efficient methodologies were reported using **31a**, **31i** and the combination of **31a** with a chiral carboxylic acid such as **131** as catalysts.[68b–d] Key to the success of these later reports is the inclusion of 4 Å MS as additive, which presumably favors the final dehydration step.

Using this approach, benzothiopyrans have been prepared by means of a sulfa-Michael/aldol/dehydration cascade[69] and access to dihydroquinolines has been achieved by using an aza-Michael/aldol/dehydration approach (Scheme 7.52).[70] In general, all these reactions proceeded with excellent yields and

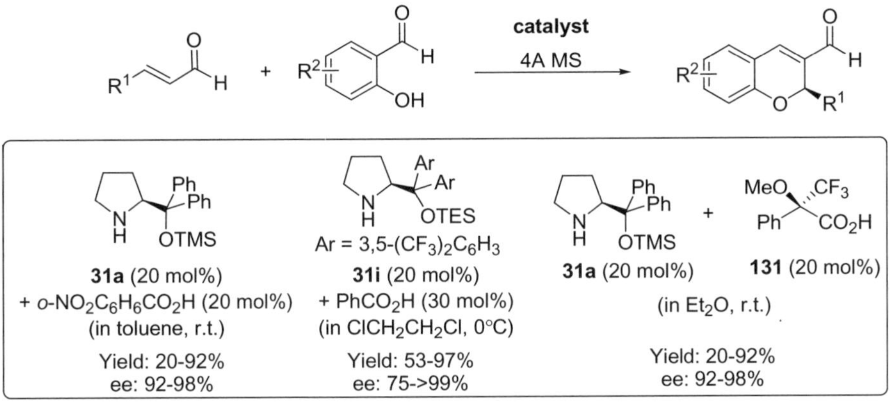

Scheme 7.51 Oxa-Michael/aldol/dehydration cascade for the synthesis of benzopyranes.

enantioselectivities and with good substrate scope regarding the enal reagent employed, tolerating the incorporation of aromatic, aliphatic and other functionalized substituents at the β-position. A related approach has also been carried out using 1H-indole-2-carbaldehyde as functionalized nitrogen nucleophile, also incorporating an adjacent formyl group ready to participate as an electrophile in the subsequent intramolecular aldol reaction. In this case, **31a** was identified as the best catalyst, which allowed the preparation of a family of differently substituted tricyclic adducts in moderate yields, albeit with excellent enantioselectivities (Scheme 7.52).[71]

On the other hand, the reaction involving enones as initial Michael acceptors has been limited to the use of cyclohexenone, showing that tetrahydroxanthenones and tetrahydrothioxanthenones could be obtained using this reaction design, although other less bulky secondary amines with improved abilities for activating enones had to be employed as catalysts.[72] Surprisingly, there has not been any report evaluating the use of primary amines as catalysts in these transformations. There is also another example, in which a 2-acetylthiophenol has been employed as functionalized Michael donor which, therefore incorporates a ketone moiety as internal electrophilic moiety for the intramolecular aldol reaction step (Scheme 7.53). In this case, as the final dehydration reaction could not take place, the final products were formed with the concomitant generation of three contiguous stereogenic centers, with remarkably high diastereoselectivities in all cases tested.[73] The use of a ketone as internal electrophile in this context has also been described in the reaction of ethyl 3-mercapto-2-oxopropanoate with β-aryl substituted α,β-unsaturated aldehydes, proceeding with moderate yields, albeit with excellent diastereo- and enantioselectivities using **31c** as catalyst (Scheme 7.53).[74]

Related to this topic, our group has also developed an interesting cascade process for the enantioselective synthesis of furofuranes in a single step by the **31a**-catalyzed reaction between dihydroxyacetone dimer and α,β-unsaturated

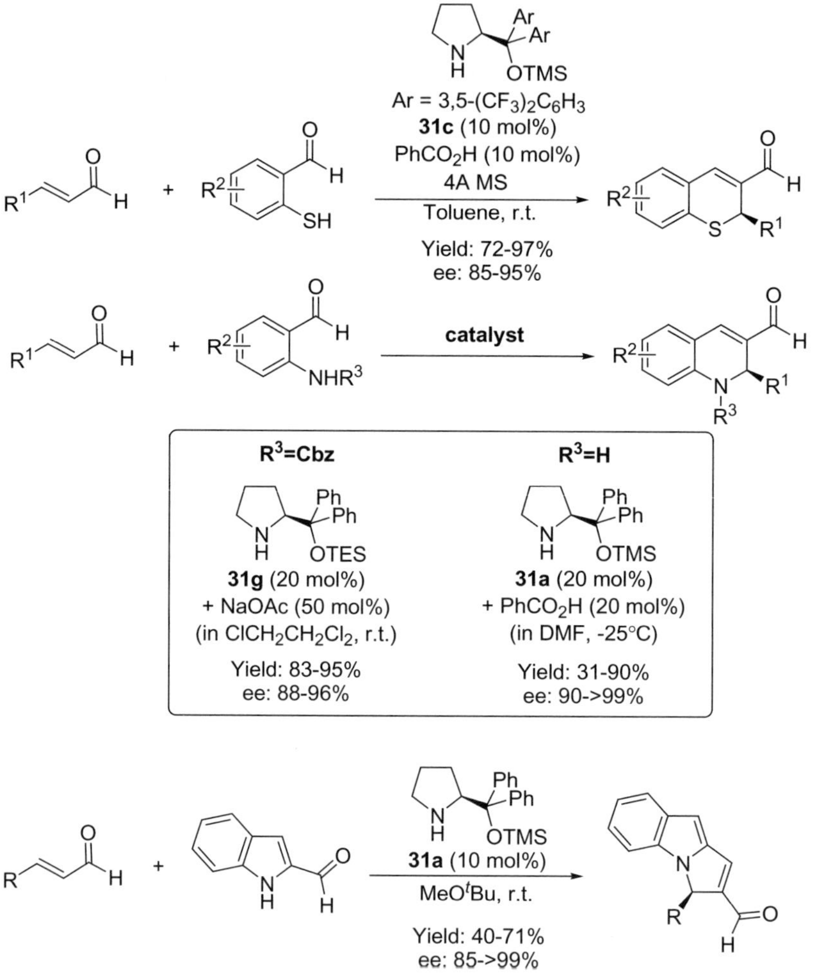

Scheme 7.52   Hetero-Michael/aldol/dehydration cascade for the asymmetric synthesis of benzo-fused heterocycles.

aldehydes (Scheme 7.54).[75] The reaction consisted of an initial oxa-Michael addition of dihydroxyacetone to the enal under iminium activation followed by a fast intramolecular aldol reaction and a final hemiacetalization, delivering the final compounds in excellent yields and as single diastereoisomers of very high enantiomeric purity. Remarkably, this sequence involved the consecutive formation of two C–O and one C–C bond and the fully stereocontrolled generation of four stereocenters, one of them a quaternary one. The fact that a high pK_a oxygen nucleophile could be employed as Michael donor to initiate the conjugate addition, in contrast with the need for the use of highly acidic phenols as O-nucleophiles described in the previous examples (see Scheme 7.51)

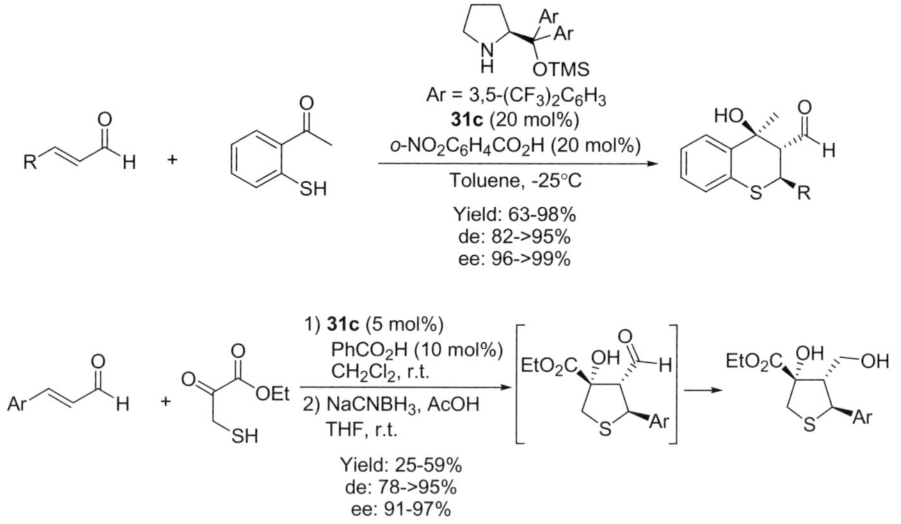

Scheme 7.53   Sulfa-Michael/aldol cascade using ketones as internal electrophiles.

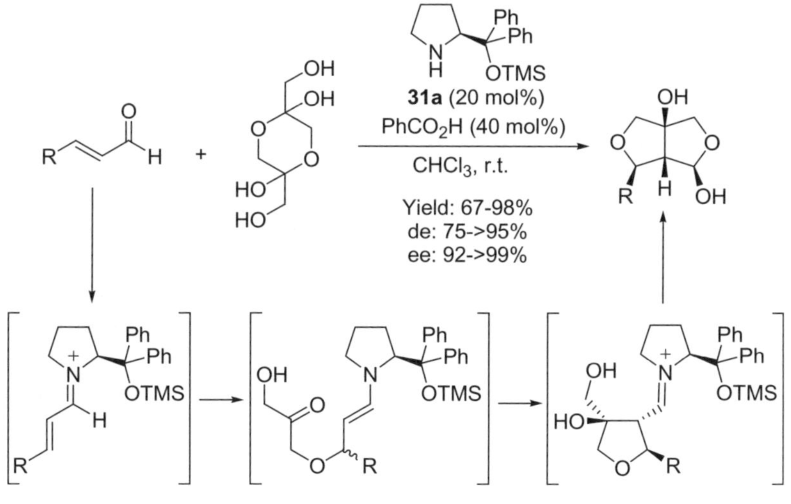

Scheme 7.54   Enantioselective synthesis of furofuranes by cascade oxa-Michael/aldol/
hemiacetalization reaction.

and also that the subsequent intramolecular aldol reaction took place with a less electrophilic ketone moiety are also unique features associated to this transformation. Moreover, the possibility of the selective manipulation of the different functionalities present at the obtained adducts was also demonstrated, allowing the preparation of a wide range of different compounds.

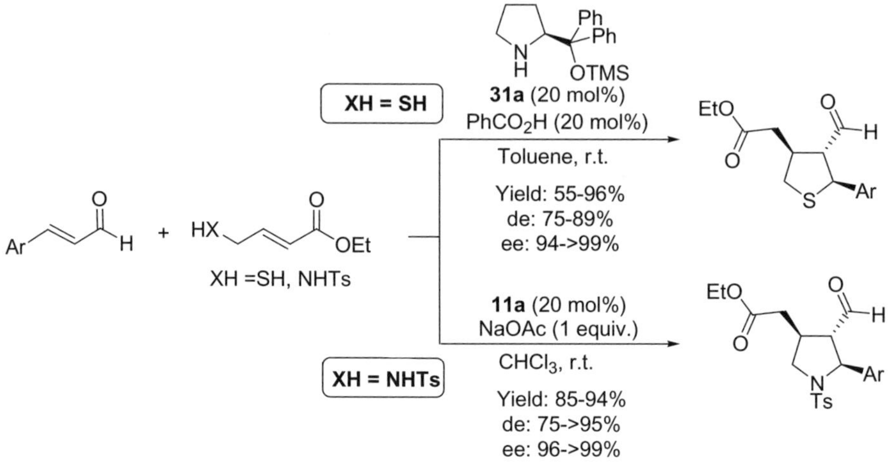

Scheme 7.55   Hetero-Michael/Michael cascade reaction for the asymmetric synthesis of tetrahydrothiophenes and pyrrolidines.

Cascade sulfa-Michael/Michael and aza-Michael/Michael reactions have also been developed for the synthesis of densely functionalized chiral tetra-hydrothiophenes and pyrrolidines (Scheme 7.55).[76] This approach consisted of the use of a 4-mercapto or a 4-amino-2-butenoate respectively as functionalized nucleophiles, which, after the initial hetero-Michael reaction under iminium activation, underwent a subsequent intramolecular Michael reaction in which the intermediate enamine moiety reacted with the enoate moiety incorporated at the functionalized reagent, resulting in the formation of a five-membered ring heterocycle. *O*-TMS-Diphenylprolinol **31a** was identified as the best catalyst for all these reactions, which proceeded in general with good yields and excellent diastereo- and enantioselectivities for a wide variety of different α,β-unsaturated aldehydes, although it appears to be limited exclusively to substrates incorporating aromatic substituents at the β-position. Importantly, the aza-Michael-initiated cascade reaction was shown to be highly dependent on the protecting group incorporated at the amine nucleophile, being limited to the use of *N*-tosyl protected substrates, and it also required the addition of one equivalent of a basic additive such as NaOAc, which was needed to assist the deprotonation of the nitrogen nucleophile.

In a different report, a cascade oxa-Michael/Michael reaction has been developed using *O*-hydroxycinnamaldehyde as functionalized Michael donor and a nitroalkene as acceptor, using **31g** as catalyst, and leading to the formation of chiral chromanes in a single step (Scheme 7.56).[77] Interestingly, it was found that the hydroxycinamaldehyde and **31g** rapidly condensed to form a stable hemiaminal intermediate after C=C double bond isomerization and intramolecular addition, which was identified and fully characterized. This intermediate is proposed to undergo oxa-Michael addition to the nitroalkene (presumably through its hydroxy-iminium open-chain tautomer) forming the

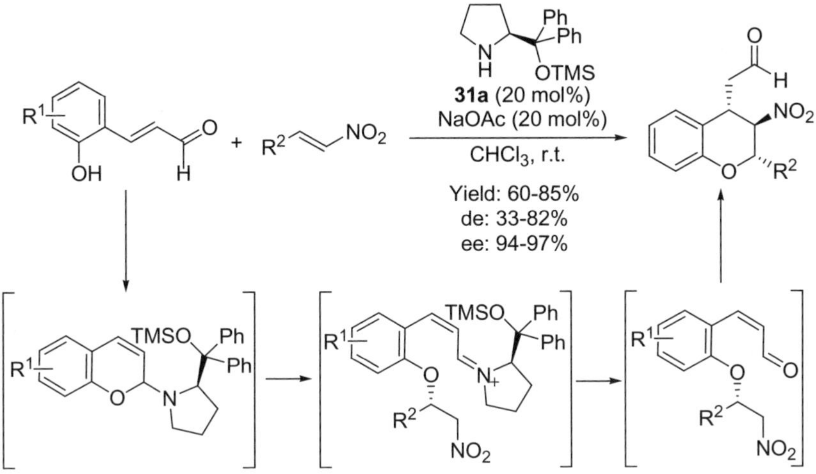

Scheme 7.56 Oxa-Michael/Michael cascade reaction for the synthesis of chromanes and the mechanism proposed.

first stereocenter and delivering an iminium intermediate, which, after release of the catalyst, would undergo diastereoselective intramolecular Michael addition under substrate control. However, other alternative pathways could also be considered involving a first non-stereoselective oxa-Michael addition followed by an intramolecular Michael addition to the intermediate iminium ion (under stereochemical control exerted by the catalyst) proceeding faster with one of the epimers, while the other would racemize in the reaction medium because of the inherent reversibility of the oxa-Michael reaction.

Simple aliphatic alcohols have also been employed as the initiating nucleophiles in a quadruple oxa-Michael/Michael/Michael/aldol cascade reaction for the synthesis of densely substituted cyclohexenes (Scheme 7.57).[78] In this case, a simple aliphatic alcohol like MeOH reacted with two equivalents of acrolein and one equivalent of a β-aryl substituted nitroalkene in the presence of catalyst **31a**, furnishing the cascade products in moderate yields but as single diastereoisomers in almost enantiomerically pure form. This reaction consisted of the first activation of acrolein by **31a** as the corresponding iminium ion followed by oxa-Michael addition and the intermediate enamine underwent a subsequent intermolecular Michael reaction with a second equivalent of the acroleine-derived chiral iminium intermediate generated before. The second enamine generated after this reaction participated in an intramolecular aldol reaction followed by dehydration and releasing of the catalyst, delivering the final cyclohexenecarbaldehyde adducts. It is also noteworthy that not only simple small alcohols like MeOH or EtOH were active in this reaction as nucleophiles initiating the cascade process but also other bulkier or functionalized alcohols were also found to be suitable substrates for this reaction.

Finally, it has to be pointed out that the enantioselective epoxidation of α,β-unsaturated compounds with peroxides can also be considered to happen *via* a

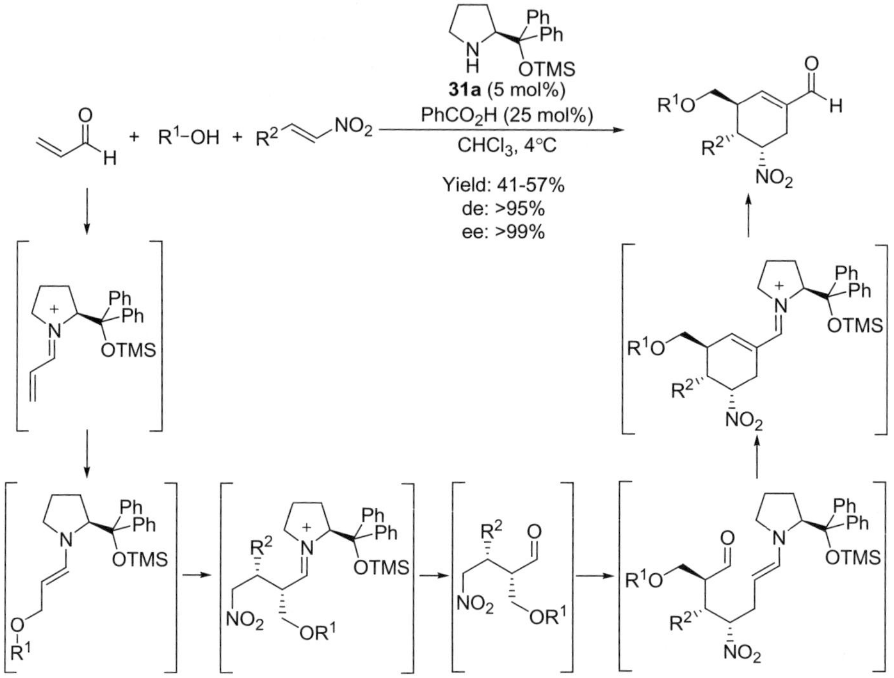

Scheme 7.57 Oxa-Michael/Michael/Michael/aldol cascade reaction for the synthesis of cyclohexenecarbaldehydes.

domino oxa-Michael/intramolecular nucleophilic substitution pathway and, therefore, the iminium activation concept has showed up as a very successful approach to face this particularly important transformation. In this context the first contribution was made by the group of Jørgensen[79] using hydrogen peroxide as the oxidant and O-TMS diarylprolinol **31c** as catalyst (Scheme 7.58). Under the best reaction conditions, a variety of both β-alkyl and β-aryl substituted α,β-unsaturated aldehydes underwent clean epoxidation delivering the final adducts with good yields, very high diastereoselectivities and enantioselectivities higher than 90% ee in all cases tested. β,β-Disubstituted enals were also surveyed as substrates but, although the reaction proceeded satisfactorily with respect to conversion, stereoselectivities were found to be somewhat lower. As an extension of this work, much more recently the same group reported a one-pot process consisting of an initial **31c**-catalyzed epoxidation of α,β-unsaturated aldehydes followed by a cascade Henry reaction/cyclization process under phase-transfer catalysis, leading to the formation of isoxazoline-*N*-oxide final products (Scheme 7.58).[80] For this second step, axially chiral ammonium salt **132** was employed as catalyst, which had already been developed by the group of Lygo as a very efficient catalyst for the alkylation of enolates.[81] As mentioned, after the epoxidation, *tert*-butyl nitroacetate, CsOH and catalyst **132** were added to the reaction mixture after complete

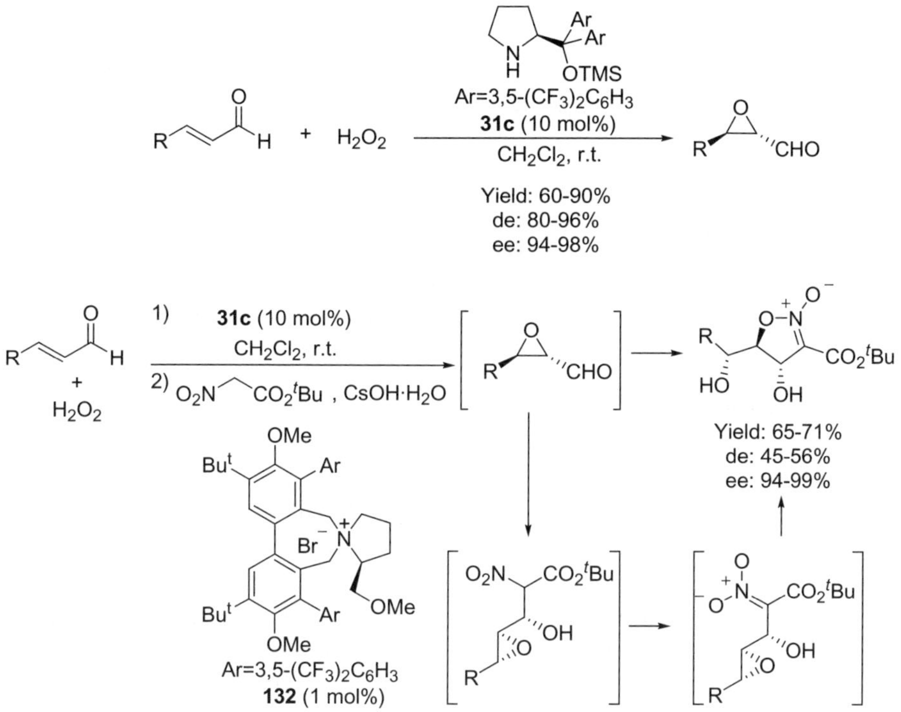

Scheme 7.58 31c-catalyzed epoxidation of α,β-unsaturated aldehydes and application to a one-pot epoxidation/Henry/cyclization reaction.

consumption of the starting material was observed and therefore promoting a subsequent nitroaldol addition to the formyl group of the epoxide followed by intramolecular epoxide ring-opening by an intermediate nitronate generated by deprotonation. The final products were obtained with good yields and as *ca.* 3:1 mixtures of diastereoisomers, although each diastereoisomer was isolated with excellent enantiomeric purity.

After the initial report by Jørgensen, several other methodologies were reported for the epoxidation of α,β-unsaturated aldehydes using different secondary amine catalysts and alternative oxidants with different success (Scheme 7.59). For example, catalyst **31c** has been employed together with sodium percarbonate (SPC) as the oxidant using water as reaction solvent.[82] In a different report, enals have been epoxidized with *p*-methyliodosobenzene hypervalent iodine reagent as oxidant in the presence of imidazolidinone **50a** catalyst[83] and, more recently, fluorine-containing catalyst **133** has also been synthesized and used in this reaction performing very well with regard to yields and enantioselectivities, although the diastereoselectivity was strongly dependent on the nature of the β-substituent of the substrate.[84]

Alternatively, the epoxidation of enones has also been covered using primary amine catalyst **28c**[85] or by applying the asymmetric counterion-directed

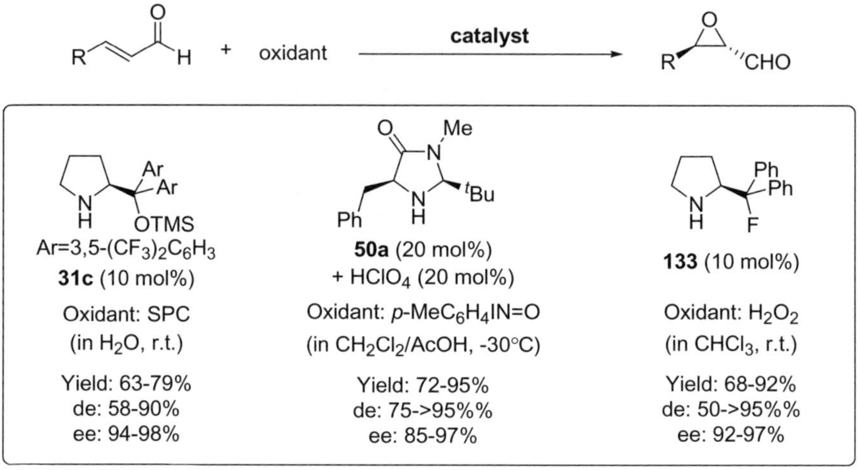

Scheme 7.59 Several methodologies reported for the epoxidation of α,β-unsaturated aldehydes using chiral amines as catalysts.

catalysis concept using the combination of **54** and chiral phosphoric acid **60a**[86] as catalytically active species (Scheme 7.60). In the first case, cumene hydroperoxide had to be employed as the epoxidizing reagent for achieving the best results and it was also found that that it was necessary to carry out the reaction at relatively higher temperatures (r.t. or 55 °C), because when the reaction was run at 0 °C only the β-peroxidation product arising from the Michael addition was observed (see Scheme 3.20 in Chapter 3). Under these conditions, a series of acyclic enones was converted into the corresponding epoxides with good yields and excellent stereoselectivities. On the other hand, the same catalyst system has been applied to the use of different cyclic enones with good results, it being noteworthy that β-substituted cyclohexenones leading to the formation of an epoxide containing a quaternary stereocenter behaved excellently, furnishing outstanding results with regard to both chemical efficiency and stereocontrol. In some other cases, the combination of diphenylethylenediamine **54** together with chiral phosphoric acid TRYP (**60a**) had to be employed.

In a similar way, the enantioselective aziridination of α,β-unsaturated aldehydes and ketones has also been achieved by a cascade aza-Michael/intramolecular nucleophilic displacement (Scheme 7.61). The reaction design involved the use of *N*-acetoxy carbamates as functionalized Michael donors, for which it was observed that, after conjugate addition, the intermediate enamine underwent intramolecular reaction with the *N*-OAc group, with the acetate moiety playing the role of an excellent leaving group. The reaction with α,β-unsaturated aldehydes required the use of *O*-TMS diphenylprolinol **31a** as the most efficient catalyst, obtaining the corresponding aziridines in good yields and a rather high diastereoselectivity.[87] On the other hand, the use of enones as substrates involved the use of primary amine **28a** together with a chiral α-amino acid such as *N*-Boc phenylglycine as Brønsted acid co-catalyst,[88] under

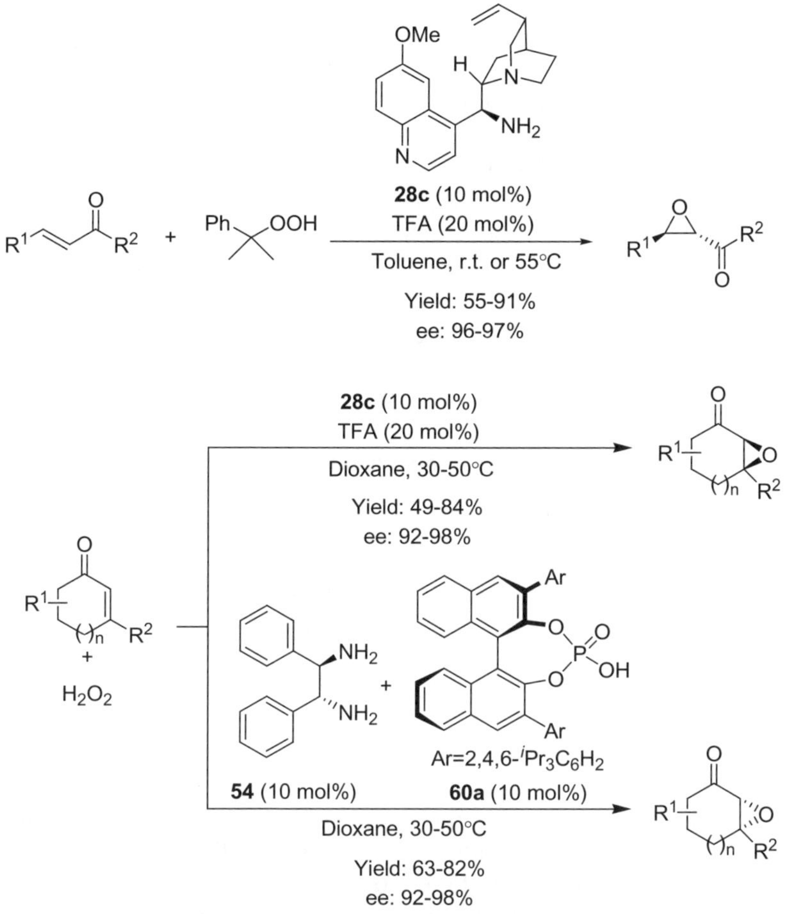

Scheme 7.60 Several methodologies reported for the epoxidation of enones.

conditions already employed by the same group with success in other Michael reactions using enones as electrophiles. Under the optimized reaction conditions, a variety of acyclic enones were converted into the corresponding *N*-protected aziridines with excellent yields and enantioselectivities and, remarkably, as single diastereoisomers.

7.4 Cascade Processes Initiated by Conjugate Addition *via* H-bonding Activation

The activation of the Michael acceptor by the formation of an H-bonded network with the catalyst has also been applied to several very efficient cascade processes, after incorporating a suitable electrophile to the reaction medium

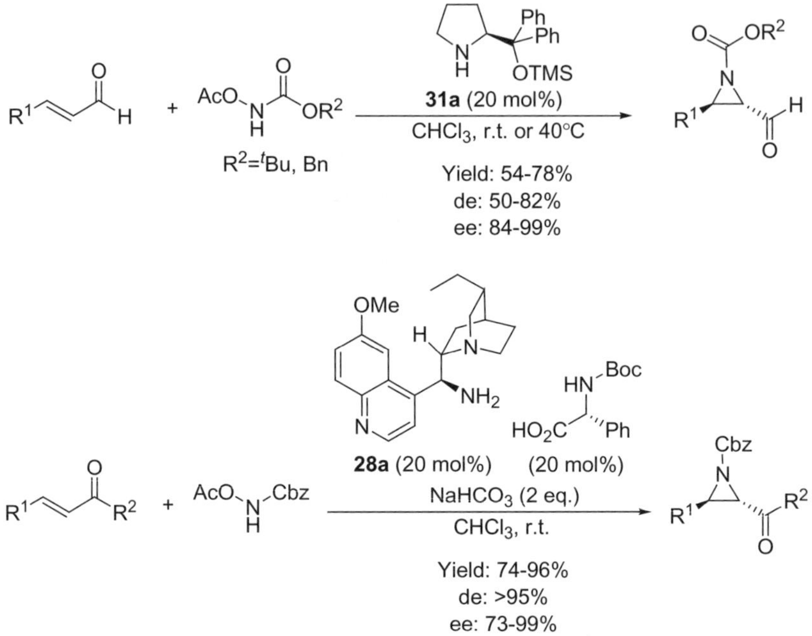

Scheme 7.61 Several methodologies reported for the aziridination of α,β-unsaturated aldehydes and ketones.

which is able to interact with the intermediate arising after the conjugate addition step. In this context, although this intermediate has the potential to remain linked to the catalyst and therefore allow the participation of the latter in the stereocontrol of the subsequent step, the weaker nature of the substrate–catalyst interaction (which on the other hand resulted in being the key feature for allowing turnover of the catalyst in the standard Michael-type reactions not proceeding in a cascade fashion) makes the second step generally controlled by the substrate, with no influence of the catalyst. On the other hand, it should also be mentioned that all the cases described in the literature for cascade processes initiated by conjugate addition proceeding under H-activation refer to intramolecular reactions, because of the already mentioned favorable entropic reasons associated to these reactions.

7.4.1 Michael/Henry and Michael/Aza-Henry Cascade Reactions

Highly substituted cyclohexanes and cyclopentanes have been prepared by means of a cascade process involving the use of a nitroalkene as Michael acceptor and an α-substituted β-ketoester incorporating a lateral β-substituent with a terminal methyl ketone moiety at the convenient position, ready to participate in an intramolecular aldol reaction after the first conjugate addition had taken place.[89] Therefore, after initial Michael-type reaction of the β-ketoester to the nitroalkene in the presence of cinchona alkaloid bifunctional catalyst **28c**, in which the first two stereocenters were generated under catalyst control, the formed Michael adduct intermediate had the potential to undergo intramolecular nitroaldol (Henry) reaction with the pendant ketone previously incorporated at the β-substituent of the β-ketoester, leading to the formation of the five- or six-membered ring final compound and generating two additional stereocenters (Scheme 7.62). Yields, diastereo- and enantioselectivities were remarkably high in all cases tested, indicating that the catalyst exerted a very efficient stereochemical control in the first step and that the cyclization step proceeded with excellent diastereoselection, the stereochemical outcome of the latter being strictly controlled by the stereocenters present at the Michael intermediate. It should be pointed out that, even though catalyst **28c** is a primary amine and consequently it might participate in the reaction *via* enamine activation, the authors point toward a bifunctional activation profile involving activation of the nitroalkene by the amino group *via* H-bonding together with the participation of the basic quinuclidine nucleus in the deprotonation of the β-ketoester (see Scheme 4.24 in Chapter 4).

On the other hand, there is a report regarding a cascade Michael reaction followed by intramolecular nitro-Mannich (aza-Henry) reaction occurring between imides derived from diethyl aminomalonate and nitrostyrenes using thiourea **68a** as catalyst (Scheme 7.63).[90] This reaction results in a formal [3 + 2] cycloaddition between these two reagents, with this aminomalonate-derived

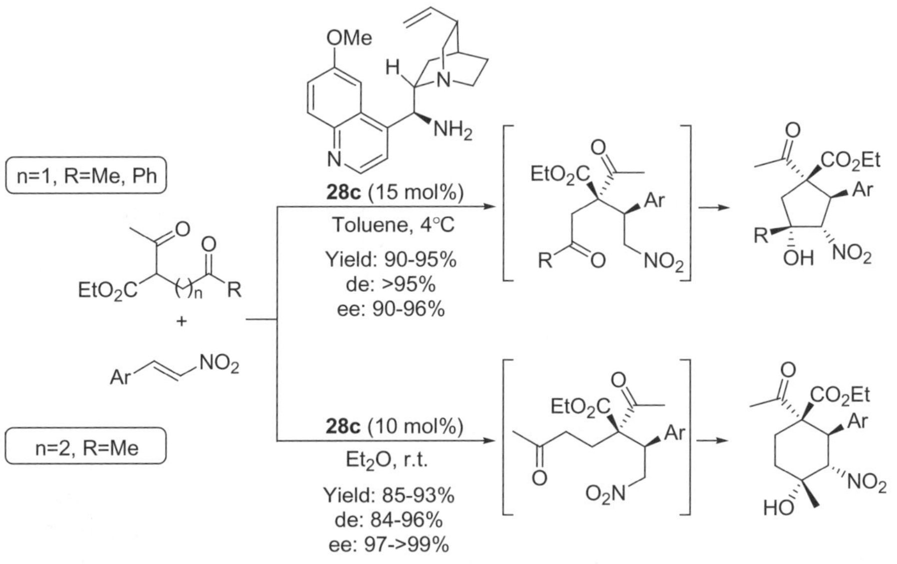

Scheme 7.62 Michael/Henry cascade reactions for the synthesis of highly substituted cyclopentanes and cyclohexanes.

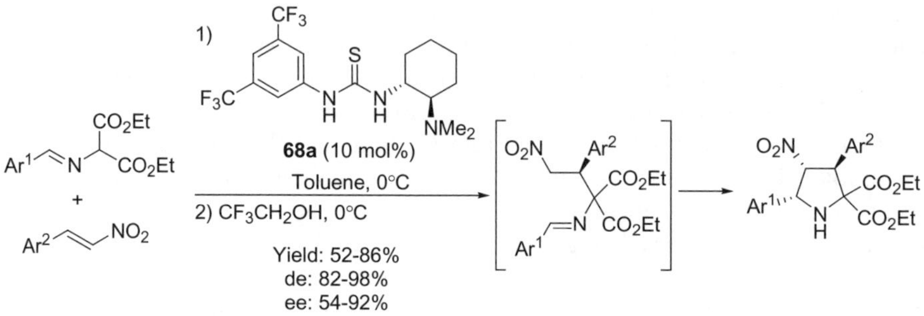

Scheme 7.63 Formal [3 + 2] cycloaddition between azomethine ylides and nitroalkenes proceeding by sequential Michael/aza-Henry reaction.

imine participating as the source of an azomethine ylide reagent and, in this case, the authors were able to isolate the intermediate Michael addition product, which was subsequently demonstrated to undergo fast intramolecular aza-Henry reaction facilitated by the presence of the catalyst, which is proposed in the simultaneous activation of the imine electrophile (by H-bonding with the thiourea moiety) and the nitroalkane (by deprotonation exerted by the dimethylamino group). It should be mentioned at this point that several other authors have carried out this reaction using other catalysts,[91] but in all these cases the reaction is considered to proceed via a concerted mechanism rather than stepwise, and that there is even also a previous example of this [3 + 2] cycloaddition between azomethine ylides and α,β-unsaturated aldehydes under iminium catalysis.[92]

7.4.2 Michael/Michael Cascade Reactions

Analogous to the cascade Michael–Henry reaction depicted in Scheme 7.61, the same group has developed an efficient synthesis of highly substituted cyclopentanes by means of a cascade process involving again the use of a nitroalkene as Michael acceptor and an α-substituted β-ketoester incorporating in this case an α,β-unsaturated ester moiety as the lateral β-substituent (Scheme 7.64).[93] In this way, after the initial Michael addition of the β-ketoester to the nitroalkene reagent promoted by 9-amino cinchona alkaloid derived catalyst **28c**, the intermediate was in a position to undergo intramolecular Michael reaction by conjugate addition of an *in situ* generated nitronate species to the remaining α,β-unsaturated ester unit. As in the previous case of Scheme 7.61, the reaction proceeded with outstanding results regarding yields and stereoselectivities.

In a similar context, γ,δ-unsaturated β-ketoesters (Nazarov reagents) have also been employed as suitable functionalized Michael donor substrates for a Michael/Michael cascade reaction with nitroalkenes using, in this case, Takemoto's catalyst **68a** (Scheme 7.65).[94] This reaction proceeded in a tandem

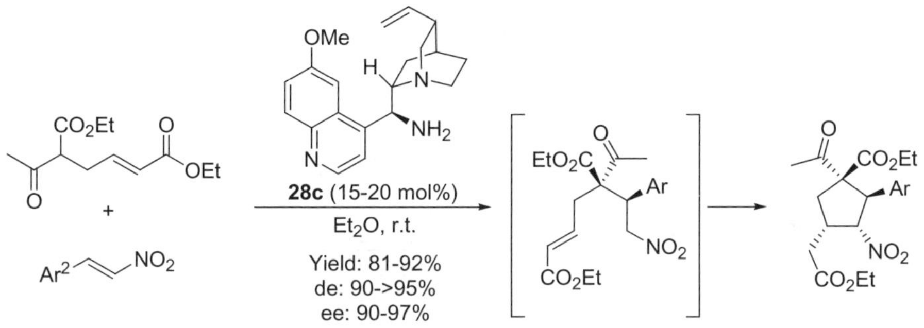

Scheme 7.64 Cascade Michael/Michael reaction for the asymmetric synthesis of highly functionalized cyclopentanes.

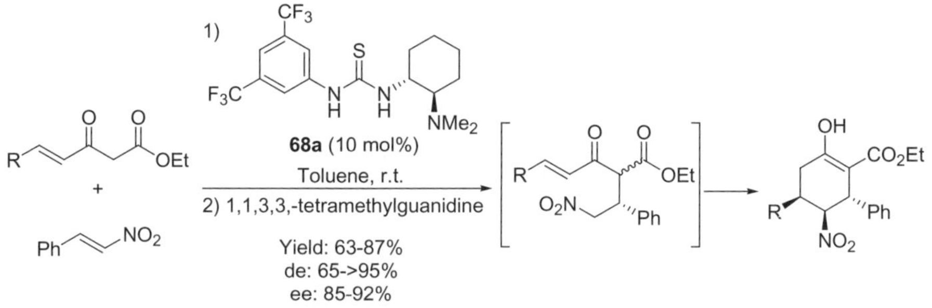

Scheme 7.65 **68a**-Catalyzed cascade Michael/Michael reaction for the asymmetric synthesis of polysubstituted cyclohexanones.

mode, with the initial intermolecular conjugate addition step proceeding in the first place with the assistance of the thiourea catalyst and requiring the incorporation of an external base such as 1,1,3,3-tetramethylguanidine (TMG) after consumption of the starting material was observed, in order to promote the second intramolecular Michael reaction by deprotonating the nitroalkane moiety and therefore generating an activated nitronate nucleophile. The reaction was tested with a variety of γ-substituted ketoesters with success, obtaining a family of different highly substituted cyclohexanes with good yields and stereoselectivities. This reaction has been subsequently applied to the asymmetric total synthesis of a natural product such as (–)-*epibatidine*, by a series of simple transformations starting from one of the adducts obtained in the tandem sequence.

On the other hand, a triple Michael/Michael/Aldol cascade sequence has been developed for the synthesis of highly substituted cyclohexanes starting from dimethylmalonate, an α,β-unsaturated aldehyde and a nitroalkene in which H-bonding catalysis and iminium activation were jointly employed for the simultaneous activation of the two Michael acceptors involved in the

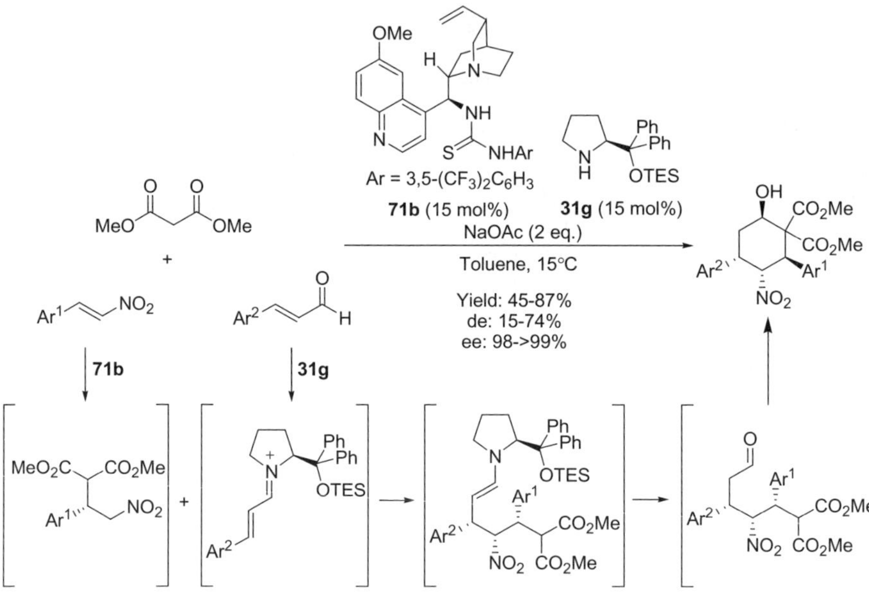

Scheme 7.66 Cascade triple Michael/Michael/aldol reaction combining H-bonding activation and iminium catalysis.

reaction (Scheme 7.66).[95] This cascade process started with the Michael addition of the malonate to the more electrophilic nitroalkene under H-bonding catalysis, leading to the formation of the corresponding Michael adduct, which was a C–H acidic enough pro-nucleophile to participate as Michael donor to the α,β-unsaturated aldehyde *via* the corresponding nitronate under iminium activation of the enal. To finish, after release of the catalyst by hydrolysis, the formed intermediate underwent intramolecular aldol reaction delivering the final cyclohexane-type adducts. After surveying several chiral Brønsted acids and secondary amines as H-bonding and iminium catalysts respectively, it was found that bifunctional cinchona alkaloid-based thiourea **71b** was the most appropriate catalyst for promoting the first Michael addition step, while *O*-TES diphenylprolinol **31g** showed up as the best chiral secondary amine catalyst to engage in the iminium-type activation required for the second Michael reaction. Under the optimized reaction conditions, a variety of different highly functionalized cyclohexanes were obtained in good to moderate yields and excellent enantioselectivities, observing in all the cases the preferential formation of one diastereoisomer out of the eight possible ones.

7.4.3 Michael/α-Alkylation Cascade Reactions

The cyclopropanation of activated olefins by cascade Michael/α-alkylation reaction using α-halomalonates or bromonitromethane as functionalized pro-nucleophiles has also been faced with the application of H-bonding catalysis.

Nevertheless, it has to be said that all the methodologies reported to date involve a tandem rather than a pure cascade process, observing that in a first stage the conjugate addition of the functionalized pro-nucleophile to the activated olefin has been performed under the stereochemical influence of the catalyst and next an external base had to be added to the reaction mixture in order to promote the subsequent intramolecular alkylation step.

For example, cupreidine **84b** has been successfully employed as catalyst in the cyclopropanation of nitroalkenes with dimethyl bromomalonate, using DABCO as the external auxiliary base required to promote the cyclization and leading to the final cyclopropanation product in excellent yields and stereo-selectivities (Scheme 7.67).[96] Alternatively, the cyclopropanation of α,β-unsaturated α-cyanoimides has been carried out using bromonitromethane as the functionalized nucleophile and Takemoto's catalyst **68a**, leading to the formation of cyclopropanes containing three stereogenic centers, one of them being a quaternary one.[97] However, in the latter case, although yields and enantioselectivities were excellent, mixtures of C-1 epimers were obtained.

There is also another version of this reaction which makes use of α,β-unsaturated selenones as substrates and α-aryl substituted cyanoacetates as Michael donors (Scheme 7.68).[98] In this case, the reaction design consisted of the initial Michael addition of the cyanoacetate to the selenone under H-bonding

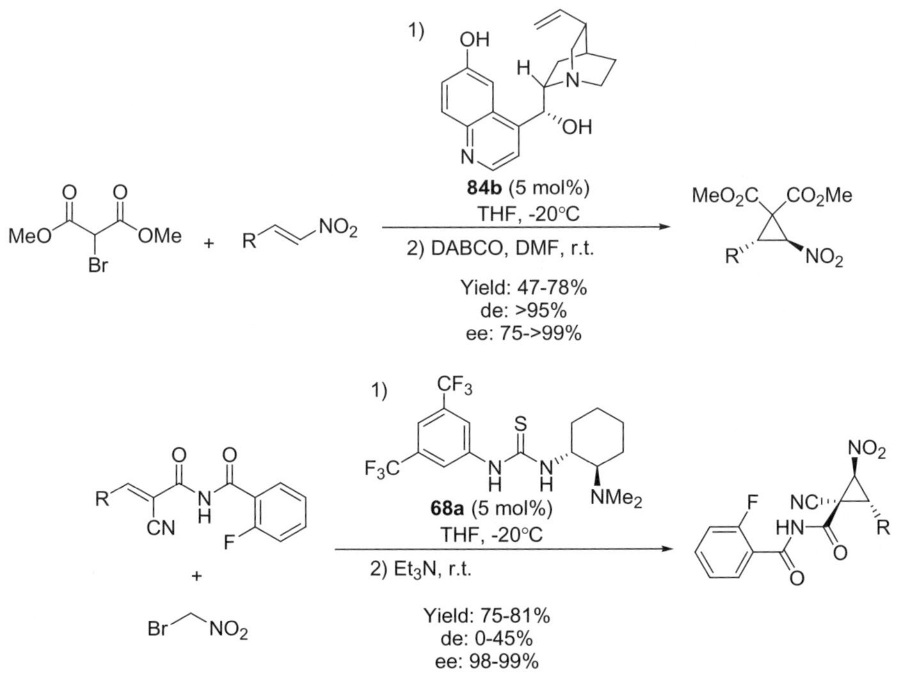

Scheme 7.67 Two examples of cascade Michael/α-alkylation for the synthesis of cyclopropanes.

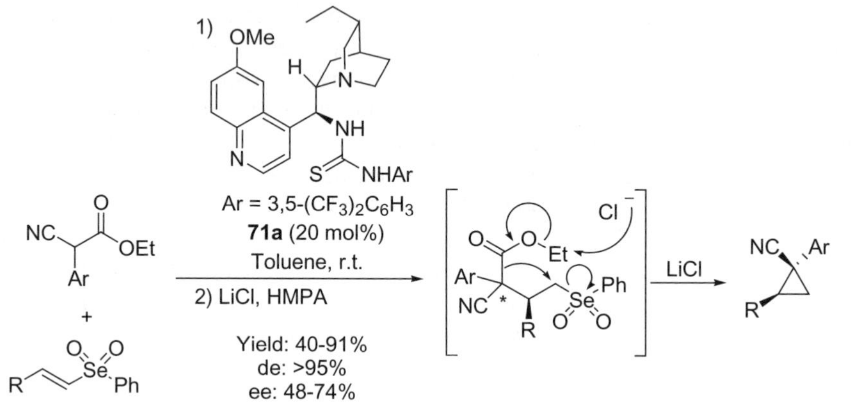

Scheme 7.68 Two examples of cascade Michael/α-alkylation for the synthesis of cyclopropanes.

catalysis and next, in the presence of another nucleophilic reagent, a de-alkoxycarbonylation process together with an intramolecular nucleophilic substitution occurred, with the selenone group itself playing the role of an efficient leaving group in this last cyclization reaction. The best reaction conditions comprised the use of cinchona-thiourea catalyst **71a** for the optimal stereocontrol in the first Michael reaction and next LiCl/HMPA was added to the reaction mixture, the chloride anion being the nucleophilic reagent promoting the decarboxylation/cyclization process. Under these conditions, a series of cyclopropanes were obtained in good to moderate yields and as single diastereoisomers on enantiomeric purities in the range of 48–74% ee.

7.4.4 Other Cascades Initiated by Michael Reactions Using Stabilized Carbon Nucleophiles

A multicomponent reaction involving an α,β-unsaturated aldehyde, a β-ketoester and a primary amine for the synthesis of Hantzsch esters in an analogous sequence to that developed by Jørgensen and depicted in Scheme 7.37 has also been developed using H-bonding catalysis for the activation of the initial Michael acceptor using chiral phosphoric acid ***ent*-60g** as catalyst (Scheme 7.69).[99] In this case, the reaction consisted of a real cascade rather than stepwise process, proceeding by the first condensation of the α,β-unsaturated aldehyde with the primary amine delivering an α,β-unsaturated imine intermediate, which, upon activation with the phosphoric acid catalyst, underwent enantioselective Michael reaction with the ketoester. Finally, this Michael adduct intermediate underwent cyclization delivering the final dihydropyridine skeleton containing a stereocenter. The reaction was found to be rather general in scope with regard to the alkoxy substituent of the β-ketoester and the enal

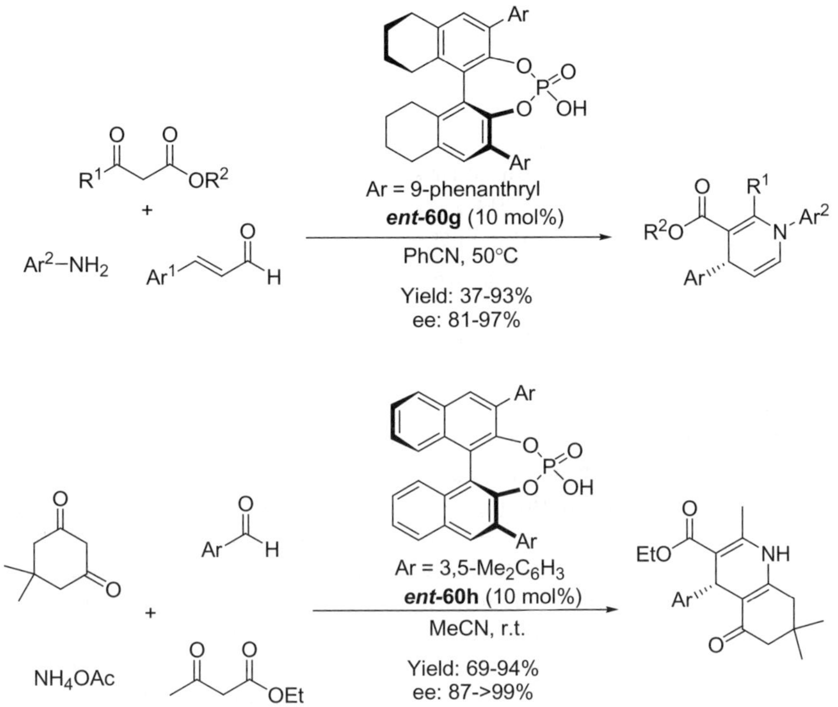

Scheme 7.69 Enantioselective synthesis of 1,4-dihydropyridines by means of cascade processes under H-activation.

reagent, although the reaction proceeded satisfactorily when the latter incorporated a β-aryl substituent, affording poorer results when a β-alkyl substituted α,β-unsaturated aldehyde was employed. There is also a related report which consists of a four component reaction between an aromatic aldehyde, 5,5-dimethyl-1,3-cyclohexanedione, ethyl acetoacetate and ammonium acetate, leading to enantioenriched tetrahydroquinolinones in the presence of phosphoric acid *ent*-**60h** (Scheme 7.67).[100] In this process, a Knoevenagel reaction had to take place in a first stage, condensing the cyclohexanedione with the aromatic aldehyde and generating *in situ* the starting material, an alkylidene-1,3-diketone in this case. This compound had subsequently to undergo condensation with ammonia, followed by a catalyst-controlled Michael addition/intramolecular condensation. Under the optimized conditions, a variety of different dihydroquinolines were obtained with excellent yields and enantioselectivities.

In a completely different context, a cascade Michael/intramolecular addition process involving nitroalkenes and isocyanoesters has been developed, which results in a formal [3 + 2] cycloaddition reaction and therefore consisting of a direct and very efficient methodology for the asymmetric synthesis of 2,3-dihydropyrroles (Scheme 7.70).[101] The reaction was catalyzed by cinchona

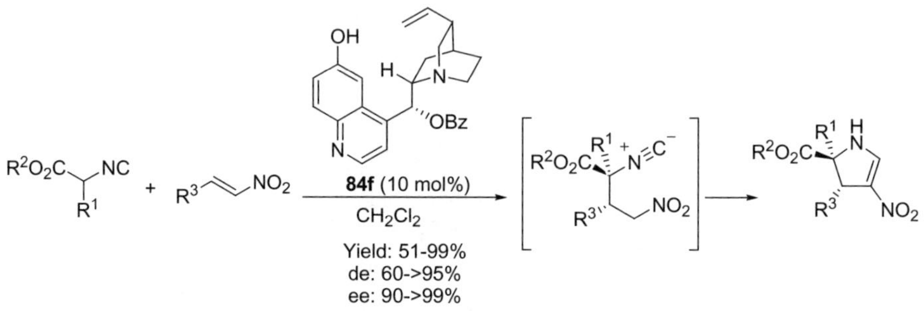

Scheme 7.70 Cascade reaction for the enantioselective synthesis of 3,2-dihydropyrroles.

alkaloid derivative **84f** and key to the success of the process was the bifunctional nature of this catalyst, which activated the cyanoester toward addition by exerting its deprotonation through the quinucludine basic site and simultaneously activated the nitroalkene through the phenolic acidic site. After this initial Michael addition step, the generated enolate-type intermediate incorporated a suitable isonitrile functionality, ready to participate as an electrophile in a subsequent intramolecular reaction, leading to the formation of the final 2,3-dihydropyrrole compounds after protonation. The reaction proceeded satisfactorily with a variety of nitroalkenes, including several β-alkyl substituted substrates as challenging Michael acceptors, although the scope of the reaction was limited to the use of α-substituted isocyanoacetates, as the functionalized Michael donors, reporting that, when the reaction was carried out with the commercially available α-unsubstituted methyl isocyanoacetate, no cycloaddition product was observed.

7.4.5 Cascade Processes Initiated by Hetero-Michael Reactions

Most of the examples of cascade reactions initiated by the conjugate addition of a heteroatom-centered nucleophile to an electron-deficient olefin previously activated by a Brønsted acid catalyst *via* H-bonding interactions have been limited to the use of sulfur-based heteronucleophiles as Michael donors. In particular, the reaction of 2-mercaptobenzaldehydes with different Michael acceptors has been the focus of many studies. This cascade process consists of an initial sulfa-Michael reaction under catalyst control, which is immediately followed by an intramolecular aldol reaction, which normally proceeds in a diastereoselective fashion under substrate control. A representative example is provided in Scheme 7.71, showing the reaction between a series of 2-mercaptobenzaldehydes with different α,β-unsaturated imides using cinchona alkaloid-thiourea catalyst **71b** and leading to the formation of chiral thiochromenes with excellent yields and enantioselectivities.[102] In several different reports, the same reaction has been carried out using alkylidenemalonates as Michael acceptors together with the same catalyst **71b**[103] or, alternatively, employing maleimides

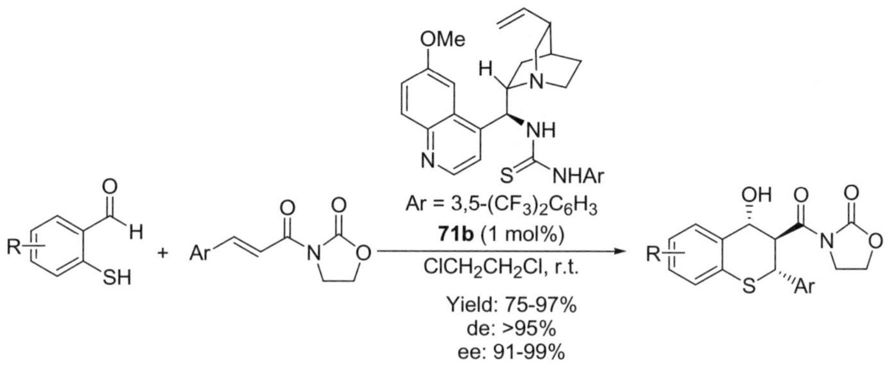

Scheme 7.71 One example of a cascade sulfa-Michael/aldol reaction catalyzed by a chiral thiourea.

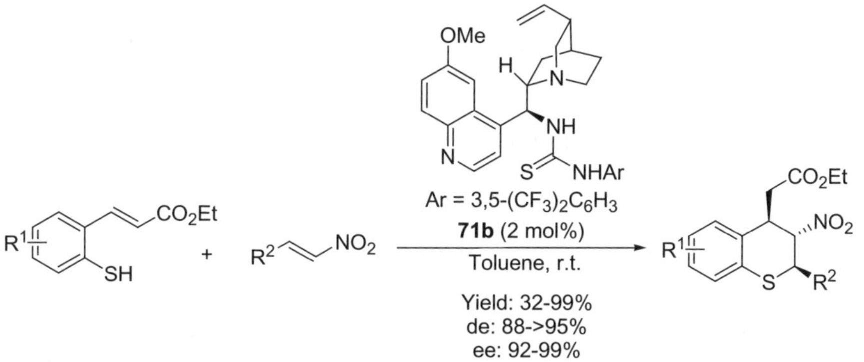

Scheme 7.72 Cascade sulfa-Michael/Michael reaction for the asymmetric synthesis of highly substituted thiochromanes.

as acceptors in the presence of Takemoto's catalyst **68a**.[104] On the other hand, simple cupreidine **84b** has been shown to perform well in the reaction between 2-mercaptobenzaldehydes and nitroalkenes.[105]

An analogous cascade sulfa-Michael/Michael reaction has also been developed using again a conveniently substituted aromatic thiol containing an *ortho*-α,β-unsaturated ester substituent, ready to undergo intramolecular Michael reaction after the first sulfa-Michael addition step had taken place (Scheme 7.72).[106] As happened in the example shown in the previous scheme, cinchonathiourea **71b** was identified as the most efficient catalyst for the reaction of differently substituted *trans*-3-(2-mercaptoaryl)-2-propenoates with a series of nitroalkenes, leading to the formation of thiocromanes in excellent yields and stereoselectivities. Interestingly, the authors demonstrated that the second intramolecular Michael reaction had to be the determining step with regard to stereochemical control, with participation of the catalyst at this point rather than in the initial sulfa-Michael addition step. Consequently, they proposed that a dynamic kinetic resolution process was operating after the first sulfa-Michael reaction, in which a racemic adduct had to be formed but which, after interaction with the catalyst, one of these enantiomers would undergo faster intramolecular Michael reaction than the other, delivering the final cyclic compound with almost full stereocontrol. The remaining enantiomer would also undergo fast racemization *via* a retro-sulfa-Michael/Michael process. This was demonstrated by treating a racemic sample of the sulfa-Michael addition intermediate in the presence of catalyst **71b**, observing the formation of the corresponding thiochromane as a single diastereoisomer of very high enantiomeric purity.

In a different approach, a cascade process initiated by an oxa-Michael reaction has also been described for the synthesis of fluorinated flavanone derivatives (Scheme 7.73).[107] In this case, a conveniently substituted 2-alkenoylphenol reagent was employed as substrate undergoing an intramolecular oxa-Michael reaction which subsequently, and in a tandem sequence, was treated with *N*-fluorobenzenesulfonylimide (NFSI) as a source of electrophilic fluorine reagent, which led to the formation of the final fluorinated flavanone derivatives. This tandem oxa-Michael/electrophilic fluorination process was carried out in the presence of modified quinidine catalyst **83h**, allowing the preparation of a wide range of different adducts in excellent yields and as single diastereoisomers of high enantiomeric purity. In this case, the authors assumed that the initial intramolecular oxa-Michael reaction was the enantiodiscriminating step, proposing a bifunctional activation mechanism for the catalyst analogous to other hetero-Michael reactions catalyzed by this type of phenolic cinchona alkaloids (see Scheme 4.23 in Chapter 4), which involved the activation of the acceptor by H-bonding with the phenolic moiety and the activation of the nucleophile by deprotonation through the quinuclidine nucleus.

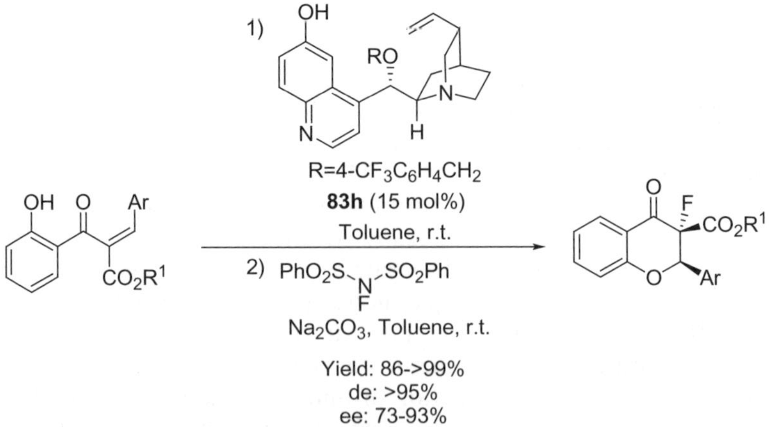

Scheme 7.73 Tandem oxa-Michael/electrophilic fluorination reaction for the asymmetric synthesis of fluorinated flavanones.

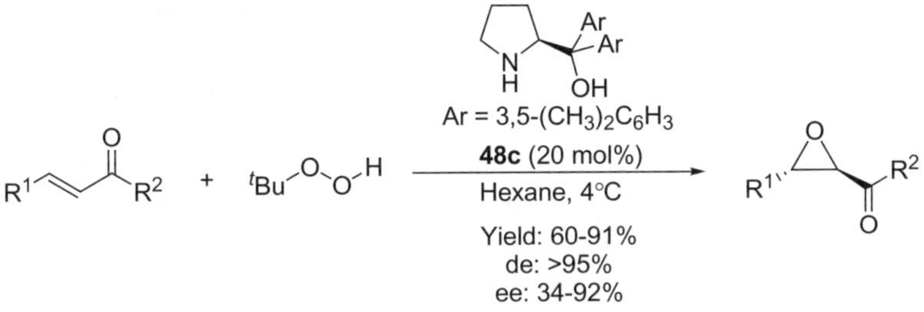

Scheme 7.74 Epoxidation of enones by cascade oxa-Michael/intramolecular nucleophilic displacement.

Finally, the epoxidation of α,β-unsaturated compounds using H_2O_2 or alkyl hydroperoxides as oxidants proceeding *via* cascade oxa-Michael/intramolecular nucleophilic displacement has also been the focus of study by several research groups using the H-bonding activation concept as the vehicle to asymmetric catalysis. In particular, several diarylprolinols have been shown to promote very efficiently the epoxidation of enones proposing that the amino alcohol catalyst had to be involved in activation of the enone by H-bonding interaction between the OH of the catalyst and the carbonyl group, while the remaining secondary amino group would contribute to the reaction by enhancing the reactivity of the peroxide by a secondary Brønsted acid/base interaction.[108] A representative example of this chemistry is provided in Scheme 7.74, in which the epoxidation of several enones was carried out successfully with *tert*-butyl hydroperoxide in the presence of diarylprolinol **48c** as catalyst.[108b] Other different β-amino alcohols[109] and a guanidine-urea bifunctional catalyst[110] have also been tested as catalysts in this context. There is also a related example in which the aziridination of enones has been carried out by a similar aza-Michael/intramolecular nucleophilic substitution process using quinine as catalyst but furnishing low to moderate enantioselectivities.[111]

7.5 Cascade Processes Initiated by Conjugate Addition *via* Phase-transfer Catalysis

Several very interesting cascade processes have also been developed under phase-transfer catalysis which are initiated by a conjugate addition reaction, although the number of reports is remarkably more limited compared to other organocatalytic cascades proceeding *via* other different mechanisms of activation. Representative examples will be presented in the following pages.

7.5.1 Michael/Elimination Cascade Reactions

The most widely studied type of cascade initiated by a conjugate addition proceeding under phase-transfer catalysis conditions involves the use of either a nucleophile or a Michael acceptor containing a good leaving group able to undergo an elimination process after the initial conjugate addition step. A good example is the formal nucleophilic vinylic substitution developed by Jørgensen and shown in Scheme 7.73. In this reaction, a β-chloro-substituted α,β-unsaturated ester is employed as Michael acceptor in the reaction with β-ketoesters using chiral cinchoninium quaternary ammonium salt **67d** as catalyst and leading to the formation of alkenyl-substituted â-keto esters containing a quaternary stereocenter in a clean and very efficient way (Scheme 7.75).[112] Important features associated to this methodology can be found not only in the excellent enantioselectivity achieved but also in the very high degree of diastereocontrol in the formation of the C=C double bond present at the final compounds, for which an outstanding Z/E-selectivity was observed in all cases tested, resulting in an overall stereoretentive vinylic substitution process at the chlorine substituent. This high diastereoselectivity was interpreted in terms of a fast elimination reaction after the initial Michael addition, in which the stereoelectronic constraints associated with the elimination process were fulfilled after a short bond rotation and therefore the elimination step had to take place much faster than a longer rotation around this C–C bond, which would lead to the formation of the opposite diastereoisomer. Interestingly, access to the E diastereoisomers from the Z adducts could also be achieved by a simple isomerization process catalyzed by tributylphosphine.

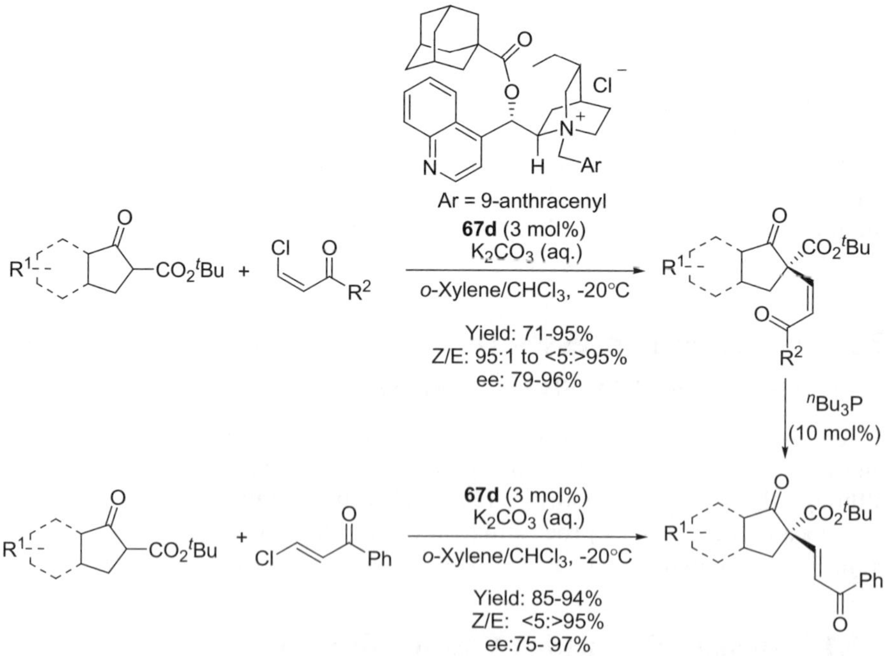

Scheme 7.75   Formal **67d**-catalyzed vinylic substitution on β-chloro α,β-unsaturated carbonyl compounds.

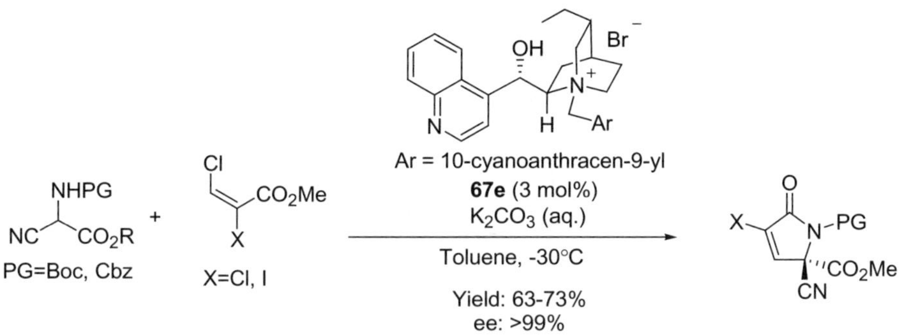

Scheme 7.76 Asymmetric synthesis of γ-lactams by cascade formal vinylic substitution/lactamization.

This vinylic substitution methodology has also been subsequently applied to the development of a very efficient protocol for the asymmetric synthesis of highly functionalized γ-lactams using *N*-protected 2-aminocyanoacetates as Michael donors, by promoting a cascade Michael/elimination/lactamization process (Scheme 7.76).[113] In this case, a different catalyst **67e** had to be employed, which led to the isolation of a series of substituted γ-lactams in good yields and enantioselectivities around 80% ee, which could be improved to >99% ee after recrystallization. The presence of a vinylic bromide or iodide at the final products allowed the authors to explore further synthetic transformations on these compounds in order to illustrate their utility as chiral building blocks, in particular carrying out several Pd-catalyzed cross-coupling reactions which allowed the introduction of an additional carbon substituent at this position of the γ-lactam framework.

On the other hand, the same authors have also employed β-chloro- or β-bromo-substituted acetylenic ketones as Michael acceptors in a Michael/elimination process, leading to the formation of β-alkynyl-substituted β-ketoesters in what can be considered a formal acetylenic substitution reaction (Scheme 7.74).[114] In this case, adamantoyl-containing catalyst **67a** initially employed in the vinylic substitution reaction displayed in Scheme 7.73 was identified as the best catalyst, leading to the formation of the final products with excellent yields and enantioselectivities (Scheme 7.77). Alternatively, the same authors have also used β-cyanosulfones as Michael acceptors in the reaction with cyclic β-ketoesters, in which the cascade process consisted of the initial Michael reaction followed by elimination, the sulfone moiety playing the role of the leaving group (Scheme 7.74).[115] Interestingly, the better abilities of the sulfone group as olefin-activating group directed the regioselectivity of the 1,4-addition process toward the β-carbon with respect to this sulfone group, which, after elimination, led to the formation of a final product resulting from the formal substitution of the α-hydrogen atom of acrylonitrile with the β-ketoester enolate, also referred to as an "*anti*-Michael" addition. The best conditions comprised the use of chiral ammonium salt **67d** as catalyst and

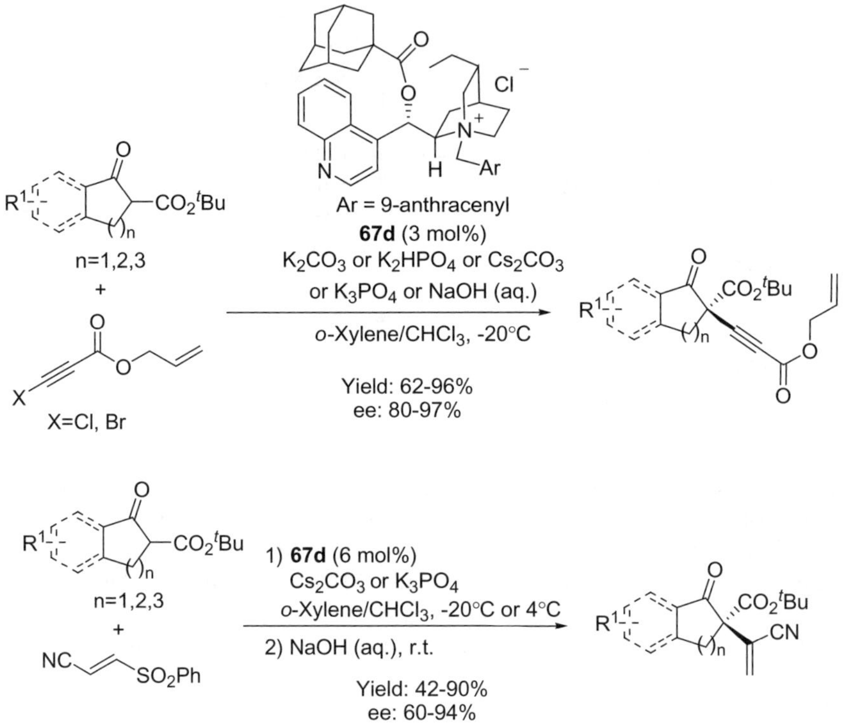

Scheme 7.77 Enantioselective alkynylation of β-ketoesters and the formal *anti*-Michael addition of β-ketoesters to acrylonitrile.

different inorganic bases depending on the structure of the β-ketoester employed and, in all cases, the elimination step had to be carried out in a one-pot fashion, by the addition of an excess of NaOH after consumption of the starting material was observed.

Finally, there is also a contribution from a different research group in which activated allylic acetates were reacted with the benzophenone imine of *tert*-butyl glycinate in the presence of cinchonidinium salt **103a** leading to the formation of 4-alkylidene substituted glutamic acid derivatives in good yields and enantioselectivities (Scheme 7.78).[116] This transformation consisted of a cascade sequence involving the initial Michael reaction under PTC conditions, using CsOH as the base required for the formation of the enolate derived from the glycinate imine reagent, which was followed by the elimination of the acetate promoted by CsOH. The reaction also proceeded with complete diastereoselection regarding the new C=C double bond formed during the elimination process, which should be attributed to the stereochemical constraints associated to the usual bimolecular mechanism associated to base-mediated elimination reactions.

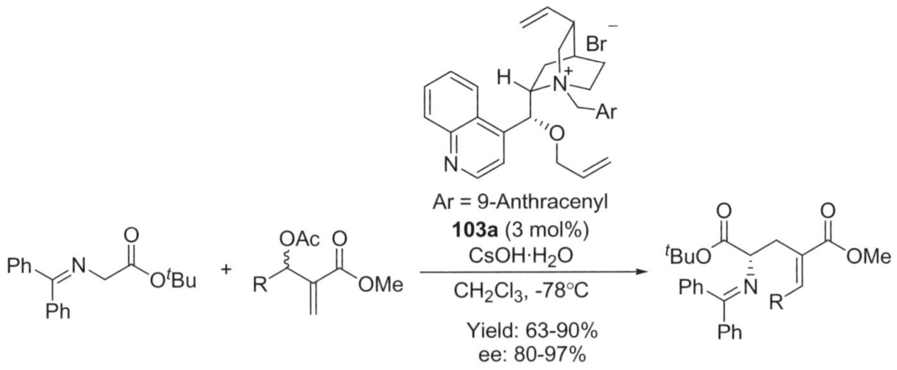

Ar = 9-Anthracenyl
103a (3 mol%)
CsOH·H$_2$O

CH$_2$Cl$_2$, -78°C

Yield: 63-90%
ee: 80-97%

Scheme 7.78 Enantioselective synthesis of 4-alkylidene glutamic acids by cascade Michael/elimination process.

7.5.2 Other Cascades Initiated by Michael Reactions using Stabilized Carbon Nucleophiles

There are a couple of examples of cascade processes starting by a Michael-type addition of a carbon nucleophile proceeding under phase-transfer catalysis conditions which deserve to be mentioned at this point. The first one consists of an enantioselective cyclopropanation of 2-bromocyclopentenone by a cascade Michael/intramolecular nucleophilic displacement in which a variety of C–H acidic carbon pro-nucleophiles such as nitromethane, cyanomethylsulfone and benzyl cyanoacetate reacted with this Michael acceptor in the presence of a quinidinium salt of type **67** (Scheme 7.79).[117] In addition, the conditions needed to be optimized for each Michael donor employed, requiring a different catalyst and inorganic base for each case. Under the best conditions, the final cyclopropanes were obtained in moderate yields and enantioselectivities, albeit as single diastereoisomers.

The other relevant report refers to a Mukaiyama–Michael/lactonization process developed for the synthesis of 3,4-dihydropyran-2-ones (Scheme 7.80).[118] Interestingly, in this case an ammonium phenoxide **103h** was employed as the phase-transfer catalyst, a most important contribution that is associated to the fact that there is no need to incorporate any external inorganic base to the reaction design, the phenoxide counterion of the ammonium salt catalyst, the basic species, being the reagent involved in the activation of the silyl enol ether nucleophile. In this reaction, the alkoxy substituent of the silyl ketene acetal was shown to play an important role with regard to stereocontrol, requiring the incorporation of bulky groups such as *tert*-butoxy, *iso*-propoxy or the rather exotic 2-*iso*-propylphenyl substituents at this position of the silyl enol ether nucleophile. After identifying the best reaction conditions, a set of different enones and silyl ketene acetals was reacted with each other delivering the target dihydropyran-2-ones in excellent yields and as single diastereoisomers of very high optical purity.

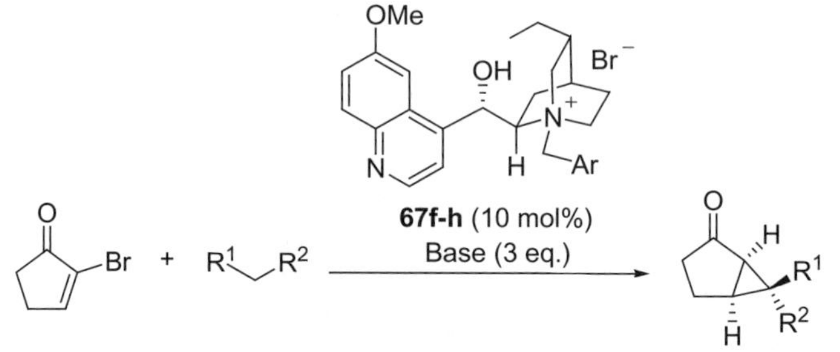

R^1=H, R^2=NO_2	R^1=CN, R^2=SO_2Ph	R^1=CN, R^2=CO_2Bn
67f: Ar = C_6F_5	**67g**: Ar = 2,4-$(CH_3)_2C_6H_3$	**67h:** Ar = 2,4-$(CF_3)_2C_6H_3$
+Rb_2CO_3 (3 eq.)	+K_2CO_3 (1.6 eq.)	+K_2CO_3 (1.6 eq.)
(in THF, r.t.)	(in toluene, 4°C)	(in C_6H_5Cl, r.t.)
Yield: 50%	Yield: 60%	Yield: 60%
de: >95%	de: >95%	de: >95%
ee: 62%	ee: 60%	ee: 83%

Scheme 7.79 Asymmetric cyclopropanation of 2-bromocyclopentenone by cascade Michael/α-alkylation.

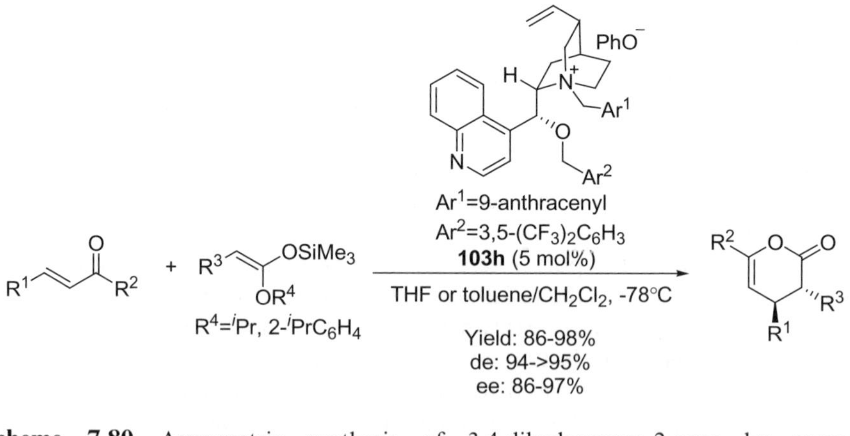

Scheme 7.80 Asymmetric synthesis of 3,4-dihydropyran-2-ones by cascade Mukaiyama–Michael/lactonization.

7.5.3 Cascade Processes Initiated by Hetero-Michael Reactions

Asymmetric phase-transfer catalyzed cascades initiated by the conjugate addition of heteronucleophiles has been employed for the enantioselective preparation of epoxides and aziridines using a suitable oxygen- or nitrogen-based nucleophile incorporating an appropriate leaving group ready to

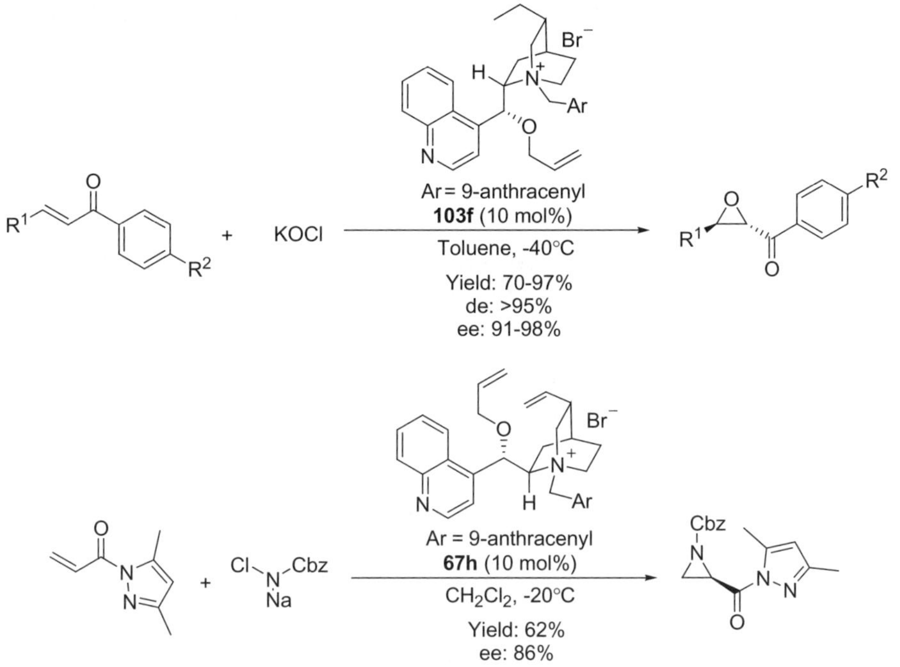

Scheme 7.81 Asymmetric epoxidation and aziridination reactions by cascade hetero-Michael/cyclization under PTC conditions.

undergo an intramolecular nucleophilic displacement after the hetero-Michael addition step. In this sense, Corey has reported the **103f**-catalyzed enantioselective epoxidation of enones using potassium hypochlorite as the oxidant (Scheme 7.81).[119] The formation of a tightly bonded ion pair between the ammonium salt catalyst and the hypochlorite anion, which approached the enone through a well-defined orientation, accounts for the highly enantioselective initial oxa-Michael addition step and the subsequent cyclization had to take place fast and in a fully diastereoselective way, affording exclusively the final *trans*-epoxides in excellent yields and enantioselectivities. The reaction was tested on a variety of aryl alkenyl ketones, but no data were reported about the performance of the system when simple aliphatic enones were employed. On the other hand, an example of a related aziridination using *N*-chloro-*N*-sodio benzyl carbamate as the functionalized nitrogen nucleophile has been reported, although limited to the use of *N*-propenoyl imides and amides as substrates. The best results were obtained on the aziridination of *N*-propenoyl-3,5-dimethylpyrazole in the presence of cinchonine-derived ammonium salt **67h** catalyst (Scheme 7.81).[120]

7.6 Cascade Processes Initiated by Conjugate Addition *via* Other Mechanisms of Activation

To end this chapter, other cascade processes initiated by organocatalytic conjugate additions proceeding *via* other types of activation mode not covered in the previous sections will be presented.

7.6.1 Cascade Processes Initiated by *N*-heterocyclic Carbene-mediated Conjugate Additions

The ability of *N*-heterocyclic carbenes to activate α,β-unsaturated carbonyl compounds *via* the formation of the corresponding Breslow intermediate, which plays the role of a homoenolate nucleophile, has also been applied to a cascade process involving a formal intramolecular Michael reaction/oxidation/lactonization, leading to the formation of complex tricyclic carbon frameworks starting from a bifunctional substrate containing an enone and an α,β-unsaturated aldehyde side chain linked to each other *via* a benzene tether (Scheme 7.82).[121] The reaction involved a complex multistep mechanism which started with the activation of the enal by the catalyst, forming the Breslow intermediate, which subsequently underwent intramolecular Michael reaction and next the generated enol-type intermediate reacted intramolecularly with the

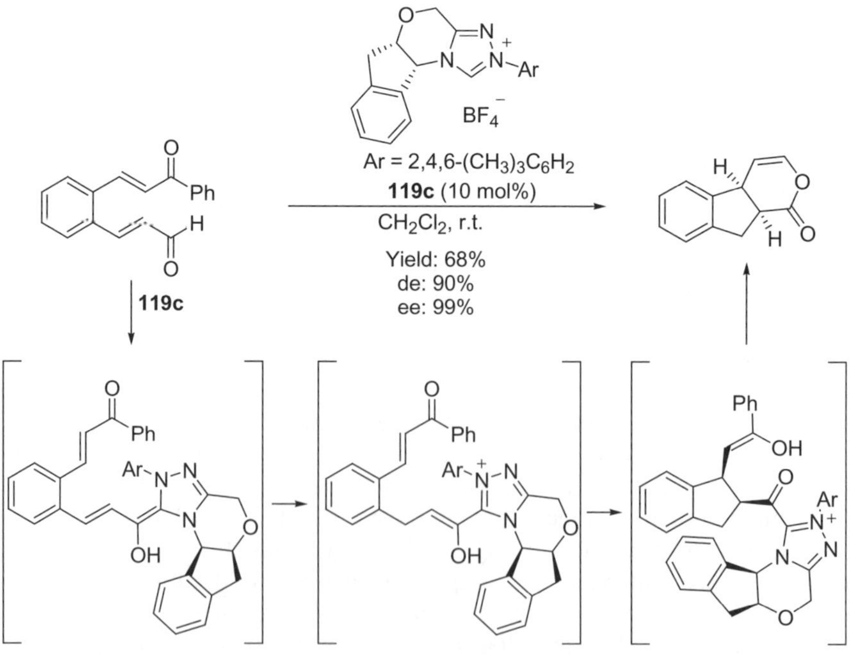

Scheme 7.82 Cascade intramolecular Michael reaction/oxidation/lactonization catalyzed by an *N*-heterocyclic carbene.

acyl-azolium moiety, releasing the catalyst and delivering the final product. An overall evaluation of the process indicates that the enal has reverted its reactivity profile from being an a^3-type electrophile to a d^3-type nucleophile and also that the formyl group had suffered a formal oxidation process appearing as an enol ester in the final compound. Under the best reaction conditions, which consisted of the use of triazolium salt **119c** as pre-catalyst, Hünig base for the generation of the carbene and CH_2Cl_2 as the solvent, the final tricyclic compound was obtained in good yield, as a single diastereoisomer and in 99% enantiomeric excess.

7.6.2 Cascade Processes Initiated by Conjugate Additions Proceeding *via* Ylide Formation

There is an excellent and very efficient organocatalytic methodology for the enantioselective cyclopropanation of α,β-unsaturated enones, which involves a different mechanism of activation which has not already been covered in the previous sections or chapters. This is based on the possibility of chiral tertiary amines to activate α-halo carbonyl compounds toward their participation as nucleophiles in a given reaction *via* N-alkylation, which delivers a nitrogen ylide intermediate after deprotonation with a suitable external base incorporated in the reaction scheme.[122] This ammonium ylide shows up as an excellent chiral nucleophile which can undergo conjugate addition in the presence of a suitable Michael acceptor, which, subsequently, after intramolecular nucleophilic substitution would deliver a final compound with a cyclopropane structure and would also allow catalyst turnover. This concept was developed by Gaunt and Ley using the simple O-methylquinidine **84g** as catalyst in an initial attempt focused on the cyclopropanation of *tert*-butyl acrylate with α-bromoacetophenone[123a] and was afterwards successfully extended to a variety of enones and α-bromoesters and amides, also using catalyst **84g** (Scheme 7.83),[123b] although dimeric cinchona alkaloid (DHQ)$_2$PHAL **124a** was also successfully employed as an alternative in the cases in which **84g** failed to give good results. Soon afterwards, the same authors also developed a very efficient intramolecular version of this reaction which also allowed the direct and easy preparation of cyclopropane-containing bicyclic carbon frameworks in a highly stereocontrolled way.[124]

Related cyclopropanations have also been reported using sulfonium and telluronium ylides as intermediates. In particular, the cyclopropanation of enones has been carried out employing an allyl bromide as the cyclopropanating reagent and sulfonium and telluronium salts **134** and **135** as pre-catalysts (Scheme 7.84).[125] These species, in the presence of a base, generated the corresponding ylide which underwent the cascade Michael/intramolecular nucleophilic substitution and it is in this second step that the real catalytically active species is released, able to interact with another molecule of the allyl bromide and thus regenerating the sulfonium or telluronium salts pre-catalysts, which can afterwards continue in the catalytic cycle. The substitution at the

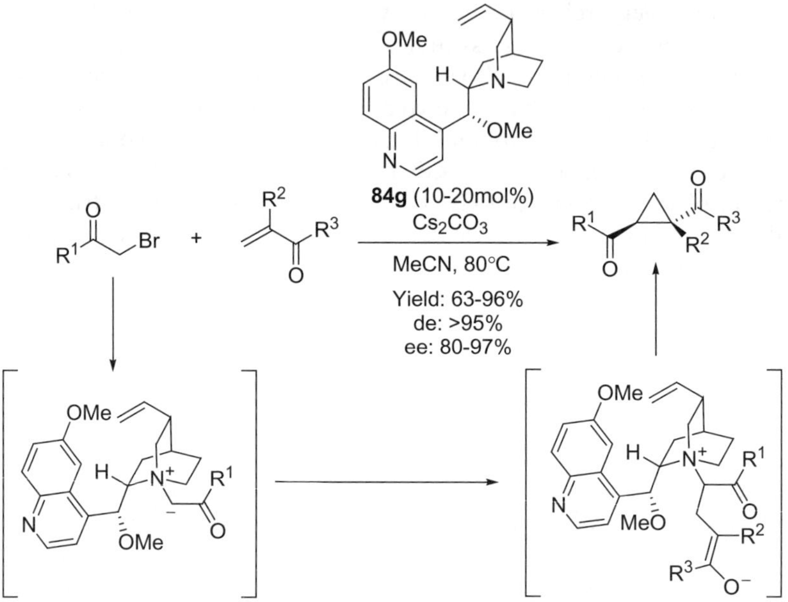

Scheme 7.83 Enantioselective cyclopropanation of enones *via* nitrogen ylides.

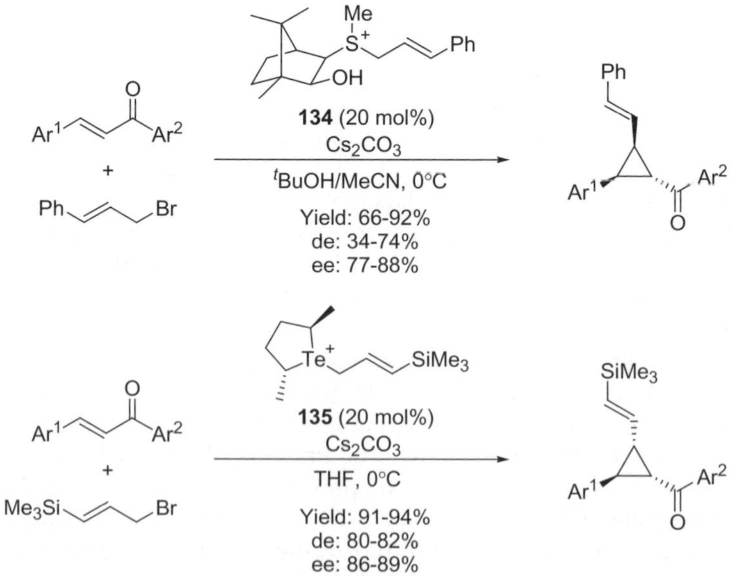

Scheme 7.84 Enantioselective cyclopropanation of chalcones *via* sulfonium and telluronium ylides.

sulfur or the tellurium atom was employed to introduce chirality, which allowed the modulation of the stereochemical outcome of the reaction. The performance of sulfonium ylide **134** was studied using a series of different chalcones as substrates, delivering the final cyclopropanation products containing three stereocenters in good yields although with moderate diastereoselectivities and obtaining typically a mixture of two diastereoisomers regarding the relative *syn/anti* configuration between the alkenyl substituent coming from the ylide and the carbonyl moiety. The major diastereoisomers were isolated with enantioselectivities in the range of 77–88% ee. The tellurium ylide-mediated cyclopropanation proceeded similarly, although in this case the major diastereoisomer formed resulted in being opposite to that obtained in the sulfonium ylide-mediated reaction.

7.6.3 Other Miscellaneous Cascade Processes

There are a few additional organocatalytic enantioselective methodologies involving a cascade process initiated by conjugate addition reaction that deserve to be mentioned and which can not be clearly classified into any of the previous sections. This is the case of the epoxidation of α,β-unsaturated carbonyl compounds by small peptides originally developed by Juliá and Colonna in the early 1980s. This methodology has become very often the reaction of choice for the synthesis of epoxides in an enantiopure form because of the high stereoselectivities obtained, the ready accessibility of the catalysts and the mild reaction conditions. In the first version, solid poly-L-alanine **136** was found to promote efficiently the epoxidation of chalcones in the presence of a H_2O_2/aqueous NaOH/toluene mixture (Scheme 7.85).[126] Further work led to the extension of the scope of the reaction to other Michael acceptors and also to the identification of other catalysts with comparable or improved performances in the reaction, which also included the development of solid supported catalyst variants and the possibility of using other different oxidants.[127]

In a different context, Tröger's base **137** has been identified as a suitable catalyst for the enantioselective aziridination of chalcones using *O*-mesitylenesulfonylhydroxylamine as the functionalized nucleophile undergoing the

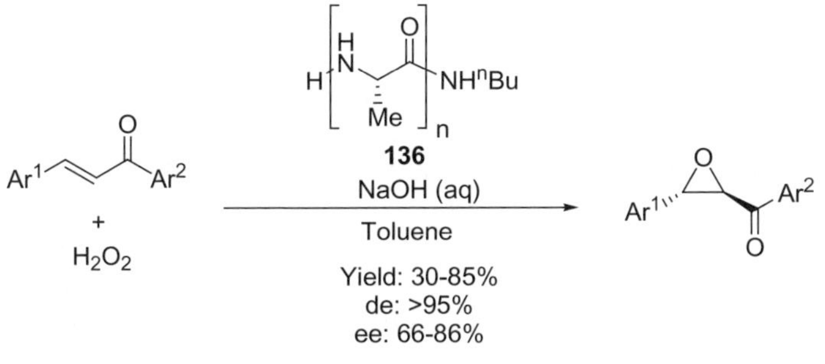

Scheme 7.85 Enantioselective Juliá–Colonna epoxidation of chalcones.

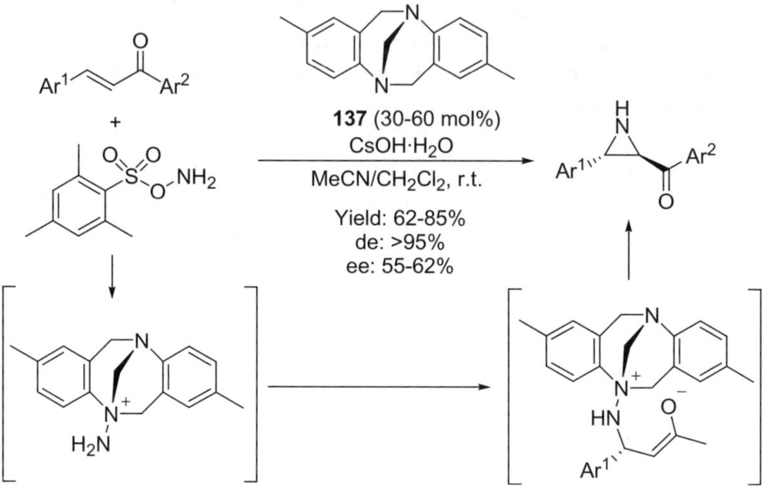

Scheme 7.86   Enantioselective aziridination of chalcones catalyzed by Tröger base.

initial aza-Michael reaction followed by ring closure (Scheme 7.86).[128] The mechanism of the reaction involved the activation of this nucleophile by the chiral tertiary amine catalyst *via* formation of a hydrazinium salt intermediate, which was proposed to be the real nucleophilic species undergoing the Michael addition and subsequent intramolecular nucleophilic substitution, releasing the catalyst ready to participate in the next catalytic cycle. High yields and a fully diastereoselective reaction were observed, although enantioselectivities remained in moderate values.

7.7   Concluding Remarks

The organocatalytic conjugate addition reaction shows up as an extremely powerful motif for initiating cascade processes which allow the preparation of complex organic compounds in a simple and easy way. Across the examples presented in this chapter, the high efficiency of the different organocatalytic approaches developed has been demonstrated, in terms of the number of bonds and stereocenters that can be simultaneously generated in a well-ordered and stereocontrolled way. Without doubt, among all the activation types by which organocatalysts can participate in a cascade process initiated by a conjugate addition reaction, the iminium/enamine manifold has been the most extensively studied and has reached an exceptional level of sophistication, enabling the assembly of many different substrates into a final complex carbo- or hetero-cyclic structure either cyclic or acyclic in nature and containing multiple functionalities and stereogenic centers. The other activation modes have also found applicability in many other different cases, showing that exceptionally efficient methodologies have also been developed in this context, which, in some cases, fill perfectly the gap left by the iminium/enamine activation profile in terms of the wider tolerance with regard to the Michael acceptor employed, which in the former case is limited to the use of enones and enals. Moreover, the possibilities of these approaches to their application in the synthesis of bioactive compounds has also been faced in several representative examples, showing the power of this methodology even for future industrial applications.

References and Notes

1. For some selected reviews see (a) N. Ismabery and R. Lavila, *Chem. Eur. J.*, 2008, **14**, 8444; (b) B. Albert and K. Scott, *Tetrahedron*, 2007, **63**, 5341; (c) C. J. Chapman and C. G. Frost, *Synthesis*, 2007, 1; (d) H. Miyabe and Y. Takemoto, *Chem. Eur. J.*, 2007, **13**, 7280; (e) N. T. Patil and Y. Yamamoto, *Synlett*, 2007, 994; (f) J. Zhu, H. Bienayme (ed.), *Multicomponent Reactions*, WILEY-VCH, Weinheim, 2005; (g) D. Tejedor and F. Garcia-Tellado, *Chem. Soc. Rev.*, 2007, **36**, 484; (h) L. F. Tietze, G. Brasche and K. Gerike, *Domino Reactions in Organic Chemistry*, WILEY-VCH, Weinheim, 2006; (i) H. Pellissier, *Tetrahedron*, 2006, **62**, 1619; (j) K. C. Nicolaou, D. J. Edmonds and P. G. Bulger, *Angew. Chem.*, 2006, **118**, 7292; *Angew. Chem. Int. Ed.*, 2006, **45**, 7134; (k) H. Pellissier, *Tetrahedron*, 2006, **62**, 2143; (l) A. Doemling, *Chem. Rev.*, 2006, **106**, 17; (m) H.-C. Guo and J.-A. Ma, *Angew. Chem.*, 2006, **118**, 362 *Angew. Chem. Int. Ed.*, 2006, **45**, 354; (n) D. J. Ramon and M. Yus, *Angew. Chem.*, 2005, **117**, 1628 *Angew. Chem. Int. Ed.*, 2005, **44**, 1602; (o) J.-C. Wasilke, S. J. Obrey, R. T. Baker and G. C. Bazan, *Chem. Rev.*, 2005, **105**, 1001.
2. (a) X. Yu and W. Wang, *Org. Biomol. Chem.*, 2008, **6**, 2037; (b) D. Enders, C. Grondal and M. R. M. Huettl, *Angew. Chem. Int. Ed.*, 2007, **46**, 1570; (c) G. Guillena, D. J. Ramon and M. Yus, *Tetrahedron: Asymmetry*, 2007, **18**, 693.
3. Y. Hayashi, T. Okano, S. Aratake and D. Hazelard, *Angew. Chem. Int. Ed.*, 2007, **46**, 4922.
4. G.-L. Zhao, P. Dziedzic, F. Ullah, L. Eriksson and A. Córdova, *Tetrahedron Lett.*, 2009, **50**, 3458.
5. B.-C. Hong, R. Y. Nimje, A. A. Sadani and J.-S. Liao, *Org. Lett.*, 2008, **10**, 2345.
6. (a) A. E. Asato, C. Watanabe, X.-Y. Li and R. S. H. Liu, *Tetrahedron Lett.*, 1992, **33**, 3105; (b) B. J. Bench, C. Liu, C. R. Evett and C. M. H. Watanabe, *J. Org. Chem.*, 2006, **71**, 9458; (c) B.-C. Hong, M.-F. Wu, H.-C. Tseng and J.-H. Liao, *Org. Lett.*, 2006, **8**, 2217; (d) B.-C. Hong, M.-F. Wu, H.-C. Tseng, G.-F. Huang, C.-F. Su and J.-H. Liao, *J. Org. Chem.*, 2007, **72**, 8459.
7. B.-C. Hong, R. Y. Nimje, M.-F. Wu and A. A. Sadami, *Eur. J. Org. Chem.*, 2008, 1449.
8. B.-C. Hong, R. Y. Nimje and J.-H. Liao, *Org. Biomol. Chem.*, 2009, **7**, 3095.

9. C.-L. Cao, Y.-Y. Zhou, J. Zhou, X.-L. Sun, Y. Tang, Y.-X. Li, G.-Y. Li and J. Sun, *Chem. Eur. J.*, 2009, **15**, 11384.

10. L.-Y. Wu, G. Bencivenni, M. Mancinelli, A. Mazzanti, G. Bartoli and P. Melchiorre, *Angew. Chem. Int. Ed.*, 2009, **48**, 7196.

11. D. Enders, M. R. R. Hüttl, C. Grondal and G. Raabe, *Nature*, 2006, **441**, 861.

12. D. Enders, M. R. M. Hüttl, G. Raabe and J. W. Bats, *Adv. Synth. Catal.*, 2008, **350**, 267.

13. M. C. Varela, S. M. Dixon, K. S. Lam and N. E. Schore, *Tetrahedron*, 2008, **64**, 10087.

14. D. Enders, M. R. R. Hüttl, J. Runsink, G. Raabe and B. Wendt, *Angew. Chem. Int. Ed.*, 2007, **46**, 467.

15. O. Penon, A. Carlone, A. Mazzanti, M. Locatelli, L. Sambri, G. Bartoli and P. Melchiorre, *Chem. Eur. J.*, 2008, **14**, 4788.

16. G. Bencivenni, L.-Y. Wu, A. Mazzanti, B. Giannichi, F. Pesciaioli, M.-P. Song, G. Bartoli and P. Melchiorre, *Angew. Chem. Int. Ed.*, 2009, **48**, 7200.

17. H. Ishikawa, T. Suzuki and Y. Hayashi, *Angew. Chem. Int. Ed.*, 2009, **48**, 1304.

18. D. Zhu, M. Lu, L. Dai and G. Zhong, *Angew. Chem. Int. Ed.*, 2009, **48**, 6089.

19. S. Belot, K. A. Vogt, C. Besnard, N. Krause and A. Alexakis, *Angew. Chem. Int. Ed.*, 2009, **48**, 8923.

20. In 2000, Barbas carried out the Robinson annulation between 2-methyl-1,3-cyclohexanedione and methyl vinyl ketone catalyzed by several chiral amines, reporting that the process stopped in many cases after the conjugate addition step. However, neither yields nor enantioselectivities were given for those cases. T. Bui and C. F. Barbas III, *Tetrahedron Lett.*, 2000, **41**, 6951.

21. (a) D. B. Ramachary and M. Kishor, *J. Org. Chem.*, 2007, **72**, 5056; (b) D. B. Ramachary and M. Kishor, *Org. Biomol. Chem.*, 2008, **6**, 4176.

22. (a) N. Halland, P. S. Aburel and K. A. Jørgensen, *Angew. Chem. Int. Ed.*, 2004, **43**, 1272; (b) J. Pulkkinen, P. S. Aburel, N. Halland and K. A. Jørgensen, *Adv. Synth. Catal.*, 2004, **346**, 1077; (c) D. Gryko, *Tetrahedron: Asymmetry*, 2005, **16**, 1377.

23. Y.-Q. Yang, Z. Chai, H.-F. Wang, X.-K. Chen, H.-F. Cui, C.-W. Zheng, H. Xiao, P. Li and G. Zhao, *Chem. Eur. J.*, 2009, **15**, 13295.

24. A. Carlone, M. Marigo, C. North, A. Landa and K. A. Jørgensen, *Chem. Commun.*, 2006, 4928.

25. P. Bolze, G. Dickmeiss and K. A. Jørgensen, *Org. Lett.*, 2008, **10**, 3753.

26. M. Marigo, S. Bertelsen, A. Landa and K. A. Jørgensen, *J. Am. Chem. Soc.*, 2006, **128**, 5475.

27. M. Rueping, A. Kuenkel, F. Tato and J. W. Bats, *Angew. Chem. Int. Ed.*, 2009, **48**, 1699.

28. J. Wang, H. Li, H. Xie, L. Zu, X. Shen and W. Wang, *Angew. Chem. Int. Ed.*, 2007, **46**, 9050.

29. (a) D. Enders, A. A. Narine, T. R. Benninghaus and G. Raabe, *Synlett*, 2007, 1667; (b) D. Enders, C. Wang and J. W. Bats, *Synlett*, 2009, 1777.
30. E. Reyes, H. Jiang, A. Milelli, P. Elsner, R. G. Hazell and K. A. Jørgensen, *Angew. Chem. Int. Ed.*, 2007, **46**, 9202.
31. Y. Hayashi, M. Toyoshima, H. Gotoh and H. Ishikawa, *Org. Lett.*, 2009, **11**, 45.
32. L. Albrecht, B. Richter, C. Vila, H. Krawczyk and K. A. Jørgensen, *Chem. Eur. J.*, 2009, **15**, 3093.
33. S. Bertelsen, R. L. Johansen and K. A. Jørgensen, *Chem. Commun.*, 2008, 3016.
34. J. L. García Ruano, V. Marcos, J. A. Suanzes, L. Marzo and J. Alemán, *Chem. Eur. J.*, 2009, **15**, 6576.
35. L. Zu, H. Li, H. Xie, J. Wang, W. Jiang, Y. Tang and W. Wang, *Angew. Chem. Int. Ed.*, 2007, **46**, 3732.
36. G.-L. Zhao, I. Ibrahem, P. Dziedzic, J. Sun, C. Bonneau and A. Córdova, *Chem. Eur. J.*, 2008, **14**, 10007.
37. (a) T.-R. Kang, J.-W. Xie, W. Du, X. Feng and Y.-C. Chen, *Org. Biomol. Chem.*, 2008, **6**, 2673 See also; (b) J. W. Xie, W. Chen, R. Li, M. Zeng, W. Du, L. Yue, Y.-C. Chen, Y. Wu, J. Zhu and J.-G. Deng, *Angew. Chem. Int. Ed.*, 2007, **46**, 389.
38. S. Cabrera, J. Alemán, P. Bolze, S. Bertelsen and K. A. Jørgensen, *Angew. Chem. Int. Ed.*, 2008, **47**, 121.
39. M.-K. Zhu, Q. Wei and L.-Z. Gong, *Adv. Synth. Catal.*, 2008, **350**, 1281.
40. P. G. McGarraugh and S. E. Brenner, *Org. Lett.*, 2009, **11**, 5654.
41. A. Carlone, S. Cabrera, M. Marigo and K. A. Jørgensen, *Angew. Chem. Int. Ed.*, 2007, **46**, 1101.
42. D. Enders, M. Jeanty and J. W. Bats, *Synlett*, 2009, 3175.
43. (a) H. M. Hansen, D. A. Longbottom and S. V. Ley, *Chem. Commun.*, 2006, 4838; (b) V. Wascholowski, H. M. Hansen, D. A. Longbottom and S. V. Ley, *Synthesis*, 2008, 1269.
44. J. Lv, J. Zhang, Z. Lin and Y. Wang, *Chem. Eur. J.*, 2009, **15**, 972.
45. (a) J. Vesely, G.-L. Zhao, A. Bartoszewicz and A. Córdova, *Tetrahedron Lett.*, 2008, **49**, 4209; (b) J. Zhang, Z. Hu, L. Dong, Y. Xuan, C.-L. Lou and M. Yan, *Tetrahedron: Asymmetry*, 2009, 355.
46. (a) R. Rios, H. Sundén, J. Vesely, G.-L. Zhao, P. Dziedzic and A. Córdova, *Adv. Synth. Catal.*, 2007, **349**, 1028; (b) I. Ibrahem, G.-L. Zhao, R. Rios, J. Vesely, H. Sundén, P. Dziedzic and A. Córdova, *Chem. Eur. J.*, 2008, **14**, 7867; (c) H. Xie, L. Zu, H. Li, J. Wang and W. Wang, *J. Am. Chem. Soc.*, 2007, **129**, 10886.
47. X. Companyó, A.-N. Alba, F. Cárdenas, A. Moyano and R. Rios, *Eur. J. Org. Chem.*, 2009, 3075.
48. R. Rios, J. Vesely, H. Sundén, I. Ibrahem, G.-L. Zhao and A. Córdova, *Tetrahedron Lett.*, 2007, **48**, 5835.
49. (a) R. K. Kunz and D. W. C. MacMillan, *J. Am. Chem. Soc.*, 2005, **127**, 3240. For a related work using a different catalyst see; (b) A. Harttika, A.

T. Slosarczyk and P. I. Arvidson, *Tetrahedron: Asymmetry*, 2007, **18**, 1403.

50. S. Lakhdar, R. Appel and H. Mayr, *Angew. Chem. Int. Ed.*, 2009, **48**, 5034.

51. (a) Y.-H. Zhao, C.-W. Zheng, G. Zhao and W.-G. Cao, *Tetrahedron: Asymmetry*, 2008, **19**, 701; (b) Y.-H. Zhao, G. Zhao and W.-G. Cao, *Tetrahedron: Asymmetry*, 2007, **18**, 2462.

52. Y. Liu, C. Ma, K. Jiang, T.-Y. Liu and Y.-C. Chen, *Org. Lett.*, 2009, **11**, 2848.

53. B.-C. Hong, R.-H. Jan, C.-W. Tsai, R. Y. Nimje, J.-H. Liao and G.-H. Lee, *Org. Lett.*, 2009, **11**, 5246.

54. P. T. Franke, R. L. Johansen, S. Bertelsen and K. A. Jørgensen, *Chem. Eur. J.*, 2008, **3**, 216.

55. J. Franzén and A. Fisher, *Angew. Chem. Int. Ed.*, 2009, **48**, 787.

56. S. P. Lathrop and T. Rovis, *J. Am. Chem. Soc.*, 2009, **131**, 13628.

57. Y. Huang, A. M. Walji, C. H. Larsen and D. W. C. MacMillan, *J. Am. Chem. Soc.*, 2005, **127**, 15051.

58. J. F. Austin, S.-G. Kim, C. J. Sinz, W.-J. Xiao and D. W. C. MacMillan, *Proc. Natl. Acad. Sci.*, 2004, **101**, 5482.

59. M. Harmata, S. K. Ghosh, X. Hong, S. Wacharasindhu and P. Kirchhoefer, *J. Am. Chem. Soc.*, 2003, **125**, 2058.

60. P. Galzerano, F. Pesciaioli, A. Mazzanti, G. Bartoli and P. Melchiorre, *Angew. Chem. Int. Ed.*, 2009, **48**, 7892.

61. J. W. Yang, M. T. Hechavarria-Fonseca and B. List, *J. Am. Chem. Soc.*, 2005, **127**, 15036.

62. A. Michrowska and B. List, *Nature Chemistry*, 2009, **1**, 225.

63. B. Simmons, A. M. Walji and D. W. C. MacMillan, *Angew. Chem. Int. Ed.*, 2009, **48**, 4349.

64. G.-L. Zhao and A. Córdova, *Tetrahedron Lett.*, 2006, **47**, 7417.

65. M. Marigo, T. Schulte, J. Franzén and K. A. Jorgensen, *J. Am. Chem. Soc.*, 2005, **127**, 15710.

66. G.-L. Zhao, R. Rios, J. Vesely, L. Eriksson and A. Córdova, *Angew. Chem. Int. Ed.*, 2008, **47**, 8468.

67. S. Brandau, E. Maerten and K. A. Jørgensen, *J. Am. Chem. Soc.*, 2006, **128**, 14986.

68. (a) T. Govender, L. Hojabri, F. M. Moghaddam and P. I. Arvidsson, *Tetrahedron: Asymmetry*, 2006, **17**, 1763; (b) S. Sundén, I. Ibrahem, G.-L. Zhao, L. Eriksson and A. Córdova, *Chem. Eur. J.*, 2007, **13**, 574; (c) H. Li, J. Wang, T. E-Nunu, L. Zu, W. Jiang, S. Wei and W. Wang, *Chem. Commun.*, 2007, 507; (d) S.-P. Luo, Z.-B. Li, L.-P. Wang, Y. Guo, A.-B. Xia and D.-Q. Xu, *Org. Biomol. Chem.*, 2009, **7**, 4539.

69. (a) W. Wang, H. Li, J. Wang and L. Zu, *J. Am. Chem. Soc.*, 2006, **128**, 10354 See also; (b) R. Rios, H. Sundén, I. Ibrahem, G.-L. Zhao, L. Eriksson and A. Córdova, *Tetrahedron Lett.*, 2006, **47**, 8547.

70. (a) H. Li, J. Wang, H. Xie, L. Zu, W. Jiang, E. N. Duesler and W. Wang, *Org. Lett.*, 2007, **9**, 965; (b) H. Sunden, R. Rios, I. Ibrahem, G.-L. Zhao,

L. Eriksson and A. Córdova, *Adv. Synth. Catal.*, 2007, **349**, 827 See also; (c) Y. Yoshitomi, H. Arai, K. Makino and Y. Hamada, *Tetrahedron*, 2008, **64**, 11568.

71. D. Enders, C. Wang and G. Raabe, *Synthesis*, 2009, 4119.
72. (a) R. Rios, H. Sundén, I. Ibrahem, G.-L. Zhao and A. Córdova, *Tetrahedron Lett.*, 2006, **47**, 8679; (b) R. Rios, H. Sundén, I. Ibrahem and A. Córdova, *Tetrahedron Lett.*, 2007, **48**, 2181.
73. G.-L. Zhao, J. Vesely, R. Rios, I. Ibrahem, H. Sundén and A. Córdova, *Adv. Synth. Catal.*, 2008, **350**, 237.
74. G. Luo, S. Zhang, W. Duan and W. Wang, *Tetrahedron Lett.*, 2009, **50**, 2946.
75. E. Reyes, A. Talavera, J. L. Vicario, D. Badia and L. Carrillo, *Angew. Chem. Int. Ed.*, 2009, **48**, 5701.
76. (a) H. Li, L. Zu, H. Xie, J. Wang, W. Jiang and W. Wang, *Org. Lett.*, 2007, **9**, 1833; (b) H. Li, L. Zu, H. Xie, J. Wang and W. Wang, *Chem. Commun.*, 2008, 5636.
77. L. Zu, S. Zhang, H. Xie and W. Wang, *Org. Lett.*, 2009, **11**, 1627.
78. F.-L. Zhang, A.-W. Xu, Y.-F. Gong, M.-H. Wei and X.-L. Yang, *Chem. Eur. J.*, 2009, **15**, 6815.
79. M. Marigo, J. Franzén, T. B. Poulsen, W. Zhuang and K. A. Jørgensen, *J. Am. Chem. Soc.*, 2005, **127**, 6964.
80. H. Jiang, P. Elsner, K. L. Jensen, A. Falcicchio, V. Marcos and K. A. Jørgensen, *Angew. Chem. Int. Ed.*, 2009, **48**, 6844.
81. B. Lygo, B. Allbutt and S. R. James, *Tetrahedron Lett.*, 2003, **44**, 5629.
82. G.-L. Zhao, I. Ibrahem, H. Sundén and A. Córdova, *Adv. Synth. Catal.*, 2007, **349**, 1210.
83. S. Lee and D. W. C. MacMillan, *Tetrahedron*, 2006, **62**, 11413.
84. C. Sparr, W. B. Schweizer, H. M. Senn and R. Gilmour, *Angew. Chem. Int. Ed.*, 2009, **48**, 3065.
85. X. Lu, Y. Liu, B. Sun, B. Cindric and L. Deng, *J. Am. Chem. Soc.*, 2008, **130**, 8134.
86. (a) X. Wang, C. M. Reisinger and B. List, *J. Am. Chem. Soc.*, 2008, **130**, 6070. This concept was also applied to the use of enals as substrates; (b) X. Wang and B. List, *Angew. Chem. Int. Ed.*, 2008, **47**, 1119.
87. J. Vesely, I. Ibrahem, G.-L. Zhao, R. Rios and A. Córdova, *Angew. Chem. Int. Ed.*, 2007, **46**, 778.
88. F. Pesciaioli, F. De Vicentiis, P. Galzerano, G. Bencivenni, G. Bartoli, A. Mazzanti and P. Melchiorre, *Angew. Chem. Int. Ed.*, 2008, **47**, 8703.
89. (a) B. Tan, P. J. Chua, Y. Li and G. Zhong, *Org. Lett.*, 2008, **10**, 2437; (b) B. Tan, P. J. Chua, X. Xeng, M. Lu and G. Zhong, *Org. Lett.*, 2008, **10**, 3489.
90. J. Xie, K. Yoshida, K. Takasu and Y. Takemoto, *Tetrahedron Lett.*, 2008, **49**, 6910.
91. (a) X.-H. Chen, W.-Q. Zhang and L.-Z. Gong, *J. Am. Chem. Soc.*, 2008, **130**, 5652; (b) Y.-K. Liu, H. Liu, W. Du, L. Yue and Y.-C. Chen, *Chem.*

Eur. J., 2008, **14**, 9873; (c) M.-X. Xue, X.-M. Zhang and L.-Z. Gong, *Synlett*, 2008, 691.

92. J. L. Vicario, S. Reboredo, D. Badía and L. Carrillo, *Angew. Chem. Int. Ed.*, 2007, **46**, 5168.

93. B. Tan, Z. Shi, P. J. Chua and G. Zhong, *Org. Lett.*, 2008, **10**, 3425.

94. Y. Hoashi, T. Yabuta and Y. Takemoto, *Tetrahedron Lett.*, 2004, **45**, 9185.

95. Y. Wang, R.-G. Han, Y.-L. Zhao, S. Yang, P.-F. Xu and D. J. Dixon, *Angew. Chem. Int. Ed.*, 2009, **48**, 9834.

96. (a) Y. Xuan, S. Nie, L. Dong, J. Zhang and M. Yan, *Org. Lett.*, 2009, **11**, 1583. Thiourea-cinchona bifunctional catalyst 71a was initially employed in the cyclopropanation of nitroalkenes with dimethyl-chloromalonate, using DBU as the external auxiliary base required to promote the cyclization but furnishing very low enantioselectivities; (b) S. H. McCooey, T. McCabe and S. J. Connon, *J. Org. Chem.*, 2006, **71**, 7494.

97. T. Inokuma, S. Sakamoto and Y. Takemoto, *Synlett*, 2009, 1627.

98. F. Marini, S. Sternativo, F. Del Verne, L. Testaferri and M. Tiecco, *Adv. Synth. Catal.*, 2009, **351**, 1801.

99. J. Jiang, J. Yu, X.-X. Sun, Q.-Q. Rao and L.-Z. Gong, *Angew. Chem. Int. Ed.*, 2008, **47**, 2458.

100. C. G. Evans and J. E. Gestwicki, *Org. Lett.*, 2009, **11**, 2957.

101. G. Guo, M.-X. Xue, M.-K. Zhu and L.-Z. Gong, *Angew. Chem. Int. Ed.*, 2008, **47**, 3414.

102. L. Zu, J. Wang, H. Li, H. Xie, W. Jiang and W. Wang, *J. Am. Chem. Soc.*, 2007, **129**, 1036.

103. R. Dodda, T. Mandal and C.-G. Zhao, *Tetrahedron Lett.*, 2008, **49**, 1899.

104. L. Zu, H. Xie, H. Li, J. Wang, W. Jiang and W. Wang, *Adv. Synth. Catal.*, 2007, **349**, 1882.

105. R. Dodda, J. J. Goldman, T. Mandal, C.-G. Zhao, G. A. Broker and E. R. T. Tiekink, *Adv. Synth. Catal.*, 2008, **350**, 537.

106. J. Wang, H. Xie, H. Li, L. Zu and W. Wang, *Angew. Chem. Int. Ed.*, 2008, **47**, 4177.

107. H.-F. Wang, H.-F. Cui, Z. Chai, P. Li, C.-W. Zheng, Y.-Q. Yang and G. Zhao, *Chem. Eur. J.*, 2009, **15**, 13299.

108. (a) A. Lattanzi, *Org. Lett.*, 2005, **7**, 2579; (b) A. Lattanzi, *Adv. Synth. Catal.*, 2006, **348**, 339; (c) Y. Li, X. Liu, Y. Yan and G. Zhao, *J. Org. Chem.*, 2007, **72**, 288.

109. (a) A. Russo and A. Lattanzi, *Eur. J. Org. Chem.*, 2008, 2767; (b) A. Russo and A. Lattanzi, *Synthesis*, 2009, 1551; (c) C. Zheng, Y. Li, Y. Yan, H. Wang, H. Cui, J. Zhang and G. Zhao, *Adv. Synth. Catal.*, 2009, **351**, 1685; (d) J. Lu, Y.-H. Xu, F. Liu and T.-P. Loh, *Tetrahedron Lett.*, 2008, **49**, 6007.

110. S. Tanaka and K. Nagasawa, *Synlett*, 2009, 667.

111. A. Armstrong, C. A. Baxter, S. G. Lamont, A. R. Pape and R. Wincewicz, *Org. Lett.*, 2007, **9**, 351.

112. T. B. Poulsen, L. Bernardi, M. Bell and K. A. Jørgensen, *Angew. Chem. Int. Ed.*, 2006, **45**, 6551.

113. T. B. Poulsen, G. Dickmeiss, J. Overgaard and K. A. Jørgensen, *Angew. Chem. Int. Ed.*, 2008, **47**, 4687.

114. T. B. Poulsen, L. Bernardi, J. Alemán, J. Overgaard and K. A. Jørgensen, *J. Am. Chem. Soc.*, 2007, **129**, 441.

115. J. Alemán, E. Reyes, B. Richter, J. Overgaard and K. A. Jørgensen, *Chem. Commun.*, 2007, 3921.

116. P. V. Ramachandran, S. Madhi, L. Bland-Berry, M. V. R. Reddy and M. J. O'Donnell, *J. Am. Chem. Soc.*, 2005, **127**, 13450.

117. S. Arai, K. Nakayama, T. Ishida and T. Shioiri, *Tetrahedron Lett.*, 1999, **40**, 4215.

118. T. Tozawa, H. Nagao, Y. Yamane and T. Mukaiyama, *Chem. Asian J.*, 2007, **2**, 123.

119. E. J. Corey and F.-Y. Zhang, *Org. Lett.*, 1999, **1**, 1287.

120. S. Minakata, Y. Murakami, R. Tsuruoka, S. Kitanaka and M. Komatsu, *Chem. Commun.*, 2008, 6363.

121. (a) E. M. Philips, M. Wadamoto, A. Chan and K. A. Scheidt, *Angew. Chem. Int. Ed.*, 2007, **46**, 3107. For a related example see; (b) Y. Li, X.-Q. Wang, C. Zheng and S.-L. You, *Chem. Commun.*, 2009, 5823.

122. For a recent review see: M. J. Gaunt and C. C. Johansson, *Chem. Rev.*, 2007, **107**, 5596.

123. (a) C. D. Papageorgiou, S. V. Ley and M. J. Gaunt, *Angew. Chem. Int. Ed.*, 2003, **42**, 828; (b) C. D. Papageorgiou, M. A. Cubillo de Dios, S. V. Ley and M. J. Gaunt, *Angew. Chem. Int. Ed.*, 2004, **43**, 4641.

124. (a) N. Bremeyer, S. C. Smith, S. V. Ley and M. J. Gaunt, *Angew. Chem. Int. Ed.*, 2004, **43**, 2681; (b) C. C. C. Johansson, N. Bremeyer, S. V. Ley, D. R. Owen, S. C. Smith and M. J. Gaunt, *Angew. Chem. Int. Ed.*, 2006, **45**, 6024.

125. (a) X. M. Deng, P. Cai, S. Ye, X.-L. Sun, W.-W. Liao, K. Li, Y. Tang, Y.-D. Wu and L.-X. Dai, *J. Am. Chem. Soc.*, 2006, **128**, 9730; (b) W.-W. Liao, K. Li and Y. Tang, *J. Am. Chem. Soc.*, 2003, **125**, 13030.

126. S. Juliá, J. Masana and J. C. Vega, *Angew. Chem. Int. Ed. Engl.*, 1980, **19**, 929.

127. For leading reviews see: (a) E. A. Colby, S. M. Davie, Y. X. Mennen and S. J. Miller, *Chem. Rev.*, 2007, **107**, 5759; (b) E. R. Jarvo and S. J. Miller, *Tetrahedron*, 2002, **58**, 2481; (c) M. J. Porter and J. Skidmore, *Chem. Commun.*, 2000, 1215; (d) L. Pu, *Tetrahedron: Asymmetry*, 1998, **9**, 1457; (e) S. Ebrahim and M. Wills, *Tetrahedron: Asymmetry*, 1997, **8**, 3163.

128. Y.-M. Shen, M.-X. Zhao, J. Xu and Y. Shi, *Angew. Chem. Int. Ed.*, 2006, **45**, 8005.

Appendix: Catalysts Referred to in the Text

RSC Catalysis Series No. 5
Organocatalytic Enantioselective Conjugate Addition Reactions: A Powerful Tool for the
Stereocontrolled Synthesis of Complex Molecules
By Jose L. Vicario, Dolores Badía, Luisa Carrillo and Efraim Reyes
© J. L. Vicario, D. Badía, L. Carrillo and E. Reyes 2010
Published by the Royal Society of Chemistry, www.rsc.org

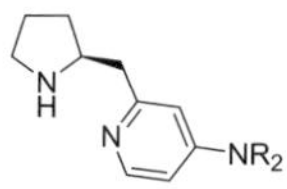

16a R= -(CH₂)₄-
16b R= Me

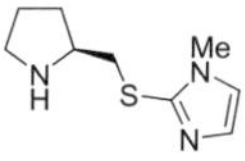

17

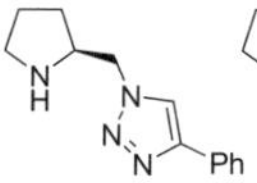

18a

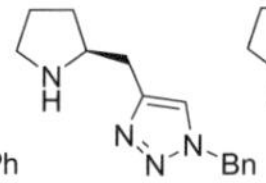

18b

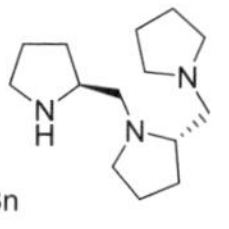

19a

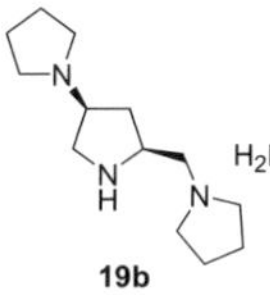

19b

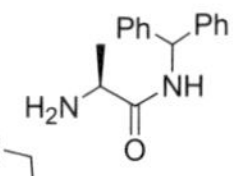

20a

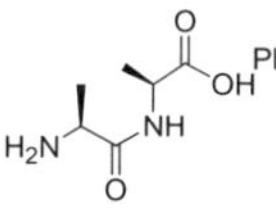

20b

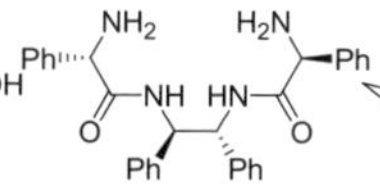

21

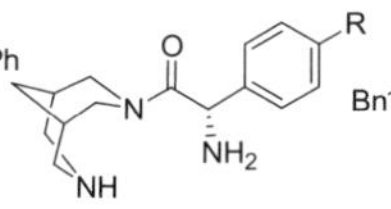

22a R=F
22b R=H

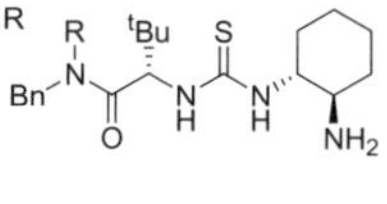

23a R=Me
23b R=H

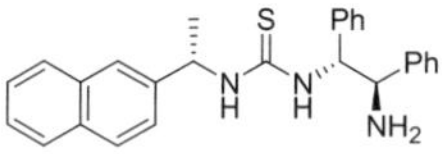

24

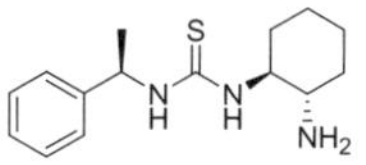

25a

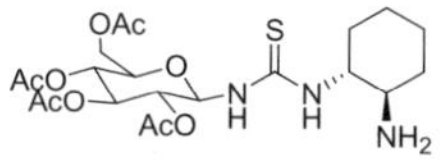

25b

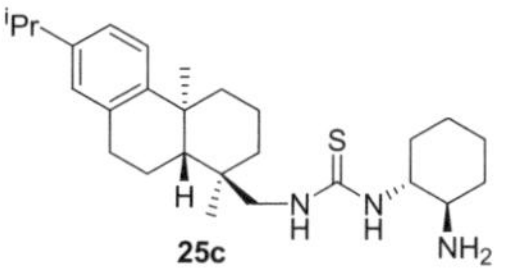

25c

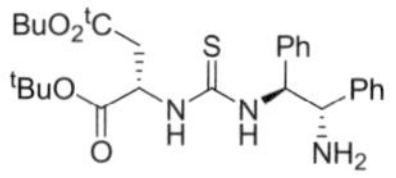

26

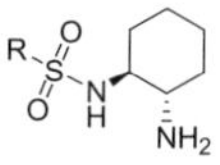

27a R=CF₃
27b R=2,4,6-(ⁱPr)₃C₆H₂

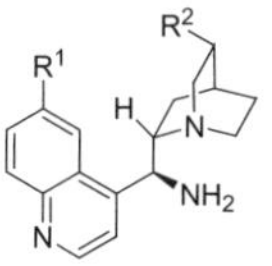

28a R¹=OMe, R²=Et
28b R¹=H, R²=vinyl
28c R¹=OMe, R²=vinyl

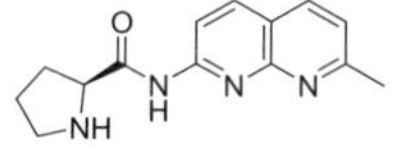

29

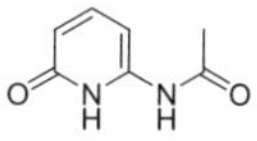

30

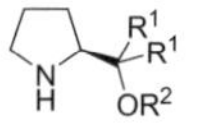

31a R¹=Ph, R²=TMS
31c R¹=3,5-(CF₃)₂C₆H₃, R²=TMS
31d R¹=Ph, R²=Me
31e R¹=n-C₆H₁₂, R²=SiPh₃
31f R¹=Ph, R²=TBS
31g R¹=Ph, R²=TES
31h R¹=3,5-ᵗBu₂-4-MeOC₆H₂, R²=TMS
31i R¹=3,5-(CF₃)₂C₆H₃, R²=TES

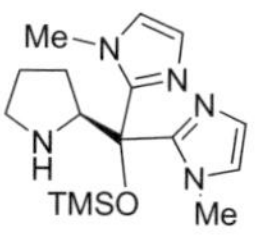

31b

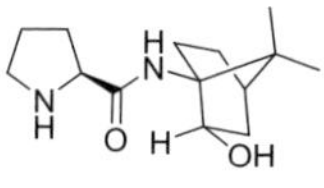

32

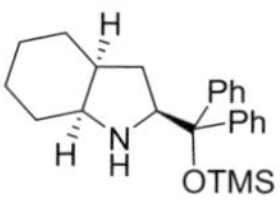

33

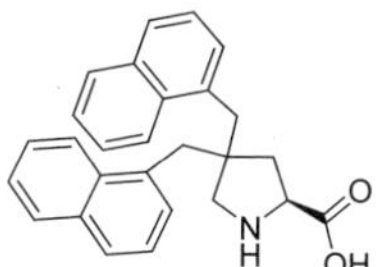

34

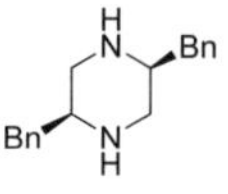

35

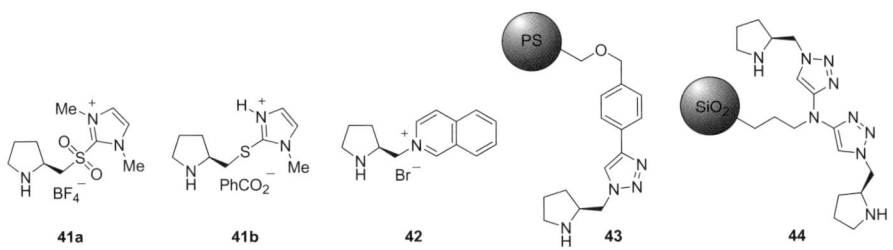

36

37a Ar = Ph
37b Ar = 3,5-(CF$_3$)$_2$C$_6$H$_3$

38

H-D-Pro-Pro-Asp-NH$_2$
39a

H-D-Pro-Pro-Glu-NH$_2$
39b

40a X = $^-$O$_3$SO^{n}C$_{12}$H$_{25}$
40b X = BF$_4$

41a

41b

42

43

44

45a Ar=3,5-(CH$_3$)$_2$C$_6$H$_3$

46a R=-(CH$_2$)$_5$
46b R=Me

47

48a R=Ph
48b R=nC$_6$H$_{13}$
48c R=3,5-(CH$_3$)$_2$C$_6$H$_3$

Ar = 3,5-(CF$_3$)$_2$C$_6$H$_3$
49

50a R^1=Bn, R^2=^{t}Bu
50c R^1=H, R^2=^{t}Bu

50b

50d

51

52

53a

53b

53c

54

55

56a R^1=OMe, R^2=vinyl
56b R^1=H, R^2=vinyl

57

58

59a: R^1=H, R^2=vinyl

60a R=2,4,6-^{i}Pr$_3$C$_6$H$_2$ (TRYP)
60b R=SiPh$_3$
60d R=9-anthracenyl
60e R=9-phenanthryl
60h R=3,5-Me$_2$C$_6$H$_3$

60c R=2,4-(CF$_3$)$_2$C$_6$H$_3$
60f R=SiPh$_3$
60g R=9-phenanthryl

Ar = 4-(^{t}Bu)C$_6$H$_4$
61

62a R^1=R^2=H
62b R^1=^{n}Pr, R^2=NO$_2$

63

64

65a

66

67a R^1=R^2=H, R^3=vinyl,
Ar = 4-CF$_3$C$_6$H$_4$, X=Br
67b R^1=H, R^2=Bn, R^3=Et,
Ar = 9-anthracenyl, X=Br
67c R^1=OMe, R^2=H, R^3=vinyl,
Ar = 3,5-(3,5-(CF$_3$)$_2$C$_6$H$_3$)$_2$C$_6$H$_3$, X=Br
67d R^1=H, R^2=adamantoyl, R^3=Et, Ar = 9-anthracenyl, X=Cl
67e R^1=H, R^2=H, R^3=Et, Ar = 10-cyanoanthracen-9-yl, X=Br
67f R^1=OMe, R^2=H, R^3=Et, Ar = C$_5$F$_5$, X=Br
67g R^1=OMe, R^2=H, R^3=Et, Ar = 2,4-Me$_2$C$_6$H$_3$, X=Br
67h R^1=OMe, R^2=H, R^3=Et, Ar = 2,4-(CF$_3$)$_2$C$_6$H$_3$, X=Br

68a R=Me
68b R=(CH$_2$)$_3$Ph

Ar = 3,5-(CF$_3$)$_2$C$_6$H$_3$
68c

Ar = 3,5-(CF$_3$)$_2$C$_6$H$_3$
69

Ar = 3,5-(CF$_3$)$_2$C$_6$H$_3$
70a R^1=OMe, R^2=Et
70b R^1=OMe, R^2=vinyl

Ar = 3,5-(CF$_3$)$_2$C$_6$H$_3$
71a R^1=OMe, R^2=Et
71b R^1=OMe, R^2=vinyl

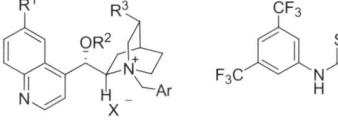

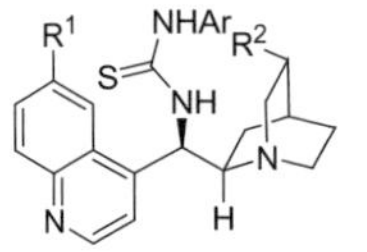

Ar = 3,5-(CF$_3$)$_2$C$_6$H$_3$

72a R^1=OMe, R^2=Et
72b R^1=H, R^2=vinyl
72c R^1=OMe, R^2=vinyl

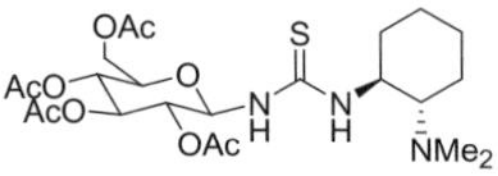

73

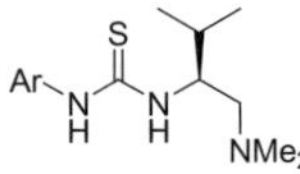

Ar = 3,5-(CF$_3$)$_2$C$_6$H$_3$

74

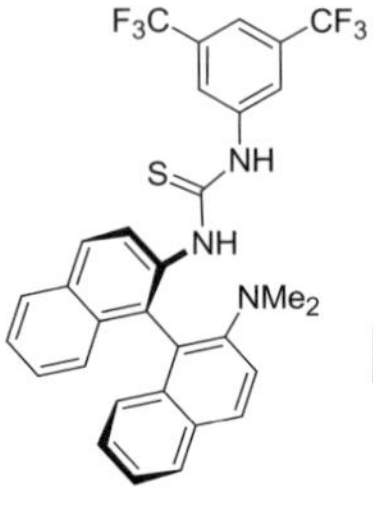

75a

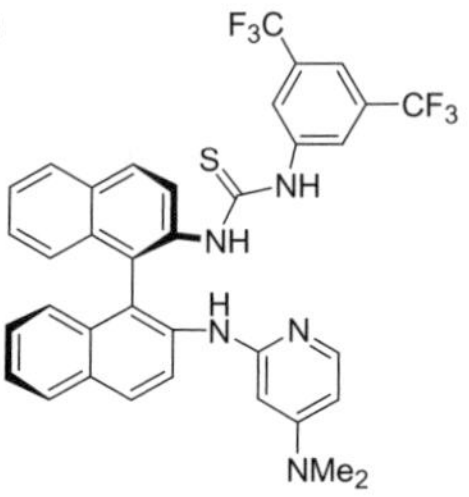

75b

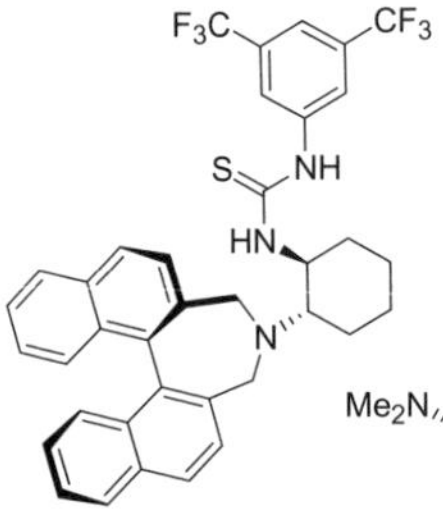

76

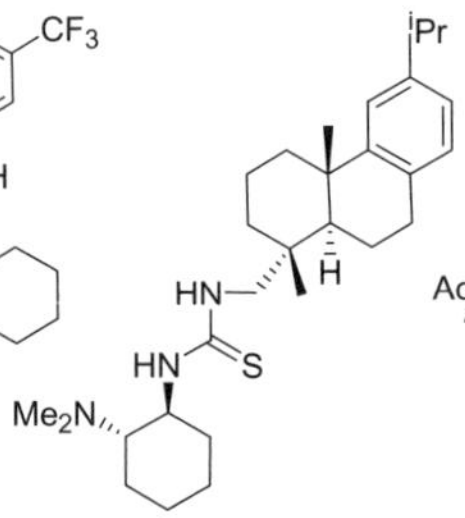

77

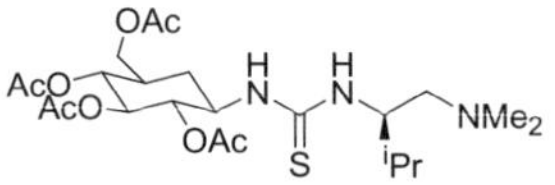

78

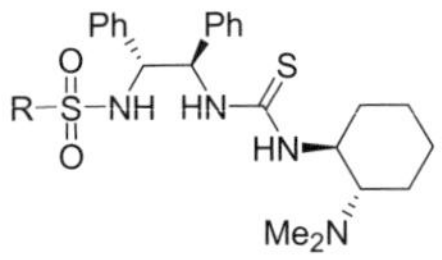

79a R=3,5-(CF$_3$)$_2$C$_6$H$_3$
79b R=OH

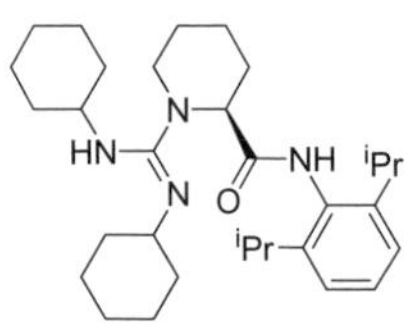

80

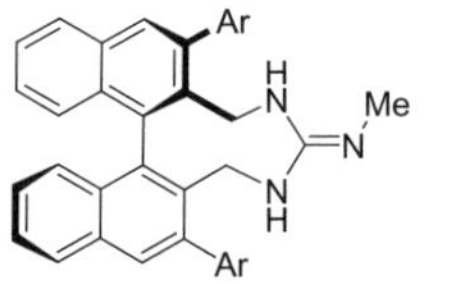

Ar = 3,5-(3,5-^{t}Bu$_2$C$_6$H$_3$)C$_6$H$_3$

81

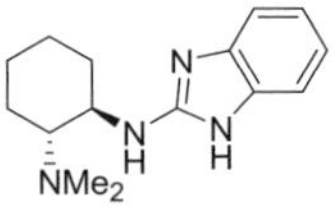

82

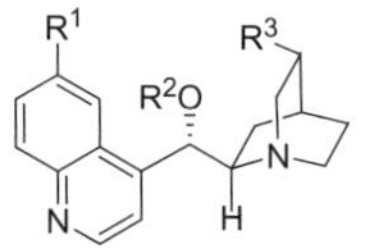

83a R^1=OMe, R^2=H, R^3=vinyl
83b R^1=OH, R^2=H, R^3=vinyl
83c R^1=OH, R^2=Bn, R^3=vinyl
83d R^1=OH, R^2=9-phenanthryl,
R^3=vinyl
83e R^1=R^2=H, R^3=(CH$_2$)$_2$SiPh$_3$
83g R^1=R^2=H, R^3=vinyl
(cinchonine)

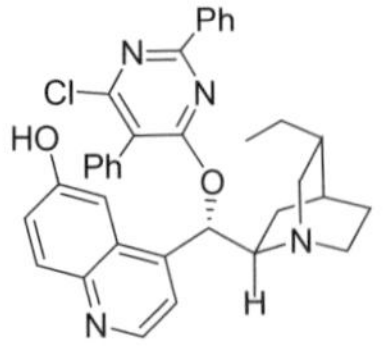

83f

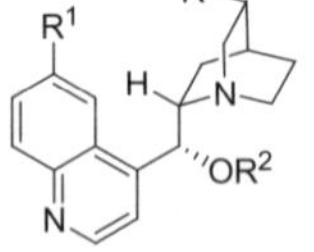

84a R^1=OMe, R^2=H, R^3=vinyl
84b R^1=OH, R^2=H, R^3=vinyl
84c R^1=OH, R^2=9-phenanthryl, R^3=vinyl
84d R^1=R^2=H, R^3=vinyl
84e R^1=OMe, R^2=H, R^3=Et
84f R^1=OH, R^2=Bz, R^3=vinyl
84g R^1=OMe, R^2=Me, R^3=vinyl

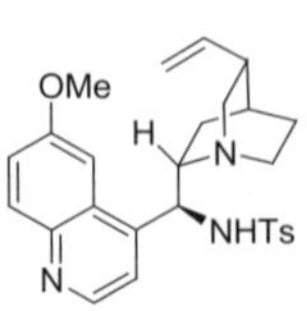

85

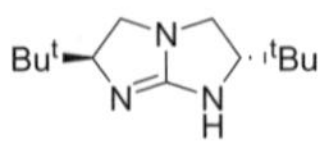

86

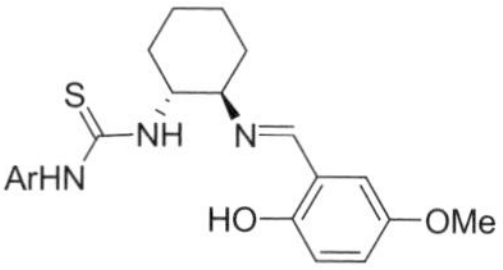

Ar = 3,5-(CF$_3$)$_2$C$_6$H$_3$

87

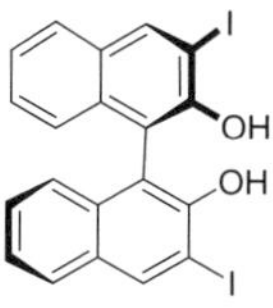

88

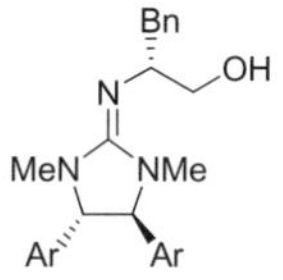

89a Ar = Ph
89b Ar = 2-CH$_3$C$_6$H$_4$

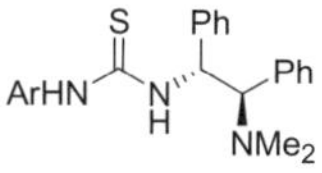

Ar = 3,5-(CF$_3$)$_2$C$_6$H$_3$

90

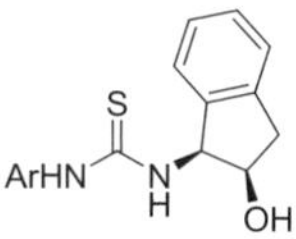

Ar = 3,5-(CF$_3$)$_2$C$_6$H$_3$

91

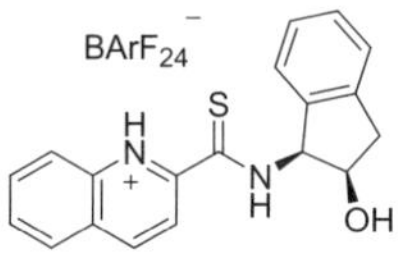

BArF$_{24}$

92

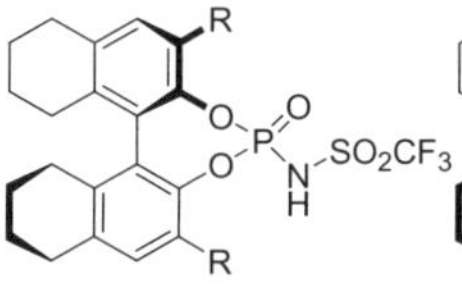

93a R=SiPh$_3$

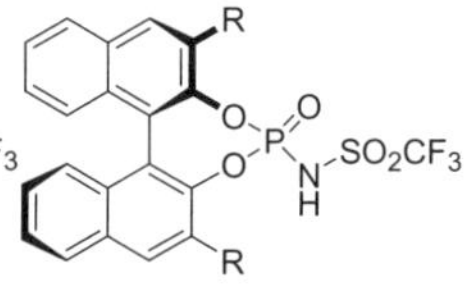

93b R = 9-anthracenyl

94

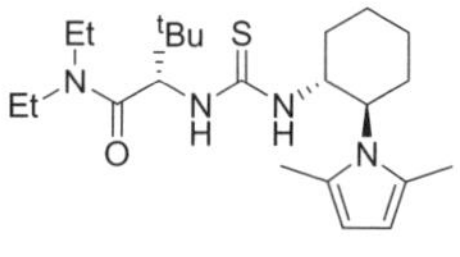

95

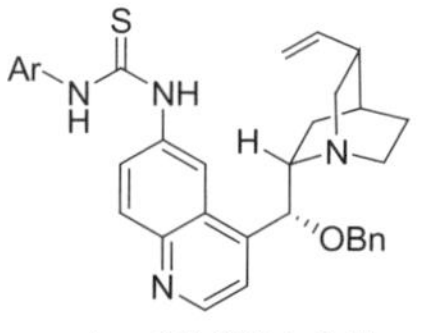

Ar = 3,5-(CF$_3$)$_2$C$_6$H$_3$
96

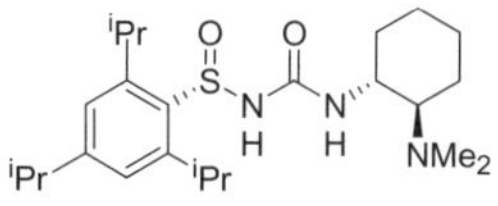

97

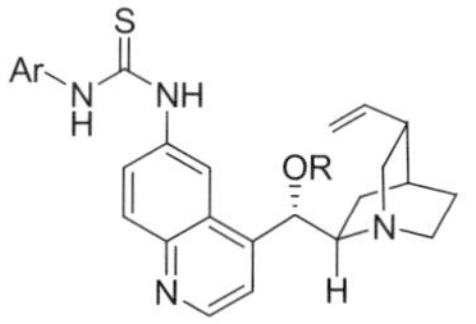

Ar = 3,5-(CF$_3$)$_2$C$_6$H$_3$
R = 9-anthracenyl

98

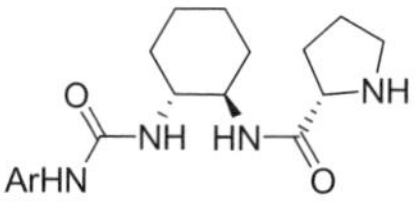

Ar = 3,5-(CF$_3$)$_2$C$_6$H$_3$

99

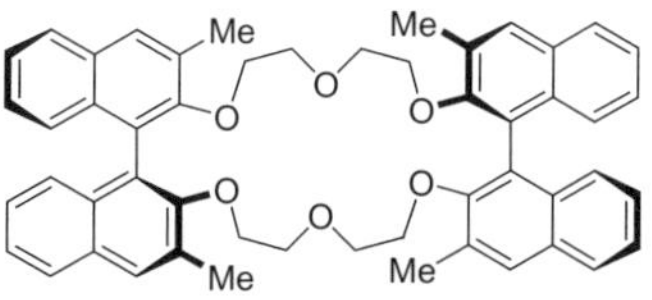

100a

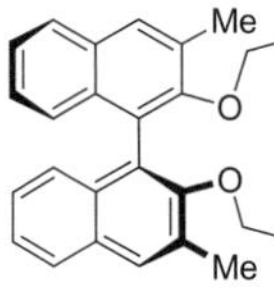

100b   **101**   **102**

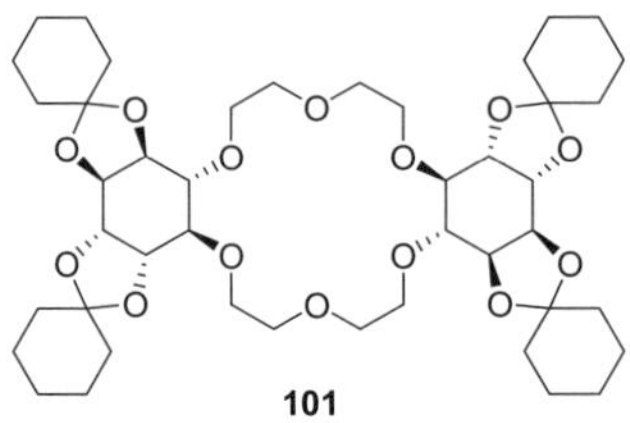

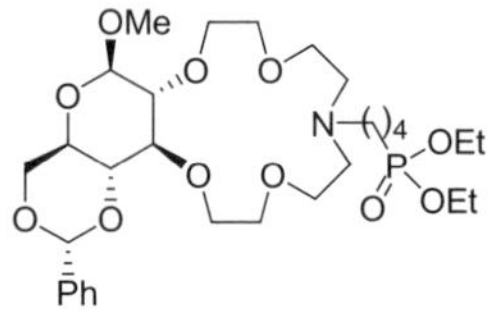

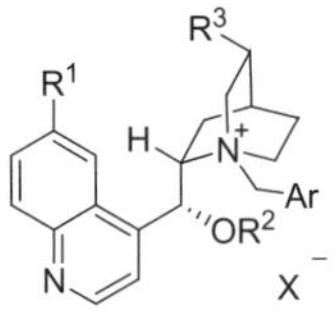

103a R¹=H, R²=Allyl, R³=Vinyl, Ar = 9-anthracenyl, X=Br
103b R¹=R²=H, R³=Vinyl, Ar = 9-anthracenyl, X=Br
103c R¹=H, R²=Bn, R³=Vinyl,Ar = 9-anthracenyl, X=Br
103d R¹=OMe, R²=H, R³=Vinyl, Ar = 9-anthracenyl, X=Cl
103f R¹=R²=H, R³=Et, Ar = 9-anthracenyl, X=Br
103g R¹=R²=H, R³=vinyl, Ar =3,5-(CF₃)₂C₆H₃, X=Br
103h R¹=H, R²=CH₂-3,5-(CF₃)₂C₆H₃, R³=vinyl, Ar =9-anthracenyl, X=PhO

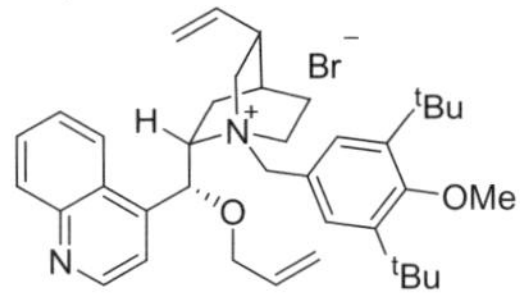

103e   **104**

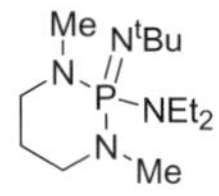

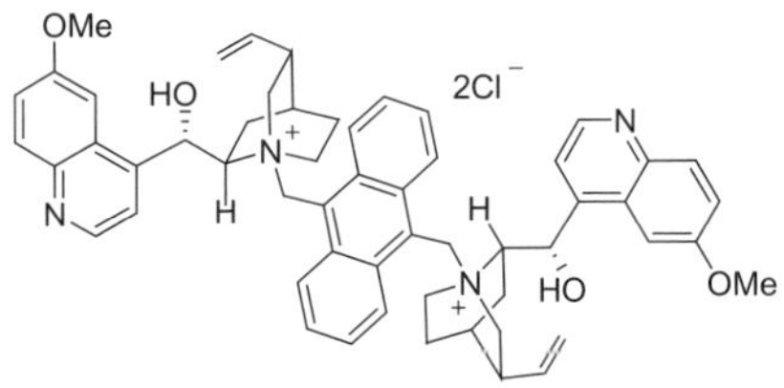

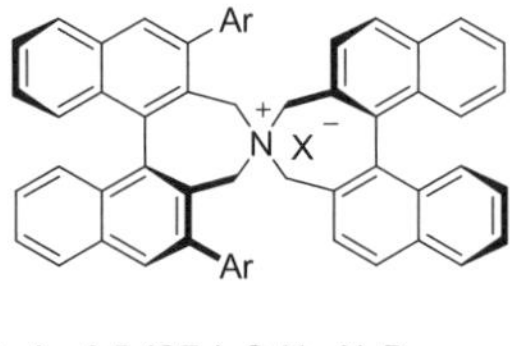

106a Ar=3,5-(CF₃)₂C₆H₃, X=Br
106b Ar=3,5-(3,4,5-F₃-C₆H₂)₂C₆H₃, X=Br
106d Ar=3,4,5-F₃-C₆H₂, X=Br
106f Ar=3,5-[3,5-(CF₃)₂C₆H₃]C₆H₃, X=HF₂
106g Ar=3,5-[3,5-(CF₃)₂C₆H₃]₂C₆H₃, X=Br

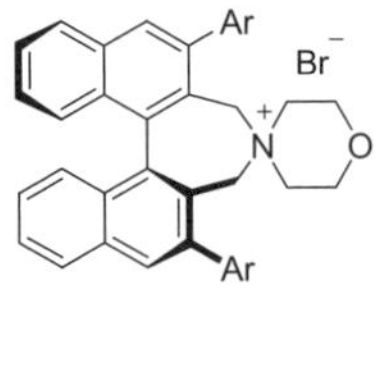

105

106c Ar=3,5-(CF₃)₂C₆H₃
106e Ar=3,5-[3,5-(CF₃)₂C₆H₃]₂C₆H₃

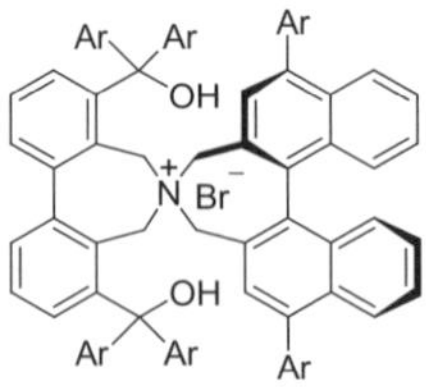

107a Ar=3,5-Ph₂C₆H₃

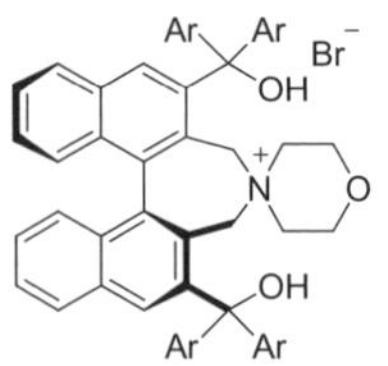

107b Ar=3,5-(CF₃)₂C₆H₃

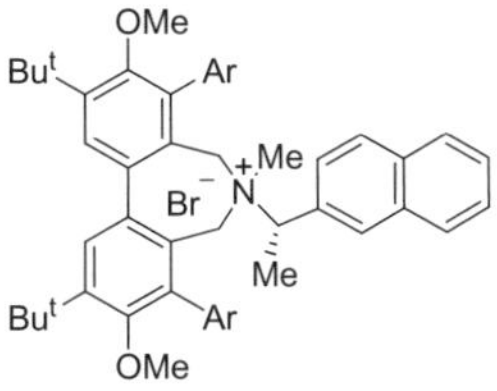

108 Ar=3,5-(CF₃)₂C₆H₃

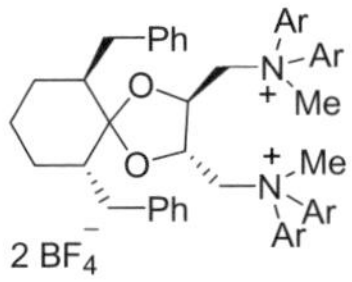

109a Ar=4-MeC₆H₄

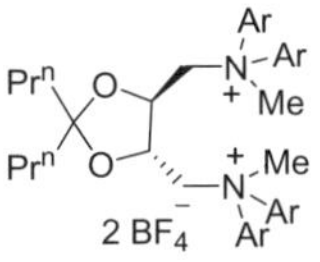

109b Ar=4-MeC₆H₄

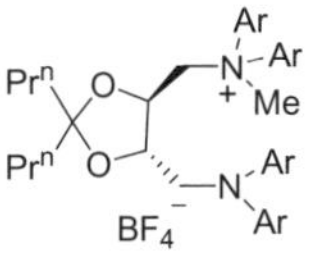

109c Ar=4-MeC₆H₄

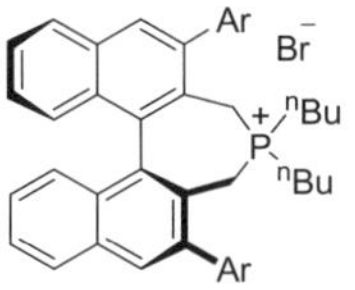

110 Ar=3,5-(CF₃)₂C₆H₃

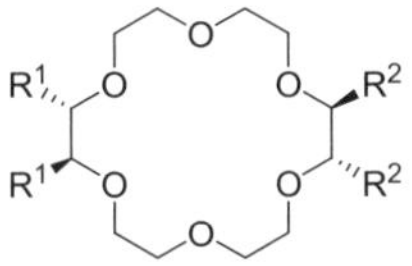

111a R¹=H, R²=Ph
111b R¹=R²=Ph

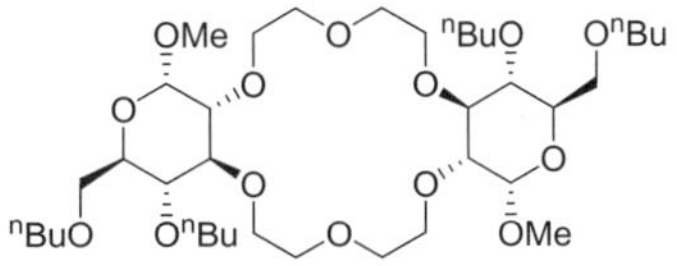

112

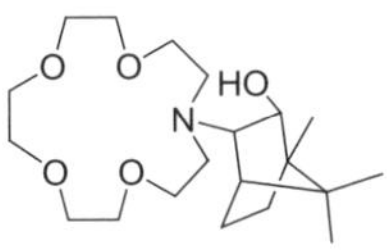

113

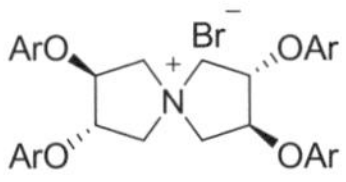

114 Ar=4-CF₃PhCH₂

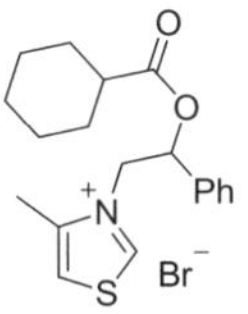

115a

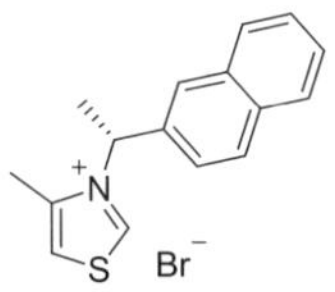

115b

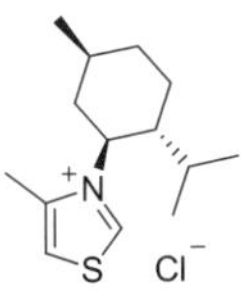

115c

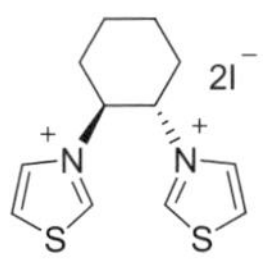

116

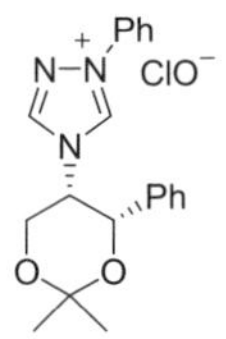

117

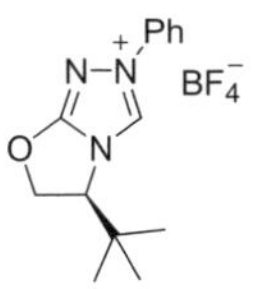

118

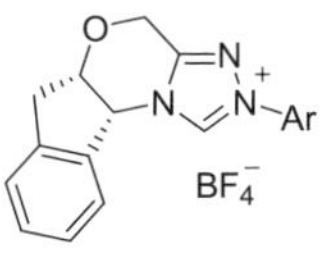

119a Ar = p-MeOC₆H₄
119b Ar = C₆F₅
119c Ar = 2,4,6-Me₃C₆H₂

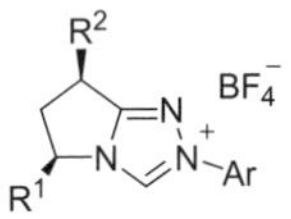

120a R¹=Bn, R²=H, Ar=Ph
120b R¹=Bn, R²=H, Ar=p-CF₃C₆H₄
120c R¹=TBSOCH₂, R²=H, Ar=Bn
120d R¹=Bn, R²=H, Ar=C₆F₅
120e R¹=ⁱPr, R²=F, Ar=C₆F₅
120f R¹=ⁱPr, R²=H, Ar=C₆F₅
120g R¹=H, R²=F, Ar=C₆F₅
120h R¹=H, R²=H, Ar=C₆F₅

121　　　　　　　　**122**　　　　　　　　**123**

124a [DHQ]₂PHAL　　　　　　　　**124b**

125a [DHQD]₂PHAL　　　　　　　　**125b** (5 or 10 mol%)

126a R=H
126b R=Me　　　　　　　　**127**　　　　　　　　**128**

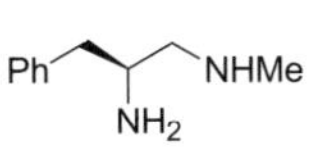

129

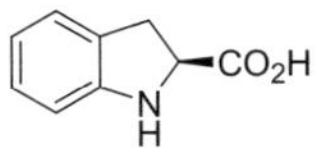

130

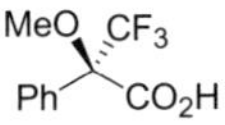

131

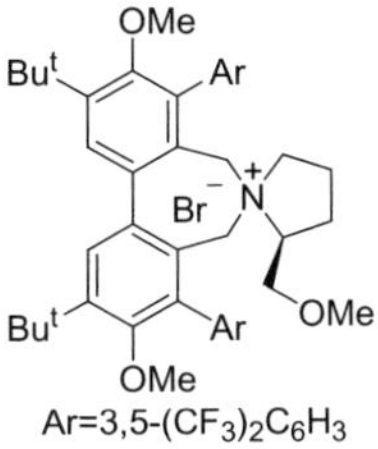

Ar=3,5-(CF$_3$)$_2$C$_6$H$_3$

132

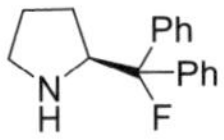

133

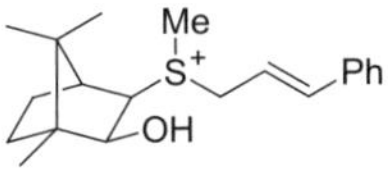

134

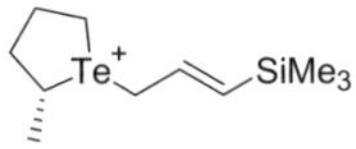

135

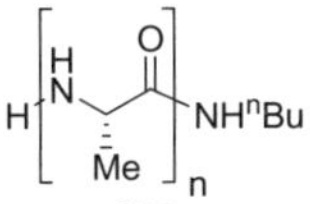

136

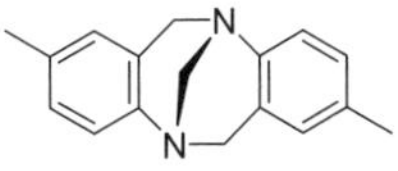

137

Subject Index